The Anaconda (U.S.A.) Electrolytic Copper Refinery.

The

Electroplating

&

Electrorefining

Of Metals

BY

Alexander Watt

Arnold Philip

MERCHANT BOOKS

ISBN 1-933998-14-8

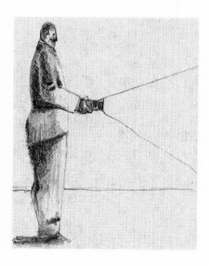

PREFACE.

I UNDERTOOK originally to edit the late Mr. Watt's well-known work on ELECTRO-DEPOSITION, and to write a further amount of about one hundred pages, but gradually (and, as I think, unavoidably) this amount expanded into two hundred and twenty pages of fresh matter. At this point the Publishers sympathetically but firmly intervened by placing their veto upon any further extension.

The present edition appears under a modified title, and has for convenience been divided into two parts—namely, Part I., or ELECTRO-PLATING, and Part II., or ELECTRO-METALLURGY. The second part, however, has included in it a description of the methods of coating iron with zinc electrolytically. It might be considered that this subject should be more correctly included in the section on ELECTRO-PLATING, but the large scale upon which the work is conducted nowadays, and the fact that at present it represents the only successfully practised process for the electrolytic treatment of zinc, have led to its being given a place in Part II. On the other hand, the whole of the subject of the electrical extraction of metals from their ores has—with the single exception of the electrolytic smelting of aluminium—been excluded from the present volume, partly because of the want of room, but largely because of the unsatisfactory state in which this portion of electro-metallurgical technology still exists.

That portion of the new matter in this edition which is original work, and which I believe to be of some importance, is contained in Chapters II. and III. of Part II., and is an attempt to systematise the electro-refining of copper—and (it may be added) of all metals, from the financial side— for a similar treatment is with very slight modifications capable of being applied to the electrolytic refining of any other metal, but owing to want of space has not been attempted here. I have to thank Mr. J. H. Brinkworth, of Bristol, very heartily for his kind assistance in checking some of the numerous calculations contained in the chapters dealing with this portion of the subject.

Two guiding principles which I have endeavoured to follow will, it is hoped, be consistently apparent in those portions of this work for which I am responsible. The first is the neglect of theory, or at least its treatment in the most superficial and sketchy manner. This is necessitated by the narrow space limits, and is also desirable as enforcing the fact that a technical text book cannot, and should not, deal with explanations of the underlying theory, but should assume them. The second principle of treatment is the constant recognition of the importance of the question of cost. Financial success is the only basis upon which technology can exist.

The alterations which have been made in this edition are chiefly as follows :—Chapters I., II., III., and V. of Mr. Watt's 1889 edition have been taken out and replaced by Chapters I. and II. of the present edition. The remainder of the volume, up to the end of Part I. of the APPENDIX ON ELECTRO-PLATING, is (with the exception of page 440) entirely Mr. Watt's work, and only differs from the 1889 edition in that the matter in Chapters XXVI., XXVII., and XXVIII., as well as Part I. of the APPENDIX, has been arranged in a somewhat different order. Part II. of the APPENDIX ON ELECTRO-PLATING is new, with the exception of those parts on cobalt, which are from the 1889 edition. The historical remarks on the electro-metallurgy of copper, consisting of 19 pages in Chapter I. of Part II. of this edition, are retained from the thirty-three pages of Chapters XXIV. and XXX. of the 1889 volume. The remaining seven chapters on electro-metallurgy are entirely new, with the exception of an account of Keith's process of electrolytically refining lead in Chapter VI., an account of Cowles' process of producing alloys of aluminium in Chapter VII., and a description of Garcia's and of Montagne's processes for recovering tin from waste tinned iron in Chapter V.

As regards the bulk of the volume, the net result (it will be seen) of the revision to which the work has been subjected is the enlargement of the text by about one hundred pages. Where necessary to bring the work up to date, new illustrative diagrams have been freely introduced.

ARNOLD PHILIP.

THE CHEMICAL LABORATORY,
H.M. DOCKYARD, PORTSMOUTH.
July 31*st*, 1902.

PREFACE

IN contemplating the present work, the Author's desire was to furnish those who are engaged in the ELECTRO-DEPOSITION OF METALS, and in the equally important department of Applied Science ELECTRO-METALLURGY, with a comprehensive treatise, embodying all the practical processes and improvements which the progress of Science has, up to the present time, placed at our command.

While the long-continued success of the Author's former work upon this subject, "Electro-Metallurgy Practically Treated"—now passing through its Eighth Edition—testifies to its having filled a useful place in technical literature, the art of which it treats has during recent years attained such a high degree of development, that it was felt that a more extended and complete work was needed to represent the present advanced state of this important industry.

In carrying out this project, the Author's aim has been to treat the more scientific portion of the work in such a manner that those who are not deeply learned in Science may readily comprehend the chemical and electrical principles of Electrolysis, the knowledge of which is essential to those who would practise the art of Electro-Deposition with economy and success. He has also endeavoured to render the work thoroughly practical in character in all its most important details ; and having himself worked most of the operations of the art upon a very extensive scale, he is enabled in many instances to give the results of his own practical experience.

ELECTRO-METALLURGY, which is now recognised as a distinct branch of electro-chemistry, has been treated separately, and those processes which have been practically adopted, such as the electrolytic refining of crude copper; are exhaustively given, while other processes, now only upon their trial, are described. In this section also will be found a description of the new process of electric smelting, as applied, more especially, to the production of aluminium and silicon bronzes.

In conclusion, the author tenders his best thanks to those who kindly furnished him with information, for the readiness and promptitude with which they complied with his requests.

ALEXANDER WATT.

LONDON, *December*, 1885.

CONTENTS.

PART I.—ELECTRO-PLATING.

CHAPTER I.

PRELIMINARY CONSIDERATIONS.—PRIMARY AND SECONDARY BATTERIES.

CHAPTER II

THERMOPILES.—DYNAMOS.—THE COST OF ELECTRICAL INSTALLATIONS OF SMALL OUTPUT FOR ELECTRO-PLATING, ETC.

CHAPTER X.

ELECTRO-DEPOSITION OF GOLD.

CHAPTER XI.

ELECTRO-DEPOSITION OF GOLD (continued).

CHAPTER XII.

VARIOUS GILDING OPERATIONS.

CHAPTER XIII.

MERCURY GILDING.

CHAPTER XIV.

ELECTRO-DEPOSITION OF SILVER.

CHAPTER XV.

ELECTRO-DEPOSITION OF SILVER (*continued*).

CHAPTER XVI,

ELECTRO-DEPOSITION OF SILVER (*continued*).

CHAPTER XVII.

IMITATION ANTIQUE SILVER.

b

CHAPTER XVIII.

ELECTRO-DEPOSITION OF NICKEL.

CHAPTER XIX.

ELECTRO-DEPOSITION OF NICKEL (continued).

CHAPTER XX.

ELECTRO-DEPOSITION OF NICKEL (continued).

CHAPTER XXI.

DEPOSITION AND ELECTRO-DEPOSITION OF TIN.

CHAPTER XXVI.

RECOVERY OF GOLD AND SILVER FROM WASTE SOLUTIONS, &c.

CHAPTER XXVII.

MECHANICAL OPERATIONS CONNECTED WITH ELECTRO-DEPOSITION.

CHAPTER XXVIII.

MATERIALS USED IN ELECTRO-DEPOSITION.

APPENDIX (PART I.) ON ELECTRO-PLATING.

APPENDIX (PART II.) ON ELECTRO-PLATING.

PART II.—ELECTRO-METALLURGY.

CHAPTER I.

THE ELECTRO-METALLURGY OF COPPER.—CHIEFLY HISTORICAL.

CHAPTER II.

THE COST OF ELECTROLYTIC COPPER REFINING.—CURRENT DENSITY AS A FACTOR IN PROFITS.

CHAPTER III

SOME IMPORTANT DETAILS IN ELECTROLYTIC COPPER REFINERIES.

CHAPTER IV.

ELECTROLYTIC GOLD AND SILVER BULLION REFINING.

CHAPTER V.

THE ELECTROLYTIC TREATMENT OF TIN.

CHAPTER VI.

THE ELECTROLYTIC REFINING OF LEAD.

CHAPTER VII.

THE ELECTROLYTIC PRODUCTION OF ALUMINIUM AND THE ELECTROLYTIC REFINING OF NICKEL.

THE ELECTRO-PLATING

AND

ELECTRO-REFINING OF METALS.

PART I.—ELECTRO-PLATING.

CHAPTER I.

PRELIMINARY CONSIDERATIONS.—PRIMARY AND SECONDARY BATTERIES.

The Electric Current.—Electricity Moving Force.—The Electric Circuit.
—Source of Electricity Moving Force.—Chemical Electric Batteries.—Magnitudes of e. m. f. of Batteries.—Polarisation.—Polarity
of Batteries.—Primary Batteries.—The Lalande Cell.--The Daniell
Cell.—Amalgamation of Zincs.—Management of Primary Batteries.—Relative Activity of Primary Cells.—Constancy of Primary
Cells.—General Remarks on Primary Batteries.—Secondary Batteries.—Care and Repair of Secondary Batteries.—Annual Cost of
Upkeep of Secondary Batteries.—Electrolytes.—Short Circuits.—
Connection of Batteries in Series and Parallel.—Ammeters and
Voltmeters.—Regulating Resistances.

IT is not the object of this treatise to enter into discussions on, and
explanations of, the theories of chemistry, electricity and magnetism,
or the methods of construction and design of dynamos. Such
questions must be studied in some one or more of the numerous
excellent text books now existing which deal especially with these
branches of pure and applied chemistry, electricity and magnetism.
The present work is intended above all to deal with the question of the
chemical action of the electric current from a practical standpoint and
all theory is as far as possible omitted, except in its simplest and most
generalised form. It must not be imagined, however, that the present
writer wishes in any way to induce the student of this particular
branch of applied electricity to consider that the theory of the subject
is unimportant. The theory is of the greatest value to guide and
direct experimental work, and it cannot be too strongly urged that

B

every practical technical worker who is possessed of little or no theory, however successful and excellent he may be in his particular branch of applied science, will find himself rendered the more capable in carrying out improvements and rectifying difficulties the greater the amount of sound theoretical knowledge he is able to obtain.

The Electric Current.—For all practical purposes the electric current may be considered as being similar to a current of water, and the electric conductors in which it can flow as being analogous to water pipes, whilst non-conductors of electricity may be represented, by analogy, by the solid materials forming the walls of the pipe through which the water flows.

Electro-Motive Force, or Electricity Moving Force.—In order that a current of water may flow through a pipe, it is necessary to have a water moving force, such as a pump, or a head of water, arranged somewhere in the course of the tube through which the water flows. And in order that the electric current may flow, it is always necessary that there shall be an electricity moving force (or as it is frequently written shortly, an e. m. f.) somewhere in the circuit.

The Electric Circuit.—If the reader will consider for one moment he will find that in all cases where we have a *continuous* flow of water there must be a closed circuit or path, round which this flow takes place. For instance, if a pump and its attached engine (the source of water moving force) is raising water from a mine, which water, as it escapes from the pump, runs back again down the shaft, we have a case of a continuous circuit. Again, in the case of a river, running continuously (or nearly so) into the sea and continuously supplied at its sources with water from the clouds (which water has been extracted from the sea by the sun); the sun is the water motive force and the closed path of flow (or the circuit as it is called in the case of electricity) is down the river bed, into the sea, up through the air from the sea, as water vapour, transport by the winds as clouds, again a downward path through the air on to the land, and once more along the river bed. In the case of the electric current we have a closely analogous arrangement. There must always be a closed circuit, and as electricity can only flow in an electric conductor, such as a metal, carbon, or a conducting liquid, the circuit throughout must consist of one or more of these materials continuously connected ; if at any point the connection between the conductors is broken by a non-conductor, such as air, ebonite, indiarubber, etc., the current must at once cease flowing.

Thus the necessary conditions for a continuous flow of electricity are :—

    1st. A source of electricity motive force or e. m. f.

    2nd. A continuous closed conducting path in the course of which the source of e. m. f. is placed.

Sources of Electricity Motive Force.—Practically there are three different types of sources of electricity motive force :—
1. The electro-chemical battery, generally known as the electric battery.
2. The thermopile.
3. The dynamo.

We shall immediately consider the relative merits and importance of these sources.

For technical purposes the prime consideration in conducting any process is economy of both time and money, and as the cost of the electric current obtained from any given source depends upon the prime cost of the generator, its rate of depreciation, and the cost of maintenance, it is of vital importance to consider these matters with every care.

In the writer's opinion the most advantageous source of electrical energy for small work up to perhaps as large a current as about three ampères, if the work is turned out at a steady rate, or as large as perhaps twice this amount if the output of work is variable and the plating plant only intermittently used, is a form of the Lalande battery manufactured by Messrs. Umbreit and Matthes of Leipzig. This refers to the cases most unfavourable to the employment of primary batteries, namely, when either cheap electrical or cheap mechanical power, or both, is to be easily obtained.   Under other circumstances, where neither electrical nor mechanical power is to be had, the importance of this form of primary battery is greatly increased, and it may be employed with advantage under the most favourable circumstances for currents up to as large as 20 or perhaps 30 ampères. In connection with these statements it may be remarked that 3 ampères is a current sufficiently large to plate satisfactorily about 500 square inches of surface with gold, or as much as about 100 to 300 square inches with silver, or to electrotype with copper, surfaces as large as about 30 to 40 square inches.   A fuller discussion of the question as to the best form of current generator to employ under different conditions is given at the end of the next chapter, after all the available current generators have been considered and their efficiency, prime cost, and upkeep noted.

It is now necessary to describe the various forms of primary electric batteries which are available for electro-plating and other electro-metallurgical work.   In the writer's opinion, none of the primary batteries, except the modified Lalande cell already mentioned, should be for a moment considered for practical work, but owing to the fact that this form of cell is of very recent introduction, and as in the larger part of the text of this book other forms of battery are continually referred to, it is necessary, in order that the reader may clearly follow

Mr. Watt's remarks, that a description of the various forms of batteries mentioned shall be given here. Moreover, as no doubt many experimentalists, amateurs and others, may have in their possession some one or more of the older forms of electric batteries, a few words on their construction and management may be found useful.

Electric Batteries.—An electric battery depends for its e. m. f. upon a chemical action occurring within it, and must have three essential parts.

I and 2. Two different conducting bodies, always solid in practice unless mercury is employed for one.

3. A liquid which can conduct electricity, but which is chemically changed by the passage of electricity through it. (Such a liquid is known as an electrolyte. This class of liquids is further considered on page 33.)

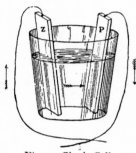

Fig. 1.— Single Cells.

The liquid is contained in some vessel made of a material upon which the electrolyte has no action, and one end of each of the two conducting solids is immersed in the liquid. The solids must be prevented from touching one another in the liquid. To each of the two ends of the conducting solids or, as they are called, elements of the battery, which are not immersed in the liquid, terminal screws are attached, and these serve to connect the battery by means of wires to the apparatus, external to the battery, through which it is desired to pass an electric current. Such a simple arrangement of materials forming a battery cell, as it is called, is shown in Fig. 1, where Z denotes a plate of metallic zinc, and P is a plate of platinum; these are immersed in a solution of sulphuric acid in water contained in a glass vessel. The binding screws are not shown, but the external conductors of copper wire are soldered directly to the ends of the zinc and platinum respectively, which emerge from the liquid.

Each particular combination of a pair of solid conductors and electrolyte solution give a perfectly definite electricity motive force at any given temperature, but if the temperature is altered, the electricity moving force of the cell is slightly altered; this alteration, however, is very slight compared with the alteration which may occur due to changing the material of either one or both of the conducting solids and the electrolyte.

Magnitude of the Electro-Motive Forces of Batteries.—The electro-motive force of any generator is measured in terms of the unit of electrical pressure or electro-motive force, which is called the *volt*. The actual value of the e. m. f. of any single cell of any electrical battery which is employed in practice varies from the highest value of about 2·2 volts in a secondary battery, down to about 1·0 volt in a Smee's cell. It must be understood that the unit of e. m. f., or electrical pressure, is analogous to the unit of head of water, or pressure of water, which is usually expressed in so many feet of head. The electrical pressure required in electro-plating work varies from about 2 volts up to as much as nearly 8 volts, whilst for ordinary house lighting work pressures usually vary from about 100 volts up to as high as 250 volts on the lamps. Most people can hardly feel a pressure of 100 direct volts, when applied to the hands across the body, whilst it is unlikely that 250 volts could give a dangerous shock, unless the hands had been previously well soaked in some conducting liquid. It must be remembered, however, that these results only apply to *direct* pressures; alternating pressures at 100 volts may produce very unpleasant sensations with some persons, and especially if the hands are moistened with a conducting liquid. These details concerning the value of different voltages, are given in order to familiarise the reader with some practical ideas of the order of the electrical pressures which are employed for most electro-chemical processes.

ELECTRO-MOTIVE FORCES OF THE CHIEF CHEMICAL BATTERIES.

Name of cell.	Approximate volts.	Remarks on voltage.
Secondary battery . .	1·8 to 2·2	constant.
Lalande (Cupron element) .	0·75 to 0·85	constant.
Daniell	1·1	constant.
Bichromate	2·1	volts fall with use.
Leclanché	1·47	volts fall with use but recover on standing.
Smee's	0·5 to 1·0	volts fall with use.
Wollaston . .	0·5 to 1·0	,, ,,
Grove	1·6 to 1·9	,, ,,
Bunsen	1·5 to 1·7	,, ,,

MINIMUM ELECTRO-MOTIVE FORCES REQUIRED FOR ELECTRO-PLATING REACTIONS.

Brass	about 4 volts.
Copper (acid bath)	0·5 to 1·5 ,,
,, (alkaline bath) . . .	3 to 5 ,,
Gold	0·5 to 1 ,,
Iron (steel facing of copper plates) .	1 to 1·5 ,,
Nickel (on various metals) . . .	1·5 to 8 ,,
Platinum	4 to 6 ,,
Silver	0·5 to 1 ,,
Zinc (on various metals) . . .	3 to 8 ,,

It will be noticed from the above given details that there is no single
cell which can give the e. m. f., which is required for many electro-
plating operations, but this difficulty may be readily overcome by
arranging the batteries in series as it is termed. This method of
arrangement will be briefly referred to later on, but in the meantime it
may be noted that by thus arranging a sufficient number of any cells
in series, any desired e. m. f. may be obtained, and such batteries of
cells of even the lowest e. m. f. have been built up until very large
e. m. fs. have been obtained, sufficiently high indeed to need great
caution in dealing with them to avoid a dangerous shock. For
all plating purposes it is therefore clear that it is a very simple
matter to connect in series a sufficient number of the single cells to give
a battery having an ample e. m. f.

Polarisation of Cells.—Most electrical batteries, when they have
been connected in a circuit with a fixed resistance, yield a current
which, although perhaps of amply sufficient magnitude at first, is
gradually found to decrease in strength. In some cases this variation
of strength is very marked, and this is the case although the resistance
of the circuit remains constant. Now the current in an electric circuit
flows, as is well known, according to Ohm's law, which states that
the resistance of any circuit, or part of a circuit, is defined as the
ratio of the total e. m. f. in that circuit, or portion of a circuit, to the
current caused to flow in the circuit. That is, if the resistance of a
circuit is represented by the letter R, and the e. m. f. by the letter E,
whilst the current is represented by the letter C, then the ratio $\dfrac{E}{C}$ is
equal to the resistance R of the circuit, or the relationship is represented
by the equation,

$$\frac{E}{C} = R, \text{ or, as it may also be written } C = \frac{E}{R}$$

If E is expressed in volts, and C in ampères, R is expressed in ohms.
This well known equation enables us to readily calculate out many
useful problems. For instance, in the present case, to explain the
diminution of the current after a cell has been running some little time
on a constant resistance R, we see that the only way to account for
the fall in current is to suppose that the value of the voltage E of the
cell has diminished, and this is precisely what investigation shows has
occurred. This decrease of the current is due to one or all of the three
following causes :—

1st. The current given out by the cell causes a chemical action to
occur in the cell, and if the cell is one containing a single fluid, such
as dilute sulphuric acid (as in the case with the Smee cell), hydrogen is
deposited on the platinised silver plate, and this coating of the plate
outside with hydrogen causes it in effect to act with the e. m. f. of a

cell made up with the elements of *hydrogen* and zinc in dilute sulphuric acid instead of *platinised silver* and zinc in dilute sulphuric ; it has already been stated that the e. m. f. of the cell practically only depends upon the nature of the elements and the liquid, and the *hydrogen* and zinc elements in dilute sulphuric have a smaller e. m. f. than the *platinised silver* and zinc in the same liquid. Therefore by Ohm's law, as the e. m. f. has been decreased, the current must decrease proportionately.

2nd. The deposition of the hydrogen upon the platinised silver plate causes the resistance of the cell to be somewhat increased, and therefore as R is increased, we again see that by Ohm's law C must decrease. This effect no doubt does exist in the Smee cell to some extent, but it is not by any means such a powerful factor in reducing the current as the first noticed cause.

The two above-described actions causing the decrease of the current given by a cell are included in what is known as the polarisation of the cell, a term which has nothing except custom to recommend it, but is always intended to indicate the actions we have just considered.

3rd. The third cause of the cell's decrease of activity is due to the fact that the chemical or chemicals in the solution of the cell become changed. This always occurs to a greater or smaller extent in all cells. For instance, in the Smee cell the dilute sulphuric acid is gradually converted into a solution of zinc sulphate. In the Daniell cell the same action takes place if dilute sulphuric acid is employed, whilst if a zinc sulphate solution is used to commence with, this solution becomes more and more concentrated. In the Bichromate cell this change of the character of the solution has a very marked effect, and it is also less, but still noticeable, in the Bunsen and in the Grove's cells. The change in the chemical character of the solutions alters one of the three things necessary for a given e. m. f., and therefore, as might have been supposed, the e. m. f. itself is altered, and always in all practical cases it is altered in such a sense as to diminish the e. m. f. This cause of alteration of e. m. f. of a cell is very small in the case of the secondary battery and the Daniell cell, it is greater in the Bunsen and the Grove's cells, and is probably most marked in the case of the Bichromate cell, whilst in the Smee it has very little effect. It must be understood, however, that this does not mean that the Smee cell has a much more constant e. m. f. than the Bichromate ; as a matter of fact their rate of fall of e. m. f. may be much the same, but in the Smee it is chiefly due to the polarising hydrogen, whilst in the Bichromate cell it is chiefly due to the alteration of the chemical nature of its exciting fluid.

Polarity of Chemical Batteries.—All the batteries, with the exception of the secondary battery, which have been enumerated in the foregoing list contain metallic zinc as one of their elements, and it is useful to note that in every case the terminal of the cell connected to the zinc is the negative terminal, that is to say, that when connected in circuit the current will flow out from the other or positive terminal of the cell and back again in at the negative or zinc terminal. As far as the writer is aware, there is no primary battery in general use which does not contain zinc, and the zinc is always the negative terminal.

In order that the references to different batteries in the following pages may be intelligible, it is necessary to briefly describe the various forms, but it must be understood that the only essentials in a given battery are the nature of its elements and its exciting fluid or fluids; the particular arrangement or form of the vessels containing the fluids, and the shape and position of the elements, may be varied almost infinitely to obtain certain advantages of packing of cells, or portability, or cheapness, or low resistance, or the contrary for special purposes. In a Daniell's cell, for instance, it is of only minor importance whether the cell is flat and rectangular, or cylindrical in shape, and whether the amalgamated zinc is immersed in zinc sulphate solution contained in the porous pot with a copper sulphate solution outside, in which the copper is placed, or the reverse arrangement is adopted, with the copper plate inside the porous pot containing the solution of copper sulphate, whilst the zinc is outside in the zinc sulphate solution; in either case the e. m. f. is

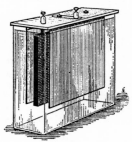

Fig. 2.—Cupron Element Cell (Lalande Battery).

the same, and the cell is a Daniell cell.

Primary Batteries.—The Lalande (Cupron element), the Daniell, the Bunsen, the Grove, the Leclanché, the Smee, the Wollaston, and the Bichromate cells are all primary batteries, that is to say, they are cells which give an electric current without having previously had an electric current passed through them to charge them.

The Lalande Cell (Cupron element).—This cell (Fig. 2) consists of two amalgamated sheets of zinc, which together form the negative element, immersed in a solution of either caustic soda (170 grams of commercial caustic soda per litre) or caustic potash (228 grams of commercial caustic potash per litre). The makers state that caustic potash or soda solution of about 19° to 21° Beaumé is employed. The positive plate of the cell is formed of a mass of cupric oxide, and it is in the mechanical construction of this plate of copper oxide that the patent for that form of Lalande cell known as the " Cupron element " exists.

When the cell is first put up it has an e. m. f. of as much as 1·2 volts. This high voltage is said to be due to the presence of oxygen in the gaseous orm present with the cupric oxide. If the cell is short circuited for a minute or two, however, the volts fall rapidly to about 0·82 volts, and then remain constant at this value until the cell is completely discharged, that is, until nearly all the cupric oxide is reduced to metallic copper, when the volts fall to slightly under 0·7.

After complete discharge the cupric oxide plate is regenerated by removing the reduced plate from the cell, washing it with water and leaving it in a dry and warm place, exposed to the air for a period of 20 to 24 hours. If a temperature of 80° to 150° Centigrade is employed, the copper is fully re-oxidised in from 20 to 30 minutes. The plate can then be replaced in the battery, and the cell can once more be used. When the caustic solution is exhausted, a yellowish-grey precipitate of zinc hydrate is thrown down. The cell can be worked after this precipitate is formed, but the e. m. f. is no longer so constant as before. The exhausted solution, which consists of caustic alkali saturated with zinc hydrate, should therefore be removed and replaced by a fresh solution. The zinc plates, which must be kept amalgamated, must from time to time be cleansed from the grey deposit which forms upon them.

The resistance of this form of cell is extremely low, the voltage is very constant (see Figs. 3 and 4), and as it gives off no noxious or corrosive fumes, it may be used in any room without any difficulty on that account. The cell behaves very much like a secondary battery with respect to its discharge voltage curve. When it is not being used, all chemical action ceases, and in this respect it is far more perfect than a secondary battery, for if the cell is kept closed up, it can be left for months, and at the end of that time its charge is as large as at the beginning. The zinc consumption is from 1·25 to 2 grams per ampère hour. The consumption of alkali is about 6 grams of commercial caustic potash, or four grams of commercial caustic soda, per ampère hour, or if the chemically pure alkalies are employed, the consumption is only half the above weights. If large batteries are employed, and much work done with this form of cell, the alkaline solutions, when saturated with zinc hydrate, need not be thrown away, as is usually done with the smaller batteries, but may be regenerated by means of the addition of a suitable quantity of sodium or potassium sulphide, according to the equation—

$$Na_2 H_2O_2 \ Zn \ H_2O_2 + Na_2 S = 2 Na_2 H_2O_2 + Zn S.$$

Although more expensive, it is rather more convenient to employ caustic potash than caustic soda for this battery, for the caustic soda is liable to form crusts of sodium carbonate, which creep up over the sides of the cell and the plate: if caustic soda is used, and this incrustation is observed, it must be removed from time to time.

The following table is given by Messrs. Umbreit and Matthes of Leipzig, the makers of this cell, summarising its chief important points, including output, weight, dimensions and price :—

Type of cell. (Trade number).	I.	II.	III.	IV.
Electro-motive force in volts .	0·85	0·85	0·85	0·85
Terminal potential difference when normal current is taken off	0·78-0·82	0·78-0·82	0·78-0·82	0·78-0·82
Terminal potential difference when maximum current is taken off	0·70-0·75	0·70-0·75	0·70-0·75	0·70-0·75
Normal current output in amperes	1	2	4	8
Maximum current output in amperes	2	4	8	16
Capacity of cell in ampère hours	40-50	80-100	160-200	350-400
Internal resistance of cell in ohms	0·06	0·03	0·0015	0·00075
Water required, in litres . .	1·2	2·3	4·4	7
Caustic soda, weight required for one charge in lbs. . . .	0·44	0·88	1·67	3·3
Caustic potash, weight required for one charge in lbs. .	0·66	1·32	2·42	4·4
Number, and dimensions of the copper oxide plates in inches, approximate	1 (4·75×4)	1 (6 × 6)	2 (6×6)	2 (8×8)
Length of cell in inches . . .	7·5	7·5	8	9
Width of cell in inches . . .	2·25	3·5	5	5·5
Height of cell in inches . . .	7·5	11	11	13·5
Weight of cell complete, in lbs.	3·3	5·82	11·55	19·8
Price in German marks (1 mark = 1 shilling, about) . .	5	9	16	27

The following are two discharge curves of a No. 1 "Cupron" element with a nearly constant current, whose mean value $= 1·55$ ampères.

The weight of sodium hydrate (commercial) was 200 grams. The external resistance between the cell terminals $= 0·43$ ohms. The internal resistance of the cell $= 0·06$ ohms. The mean terminal voltage during discharge $= 0·76$ volts. The ampère hour capacity $= 53·5$.

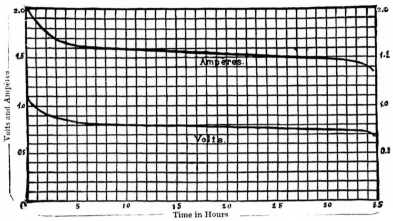

Fig. 3.—Strong Current Discharge Curve.

The following are the discharge curves of a No. 1 "Cupron" element with a nearly constant current of mean value $= 0 \cdot 15$ ampères. The external resistance $= 5 \cdot 34$ ohms. The internal resistance $= 0 \cdot 06$ ohms. The mean terminal voltage $= 0 \cdot 80$ volts. The ampère hour capacity $= 60$. The weight of sodium hydrate employed was, of course, the same as in the last case (Fig. 3).

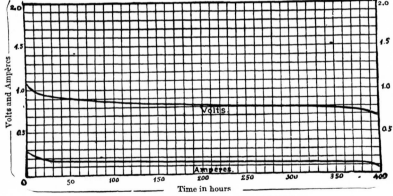

Fig. 4.—Weak Current Discharge Curve.

Daniell's Cell.—This consists of a rod of amalgamated zinc immersed in either dilute sulphuric acid (eight of water to one part acid), or a dilute solution of zinc sulphate contained in a pot of unglazed and porous porcelain. This pot stands in an outer vessel of glazed earthenware containing a saturated solution of copper sulphate, which should contain a little free sulphuric acid, and is often provided with a shelf partially immersed in the liquid, upon which crystals of copper sulphate may be placed, so that the strength of the copper sulphate solution may be preserved in spite of the constant removal of copper from it, due to the action of the cell. A sheet of metallic copper is bent round the porous pot, and stands immersed in the copper sulphate solution. The cell terminal screws are attached to the zinc and copper plates respectively, the zinc being the negative pole. When the cell is not in use for some time the porous pot should be lifted out. The level of the

zinc sulphate solution or sulphuric acid solution, according to which is employed, should be kept an inch or so above the level of the copper sulphate solution. This cell gives a remarkably constant e. m. f., and therefore a very constant current. It does not polarise. The zinc sulphate solution gradually gets stronger, and must from time to time be diluted by removing some of the liquid, and filling up with water, or dilute sulphuric acid. The copper sulphate solution gets weaker in copper, and its strength must be kept up by adding fresh crystals of copper sulphate, either placed on the shelf described above, or by suspending a muslin bag containing crystals immersed in the solution near the top. A two-pint cell will at the most give a current of not greater than about $\frac{1}{4}$ ampère in practice, even when dilute sulphuric acid is employed

Fig. 5.
Daniell's
Cell.

with the zinc. In Fig. 5 is shown a view of a Daniell cell, in which the outer glazed earthenware pot, described above, is replaced by a solid copper external pot to which is attached the positive terminal of the cell.

I have been at some pains to find out the particular form of Daniell cell which may be most cheaply and satisfactorily made, in order to ascertain how far this form of cell will compare favourably with the Cupron element for use in electro-plating, more especially for electro-silvering and gilding. Mr. F. Lyne, silversmith and electro-plater, of 5, Perry Road, Bristol, has shown me a Daniell cell which he uses, and in my opinion it is as cheap and serviceable a form of Daniell as can be obtained, and I am indebted to this gentleman for the details as to cost, etc., which are here given. The outer vessel or containing pot of the cell is made of a glazed earthenware cylindrical vessel, known in the pottery trade as a dyer's pot, it has a capacity of about four gallons, and is about 12 inches high, it costs four shillings. The copper cylinder which stands inside this outer pot is fourteen inches high, and is made by bending up a rectangular sheet of metallic copper, measuring 14 inches by 22 inches, into a roughly cylindrical form. The thickness of the copper sheet need only be sufficiently great to permit it to stand stiffly after it is bent up. The cost of this copper, which weighs roughly about 3 lbs., is about two shillings and threepence. Inside the copper cylinder is placed an unglazed or porous pot, 13 inches high, and having a diameter of about 5 inches. These porous pots cost twelve shillings per dozen. The zinc element which stands inside the porous pot consists of a cylindrical rod of zinc about 12 inches long and 2 inches diameter. It weighs about 8·5 pounds, and contains 8 pounds of zinc and 8 ounces of mercury. The mercury is added to the molten zinc just before casting, but this addition should be made when the zinc is very nearly cold enough to solidify, otherwise most of the mercury is volatilized and lost. The addition should also be made under a chimney hood, so that any mercury vapour formed may as far as possible be carried off, for it is poisonous. In my opinion the mercury can be more easily and safely added to the molten zinc if beforehand it is allowed to soak with about a pound of granulated zinc, which has been moistened with dilute sulphuric acid (one of acid to three or four of water), and after this amalgamation has been fairly completed, and the acid poured off, and the resulting amalgam thus obtained washed and *dried*, the dried amalgam can be added to the remaining melted 7 pounds of zinc, at a moderate temperature, with less danger of loss of mercury by volatilization. In any case the mercury and zinc alloy obtained is cast into rods of the form above stated. The cost of the zinc is about one shilling and twopence, and the mercury about one shilling and sixpence. The outer glazed pot contains a saturated solution of copper sulphate, to which an addition of about 2·5 per cent. by volume of sulphuric acid is made, and an equal or rather smaller amount of nitric acid. The amount of this copper sulphate solution is about 2½ gallons, each gallon contains about 2 pounds of crystallised copper sulphate, which costs about twopence per pound. The solution in the porous pot consists of dilute sulphuric acid, one part of acid to ten parts of water. The total cost of the solutions is about one shilling and eightpence. The brass terminals, of which one is soldered on to the edge of the copper plate, and the other has the zinc cast on to it, cost about one shilling. The total cost of this cell is therefore about fourteen shillings to make and charge complete. The resistance of the cell is slightly under 0·75 ohms. Six of these cells in parallel, when short-circuited, give a current of nine ampères. The e. m. f. is of course close to 1·1 volts, and the maximum current one cell can give is about 1·5 ampères. The total weight of the cell complete is over 50 pounds, and its output is less than that of two No. 1 Cupron elements, which have in series a voltage of 1·5 volts, and give a current of about 1·5 ampères, weigh under seven pounds complete, have an internal resistance (the two in series) of under 0·2 ohms, and finally take up a space 7·5 inches high and about 7·5 × 5 square inches standing room, whereas the Daniell cell just described takes up a space of about 144 square inches standing room, and is over 12 inches high, whilst finally the Cupron element cells cost (in Germany) ten shillings, as against a cost of about fourteen shillings for the Daniell cell, which has, however, a smaller output. There can therefore be little doubt as to which is the more advantageous cell to employ.

Smee Cell.—The Smee Cell consists of two amalgamated zinc

Fig. 6.

plates arranged on either side of a thin sheet of platinised silver. The zincs are connected together to the negative terminal of the cell, and the platinised silver is connected to the positive terminal. There is no porous pot, and the plates, which are supported at the top by a piece of wood or ebonite, to which they are attached by the terminal binding screws, are separated from one another below by a wooden frame or distance-piece, and the whole of this arrangement is immersed in dilute sulphuric acid (eight parts water and one part sulphuric), which is contained in an external glass, or glazed earthenware pot. A form of the cell is shown in Fig. 6. This cell, which has an e. m. f. varying from 1 volt to 0·5 volt, has a low internal resistance, and will, for the same size of positive plate, give a larger current than the Daniell, but the current is nothing like so constant.

Grove's Cell.—This cell consists of an external flat pot of glazed earthenware inside which is another cell of a similar shape, but made of porous or unglazed porcelain. A flat plate of zinc is bent in such a form that the porous cell may be placed within its folds, by means of which arrangement a surface of zinc is exposed to each side of the inner cell. A plate of platinum foil is inserted in the porous pot, and is of sufficient length to be attached to the projecting end of the zinc plate of the next cell (when arranged in a battery), or to a piece of ebonite or pitch-coated wood when used singly, by means of a binding screw or clamp. The inner porous pot, containing the platinum element, is filled with strongest nitric acid, and the outer, in which the zinc is placed, is filled with dilute sulphuric acid (one part sulphuric acid to eight parts water). In Fig. 7, on the right, is shown a cell in which A *a* is the bent zinc plate, C C are the platinum plates of this cell and the next one to it, and B is the porous pot. On the left of Fig. 7 is shown a battery of four of the cells contained in one common external glazed earthenware cell D, the alternate zincs and platinums being connected by brass clamps. This cell has a high e. m. f. of about 1·9 volts; it, however, gradually falls when a current is generated, owing chiefly to the weakening of the nitric acid in the porous pot, due to the chemical action taking place. The cell has a low resistance, and will give a larger current per square inch of positive

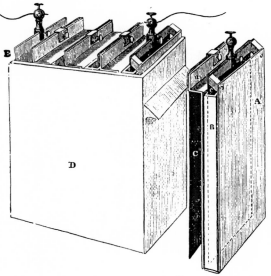

Fig. 7.—Grove's Cell.

plate than either the Daniell or the Smee. The objections to this cell are chiefly that it is expensive, it gives off corrosive and unpleasant fumes, and the nitric acid if spilt is liable to do much damage to any substance on which it falls.

Bunsen's Cell.—This cell is precisely similar in its constituents to the Grove cell, except that the platinum is replaced by gas retort carbon, which therefore makes it a much cheaper form of cell. Mr. Watt, in his original edition of the present work, praises it as being " one of the most useful batteries for the practical purposes of the electro-metallurgist.'' One form of this battery is shown in Fig. 8. In this particular cell the outer vessel is a cylindrical stoneware jar capable of holding about 4 gallons (but, of course, smaller cells are made). A plate of stout sheet zinc is turned up in the form of a cylinder *A*, and this is well amalgamated with mercury. A suitable binding screw is attached to this cylinder to receive the conducting wire. A porous cell about $3\frac{1}{2}$ inches in diameter is placed within the zinc, and in this a block of gas retort carbon is stood, and is furnished

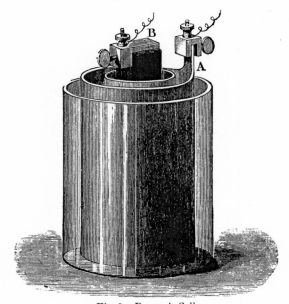

Fig. 8.—Bunsen's Cell.

with a suitable clamp *B* for attaching a conducting wire. The porous cell is then nearly filled with strong nitric acid, and the outer vessel is filled to the same height with dilute sulphuric acid—about 1 part of acid to 8 parts of water. This battery, like the Grove, emits noxious fumes, and must be kept either in a well-ventilated cupboard or outside the windows if there are any substances, such as metals, etc., which these fumes might damage in the room in which the current is being employed. The reason that a cylindrical porous pot is employed in this battery instead of the flat form used in the Grove, is owing to the fact that though platinum is cheapest in a thin plate carbon is most expensive and is much more readily obtained in the form of a rectangular rod. The carbon rods are cut from retort carbon, and this is sometimes rendered still more dense by immersing in sugar solution and then heating to a high temperature repeatedly.

The original form of the battery and its dissected parts are shown

Fig. 9.—Bunsen's Cell.

in Fig. 9. The carbon block in the Bunsen cell is more or less porous, and absorbs the nitric acid in which it is immersed by capillary attraction, in the same manner that a lump of sugar sucks up tea or other liquid into which one end is dipped. This acid will act on the brass clamp shown at B, Fig. 8, but to prevent this the outer end of the carbon may be made hot and then dipped into hot paraffin wax ; this will block the pores at the top, but, as the electricity travels along the solid carbon and not through the pores, if the outside of the block is scraped free from paraffin the clamp can be screwed on and good metallic connection obtained. This cell slowly drops its e. m. f. like the Grove, and due to a similar cause.

Bichromate Cell. — The bichromate cell, Fig. 10, is usually met with as a single fluid cell, and consists of two plates of gas retort carbon forming together the positive element, and placed between them, but not touching them, is a single plate of zinc, which is the negative element of the cell. The exciting fluid is made by making a saturated solution of potassium bichromate and adding to 10 parts by volume of the solution, about 1 volume of strongest sulphuric acid.

Fig. 10.
Bottle form of
Bichromate
Battery.

The acid must be added gradually, and with constant stirring, or the heat set at

c

liberty may crack the glass vessel in which the mixture must be made. The zinc plate is attached to a brass rod, which is held in position by a thumb-screw. When the battery is not in use this screw must be slackened and the zinc raised out of the liquid by means of the rod, and must be held in this raised position by again tightening up the thumb-screw. When the battery is being used the zinc must be lowered and *the thumb-screw again tightened*. The neglect of this precaution is not infrequently a cause of considerable trouble, as the battery will then give no current. The bichromate battery has a high e. m. f., which is fairly constant, but in time falls, owing to the chemical alteration of the exciting fluid. The exciting fluid is, when fresh, of a dark orange colour, but becomes after it has been used of a darker and darker brown, and then greenish brown colour, and finally quite dark green ; before this complete change takes place, however, the e. m. f. of the cell will have fallen considerably, and the solution should be renewed. As the solution is altered it deposits hard dark-coloured crystals of potassium chrome alum, which must be removed from time to time. The chemical action in the battery goes on whether it is being used or not, if the zinc is immersed in the solution, and it is on this account that the zinc plate is so made that it can be readily withdrawn directly the cell is out of use. The resistance of a bichromate cell is low, and it will give about as large a current as a Bunsen cell for the same area of the positive element immersed in the exciting fluid. A double fluid bichromate battery is also made, in this the carbon plate is placed alone in the bichromate solution and the zinc element, which must now be amalgamated, is placed in a separate porous pot with dilute sulphuric acid (one volume of concentrated acid to ten of water). In this form of the cell the zinc need not be withdrawn when the cell is not in use, but if it is to remain out of use for some time it is better, as in the Daniell cell, to remove the porous pot and its contents until the cell is again required. One marked advantage of the bichromate cell over other cells having high e. m. f. is due to the fact that it does not give off corrosive fumes.

The Leclanché Cell.—This battery consists of a positive carbon element surrounded by some paste or conglomerate of manganese dioxide and carbon. The carbon plate and its surrounding carbon and manganese composition stands in a solution of ammonium chloride, which is kept nearly saturated. The solution flows freely through the porous pot into contact with the carbon plate and its surrounding manganese dioxide, the cell being a single fluid cell. The negative element of the cell is a zinc rod. This cell gives off no objectionable fumes, and has a maximum e. m. f. of 1·43 volts about, but after use for a short time it polarises, and its voltage falls considerably ; if allowed to remain

idle, however, for a short time, it quite recovers its original e. m. f.
The exciting liquid is but slightly poisonous, due to the zinc which
dissolves in it, and it is non-corrosive; it is not, however, very suitable
for any electro-plating work.   All the many forms of what are known
as *dry* cells are variations on the Leclanché cell, in which the ammonium chloride solution is made into a thick paste with some inert
powder, such as plaster of Paris, mixed with some calcium chloride.

Amalgamation of Zinc Plates.—If a plate of ordinary commercial
metallic zinc, containing perhaps 2 per cent. or so of impurities, is placed
in dilute sulphuric acid, it immediately commences to dissolve, large
quantities of hydrogen gas being given off at its surface, and zinc
sulphate is formed which dissolves in the liquid.   This chemical action
is due to what is called *local action,* caused by the presence of the impurities in the zinc, for if these are removed, and quite pure zinc,
obtained by distillation, is used instead of the impure commercial zinc
no such chemical action occurs, or at any rate it is extremely slow.
Perfectly pure zinc may be employed as one of the elements of any
electric battery, and the battery will act perfectly, but when not in use
the corrosion of the zinc will cease.   The cost of this pure zinc is, however, very high, and it has been found that if the surface of impure
commercial zinc is coated with a sheet of mercury, or rather an amalgam
of zinc and mercury, the e. m. f. of the battery is not affected, and
the battery acts as satisfactorily as before, but the local action is completely stopped, and when the battery is not being employed to give
current the zinc does not dissolve.   The coating of the zinc plates
with mercury, or amalgamation as it is termed, is performed by rubbing
the plate with a rag tied on to the end of a stick in a little dilute sulphuric
acid (one of acid to ten of water), which may conveniently be placed in
a deep saucer, and at the bottom of the saucer, under the acid, must be
placed a little mercury, which must be pushed up over the acid-cleaned
zinc plate : the mercury will be found to wet the zinc, and leave it
with a bright silvered surface of zinc mercury amalgam.   Only the
smallest amount of mercury possible to thus completely silver over the
plate must be employed, as an excess of mercury merely causes the
plate to become rotten.   When in use, if blackish spots appear on the
zinc plate it must be again further amalgamated.   The mercury
employed for amalgamating must be kept by itself in a separate jar or
bottle, as it contains dissolved zinc, and must be on no account mixed
with mercury it is desired to keep pure.   Dirty or blackish zinc plates
are conveniently scrubbed with a flat piece of pumice before amalgamating.

Management of Primary Batteries.—The screws and connections
must be kept scrupulously clean, and the zinc plates must always be
properly amalgamated.   The solutions in the battery must be renewed

from time to time, as they are seen by inspection to be becoming run down, or if the cell does not act sufficiently energetically. Another frequent cause of a battery's failure to act is the contact, however slight, of one of the elements with the other inside or outside the liquid, an accident which is known as a short circuit. A loose or corroded attachment between the battery terminals and the active elements, or a loosely screwed up wire in the terminal may also cause the battery to cease to work entirely. In Bunsen batteries the upper ends of the carbons and the brass clamps should be coated with varnish after they have been screwed up, in order to avoid action on the brass by the nitric acid.

The copper plates of the Wollaston and the silver plate of the Smee batteries must be kept clean, and if accidentally spotted with mercury from contact with the amalgamated zinc plates, the sheet of metal should be heated in a flame to expel the mercury, and then should be pickled in dilute sulphuric acid, and scoured after rinsing. The zinc elements in Daniell cells should not be permitted to touch the porous cells at the bottom, or a deposit of copper may take place both inside and outside the cell and render it useless. Porous cells often crack from this cause. When porous cells have been used, and are laid aside until again required for use, they should first be well rinsed in rain or distilled water, and then filled with distilled water. They should never be allowed to become dry, or otherwise any sulphate of zinc or copper remaining in their pores will crystallise, and probably in so doing crack the pot in many places. If when a porous pot is removed from a Daniell cell, in which the acid is weak or is entirely replaced by zinc sulphate, it is rinsed out and stood in hard water, that is water containing calcium carbonate, a green deposit, or precipitate of cupric, and lime carbonates will be formed in the pores of the pot, and this deposit will, in the course of time, very greatly increase the resistance of the cell. If distilled water or very soft water is used this trouble will not occur, but if hard water is the only available variety, it should be slightly acidified with sulphuric acid before it is poured into the cell, and the cell and contents stood in a sink, then the slow oozing of the acid water through the cell's pores will remove the copper and zinc salts without precipitating them. If a cell whose resistance has been raised by the deposit of the basic carbonates, as described above, is washed or soaked in dilute sulphuric acid, it is often found that the cell becomes cracked all over and perfectly useless, caused by the chemical action which is set up in its pores.

Relative Activity of Primary Batteries.—The following experiments roughly indicate the relative activity of different kinds of primary batteries. The zinc plates were the same in each battery, and in each battery the positive plates had double the area of the zinc plates, and there was in each case the same distance between the positive and negative plates. The currents obtained for each battery so arranged were passed through solutions of copper sulphate of the same strength, with the electrodes of copper of the same size and equal distances, each during the period of one hour. The following results were obtained:—

Grove battery . . 104 grains of copper deposited.
Daniell battery . . 33 　　,,　　　　,,
Smee battery . . 22 　　,,　　　　,,
Wollaston battery . 18 　　,,　　　　,,

Constancy of Batteries.— The activity of most batteries gradually alters if they are left unadjusted, so that one kind of battery may be useful for a short period, and another kind if the action is to be sustained for any length of time. This is illustrated by the following table, showing the weight of copper deposited, the conditions being the same as in the last experiment:—

	One hour.	Two hours.	Three hours.	Four hours.	Five hours.	Six hours.	Seven hours.	Total.
Grove battery	104	86	66	60	54	49	45	464 grs.
Single-cell .	62	57	54	46	39	29	24	311 ,,
Daniell . .	33	35	34	32	32	30	31	227 ,,
Smee . . .	22	16	14	11	12	11	10	96 ,,
Wollaston .	18	14	15	12	11	10	10	90 ,,

In a second experiment of a similar nature, larger plates were used in the batteries, and proportionately larger electrodes in the copper sulphate solution, and each battery was kept in action until one pound of copper was deposited, the acid being renewed and the zincs brushed every twenty-four hours. The time taken to effect this is shown in the following table:—

Grove battery . . 19½ hours. 　　Smee battery . 147 hours.
Single-cell . . 45 　,, 　　Wollaston . . 151 　,,
Daniell . . . 49 　,,

Binding Screws.—These useful and necessary appliances are usually made from cast brass, and may be obtained in a great variety of forms. A few examples are shown in the accompanying engravings. Fig. 11 is used for connecting the platinised silver of a Smee battery to the wooden cross-bar, or for casting in zinc bars for Daniell's battery: Fig. 12 is used as a connection for a zinc or flat carbon plate; Fig. 13 is a binding screw for zinc plates, or for the cylinders

of a Bunsen battery ; Fig. 14 is for uniting the poles of dynamos with
leading rods : Figs. 15 and 17, are for connecting flat copper bands to

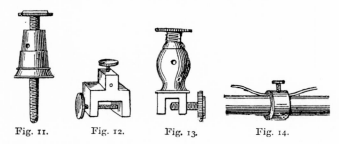

Fig. 11.   Fig. 12.   Fig. 13.   Fig. 14.

zinc and platinum plates, as in Grove's battery : Fig. 16 is a clamp
for large carbon blocks, for uniting the zincs of a Smee, or the copper
plates of a Wollaston battery.

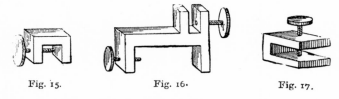

Fig. 15.   Fig. 16.   Fig. 17.

General Remarks on Primary Batteries.—With the exception of
the "Cupron element," primary batteries require much more care
and attention to keep in proper working condition than any other form
of generator. Their resistance is as a rule large, and varies with their
output, and, as has just been shown, their activity diminishes very
seriously with the time they are left in circuit. The ordinary forms of
primary batteries are therefore on all grounds, including cost, the least
advantageous form of source of e. m. f., and should if possible never
be employed for electro-technical work ; and, unless already possessed
by the experimenter, the writer strongly advises him not to purchase
them, but either to invest in some form of the Lalande or "Cupron
element," cell, or in a dynamo or secondary battery, according to the
circumstances of the work it is desired to undertake, a discussion of
which considerations will be found at the end of the next chapter.

Secondary Batteries.—Secondary batteries, which are made in a
large number of different forms, always consist (at least in all forms
used commercially up to the present) of a negative element of metallic

lead and a positive element of lead peroxide, supported on some form of lead frame-work. There is always one more negative plate than the number of positive plates present in a cell. The exciting fluid is a solution of sulphuric acid in water, having a specific gravity which varies from 1·170 when the cell is discharged, up to 1·215 when it is fully charged.

The voltage of a secondary battery, when fully charged, should be 2·2 volts, measured whilst the cell is giving a discharge current of about half its normal charging current, and the cell may be used without re-charging until its voltage drops to not lower than 1·8 volts, measured whilst the cell is giving a discharge current of about half its normal charging current.

The voltage of a secondary battery is for all practical purposes very constant, and during the greater part of its discharge is very close to 2 volts. The resistance of a secondary cell is very much lower than that of any other form of cell of equal current output. The current output of any cell is always stated by the maker, but in each case the actual current output at which the cell is run must depend upon the number of hours during which it is required to be used ; for instance, a single plate " Chloride " secondary battery, manufactured by the Chloride Electrical Storage Syndicate of Clifton Junction, Manchester, which costs well under twenty shillings complete, may be discharged for one hour at the rate of 30 ampères, but if it is wished to run it for three hours only 15 ampères must be taken from it, whilst for a six hours discharge, the rate may only be 9 ampères, and if discharged at a uniform rate for nine hours, the current taken out must not be greater than 6¾ ampères. The normal charging current for this cell is stated to be 8 ampères, and the maximum charging current must not be greater than 15 ampères. A six-plate cell of this type is shown in Fig. 18, and Figs. 19 and 20 give views of the negative and positive plates respectively ; Fig. 21 gives a view of three of the Electric Power Storage Company's secondary cells arranged in series on a stand. This particular size contains five positive plates. When the voltage of a secondary battery cell has fallen to 1·8 volts, as measured by a voltmeter whilst the cell is discharging at the rate of about one half its normal charging current (that is to say, in the chloride cell we have been considering above, whilst the cell is discharging at the rate of about 4 ampères), the cell must not be used any more until after re-charge, otherwise it will be more or less permanently damaged. The cell can be re-charged, however, and when its voltage has risen to about 2·2 (as measured with the discharge current stated above), it is completely re-charged, and can be used again and again under these conditions, with alternating charge and discharge, for a very long period if proper care is taken of it. The chief necessary precautions which

Fig. 18.—Chloride Secondary Cell, having Six Positive Plates
in Glass Box.

must be taken in order to keep a secondary cell in good condition are
as follows :—

1. Never leave the cell for any long period in the discharged, or only

partially charged condition. That is if it becomes necessary to leave it without discharging it for, say, a week, see that it is charged up fully to begin with, and disconnect all leads from it in order that any leakage may be as much as possible reduced.

2. Never discharge below the voltage limit of 1·8, measured as specified above.

Fig. 19.—Single Negative Plate of Chloride Secondary Cell.

3. Never allow the acid in the cell to evaporate below the top edges of the battery plates. The acid in a cell always tends to decrease, due partly to evaporation, and partly to what is called spraying. Spraying is the name given to the spray carried off by the hydrogen and oxygen gases liberated in the liquid when the cell is charged, and is especially noticeable towards the end of the charge. In order to replace such lost acid, the cell must from time to time be filled up with either distilled water or rain water, until the level of the liquid is about one to

one and a-half inches above the top of the edges of the plates. As water
has a less density than the acid liquid in the cells, the added water will
tend to remain as a layer at the top of the cell, floating upon the
underlying denser acid. This weak acid at the top tends to damage
the tops of the plates, and it is therefore a good practice to mix the
liquid in the cell, after adding the water, by blowing through a glass
tube, pushed down to the bottom of the cell, and having a piece of india-
rubber tubing attached to it for a mouth-piece.

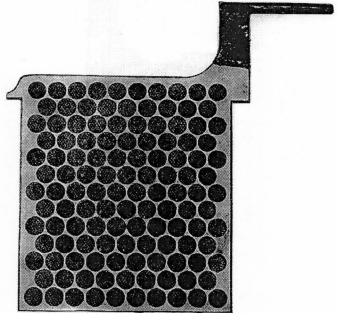

Fig. 20.—Single Positive Plate of Chloride Secondary Cell.

It is perhaps hardly necessary to caution the reader against getting
the acid into the mouth : the result will be disagreeable in the highest
degree. As the spraying of the acid liquid removes not only water
but acid, and as the directions above given for the making up the loss
only involve the addition of water, it is clear that the acid liquid in
the cell must gradually become weaker : this weakening certainly does
occur, but only slowly, and when it becomes detectable sufficient fresh

acid must be added to make good the loss. To do this it is advisable to keep a mixture of about 3 parts by volume of sulphuric acid (s. g. $= 1\cdot85$) with 5 parts by volume of distilled water or rain water.

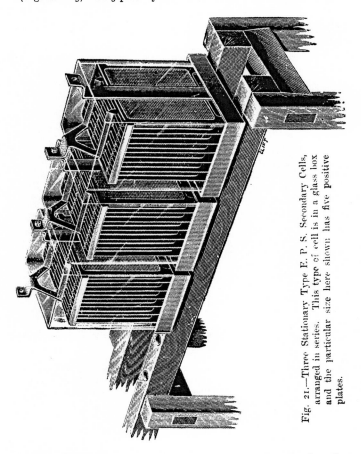

Fig. 21.—Three Stationary Type E. P. S. Secondary Cells, arranged in series. This type of cell is in a glass box and the particular size here shown has five positive plates.

This mixture has a specific gravity of about $1\cdot28$, and when the gravity of the acid in the secondary cell is found immediately after a full charge to be as low as $1\cdot19$, the stronger acid solution must be added gradually, a little at a time, until the strength of the acid in the cell rises to 1·21 once more.

4. Never remove the negative, or lead (that is the grey-looking) plates, from their acid solution, or, at any rate, do not let this be done for more than a very short period when it becomes necessary, as is sometimes the case, to straighten plates damaged by accidental short circuit or other cause.

5. The positive or reddish-chocolate coloured plates, may be removed from the acid safely when necessary, but should not be left out longer than can be helped. No attempt, however, must be made under any circumstances, unless the cell has already been accidentally completely

Fig. 22.—E. P. S. Portable Q Type Secondary Battery. (In the particular battery shown here four separate cells are connected in series, each separate cell having three positive plates.)

short circuited, to remove either the positive or negative plates from a cell whilst it contains acid, otherwise a serious short circuit will almost certainly occur. The best method of taking to pieces and overhauling a cell will be mentioned later on.

6. Be very careful to see that no leakage occurs outside the cell from the positive to the negative terminal, a common but often unnoticed cause of leakage is the acid soaked wooden cover of the portable or enclosed form of secondary battery. A partly dissected view of a portable E. P. S. secondary cell is shown in Fig. 22. Do not have a metallic handle on the lid of such a cell, or if it is placed on the cell by the makers, ask them to remove it and place metallic handles, one on

each of the sides of the cell, and have a sufficiently strong leather sling-handle soaked in paraffin wax attached to these. This sling-handle must be large enough to permit of its being readily pushed on one side to allow the cover of the cell to be removed when desired.

7. Be very careful that in charging a cell you connect the positive pole of the cell to the positive pole of the charge apparatus, and the negative to the negative. The polarity of the terminals can readily be found by inspection, for the positive terminal is attached to the lead peroxide plate, which has always a more or less marked dark chocolate colour, whilst the negative is of a darker or lighter cool grey colour. In cells which are fitted with covers, however, the error of mistaking the polarity is rather easily made if care is not taken when the lid is replaced after inspection or adjustment, for the *lid* of these cells has the polarity marked upon it, and it is sometimes not marked on the emergent lugs of the plates ; consequently if the lid is placed on in reverse position the poles, as judged by inspection, are apparently the reverse of what they really are ; this difficulty can be got over by never allowing anyone but a reliable and responsible person to remove the cover of the cell.

8. Do not short circuit the cell, that is do not place a very small or zero resistance between its poles ; the cell will under these circumstances give a very large current and will thereby have its useful life much shortened. Some people make a practice of what is called sparking a cell, that is, rapidly drawing a wire attached to one terminal over the other terminal of the cell in order to see if a spark is given. This spark is taken to indicate by its brightness the more or less complete charge of the cell. The proper test to use to ascertain this, is a small cell-testing voltmeter. Such an instrument costs only about thirty shillings, it need not read to higher than about 3 volts, and should read with an accuracy of not less than 0·2 volts per division. A good pocket instrument of this kind (Fig. 23) is put on the market by O. Berend & Co., Dunedin House, Basinghall Avenue, London, E.C., and good forms are made by many other manufacturers of electrical instruments.

If by any accident any material of any kind falls in between the plates, if it is made of a conducting substance, it will probably produce a short circuit, and the cell must be taken to pieces to remove it. If, however, it is not of a conducting material and it is desired to remove it, a rod of wood, glass, or other non-conducting material must be employed to fish it out. A rod of conducting material must on no account be introduced between the plates of a charged cell for any purpose whatever.

9. To keep cells in good condition it is very advisable to rule out a book in columns, two columns to each cell, one for the volts and the other

for the specific gravity, and record the voltage of each cell (measured with the precautions already stated) and the specific gravity of its solution about twice a week.   Such a record is most useful for reference and shows at a glance the conditions of the cells.   The specific gravity of the solution should be taken by means of a hydrometer (Fig. 24) such as is sold by most cell manufacturers at a price of about two or three shillings each.   In many small secondary batteries it is impossible

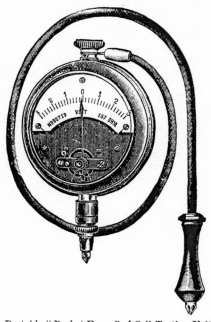

Fig. 23.—Portable " Pocket Form " of Cell-Testing Voltmeter.

to use a hydrometer directly in the liquid of the cell, as there is no room for it ; under these circumstances a sufficient quantity of the liquid should be carefully withdrawn from the cell by means of a pipette and transferred to a cylindrical tube of sufficient length and width to permit the hydrometer to float.

10. Great care should be taken to prevent any copper or brass material (as, for instance, salts and oxides from corroded connecting screws) from falling into the battery acid, as the presence of salts of

copper and, indeed, most other substances, have a very prejudicial effect upon the battery action, and all possible sources of contamination should be as far as possible avoided.

11. All connections to the battery—as, for instance, brass bolts, nuts, washers, etc.—must be kept carefully clean, and when in use must be tightly screwed up. Do not clean up the metal parts whilst they are in position over the acid, for it may thereby be contaminated.

If the secondary battery is necessarily to be carried about as, for instance, when it has to be charged up on other premises, it is advisable to purchase a portable form of battery, but if the battery can be charged whilst in a fixed position it is better not to purchase a portable form of cell. Quite apart from the fact that portable cells are liable to damage somewhat with carriage, the non-portable type gives a better cell, for there is better insulation and therefore less liability of leakage, which is itself a very fruitful source of deterioration in a cell.

Buckling of Cell Plates.—If a secondary cell is badly short circuited, or if discharges are taken from it which are above its proper capacity, the plates are liable to buckle or bend, and the paste or plugs of peroxide of lead may fall out, and either or both of these accidents may completely short circuit a cell internally. This will become evident by the fact that the voltage of the cell reads zero on the voltmeter and the cell will give no current, nor can it be charged up so as to yield a current. Under these circumstances it is absolutely necessary to take the cell to pieces and bend the plates straight once more, or remove the plug of lead peroxide which is causing the short circuit. This operation may be performed as follows : first pull out gently the ebonite, glass or celluloid insulators or combs, which will be found to be separating the positive plates of the cell from the negative ; when these separators are all removed lift out the whole of the positive plates and carefully straighten them by gentle bending where necessary. This bending must be very carefully performed, and is best effected by placing a wooden board of suitable thickness between each two positive plates, and gently pressing the whole pile until all the plates are bent flat and parallel to one another. The negative plates are not often found to be buckled, but if by any chance they are bent they must be lifted out of the acid and

Fig. 24.
Cell-
Testing
Hydrometer.

straightened in the same manner as the positives; this operation must be performed without undue force, but as rapidly as possible, for the exposure to the air damages the negatives, and the straightened plates must be returned into the dilute acid as promptly as possible. The rapidity and violence of the chemical action of the air upon the moist negative plates is shown by the fact that, after a very short exposure to the air, they will become hot and the dilute acid on their surfaces will rise in clouds of steam. When the plates have been straightened as described above, they are returned to the acid from which, however, any plugs of paste or sediment of lead peroxide should be previously removed. The plates sometimes rest upon glass, celluloid, ebonite, or even wooden racks at the bottom of the cell, and care must be taken that these are in their proper position when the plates are replaced. The insulators should next be re-inserted between the positive and negative plates, the level of the acid made up if necessary with dilute 1·21 s. g. acid, and the cell is ready for re-charging. It should be charged for a considerable period with a current which should be about that of the minimum charging current specified by the makers of the cell. This charge may with advantage last 30 hours, but usually this is inconvenient, but it should in any case be made as long as possible. Charging and discharging alternately at low rates of current output over long periods of time, and taking care never to discharge below 1·8 volts, is a method of treatment which tends to get the plates of a repaired short circuited cell into good condition.

Treatment of Cells in Bad Condition.—If a cell is found, when charged in series with others of the same size and make, not to effervesce so copiously when the charge is being completed, and if this cell drops its volts on discharge after it has been in use for a short time, whilst similar cells working at the same output keep their voltage for a longer period, it is clear that the cell in question is getting in an unsatisfactory condition. It is said to have a small ampère hour capacity. In order to restore it to condition it is advisable to charge it up with the other cells, but not to discharge it or, at any rate, to only slightly discharge it : this treatment continued over some half-a-dozen charges will usually do much to get the cell back into condition once more.

If the electrolyte of the cell gets into bad condition by reason of copper salts or other contamination finding its way into the liquid it may become necessary to change the liquid ; this may be done by syphoning it out and having at hand some previously prepared cold dilute acid of the correct gravity, which can be rapidly added in order that the negative plates may remain uncovered for as short a period as possible.

If by any chance the electrolyte of the cell is spilt the loss must be made up, not by the addition of distilled water or rain water, as when evaporation is to be made good, but a sufficient quantity of cold dilute sulphuric acid of 1·19 to 1·21 s. g. must be added.

Correct Acid Strength for Secondary Batteries.—The proper strength of the dilute sulphuric acid to be employed for secondary batteries is usually stated by the maker of the cell, but it is always closely in the neighbourhood of 1·21, measured at the ordinary air temperature. This strength of acid may be made by gradually pouring two parts by volume of strong sulphuric acid (1·85 s. g.) into five parts by volume of distilled or rain water. Heat is liberated, and the acid must be added in a thin stream to the water, which is kept well stirred by a glass rod or a wooden stick. The mixture is best made in a glass vessel, but care must be taken that the acid is *gradually* added, otherwise the rise of temperature will crack the glass; after this mixture has been made it must be allowed to cool, its specific gravity carefully measured and adjusted by the addition of more water or more acid according as the gravity is above or below the desired value of 1·21.

Annual Cost of Secondary Batteries.—If a secondary battery is properly attended to, it may be taken that the annual charge for interest, depreciation, and repairs is not less than twenty per cent., reckoned on the prime cost of the battery, but this charge may be considerably increased if proper care is not taken to superintend the cells. A portable battery has usually a higher rate of depreciation than a fixed battery. Makers will under certain conditions, which are, however, somewhat strict, enter into maintenance guarantees with purchasers of their battery. The Chloride Company, for a fixed battery, will enter into such a contract for five years for an annual payment of 12·3 per cent. of the value of the battery, but for longer periods this firm requires a larger annual payment, and if we allow interest charge at 5 per cent. the total charge will not be appreciably less than 20 per cent.

Electrolytes.—Liquids may, in so far as their electric behaviour is concerned, be divided into two main classes, namely : liquids which will allow the electric current to pass through them, and liquids which will not allow the current to pass through them. The second of these classes, namely, those which will not permit the electric current to flow through them, are of course insulating liquids, such as, for instance, paraffin oil, turpentine, resin oil, and generally many organic compounds known as hydro-carbons and fatty acids : with this class we have practically nothing to do in electrolytic refineries, although it is true that insulators for stationary secondary batteries, and insulators for the support of electric conductors, are frequently filled with some

such liquid non-conductor. The first class of liquids, namely, those
which permit a current of electricity to flow through them, may be
again divided into two other sub-classes : first, liquids which will permit
an electric current to flow through them, but which suffer no alteration of composition or chemical change due to the passage of the current. Such liquids are usually molten metals or molten alloys, such as
mercury, or molten iron, brass, solder, etc. The second sub-class consists, however, of liquids which allow the current to flow through
them, but at the same time the current causes a chemical alteration or
decomposition of the liquid. Such liquids are called electrolytes, and
usually consist of an aqueous solution of a salt, an acid, or a base, or
consist of a molten salt, acid, or base. Examples of these electrolytes
are :—a solution of copper sulphate or of common salt in water ; a solution of sulphuric acid or hydrochloric acid in water ; a solution of caustic
soda or potash in water ; liquid fused zinc or sodium chlorides ; fused
phosphoric acid, or fused sodium or potassium hydrates. The essential
difference, which theory and many experiments show exists between
liquids which are electrolytes and liquids which are not electrolytes, is
that in the non-electrolyte solution the molecules of matter of which
the substance dissolved in the liquid is made up are all of the same
nature and chemical composition, whilst in an electrolyte the
solution consists of a mixture of two or more different kinds of
molecules. Thus, although common salt or sodium chloride in its
solid state is probably built up of only one kind of molecule or minute
unit, each of which molecules consists of the same combination of sodium
with chlorine, yet if sodium chloride is fused we have a solution of sodium
molecules and of chlorine molecules in a solvent of sodium chloride, whilst
if common salt is dissolved in water, we are dealing with a solution of
sodium molecules, and chlorine molecules, and sodium chloride molecules
in water. This is the case with all electrolytes, and the electric current, or at
any rate the electric difference of pressure at the two electrodes or conducting terminals from the current generator, which are immersed in the
liquid, simply act by attracting these different molecules, into which the
original compound has split up, into different directions. The electrode
attached to the negative pole of the generator always attracts the
metallic element's molecules in an electrolyte, and these molecules are
sometimes known as the cathions, the electrode which attracts them is
called the cathode. The electrode attached to the positive pole of the
generator attracts the non-metallic element's molecules from the electrolyte ; these molecules are sometimes called the anions, and the
electrode which attracts them is called the anode.

Short Circuits.—When an electric generator has its terminals
joined by a conductor possessing very small or negligible resistance,
the generator gives out the largest current which it is capable of yield-

ing, and in many cases, especially in the case of a secondary electric battery, or any cell having a very low internal resistance, or a dynamo under the same condition of low internal resistance, which always exists in commercial machines, much damage may be done to the generator: the cell or dynamo, as the case may be, being often ruined unless the " short " lasts a very short time indeed. The danger of such a " short circuit " occurring is avoided as far as possible by placing fusible cut-outs in circuit, which will be melted by the current flowing due to the " short circuit " if this lasts for more than a very limited time, thus interrupting the " short " a..d preventing the damage to the generator.

The term " short circuit " may also be applied to the case where a current which is flowing through, say, an electrolytic bath, by reason, let us imagine, of some piece of metal falling across the metal terminals of the bath, flows through this metallic path thus provided, which has

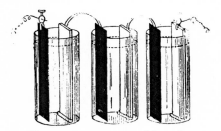

Fig. 25.—Battery arranged in series

a very low resistance, instead of through the higher resistance path of the electrolyte, in which its function is to produce some chemical change. Thus, whilst this short circuit lasts the electrolytic vat, which is short circuited, is useless, its duty of providing so much metal or other material per hour is completely interrupted until the short circuit is removed. Lastly, a current generator may be internally short circuited, as has been described in the case of a secondary battery in which a plug of paste has fallen between the positive and negative plates, thus electrically connecting them inside the cell terminals. This is called an internal short circuit.

Connection of Batteries in Series and Parallel.—As has already been remarked, batteries of electrical cells may be arranged connected in series to give a large voltage. This arrangement is shown in Figs. 25 and 27, where the positive of one cell is connected to the negative of the next, and so on, until the desired number of cells have been connected, and the final unconnected positive and negative of the end cells form the terminals of the battery. The electro-motive force of such a battery is equal to the sum of the electro-motive forces of all the cells in the series. Usually only one kind of cell is thus

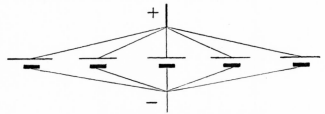

Fig. 26.—Diagram of Cells arranged in Parallel.

connected to form a battery, and if so the e. m. f. of the battery is equal to that of one cell multiplied by the number of cells in the series.

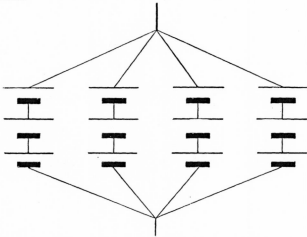

Fig. 27.—Diagram of Cells arranged in Four Groups in Parallel, each group consisting of three in series.

Cells may also be connected in parallel, as shown diagrammatically in Fig. 26, where all the positives are joined together, forming the common positive terminal of the battery, and also all the negatives are

joined together, forming the common negative of the battery. The
e. m. f. of the battery thus connected is that of only one of the

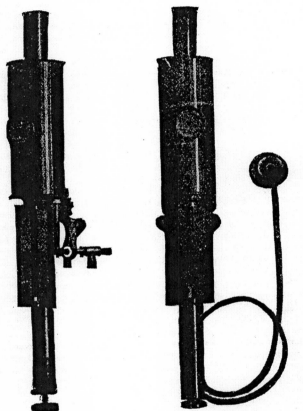

Fig. 28.—Atkinson's
Ampère-meter or Voltmeter.

Fig. 29.—Atkinson's Ampère-
meter or Voltmeter fitted with
flexible leads and connecting plugs

constituent cells, but it is capable of giving a current which is
larger than one cell can give, in proportion to the number of
cells thus connected in parallel. The cells of a battery may be

arranged partly in parallel and partly in series ; thus Fig. 27 shows a diagram of twelve cells arranged in four groups, each of three in series with each other, the groups being then connected in parallel. In the diagrams the usual convention of a thick line for the negative element and a thin line to denote the positive element is adopted. Large and small cells of any given kind may be arranged in parallel with each other, but cells of different kinds should not be so connected, that is, a Bunsen cell should not be employed in parallel with a Daniell cell. Also in putting groups of cells in series into parallel with each other, as in Fig. 27, the groups must not only be all of the same kind of cell, but each group must contain the same number of cells.

Ampère-meters and Voltmeters.—There is a very large number of these instruments of very different prices and construction now upon the market. The great point is that they shall be accurate, and also that they shall be fairly dead beat, that is, that the needles or indicators shall not oscillate about much when the current is changed, before they settle down to the position required by the changed current.

A cheap and rather novel form of ampère-meter and voltmeter which appears to fulfil these conditions is made by Messrs. Atkinson, of Cardiff. Two views of this instrument are shown in Figs. 28 and 29. Fig. 29 shows an instrument provided with a flexible lead and plug connector for connecting on to different points. These ampère-meters and voltmeters are identical in general construction and appearance ; they are intended to be attached to the wall, but there is no reason why they should not be made portable by attaching them to suitable stands. For many purposes, especially in a small electro-plating shop, it is an advantage, or indeed almost necessary, that the instruments should be portable, that is, that they shall be capable of being moved from one part of the building to another, and quickly and safely set up at any desired point. The price of these instruments is about £1.

Regulating Resistances.—For adjusting the current to any desired value from any given source of constant e. m. f. it is very desirable to have some form of regulating resistance. Several forms of this class of apparatus are described and figured at the end of the next chapter.

CHAPTER II.

THERMOPILES.—DYNAMOS.—THE COST OF ELECTRICAL INSTALLATIONS OF SMALL OUTPUT FOR ELECTRO-PLATING, ETC.

The Thermopile.—The Gülcher Thermopile.—The Cox Thermopile.—The Clamond Thermopile.—The Dynamo.—Points to be Considered in Buying a Dynamo.—Care of Dynamo.—Driving Belts.—Starting and Stopping a Dynamo.—Cost of Dynamos.—Cost of Motor-Dynamos.—Specification for and Choice of Motor-Dynamos.—Safety Precautions with Motors of Motor-Dynamos.—Choice of Electric Generators of Small Output, under Various Circumstances.—Comparison of Costs of Primary Batteries, Secondary Batteries, Dynamos, and Motor-Dynamos.—Costs of Gas Engines, Steam Engines, Oil Engines.—Gas Engines Run on Producer Gas.—Regulating Resistances.—Determination of Polarity of Generators.

Thermopiles.—The next form of electric generator which we have to consider is that form of apparatus known as a thermopile. An apparatus of this nature may be briefly described as one which converts the heat-energy of some burning fuel into the energy of the electric current. At one time it appeared as though for all ordinary purposes this form of apparatus must completely displace the electric battery in which chemical reactions are the source of the e. m. f.; and, indeed, the undoubted fact that in the thermopile the energy of the heat of combustion of a fuel is directly converted into the energy of the electric current without any complicated intermediate apparatus such as the steam-boiler and the steam-engine, or the gas-engine, with the dynamo, seemed to indicate that the thermopile might eventually quite supersede the dynamo. These attractive dreams have not, however, up to the present been realised in practice, owing to the low efficiency of the thermopile, and chiefly to its high prime cost and the care necessary to keep it in satisfactory working order. Until the advent of the "Cupron element" primary cell the thermopile might probably have had a fairly large field for small plating work and especially for laboratory work, but the present writer considers that, unless under very special circumstances, the thermopile would be better replaced by either "Cupron elements," secondary batteries, or dynamos. It is,

however, desirable to describe the forms of thermopile at present
employed, more especially as it is still quite possible that cheaper,
stronger, and more efficient forms may be developed in the future, and
if this improvement could be carried sufficiently far, it is certainly
possible that the dynamo might be largely or altogether displaced, but
up to the present there does not seem great *probability* of this desirable
end being attained.

The thermopile consists essentially of a series of junctions of different
metals, alloys, or other conducting solids, which are kept at a fixed
high temperature, and another set of alternately arranged similar
junctions, which are maintained at a fixed low temperature. The

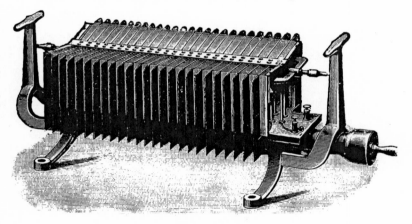

Fig. 30.—Gülcher's Thermopile.

theory of the action is quite unnecessary for practical purposes, and as
we are here merely considering the thermopile as a current generator
we shall say nothing on the theoretical aspects of the matter, which
may, however, by the aid of a moderate amount of mathematics, be
obtained from most treatises on theoretical electricity.

In all forms of thermopile, then, with which the author is acquainted,
the hot junctions are heated by means of a gas burner, or a coke or coal
furnace, and the cool junctions are kept cold either simply by direct
exposure to the air or by the circulation of a current of cold water from
a neighbouring tap.

The Gülcher Thermopile.—This apparatus, a view of which is
shown in Fig. 30, is manufactured by Julius Pintsch, Berlin, and

is listed by Franz Müller—Dr. Geissler's successor—of Bonn am Rhein, Germany, at a price of 190 marks for the largest size, with 23 marks extra for a gas pressure regulator.

Dr. G. Langbein, of Leipzig-Sellerhausen, Germany, also lists this thermopile at the same price, whilst the same size of thermopile may also be purchased of Messrs. O. Berend & Co., of Dunedin House, Basinghall Avenue, London, E.C., at a price of £14 5s., or with a gas pressure regulator extra, the cost is £15 15s.

It is claimed for these thermopiles that they may be employed continuously, and that their efficiency does not vary, neither does the apparatus deteriorate. They yield a constant electro-motive force which may be slightly varied by turning the gas up or down, and the internal resistance of the pile is not large. A thermopile, of course, causes no polarisation troubles like a primary battery.

The following details as to price, etc., of this type of generator are given by Messrs. O. Berend & Co. :—

	Size.		
	No. 1.	No. 2.	No. 3.
Number of elements . . .	26	50	66
E. M. F. in volts	1·5	3	4
Current strength obtainable when the external resistance is equal to internal	3	3	3
Internal resistance in ohms .	0 25	0·50	0·65
Approximate consumption of gas in cubic feet per hour . .	2	4·9	6·4
Price of thermopiles . . .	£6 7 6	£12 0 0	£14 5 0
Price of gas regulators . .	£1 10 0	£1 10 0	£1 10 0

With reference to the foregoing table it must be remarked that when the thermopile is running with the currents there stated, that is, when the external resistance is equal to the internal resistance of the generator, it is working at its greatest rate, that is, it is giving out electrical energy to the outside circuit at its greatest rate.

The following instructions are given as to the use of this thermopile :—

"First place the tubular porcelain chimneys in their places, on top of the elements, by inserting the projecting mica tube in the holes until the porcelain rests on the metal ; in this position they must remain. Thereupon connect the thermopile by a rubber tube to the gas supply, open the tap, and after a lapse of about half a minute— the time it will take for the air to escape from the tubes—set light to the gas escaping from the tubular elements at the mouth of the porcelain chimneys. (See that all the burners are alight to prevent escape of gas, and to ensure full efficiency of the thermopile.) When the burners are lighted the thermopile is ready for use, and requires no further attention. In 8 to 10 minutes after lighting up, the pile is sufficiently heated to give off a *perfectly constant e. m. f.* The consumption of gas given in the table is calculated upon a gas pressure of 30 m/m = 1·18 in. water column. Before packing, each thermopile is carefully tested, and adjusted to the highest admissible gas pressure, viz., 50 m/m = 1·96 in. water column. As the gas pressure varies considerably, it is strongly advisable to always insert a gas pressure regulator as quoted in list, but in most instances the gas companies will be able to inform users of their normal pressure, and when such normal pressure does not surpass the maximum, it is not necessary to use a regulator. Simply turn off the gas tap for putting the thermopile out of work.

"CAUTION.—*Do not interfere with the gas inlet on the thermopile; any increase of this inlet will ruin the instrument.*

The thermopile must be kept in a dry place, and should be protected from acid vapours, which would attack the metal. It is advisable to fix the pile to the wall on a horizontal bracket at a height sufficient to protect it from interference."

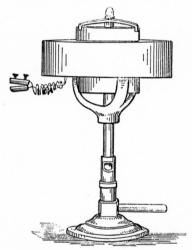

Fig. 31.—Cox Thermopile for gas without water-jacket.

The Cox Thermopile.— This thermopile, invented by Mr. H. Barrington Cox, is manufactured by the Cox Thermo Electric Company, Limited, at the Cox Laboratory, St. Albans, Herts, England.* Several general views of this form of thermopile are given in Figs. 31, 32, 33.

The hot junctions are heated by means of a Bunsen gas flame as in the Gülcher pile, but in the larger sizes of the Cox thermopile the

---

* I hear that the English Company has been wound up, but the American business is, I believe, still carried on at 126, Liberty Street, New York.—A. P.

cold junctions are kept cool by means of a stream of water from the ordinary supply tap, whereas in the Gülcher pile, air cooling alone is employed. The following directions for use are given by the makers of this thermopile :—

1. Directions for use of smaller-sized generator, heated with gas burner, and not supplied with water cooling arrangements (Fig. 36).

The generator with its gas burner should be placed upon some convenient support, such as a table or stand, where the apparatus operated can be conveniently placed. Use a sufficient length of rubber tube to connect the burner to the nearest gas jet in the room. See that the deflectors, which are shown at D in Fig. 35, are as nearly as possible in the centre of the generator ; this is necessary in order that the heat from the burner shall be supplied equally to all parts of the machine. The burner may be lighted at either the top of the tube or at the top of the generator. The deflectors require to be heated up to a cherry-red heat, and sufficient gas must be supplied to do this The gas flame should never impinge upon, or strike against the interior of the element, that is to say against the inside of the disc, which is the part that produces the electricity. This last precaution is extremely important. The pile may be run continuously,

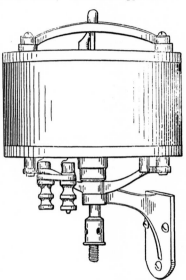

Fig. 32.—Cox Thermopile for gas, fitted with water-jacket and bracket support.

and provided the gas supply is steady, the current will not vary in the slightest degree, but, it it is required, the strength of the current may be increased or decreased by regulating the amount of gas supplied. More gas will give more current, less gas less current. Short circuiting the generator does not injure it in any way. It should be remembered that the element under working conditions becomes hot, and must not be handled without using the fire-proof asbestos pads provided, or some other suitable means. To stop the pile acting simply turn off the gas.

Heat in the form of stored energy will operate the apparatus for some minutes after the gas is put out. The generator should be allowed to thoroughly cool before it is handled.

2. Directions for installing and using the larger sized generators, heated with gas and supplied with water circulator.

Screw the machine upon the wall in a vertical position and in such a place that it may be as free as possible from draughts of air. Connect the gas burner A, Figs. 34 and 35, on the pile by means of a rubber tube to a convenient gas supply, also connect the water tubes E E in the same manner with the water supply, taking care that the water enters the pile at the bottom opening, following the direction of the arrows so as to leave at the top as shown in the drawing. Remove the string and tag which is tied about the deflector tripod C at the top of the machine, and see that the deflector rod D rests in the centre of the burner B. Also note that all screws are tight, especially the connections. The use of small lead or compo piping is recommended instead of rubber tubes. A lead or compo pipe is found to be more satisfactory, especially if a pipe is to be kept in constant or practically constant use.

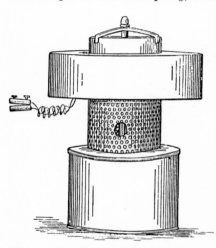

Fig. 33.—Cox Thermopile for Spirit Lamp, without water-jacket.

To start the generator, turn on the gas and water and light the gas, the flame from which should appear as in Fig. 34. Observe that the water is running freely through the apparatus, though only a small current is necessary. The correct quantity is determined by the temperature of the waste water, which should be kept at a temperature of from 70° to 80° Fahrenheit. The temperature must on no account be permitted to exceed 150° Fahr. For ordinary purposes the temperature of the cooling water can be determined with sufficient accuracy by placing the hand upon the casing, which should not feel more than slightly warm. This, of course, must be done when the pile has been lighted up for some ten or fifteen minutes, and has had time to reach its full temperature. Too much water is useless. The actual temperature of the waste water may of course be obtained by placing a thermometer bulb in the issuing stream. An electric current may be taken from the pile directly it is lighted, but the output steadily increases with the time until the maximum is reached in about ten minutes.

To stop the pile, turn off the gas and water, but do not disconnect the water pipes. Always allow the water jacket of the pile to remain full.

The same precautions in lighting the gas and regulating the height of the gas flame apply to the water-jacketed thermopile as to the

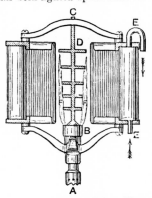

Fig. 34.—Section of Cox Thermopile having water-jacket. Flame burning correctly.

smaller unjacketed form. The appearance of the flame should be like that shown in Figs. 34 and 36. In Fig. 35 a gas flame is shown burning incorrectly, and in such a manner as to damage the pile. The correct form of flame can always be obtained by turning the gas up or down. The pile must never be used without the circulating water running. These are the only points which require attention, and by taking these precautions the natural life of the machine is indefinitely long. Negligence in regard to their observance may spoil the pile in a few hours. To observe the gas-flame look from the bottom of the pile upwards. One form of the Cox generator is made to be run with a spirit-burner, Fig. 33.

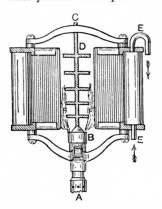

Fig. 35.—Section of Cox Thermopile, showing flame burning incorrectly.

All forms are claimed by the makers to be economical and

strong. The following details as to price and output have been supplied :—

Cox Generator for Gas with Water-Jacket.—

Volts on open circuit.	Short circuit current.	Approx. internal resistance.	Price. £ s. d.
3·0	4·0 ampères.	0·7 ohms.	5 0 0
4·5	3·5 ,,	1·2 ,,	5 10 0
5·0	3·5 ,,	1·5 ,,	6 0 0
8·5	2·5 ,,	3·5 ,,	10 0 0

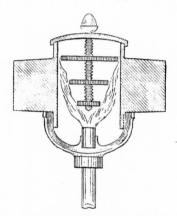

Fig. 36.—Section of Cox Thermopile for gas, without water-jacket.

Cox Generator for Gas or Spirit-Burner, without Water-Jacket.—

Volts on open circuit.	Short circuit current in ampères.	Price, complete with burner, etc. £ s. d.
3	1·3	1 0 0
1·3	3	1 0 0

Prices for separate items :—

	s. d.
Element	15 0
Deflectors	1 6
Spirit or gas-burner and stand	3 6

The Clamond Thermopile.—This pile was originally made by the Thermo Generator Company in or about 1876, and the following are details concerning their cost and output :—

Volts.	Internal resistance : ohms.	Short circuit current.	Gas consumption: cubic ft. per hour.	Price. £ s. d.
2	1·0	2 ampères.	—	3 0 0
3	1·5	2 ,,	3	4 0 0
6	3·0	2 ,,	5	6 10 0
12	6·0	2 ,,	9	13 0 0
34	12·15	2·7 ,,	37	32 0 0

Piles having a lower resistance were also made, of which the following results were given :—

Volts.	Internal resistance.	Short circuit current.	Coal gas consumption in cubic feet per hour.	Price. £ s. d.
3	0·6 ohms.	5 ampères.	6	8 0 0
10	2·0 ,,	5 ,,	23	20 0 0
20	4·0 ,,	5 ,,	37	32 0 0

The piles were spoken well of by Sir W. H. Preece, and were by his directions employed in the Government Telegraph Service, but owing to the fact that the junctions after a time were found to fuse by coming into direct contact with the burning gas, and to break by too sudden cooling and heating, and also because moisture and products of combustion apparently in time corroded the junctions, the use of the piles was discontinued ; but Sir W. H. Preece stated that the piles were far more compact than ordinary telegraph batteries, and expressed his belief that the difficulties mentioned above could be overcome. The Company, however, collapsed, and were unable to supply a complete set of thermopiles contracted for to the Postal Telegraph Department in 1876. Whether the pile is manufactured under the name of Clamond or any other title the present writer is not aware, but apparently, with a few not very complicated modifications, this pile could be made as successful as any others on the market.

The following is a description of the original Clamond pile given by Mr. Latimer Clark at a meeting of the Society of Electrical Engineers in 1876, an account of which, accompanied by an interesting discussion, may be found in the *Journal* of that Society, vol. v., p. 321, *et seq.*

Mr. Latimer Clark states that :—

"The mixture employed by Clamond consists of an alloy of 2 parts antimony and 1 of zinc for the negative element, and for the positive element he employs ordinary tinned sheet-iron, the current flowing through the hot junction from the iron to the alloy. The combination is one of great power. Each element consists of a flat bar of

the alloy from 2 inches to 2¾ inches in length, and from ⅜ to 1 inch
in thickness. Their form is shown in Fig. 37, by which it will be
seen that, looking at the plan, they are spindle-shaped or broader in
the middle than at the ends. The sheet-tin is stamped out in the
form shown in Fig. 38 ; the narrow portion is then bent in the forms
shown, in which state they are ready for being fixed in the mould.
The melted alloy is poured in, and, before it has cooled, the mould is
opened and the bars removed with the lugs securely cast into them.
The mould is heated nearly to the melting point of the alloy, and 10
or 12 bars are cast at one time. A little zinc is added from time to
time to make up for the loss due to volatilisation. The alloy melts at
about 500° Fahr. ; it expands considerably on cooling. The more
frequently the alloy is recast the more perfect becomes the mixture,
so that old piles can be reconverted with advantage and with little
loss beyond that of the labour. The alloy is extremely weak and brittle

Fig. 37.—Elements of Clamond's
Thermopile connected in series.

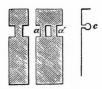

Fig. 38.—Sheet Tin Stampings
for Clamond Thermopile.

and easily broken by a blow—in fact, is scarcely stronger than loaf
sugar.

"The tin lugs are bent into form, and the bars are arranged in a
radial manner round a temporary brass cylinder, as shown in Fig. 37,
a thin slip of mica being inserted between the tin lug and the alloy,
to prevent contact, except at the junction. The number of radial
bars varies with the size of the pile, but for the usual sizes eight or
ten are employed. As fast as the bars are laid in position, they are
secured by a paste or cement formed of powdered asbestos and soluble
glass, or solution of silicate of potassa ; flat rings are also formed of the
same composition, which possesses considerable tenacity when dry ;
and as soon as one circle of bars is completed, a ring of the dry asbestos
cement is placed upon it, and another circle of elements is built upon
this, and so on until the whole battery is formed. Cast-iron frames
are then placed at top and bottom of the pile, and drawn together by

screws and rods, so as to consolidate the whole, and in this condition the pile is allowed to dry and harden. Looked at from the inside, the faces of the elements form a perfect cylinder, within which the gas is burned. The inner face of each element is protected from excessive heat by a tin strip or cap of tin bent round it; before it is embedded in the cement the projecting strips of tin from the opposite ends of each pair of elements are brought together and soldered with a blowpipe and soft solder. The respective rings are similarly connected, and the whole pile is complete, except as regards the heating arrangements. The positive pole of these piles is always placed at the top. Cumming was the first to use this stellar arrangement of couples. The pile is usually heated by gas mingled with air, on the Bunsen principle; gas is introduced at the bottom of a tube of earthenware, which is closed at the top, and is pierced with a number of small holes throughout its length, corresponding, approximately, in number and position with the number of elements employed. Before entering this tube, the gas is allowed to mix with a regulated proportion of air, by an orifice in the supply tube, the size of which can be adjusted; the mixed gases escape through the holes in the earthenware tube, and there burn in small blue jets, the annular space between the gas tube and the elements forming a chimney to which air is admitted at bottom, the products of combustion escaping at the top. In order to prevent injury from over-heating, and to diminish the consumption of gas, M. Clamond has introduced a new form of combustion chamber, by which he obtains very great advantages. This form is shown in Fig. 39. The mixture of air and gas is burnt in a perforated earthenware tube, as before described, but instead of extending the whole height of the battery, it only extends to about one-half of its height. The earthenware tube is surrounded by an iron tube of larger diameter, which extends nearly to the top of the battery, and is open at the top. Outside this iron tube, and at some distance from it, are arranged the elements in the usual manner. A movable cover fits closely over the top of the pile, and a chimney is connected to the bottom of the pile. Leading off from the annular space between the iron tube and the interior faces of the elements, the air enters at the bottom of the iron tube, and the heated gases, passing up the tube, curl over at the top, and descend on its outside, escaping eventually by the chimney. The elements are heated partly by radiation from the iron tube, and partly by the hot gases which pass, outside the tube, downwards towards the chimney. By this arrangement, not only is great economy of gas effected, the consumption, as I am informed, being reduced by one-half; but the great advantage is obtained that the jets of gas can never impinge directly on the elements, and it is thus scarcely possible to injure the connections by over-heating. In the event of a bad connection occurring, it is easy to find out the imperfect element, and throw it out of use by short-circuiting it over with a piece of wire, and the makers have no difficulty in cutting out a defective element and replacing it by a sound one. Coke and charcoal have also been employed as a source of heat, with very great economy and success : in fact, there are many countries and places where gas would not be procurable, but where charcoal or coke could be readily obtained.

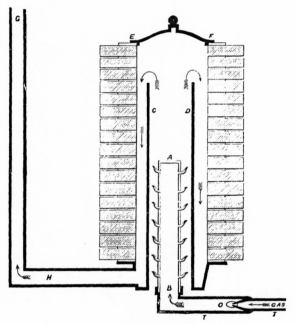

Fig. 39.—Section of Clamond's Thermopile.

The tension produced by Clamond's thermo-elements is such that each twenty elements may be taken as practically equal to one Daniell's cell, or about one volt."

It is stated that a Clamond pile of 100 bars, with the consumption of 5 cubic feet of gas, deposits about 1 ounce of silver per hour, and the same machine, arranged in multiple arc (that is, for *quantity*) will deposit about 1 ounce of copper in the same time : 400 large bars, consuming 2 lbs. of coke per hour, will deposit about four times the above quantities in the same period of time.

Messrs. L. and C. Wray introduced some important improvements upon the form of Clamond's thermopile as described by Mr. Latimer Clark, the descriptions of which were given in the discussion on Mr. Clark's paper (*loc. cit.*), but, as far as I am aware, there are no published details concerning the output, price, gas consumption, and life of any other thermopiles of sufficient magnitude for any electro-chemical work beyond those given above.

The Dynamo.—The third and last, and, for the larger operations, the most important current-generator is the dynamo. As was the case with the other forms of current-generators, no attempt will be made here to treat of the theory of the dynamo, nor its construction. The theory is well understood and the art of construction has reached a high degree of perfection. Particular information on these points does not concern the electro-metallurgist, who is merely interested in the cost, efficiency, durability, and method of treatment of the finished machines. The large number of modern treatises on the theory and construction of the dynamo now available make it an easy matter to obtain all details for those anxious to be acquainted with these subjects, but (let it be remembered) such knowledge is in no way *necessary* to the practical electro-metallurgist.

The dynamo then, in so far as we are concerned, is, like the electro-chemical battery and the thermopile, merely a source of e. m. f., which when connected up in a conducting circuit will yield a larger or smaller current according to the largeness or smallness of its e. m. f. and the smallness or largeness of the resistance placed in circuit.

There are several forms of dynamos available for obtaining an electric current, but there is only one form which is of any practical use for the electro-metallurgist, and that is the particular form of machine known as the shunt wound dynamo. Other forms of dynamos may possibly be employed, but they possess no advantage in cost or efficiency, and even when used by experienced persons for electro-chemical work, are always liable to cause more or less serious trouble from time to time. The soundest advice, then, to anyone about to purchase a dynamo for electro-chemical work of any other form than the shunt wound machine, is " Don't."

The dynamo is so largely employed nowadays that it is difficult to imagine that anyone who is in any way connected with technical work is not familiar with at least the general appearance of the machine. It is not possible to tell by mere inspection the efficiency and output of a machine, and the general details of equally good dynamos may be very different. Every machine, however, must contain the following ten important parts (see Fig. 40) :—

1. Field-magnets, M, made of steel or iron (of almost any shape).
2. Field-magnet windings, M', consisting of insulated copper-wire coils carried round the field magnets.
3. Armature shaft or spindle, E E, which rests on and runs in—
4. The armature shaft bearings.
5. The armature, A A.—This consists of a collection of copper wires or bars attached to a cylindrical core of iron plates. The complete structure is keyed firmly to the armature shaft.
6. The commutator, c.—This is a cylindrical assemblage of copper or occasionally iron segments, having an insulating layer between each segment. It is connected by wires to the armature, and is keyed on the armature shaft at one end of the armature.
7. The brushes, b b.—These are copper-wire gauze or sometimes carbon blocks, which are held in what are called brush-holders, which in their turn are held on the rocker. There are at least two brushes and may be more, but there is always an even number. The brushes are held pressing upon the commutator by means of hold-on springs, and when the dynamo is running the commutator slides beneath the brushes, with which it is always kept in close contact by the brush springs. The brushes are normally held in a fixed position whilst the machine is running, but can, if it is desired, be moved forward or backward a certain distance by rotating the whole of the rocker to which the brushes are attached. The rocker is carried on one of the fixed bearings of the armature spindle, on which it can rotate, but it can be fixed at any desired position and rendered immovable by means of a pinching or clamping screw.
8. The machine terminals.—These are usually fixed to a small wooden panel on the side or top of the field-magnets of the machine, and it is from these terminals that the current is taken off from the machine by means of conductors running to the vats direct, or more usually *viâ* a switchboard. The terminals are also connected one to each of the brushes of the machine, and also one to each of the ends of the field-magnet windings.
9. The pulley, P.—If the machine is belt-driven this is keyed to the armature shaft at the opposite end of the armature to the commutator. If the machine is not belt-driven there is a coupling for direct coupling to its engine.
10. In many cases there is also a fly-wheel keyed on to the armature shaft, especially when very steady speeds are required, but this fly-wheel may be, and generally is, absent. A view of an electro-plating dynamo made by the General Electric Company for 10 volts and 280 ampères is shown in Fig. 41. The price of this machine is £40.

A dynamo is really only a piece of apparatus for converting the mechanical energy of some steam- or gas-engine into the energy of the

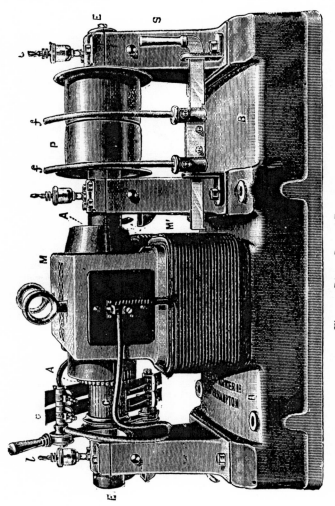

Fig. 40.—Direct-Current Dynamo.

electric current, and by itself can give out no energy unless it has mechanical energy constantly given to it by some engine, such as a steam-engine, gas-engine, oil-engine, electric motor, or a water or wind turbine. In setting up a dynamo it is therefore necessary to see that a source of a sufficient amount of mechanical energy is first provided.

The number of makers of good and reliable dynamos now to be found is large, and it is difficult to mention any particular machine or machines as being specially good In order, therefore, to purchase and instal a suitable dynamo and engine for any purpose involving more than a

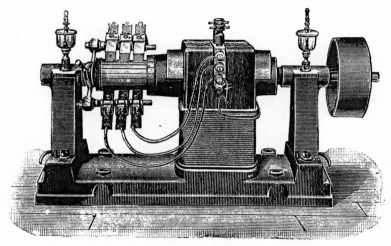

Fig. 41.—G. E. C. Electro-plating Dynamo, 10 volts, 280 ampères.

small outlay, it is, if the person desirous of obtaining the plant is not an electrical engineer himself, very desirable to obtain the assistance of a consulting electrical engineer, who will draw out specifications and obtain tenders for supply, delivery, and erection ; by this means usually a good deal of money and time is saved, and satisfaction obtained at an outlay of a small percentage of the total cost of the plant installed. For small plants, however, and for persons having some knowledge of mechanical, if not of electrical, engineering, it may be worth while obtaining the dynamo direct, and in this case it is advisable to write, stating the following details, to a number of firms manufacturing dynamos, and from their quotations a little consideration will enable the purchaser to select the machine whose price, output, etc., are best suited to his needs and pocket :—

Details concerning Shunt Dynamos for ————— —————

Power available to run dynamo, ————— brake-horse-power.
Current required not less than ———— ampères.
Volts required, about ———— volts.
Speed at which driving engine runs, ———— revs. per min.
The machine must not spark under any load up to full load **when** once the brushes are properly adjusted.

The manufacturer is asked to give the following details concerning the machine offered, for the output above stated :—

Volts on machine on open circuit at normal speed.
Volts on machine at full load current, at normal speed.
State what full load current is.
State whether voltage of machine can be varied by field rheostat, and, if so, between what limits.
State normal speed of machine.
State diameter and width of face of pulleys supplied.
State power required to drive dynamo at its full load at normal speed.
State power required to drive dynamo, when giving the number of ampères specified as being required, at normal speed.
State resistance of machine from brush to brush hot.
State temperature rise of armature and of field-magnets after running for six hours at the full load current at normal speed.
State direction of rotation.
State price of machine delivered at ———— ———— with fast and loose pulleys, belt fork, and striking-gear complete, and if of size requiring above 1½ horse-power to drive it, also state separately extra cost of foundation rails and belt tightening screws and field rheostat, if the latter is advisable.

Other things being equal, the machine with the smallest resistance from brush to brush, and the least difference between its voltage at zero load and full load will be found the most satisfactory, whilst a low speed and a large output are also good points which, however, unfortunately, usually involve correspondingly good prices. The same remark is true of a low temperature rise of both armature and field-magnets after a prolonged run at full load (the rise should not be greater than about 70° Fahr.), and also the smallness of the power required by the machine to drive it at full load. It is particularly necessary that the power required by the dynamo to drive it at full load shall be stated in brake-horse-power, as other units of horse-power are misleading. It is desirable that the speed should not be greater than about 1,500 per minute.

Care of a Dynamo.—When once a dynamo is erected and running, it is necessary to understand that by constantly observing certain simple precautions (which are much more readily carried out than is the case with the similar precautions required when employing primary or secondary batteries, or indeed with thermopiles) the life of the dynamo can be indefinitely extended, and, as a matter of fact, the repairs and depreciations incidental to running a properly looked after dynamo is not a quarter of the amount expended on a secondary battery which is equally carefully attended to, whilst the time and trouble necessitated in carrying out these precautions with the dynamo is very much less than is the case with cells.

It is, if possible, very desirable that a dynamo employed for electro-plating shall not be used in the same room as that in which the plating tanks are situated, for otherwise the vapours from the vats, more especially if these are run hot, are liable to condense on the machine and rust it, and reduce its insulation resistance until at last some short circuit of the machine may occur. It must be carefully remembered that probably the most active agent in deteriorating a dynamo is damp.

The bearings should be kept well oiled, and for small machines a light mineral oil may be employed : it is, however, necessary to guard against the lubricating oil creeping over to the armature and commutator. This mishap is provided against in well-designed machines by having a V-shaped ring shrunk on the shaft just outside the journals, which, as the oil creeps over it, throws it off on to a curved oil hood cast on the bearings.

Probably the most important point in the wear and tear of a dynamo is that involved in the commutator, and every effort must be made to keep this small. The copper brushes bearing on the rapidly rotating bars of the commutator are gradually worn down, and in their turn the brushes gradually wear more or less severely the surface of the com-mutator. The wear may be kept small by seeing that the surface of the commutator is very lightly coated with a thin film of vaseline, and the brushes must be carefully trimmed. The most destructive agent in the wear of the commutator is not, however, the friction between the brushes and the commutator surface, but is due to the sparking which occurs at the brushes. This sparking tends to pit the surface of the metal where it occurs, and once this pitting has taken place the action proceeds at a constantly accelerating rate. The great thing, then, is to avoid as far as possible all sparking. As far as the attendant is able to do this it can be done by seeing that the brushes are carefully filed to a flat bearing surface. For this purpose it is advisable to make a brush clamp of cast iron, as shown in Fig. 42.

The angle of the slope at the top of the clamp may be set for each particular dynamo and particular thickness of brush in the following manner :—Draw two parallel lines on a sheet of paper, Fig. 43, having the distance between them equal to the thickness of the brush which is to be employed. Then take a compass, and with one point at A, set off an arc whose radius is equal to the thickness of about $1\frac{1}{2}$ to $1\frac{3}{4}$ commutator segments, cutting the other line at the point B. Join A to B and this line gives the angle of slope of the top surface of the required brush filing clamp. To file up the brush place it in the groove G of the clamp and place the covering piece, C, over the brush, and then place in a vice and file the top end of the copper with a fine file with an outward stroke only, until the copper surface lies even with the sloping cast-iron surface of the top of the clamp. The brush is then fastened in the brush-holder, so that the sloping surface lies closely on the surface of the commutator. If the machine is a two-pole machine the points of contact of the brushes are arranged at diametrically opposite points on the commutator.

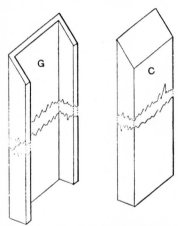

Fig. 42.—Brush Trimming Clamp.

The pressure with which the springs hold the brushes in contact with the commutator is usually adjustable, and should be arranged to be not too heavy or the brushes cut the commutator severely, and not too light or the vibration of the running machine causes the brushes to jump slightly, thus causing bad sparking. When sparking is observed at the commutator the brush rocker must be slowly rotated backwards and forwards until some point is found at which the sparking is absent or at a minimum, which should mean a very slight amount of sparking indeed. If this is not the case, at the earliest opportunity the brushes must be removed and trimmed as already described, and again adjusted. If the sparking still continues, and the machine has previously worked sparklessly, or nearly so, it is probable that either the machine is being overloaded, in which case sparking is often quite unavoidable, or some breakage has occurred in the armature windings. Do not overload a machine so as to cause sparking, for to run a machine with sparking brushes is the most extravagant method of running it possible. But if the sparking is not due to this, but has set in and cannot be stopped by any of the means noted above, call in an electrician to put the trouble right. Some machines, owing to faulty design, will always spark, and it is particularly necessary in purchasing a machine to ascertain by experiment that the machine will run at its full or any lower load, with properly adjusted brushes, without any sparking. If a machine does not spark at the brushes the commutator quickly assumes a dark blackish-brown polished appearance, which is very characteristic of a properly run dynamo. If in spite of all the precautions mentioned above, or because of their neglect, a dynamo which did not spark when first set up commences to spark, and, as a result, the commutator gets out of truth and is no longer truly cylindrical, but is worn down at certain points, either due to the sparking or to the cutting effect of too large a brush pressure, it is advisable to adjust a slide-rest parallel to the arma-ture spindle, and with a very sharp and fine-nosed tool turn down the surface of the commutator until it is once more truly cylindrical. Care must be taken in performing this operation : first, that a very light cut is taken ; second, that the rest is parallel to the armature spindle, or otherwise the surface of the commutator, after turning up, will be conical instead of cylindrical ; third, before starting the machine again after turning up the commutator examine the commutator very carefully to ascertain whether any metallic copper has dragged across the insulation between the commutator bars. If it has, it must be carefully removed. It is to avoid the danger of this occurring that a sharp cutting tool and a light cut are necessary in the turning operation. Running the dynamo with such a piece of copper connecting two neighbouring commutator segments may cause very serious damage indeed to the armature and necessitate lengthy and expensive repairs. It is only in the case of a bad form of dynamo, or very serious neglect of the proper precautions, that it should become necessary to have recourse to turning the surface true again. In any case it is evident that only just sufficient metal should be removed by the turning process to get to the bottom of the cavities worn in the commutator.

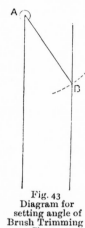

Fig. 43
Diagram for setting angle of Brush Trimming Clamp.

The armature shaft may be rotated by hand during the turning, and a little milk or soap and water may be used as a lubricant. The tool should be constantly re-sharpened. If it is noticed that the commutator is getting rather rough, but before the trouble has gone so far as to produce real inequalities of any size on its surface, necessitating re-turning to remove them, the roughness may be conveniently removed by means of a cylindrical segment cut in a thick piece of board whose internal radius is equal to the external radius of the commutator. If this cylindrical recess is lined with a piece of the finest glass paper and the dynamo is run at its usual speed, the brushes being removed, then, if the cylindrical recess lined with glass-paper is pressed lightly upon the revolving commutator, it will rapidly smooth it up bright. This treatment may also be used after the commutator has been re-turned, in order to give it a last smooth surface.

The dynamo must be kept as free as possible from dust, and if the presence of dust is unavoidable, then the dynamo itself must be closed in in some suitable cover. In any case, dust which gradually collects, and especially the copper dust arising from the wear of the brushes and the commutator, may be conveniently removed from the end windings of the armature and other places where it collects, by means of blowing with a pair of bellows. The foundations of the dynamo must be firm, and the machine bed-plate must not move at all perceptibly when the dynamo is being driven.

Driving Belts.—Concerning driving belts Mr. Watt remarks as follows : Most users of dynamo machines, polishing lathes, and other machinery driven by steam-power or gas-engines, will have experienced some trouble from the breaking, slipping, or slackening of the driving belts. Since a few hints upon these matters may prove acceptable, we give the following extracts from an interesting and thoroughly practical paper, by Mr. John Tullis, of St. Ann's Leather Works, Glasgow.*

"*Main Driving Leather Belts* should be manufactured so that when the joint is made while the belt is in its place, it ought to present the appearance of an endless belt. After having been taken up once or twice during the first year, good belts such as these require very little attention during the subsequent years of their long life. If the belt is driving in a warm engine-room, it ought to get a coating of curriers' dubbing three times a year. All belts having much work to do ought to present a clammy face to the pulley, and this condition can be best maintained by applying one coating of dubbing and three coatings of boiled linseed oil once a year. This oil oxidises, and the gummy surface formed gives the belt a smooth, elastic driving face. A belt looked after in this way will always run slack, and the tear and wear will be inconsiderable. On the other hand, dry belts have to be kept tighter,

---

* *Scottish Leather Trader*, July, 1885.

because they slip and refuse to lift the work. The friction of the running pulley 'burns the life' out of the belt while this slipping is going on."

Fixing the Belt.—As to which side of the leather ought to be placed next the pulley, Mr. Tullis says, "It is well known that by running the *grain* or smooth side next the pulley, there is considerable gain in driving power. However, by using boiled linseed oil, as before mentioned, the *flesh* side will soon become as smooth as the grain, and the driving power fully as good. A belt working with the grain side next the pulley really has a much shorter life than the belt running on the flesh side. The reason is, the one is working against the natural growth of the hide, while the other is working according to nature. . . . If you take a narrow cutting of belt leather, pull it well, and lay it down, you will at once observe that it naturally curves flesh side inwards. Nature, therefore, comes as a teacher, and tells us to run the flesh side next the pulley, and practice proves this to be correct."

Jointing Belts.—"Whether the belts are new or old, a properly made joint is of the first importance to all users of belting. . . . A well-made butt joint, with the lace holes punched in row of diamond shape, answers the purpose fully as well as any. Care should be taken that the holes do not come in line across the belt. A good lace, properly applied, with all the strands of the lace running lengthwise of the driving side of the belt, will last a long time and costs little. If a lap joint is made, time should be taken to thin down the ends of the lap. Joints of this sort should be made to the curve of the smallest pulley over which the belt has to work."

Accumulation of Lumps on Pulleys and Belts.—"Dust should never be allowed to gather into a cake either on pulley or belt, for if so, the fibre of the leather gets very much strained. The belt is prevented from doing its work because this stranger defies the attempts made by the belt to get a proper hold of the pulley."

Belts and Ropes coming off Pulleys.—"When a bearing gets heated, the shaft naturally becomes heavy to turn. The belts or ropes, having already the maximum of power in hand they are designed to cope with, refuse this extra strain, and will leave the pulleys at once, or break. This accident directs the attention of those in charge to the belts or ropes, when time is taken up in consulting as to what is to be done. Meanwhile the cause of all the trouble gets time to cool, and the source of annoyance is never discovered. Before a new start is made, all bearings are well lubricated. All goes smoothly, yet some one is blamed for the break down."

The above hints, coming as they do from an experienced manufacturer of leather, who is also an extensive user of belting, should be valuable to those who, though constantly using driving belts, may be unacquainted with the principles of their action.

To Start the Dynamo.—All that is necessary is to run up the speed of the engine, see that the brushes of the dynamo are up, and the switches open, then throw over the belt from the loose to the fixed pulley, and when the machine is running adjust the lubricators to give about three or four drops of oil per minute, or if the lubricators are not sight-feed, but have worsteds or other adjustments, see that these are properly set ; then, having noticed that the machine is running in the correct direction, which may be done by touching the commutator with the finger tip, put down the brushes, and when the machine has excited, and the outside circuit is ready, close the switches and the current can then be adjusted to its correct value, either by means of a regulating resistance in series with the machine and the plating baths, or by means of what is occasionally convenient, a regulating resistance or rheostat in the shunt windings of the dynamo, or by both of these devices.

To Stop the Dynamo.—The main switch is first opened, the belt is then thrown on to the loose pulley by the belt fork, and the brushes are raised, and the lubricators stopped running. The brushes should always be raised before the machine comes to rest, but not while running full speed.

Prices of Small Dynamos.—Small dynamos for electro-plating can be obtained for as low a price as about £10, up to very large prices for machines of large output. In order to give some rough idea of the prices at which the smallest machines may be purchased, the following details are given of machines listed by the General Electric Company, Ltd., 69, Queen Victoria Street, London, E.C., in 1895. Prices are probably somewhat higher now (1901). One of these machines is shown in Fig. 41.

Catalogue No.	200	201	202	203	204	205
Current in ampères . . .	32	56	72	96	96	280
E. M. F. in volts . . .	5	6	6	6	10	10
Diameter of pulley in inches .	3	3	3	4	4	6
Width of pulley face in inches .	2	2	2	$2\frac{1}{2}$	3	4
Approx. revs. per minute .	2,400	1,400	1,200	1,200	1,000	750
Approx. brake-horse-power at full load . . .	$\frac{1}{4}$	$\frac{1}{2}$	$\frac{3}{4}$	$1\frac{1}{4}$	$1\frac{1}{2}$	5
Price	£8	£16	£19	£24	£30	£40

The prices of small dynamos, with suitable direct current motors for driving them on the same shaft, Fig. 44, or, as they are called, motor-dynamos, which are listed by the same firm in an 1899 list, are as follows :—

PRIMARIES WOUND FOR 100, 115, and 230 VOLTS.

Output in volts.	Efficiency per cent.	Price. £ s. d.			Over-all dimensions, in inches.
80	40	17	10	0	14 × 9 × 7
175	43	30	0	0	20 × 11 × 9
375	50	36	10	0	24 × 13 × 10
800	62	55	0	0	26 × 15 × 14

Another form of motor-dynamo, known sometimes as a dynamotor, in

Fig. 44.—1 H.P. Motor-Dynamo.

which the motor and dynamo windings are on the same armature, running in a single field, Fig. 45, is listed at the following prices :—

Output in volts.	Efficiency per cent.	Price. £ s. d.			Over-all dimensions, in inches.
35-42	42	12	10	0	12 × 9 × 7
80	45	22	0	0	16 × 11 × 9
175	52·5	28	0	0	19 × 13 × 10
500	62	43	0	0	20 × 15 × 14

The voltage obtained from these motor-dynamos or dynamotors (as those in which there is a common field-magnet to the motor and the dynamo armature are sometimes called) can be made of any value desired by the purchaser for special plating work. The dynamotor is not so suitable for electro-plating work as the motor-dynamo, because the voltage from the dynamo side cannot be so cheaply and readily regulated as is the case with the motor-dynamo. In purchasing a machine of this type it is important to obtain precise details beforehand of the volts taken by the motor, and the volts delivered by the dynamo portion. Also the variation in the voltage of the dynamo portion from zero load right up to the full load current (which must

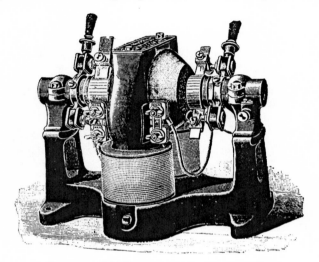

Fig. 45.—1 H.P. Dynamotor. For Electro-plating.

always be stated), should be guaranteed. When inquiring for a motor-dynamo, the following details should be given to the manufacturer :—

1. Maximum current required from dynamo armature.
2. Voltage required.
3. Voltage of circuit from which the motor portion of the machine is to be driven.
4. Distance of point at which motor-dynamo is to be installed from the point where power supply leads enter the building (approximately).
5. Work which the dynamo is to be employed for, e.g., plating, metal refining, etc.

The motor-dynamo is not to spark either on the motor or the dynamo side under any load up to full load when once the brushes are properly adjusted. The dynamo side must be shunt wound.

The manufacturer should be required to give the following details :—

1. Normal speed at specified voltage of power supply circuit.
2. Normal dynamo voltage at specified normal speed at zero load.
3. Normal dynamo voltage, motor being run off specified power supply voltage, when the dynamo is loaded to specified full load.
4. Normal full load current on motor at specified power supply voltage.
5. Normal full load current from dynamo when run at normal specified speed.
6. Temperature rise of field-magnets and also of the armature or armatures after the machine has been running at full load for six hours.
7. Price of motor-dynamo or dynamotor delivered and installed in purchaser's premises, the price to include starting-gear, switch, and safety cut-outs on the motor side. Full details of the starting apparatus, main switch, and safety cut-outs, if any, which are included, must be given.
8. State if it is possible to regulate the voltage of the dynamo output, and if so, between what limits ; also state whether the apparatus necessary for this voltage regulation is included in the price given under (7), and if not, give the price for this regulator installed complete, separately.

Other things being equal that motor-dynamo or dynamotor is the most satisfactory which :—

1. Takes the smallest current at full load from the supply main at the specified power supply voltage.
2. Gives out the largest current from the dynamo side at the specified plating voltage.
3. Has the smallest variation of volts on the dynamo side as the load is altered from zero to full load.
4. Has the smallest temperature rise under the conditions stated above.
5. Runs at the lowest speed.

It is desirable that the bearings shall all be fitted with automatic ring lubricators with oil level sight tubes, or in any case sight feed lubricators should be employed.

It is also very desirable that the motor side of the machine shall be fitted with a double pole switch, a starting switch, double pole fuses, and also an automatic zero current cut-out. A very convenient and largely used form of switch, which combines a starting switch and a zero current automatic cut-out, is manufactured and sold by the Sturtevant Engineering Company, Limited, of 75, Queen Victoria Street, London, which is known as the Cutler-Hammer Plain Motor-starting Rheostat, with automatic zero current release, Figs. 46 and 47. It is made at a price from £1 10s. for ¼ B. H. P. motors; £2 for 1½ to 2 B. H. P. motors; £2 10s. for a 3 B. H. P. motor; £4 for a 5 B. H. P. motor, up to £16 for a 50 B. H. P. motor.

It is, in the writer's opinion, quite essential that motors shall be furnished with the safety and starting gear specified above, if trouble with the motor or the supply mains is to be avoided. The whole cost of the double pole switch, double pole fuse holder and fuses, and including the starting switch and automatic magnetic zero current cut-out

Fig. 46.—Motor-starting Switch.

mentioned above, should not cost more than £2 15s. for a 1½ to 2 B. H. P. motor, not including the cost of fixing. For a motor of about ¼ to 1 B. H. P., the cost will be about 10s. less.

Having now considered the various available sources of the electric current with as full detail as the limits of this treatise permits, we must turn to the question of the most suitable source to select under given conditions of work, always bearing in mind the important questions of prime cost, depreciation, maintenance, and cost of energy per given output of the electric generator.

F

By the cost of energy we therefore include the cost of attendance
together with the cost of the chemicals consumed in a battery, the
cost of the gas or fuel burnt in a thermopile, and the cost of gas
burnt in a gas-engine, oil in an oil-engine, or coal or other fuel under
the boiler supplying a steam-engine; or lastly the cost at which
electrical energy can be obtained from some source of electric supply.
This source of electric supply is usually the public electric power or
lighting supply mains, but may also include electric power obtained from
a neighbouring premises or the electric power obtained in a charged

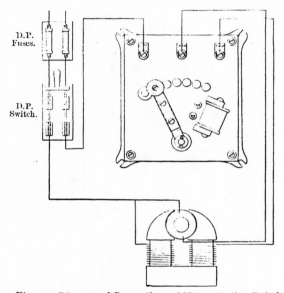

Fig. 47.—Diagram of Connections of Motor-starting Switch.

secondary battery supplied from some neighbouring generator.   The
possibility of obtaining power from either water or wind turbines is at
present so limited (and to all appearances will remain so) that these
sources of energy will not be considered in any detail here.

Electric Generators for Small Outputs.—In any particular case
where it is proposed to set up some form of electric generator for a
small amount of electric current for electro-plating or other small
electro-metallurgical operations, the circumstances under which the
installation is to be made are, of course, definitely known, and there

must be some definite solution of the question as to what particular generator to instal. In giving general advice, however, it is difficult to consider every possible condition which may arise in practice, and it has therefore been thought wise to select some few definite cases under definite stated conditions, and give the solutions for these, which, in the writer's opinion, appear most satisfactory, leaving to the reader the task of applying one or other or a combination of any of these solutions to his own case.

It must be borne in mind, that throughout this book the prices of apparatus which are given are only stated in order that some idea of the nature of the outlay involved in a given installation may be obtained. Prices vary from year to year, and a correct estimate can only be made after having obtained a tender to a definite specification. The prices given here, however, are always taken from some stated list.

Case I.—No power, either electrical or mechanical, is available, and it is not possible to get secondary batteries charged in the neighbourhood. Under these circumstances it is probably advisable to employ the " Cupron " element up to a current output as large as 20 ampères at 5 volts, which will cost probably about £24 10s., or if the work is not very steady in character, but is required a good deal at one time, with periods of little or no work intervening, it is probably wise to use the " Cupron " element for currents up to as high as even 30 ampères at 5 volts. The former battery, costing £24 10s., may be still used for this current, as the normal current output of such a battery is 24 ampères, and the maximum which may be used is 48 ampères, although it is advisable to run the battery with currents as near the normal as possible, if it is desired to obtain the largest ampère-hour efficiency. If the current required is larger than 30 ampères, then either a small steam-engine, boiler and dynamo, or a gas-engine and dynamo should be purchased. There is not much difference in the prime cost of such a combination for small powers, but the room required and also the small attention needed make it desirable to purchase the gas-engine combination. The total cost f. o. r. of the dynamo yielding 32 ampères at 5 volts, is listed by the General Electric Company, in 1895, at £8 (see p. 61), and the gas-engines may be purchased for the following prices, taken from the Stockport Gas-Engine List of 1900.

Brake-horse-power.	1½	2½	2¾	4	4¾
	£ s.	£ s.	£ s.	£ s.	£ s.
Price . . .	36 10	42 0	50 0	58 10	68 0

The cost of the necessary water-tanks is included in these prices. As these machines are fitted for constant speed for electric lighting with two fly-wheels, it is probable that cheaper engines might be purchased. The difference would not, however, be much. The prices also include, standard-size pulley, patent anti-pulsating gas bag, exhaust boxes and the usual set of spares with spanners. These engines are made by J. E. H. Andrew & Co., of Reddish, near Stockport. A gas-engine dynamo plant therefore, to yield 32 ampères at 5 volts, can be purchased for about £44 10s., exclusive of erecting, and one to yield 56 ampères at 5 volts for £52 10s., exclusive of erecting. The cost of the " Cupron " element for various current output is worked out in the following table :—

Volts.	Current.		Price.
			£ s. d.
5	normal 8	maximum 16	8 2 0
5	„ 16	„ 32	16 4 0
5	„ 32	„ 64	32 8 0
5	„ 40	„ 80	40 10 0
5	„ 48	„ 96	48 12 0
5	„ 56	„ 112	56 14 0

It must be remarked that the prices stated for the " Cupron " element, as given above, are those given on a German list, and they would, no doubt, be more expensive delivered in England. The reason that the writer would advise a gas-engine dynamo set costing say £45, and capable of yielding 32 ampères at 5 volts, rather than a " Cupron " element set of cells having precisely the same normal output as the dynamo, is because the depreciation and upkeep of the " Cupron " elements would be higher than is the case with a properly looked after gas-engine and dynamo. A cheap vertical steam-engine and boiler, packed and f. o. r., is listed by the General Electric Company, at the following prices :—

Brake-horse-power.	$\frac{1}{2}$	1	$1\frac{1}{2}$	2	$2\frac{1}{2}$	3
Price . . .	£36	£43	£49	£57	£57	£82

It is therefore evident that for very small powers the steam-engine has about the same cost as the gas-engine, but that as the powers increase the gas-engine has the lower prime cost. For any powers up to about ten brake-horse-power, the present writer considers, partly for reasons already pointed out, that the gas-engine run on town gas is better for electro-plating factories than a steam-engine of the same output.

Case II.—If, as in the last case, there is no electrical or mechanical power available, and no means for charging a secondary battery, and there is further no supply of coal-gas available, then a steam-engine dynamo combination with boiler may be purchased to yield 32 ampères at 5 volts for about £44, and a similar combination to yield 56 ampères at 5 volts for about £49, and when the current required is under about 20 ampères for steady work, or 30 ampères for variable work, the "Cupron" element should be employed instead of steam-dynamo set.

Case III.—If sufficient power is available, as, for instance, from some engine on the premises which is only partially loaded, it is probably better to purchase a small dynamo costing, say, £8 or £10, when the probable quantity of current required is found to be greater than about 3 or 4 ampères. Remembering that such a dynamo will yield as much as 32 ampères. If a current of less than about 3 or 4 ampères is required it is probably best to purchase 5 No. II. "Cupron" elements, costing £2 5s., and yielding, when in series, 2 to 4 ampères at 5 volts. (*See p.* 11.)

Case IV.—If an electric supply company can supply a continuous current on the premises at any reasonable rate, then it is probably advisable to purchase a suitable motor-dynamo or dynamotor directly it is seen that the current required will be as large as 6 ampères or so, for "Cupron" elements for this output at 5 volts would cost about £3 2s., whilst a dynamotor having an output of 8 ampères would cost about £12.   Before, however, deciding to instal an electric motor, a clear understanding must be arrived at as to the conditions of charge for power under which it will be placed. The charge is reasonably low, in many cases about 2d. per Board of Trade unit ; which unit, for a motor-driving dynamo, may be looked upon as one horse power used throughout one hour, or half that power used steadily for twice the time and so on.   But in some one or two cases with which the writer is acquainted a very large charge is also made per annum or per quarter by the supply corporation for each horse-power or fraction of a horse-power of the motor installed.   This appears to be an arrangement which must, one would imagine, have a good deal of effect in discouraging the use of electricity for small power units, a use which is generally understood to be the great object of supply companies to foster in order to increase their dividends.   This heavy charge per b. h. p. of power taken by the motor installed per annum is precisely the same whether any power is used or not, and although not usual, must, where imposed, be carefully considered when deciding on the installation of a dynamo driven by electric power. The care required to keep a motor-dynamo or dynamotor running satisfactorily is slight, and the details are the same as those necessary when running a dynamo, described on pages 51 to 61. The cost of about £1 10s., or perhaps less, must be added to the above price for the motor-dynamo to include a satisfactory arrangement of the main switch, starting switch, automatic zero current cut out, and fuses, as previously specified for any electric motor. (*See p.* 64.)

Case V.—If portable secondary cells can be charged in the neighbourhood at a moderate rate then the secondary battery is cheaper in prime cost, and probably in depreciation, than the " Cupron " element. For instance, a portable secondary battery cell, manufactured by the Electric Power Storage Co., of 4, Great Winchester Street, E.C., having a capacity of 50 ampère-hours, and giving a safe maximum discharge at the rate of 9 ampères at 2·2 volts, and which can, therefore, be kept up over 5 hours, weighs 21 lbs. complete, and costs 15s., whereas 3 No. IV. " Cupron " elements, which give a safe discharge at normal current of 8 ampères and 2·4 volts, can be run for about nine times this length of time, namely, about 45 hours, before they require the copper oxide regenerating. Now each charge of the secondary may, perhaps, be made at 1s. per charge, and, consequently, the cost of the secondary battery and nine charges is 24s., whereas the price of the 3 No. IV " Cupron " elements is about £4 5s. The " Cupron " elements can, of course, be regenerated at a small cost, but there can be no doubt that secondary batteries are cheaper, both in prime cost and up-keep, than " Cupron " element batteries if the battery-charging can be performed in the neighbourhood at a reasonable cost and with reasonable care of the cells. It must, indeed, be remembered that much damage may be in time done to secondary batteries by charging them with currents much above their specified charging currents, and as the batteries sent in, to anyone undertaking charging, usually vary much in size and the current they require, there is a good deal of temptation to connect them in series and charge them all with the same current. If the maximum current employed, however, is within the smallest specified current for any cell in the batch no harm will be done to any of the cells charged at too small a current, the only trouble being that they will require a longer charge.

Below is given a table of the details and prices of the Electrical Power Storage Co.'s Q type portable cells, also shown in Fig. 22, from their list dated 1898.

Q Type Portable Secondary Batteries.

The prices given below are for cells in lead-lined wooden boxes.

Description of Cell.		Working Rate.		Capacity Ampère Hours.	Approximate Dimensions over handles and terminals.			Acid for each Cell.	Weight complete with Acid.	Price.	
Number of Cells.	Number of negative Plates.	Charge Ampères	Maximum Discharge Ampères.		Length.	Width.	Height.	Weight in lbs.			
					ins.	ins.	ins.		lbs.	s.	d.
Single Cells.	3 Q Plates	·5 to 1·3	1·3	7	2	$5\frac{5}{8}$	$9\frac{1}{4}$	·9	6	6	6
	5 ,,	1 ,, 2·6	2·6	14	$2\frac{3}{4}$	$5\frac{5}{8}$	$9\frac{1}{4}$	1·5	8	8	0
	7 ,,	2 ,, 4	4	21	$3\frac{3}{8}$	$5\frac{5}{8}$	$9\frac{1}{4}$	2·2	$10\frac{1}{2}$	9	6
	11 ,,	4 ,, 6·5	6·5	35	5	$5\frac{5}{8}$	$9\frac{1}{4}$	3·4	16	12	9
	15 ,,	7 ,, 9	9	50	$6\frac{5}{8}$	$5\frac{5}{8}$	$9\frac{1}{2}$	4·5	21	15	3
	21 ,,	10 ,, 13	13	70	9	$5\frac{5}{8}$	$9\frac{3}{4}$	6·5	28	19	0

Case VI.—Supposing that both mechanical power and electrical power are available. Then for small currents up to about 10 ampères the "Cupron" element may be employed, or if cheap secondary battery charging is available, secondary batteries may be conveniently used probably up to as much as 40 or 50 ampères, but above this amount there is no doubt but that a motor generator costing £36 10s. and yielding 75 ampères at 5 volts is the best thing to invest in, for the up-keep and depreciation of a motor generator is very small and the power used cheap, whilst the up-keep and depreciation of secondary batteries is about 20 per cent. on their prime cost per annum, whilst power obtained by external charging is always expensive.

The foregoing discussions of several definite cases are, however, incomplete, for they do not and cannot take account of several factors which must always be considered when deciding the best course to pursue. For instance, a very essential point in commencing a business may be to keep the capital outlay small, and again, the knowledge as to whether a given output is likely to remain steady or to increase, or whether the business, although averaging a steady output during the year, is likely to vary very much from time to time during that year, are all points with many others to which due weight must be given. It is, however, to be hoped that the cases treated of and the information as to prices given may prove of assistance in guiding an intending purchaser of electric generators.

Gas Engines run on Producer Gas.—Nothing has been said

hitherto of the use of gas-engines run on producer gas, and for this reason, it is not, in the writer's opinion, an economical plan to fit a gas-engine with a gas producer plant unless the output of the engine is well over 20 b. h. p., and, therefore, although producer gas is an enormous saving for plants of larger output it is not suited to the small plants we have hitherto considered.

Oil Engines.—Oil engines are considerably more expensive than either gas or steam engines of equal output, and they cannot, therefore, compete with these generators unless coal gas is not to be had and one or more of the following conditions exist :—

1. Coal is very dear, or there is no facility for its delivery and storage.
2. Space is of great importance.
3. A boiler and boiler-furnace and flue are, for any reason, objectionable.

As it is possible that such cases may arise, the following prices may prove of use.

Brake horse-power	$1\frac{1}{2}$	$2\frac{1}{2}$	$3\frac{1}{2}$	5	$6\frac{1}{2}$
Price, including water vessel	£ s. 80 15	£ s. 93 10	£ s. 107 5	£ s. 129 0	£ s. 145 0

The above prices do not include connections to water vessel, foundation bolts, piping to silencer, nor pulley.

The cost of fuel for these engines is stated to be under one half-penny per brake horse-power hour. These prices are given in the General Electric Co.'s list for 1895, but apply, probably, to Tangye's or Priestman's oil engines. The price of either make is about the same.

Advantages of Portable Cells.—Primary and portable secondary battery cells have two distinct advantages over dynamos, and these are :—

1. They can be brought close to the work it is intended to treat, thus saving long leads.
2. They can be combined by connecting either in parallel or series so that the output from them may be varied considerably with regard to the volts in circuit, which is sometimes useful for different kinds of work.

These advantages are, however, not very marked unless the work is of an irregular character. Moreover, the variation of the voltage of the dynamo may be secured either by a shunt field regulating rheostat, or by means of a regulating resistance introduced into the circuit in series with the plating vat and the dynamo.

Regulating Resistances.—When using a dynamo or batteries for electro-plating, it is frequently necessary to vary the current employed, whilst the e. m. f. of the current generator is constant. Thus, supposing that we have a dynamo driven by a gas engine whose speed we cannot alter appreciably and that work is in course of being put into the plating vats, it is clear that the work first put in may have a very much larger number of ampères per square foot than will be the case with the work which is put in last, it may, therefore, be desirable to somewhat reduce the current on the work first put in the bath, and this may be done by means of an adjustable resistance placed in circuit with the leads running from the dynamo to the bath. Again, the work plated by the same dynamo may be of quite a different type on different days, silver at one time, gold at another, or sometimes alkaline and sometimes acid copper baths and sometimes nickel, or it may happen that a silver, a copper, and a nickel plating vat may all be taking current from the same dynamo at the same time. Under each of these conditions the e. m. f. required and the current density will be different, but if the dynamo has been purchased to yield the highest electrical pressure required, any lower pressure at the terminals of the plating-vat, and any desired current may be obtained at will by placing a suitable regulating resistance in circuit as described. What has been said about the regulation of dynamos applies equally to electric batteries, except that the batteries may be employed, so to speak, to build up any desired e. m. f. by connecting them suitably in either parallel or series; notwithstanding the facility which batteries offer in this respect, it is, nevertheless, most desirable that the flexible method of controlling the current, which is afforded by the use of a regulating resistance, should be employed if the best results are to be obtained.

An electro-plating plant, therefore, should be supplied with generators capable of yielding an e. m. f. and a current equal to the greatest calls which are to be made upon the establishment, but to properly manipulate these quantities in order to get the best results under the varying conditions of solution, nature of deposited metal, nature of surface upon which deposit is to be made, and the amount of cathode surface at any time in the plating vats, it is most desirable (even when, as should always be the case, a shunt regulator is supplied with the dynamo) that some current-regulating device should be provided. One of the most generally useful of these is the carbon plate resistance. It is shown in Fig. 48, and consists of a series of carbon plates, c, in an insulated support, fitted with a screw H, at one end. The current passes through the carbon plates in series, and the resistance of the pile varies with the pressure with which the plates are squeezed

together, and this is varied by means of the large hand screw at one
end of the apparatus. The cost of such a regulatable carbon resistance
depends, of course, upon its size. One which will take 50 ampères
continuously without heating to redness is made by Messrs. Parfitt &
Webber, of Denmark Street, Bristol, for a price of £2. It contains
50 carbon plates, each about 3 ins. square and ¼-in. thick. These are
insulated from the iron rods at the sides and bottom by means of thick
sheets of asbestos millboard, A. Similar asbestos sheets protect the cast
iron end-plates from coming into contact with the carbons or the gun
metal terminal plates, E E. To these plates the terminals are screwed.

The resistance of such a carbon plate resistance may always be halved
or reduced to any value, at a given tightness of the carbon plates,
by means of a simple device—namely, by taking out the gun-metal
terminal plate and inserting it between the carbons at any other point
than the extreme end of the pile where its normal position is. The
resistance, therefore, of this apparatus may be reduced from that of
the whole pile of carbon plates loosely pressed together down to that
of one carbon plate only, tightly clipped between the two end or

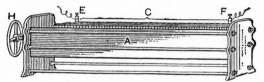

Fig. 48.—Adjustable Carbon Plate Resistance.

terminal metal plates of the apparatus. There is no harm whatever
to be feared from running such a carbon resistance with a current so
large and for so long a time that the carbon plates are at a dull red
heat, provided that the resistance is fixed in such a position that there
is no danger of fire being caused by it. If it is stood on sheet iron
which is supported on bricks, there can be no danger at all on this
account, even if the bricks rest upon wood.

Another form of carbon resistance which is a good deal cheaper than
the one last described is made of electric light carbons, and is shown in
Fig. 49. It fulfils the conditions that such a resistance for electro-
plating requires, namely, that it is cheap, strong, of considerable range
of adjustment, and not easily damaged by moisture and fumes. It
consists of a fixed rack, A, and a movable rack, B, both of cast iron.
The movable rack B, can be moved nearer to or farther from A, and
can be clamped in any position on the board C, by means of a clamping
screw, not shown in the figure, which passes through the hole D in
the metal projection at the right hand side of B. This screw exerts

pressure on the strip of tin, E, which is let into the board, but does not come nearer than about 2 to 3 inches from the rack A. The depressions in the rack, the number of which may be as large as is desired, but which the writer finds may be conveniently two, are made of a curved outline, and are meant to take ordinary electric light carbons of from 10 to 15 millimetres diameter. A good contact between these carbons and the metal racks is obtained by means of two cast iron clamping bars, F F, only one of which is shown in the figure, these are about one inch wide. To clamp the carbons annealed copper strips, one inch wide, $\frac{1}{20}$th of an inch thick, with holes at the ends, are slipped on the $\frac{1}{4}$ inch stud screws G G G G, the carbons are then laid in position over these strips, and two more similar strips of copper are placed over them. The clamps are finally placed in position, and are screwed down with sufficient pressure by butterfly nuts, screwing on to the

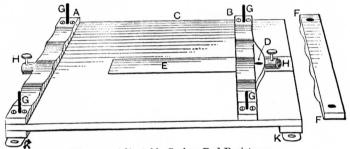

Fig. 49.—Adjustable Carbon Rod Resistance.

stud screws. The length of all the copper strips between the holes in these ends is less than the distance measured along the curved surface of the iron racks and clamps, and hence these copper strips are pulled into tight contact with the carbon rods when the butterfly nuts are screwed down. The terminals of the resistance are at H H. The resistance may be either employed horizontally as shown, or may be hung on the wall by mirror plates or hooks at K K. This resistance is also made by Messrs. Parfitt and Webber, of Denmark Street, Bristol. One with four carbons costs 15s. complete, whilst two, with two carbons each, would cost about 11s. or 12s. each. The four-carbon form* gives a range from one carbon a foot long as the highest resis-

* The resistance of an electric light carbon, having a diameter of about 14 m.m. and not soft cored, I have found to be about 0·142 ohms per foot length when carrying a current of 30 ampères in air. The temperature of the carbon under these conditions was just sufficiently high to blacken paper pressed against it.—A. P.

tance down to four in parallel, which may be shortened up by shifting the rack B to a length of not more that about 4 inches. The two double-carbon form gives, however, double this range, *i.e.*, from a maximum resistance of two foot long carbons in series down to the

same value for low resistance as the four-carbon form. The prices given above for these resistances include slate boards, which in that case do not require the strip of tin E, which is intended to protect the wood from the end of the screw. Each carbon will carry about 30 ampères without becoming sufficiently hot to set light to paper.

A variable liquid resistance, sold by the General Electric Company, of 69, Queen Victoria Street, is shown in Fig. 50. It is known as Lyons' Variable Liquid Resistance, it will carry from 50 to 80 ampères. It consists of a stoneware jar 24 inches high and 8 inches in diameter. Each jar is in a framework, to which is attached a raising or lowering gear for the purpose of inserting a cone of lead in the so-called non-ascetic liquid of secret composition, probably a solution of some neutral

Fig. 50.—Lyons' Variable Liquid Resistance.

salt. The price of the cheapest of these resistances is high, being over £6. Cheap adjustable liquid resistances can, however, be rigged up on a somewhat similar principle for temporary purposes by means of a large jam jar and a couple of lead plates in sodium sulphate solution.

Adjustable wire resistances, sold by Messrs. O. Berend & Co., of Dunedin House, Basinghall Avenue, London, are shown in Figs. 51 and 52. These resist-
ances cost from 16 to 19 shillings each in the form of Fig. 51, carry-ing a current of from 4 to 15 ampères : whilst resistances of the form shown in Fig. 52, in which the steps are not so gradual as in the resist-ances shown in Fig. 51, the price varies from 25 to to 45 shillings, according to size, the current car-ried varying also from

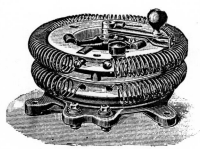

Fig. 51.—Adjustable Wire Resistance.

4 ampères in the smallest to 20 ampères in the largest size.

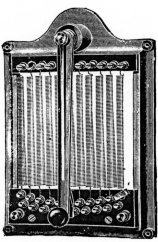

Fig. 52.—Adjustable Wire Resistance.

Determination of the Polarity of Generators.—It is not infrequently necessary to ascertain the polarity of a dynamo, or even a battery, and this may be done, in the case of a dynamo, either by passing a current from it through a solution of one part of strongest sulphuric acid and three parts of water contained in an ordinary jam jar from terminals of lead scraped bright. The greatest care must be observed, at least in testing larger machines in this way, to prevent the two lead strips coming into con-tact. They may be conve-niently kept at a distance of about three inches from one another. The lead strip at-tached to the positive pole of the dynamo will become of a dark brown or chocolate colour in about 5 to 10 minutes. A quicker test, which may be used with either batteries or dynamos, is to employ pole finding paper made with phenolphthalein. This paper is white, but if it is moistened and laid on a table, and the two wires from the terminals of the dynamo or battery are pressed upon it at a distance of an inch from each other, a deep crimson stain will be shown under the negative wire. This action takes place at once. Such pole papers may be purchased at a cheap rate made up into little books.

General Arrangement of Plating Vats.—Fig. 52A gives a general perspective view of two vats in a plating shop, arranged in parallel, with their ampère-meters and regulating resistances in circuit,

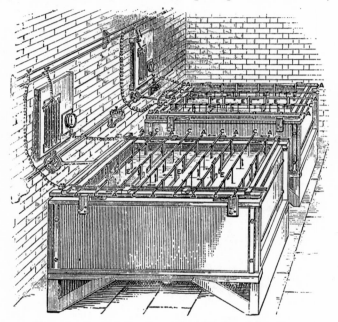

Fig. 52A.—General Arrangement of Electro-plating Vats.

and a shaft along the side of each vat which, by means of eccentrics, moves an oscillating frame, causing the work to constantly keep in gentle movement, a movement which is found to be very conducive to uniform plating deposits. A are the anode rods and C the cathodes.

It need hardly be said that these vats are intended for nickel, copper or silver plating, but not for gold work, which is done in much smaller vats, usually of stoneware or china. A somewhat similar view is shown in Fig. 108.

CHAPTER III.

HISTORICAL REVIEW OF ELECTRO-DEPOSITION.

Announcement of Jacobi's Discovery.—Jordan's Process Published.—Jordan's Process.—Spencer's Paper on the Electrotype Process.—Effect of Spencer's Paper.—Vindication of Jordan's Claim.—Mr. Dircks on Jordan's Discovery.—Sir Henry Bessemer's Experiments.—Dr. Golding Bird's Experiments.—Origin of the Porous Cell.

Long before the art of Electro-deposition was founded upon a practical basis, it was well known, experimentally, that several metals could be deposited from their solutions upon other metals, by simply immersing them in such solutions ; but this knowledge was of little importance beyond the interesting nature of the results obtained. The schoolboy had been accustomed to amuse himself by producing the ever-popular "lead tree," by suspending a piece of zinc attached to a copper-wire in a solution of sugar of lead, or the "silver tree," with a solution of nitrate of silver and mercury ; or he would coat the blade of his penknife with copper, by dipping it for a moment in a weak solution of sulphate of copper (bluestone). But these, and the like interesting facts, were of no practical value in the arts. It was also known that articles of steel could be gilt by simple immersion in a dilute solution of chloride of gold (that is gold dissolved in aqua regia), or still better, in an ethereal solution of the chloride, and this simple process was sometimes adopted in the ornamentation of engraved articles, in imitation of the process of *damascening*. The eyes of needles were also gilt by a similar process, and "golden-eyed needles" became popular amongst the fair sex. With this exception, however, the deposition of metals, even by simple immersion in metallic solutions, was regarded as interesting and wonderful, but nothing more.

As far back as about the year 1820, the author's father covered the "barrels" of quill pens with silver, by first steeping them in a solution of nitrate of silver, and afterwards reducing the metal to the metallic state in bottles charged with hydrogen gas, the object being to protect the quills from the softening influence of the ink.

In the year 1836, Professor Daniell made known his *constant battery*, and in the same year, Mr. De la Rue constructed a modification of this battery, in working which he observed that "the copper-plate is also covered with a coating of metallic copper which is continually being deposited; and so perfect is the sheet of copper thus formed, that, being stripped off, it has the counterpart of every scratch of the plate on which it is deposited.* Although this interesting observation did not lead to any direct application at the time, it is but reasonable to presume that in the minds of some persons the important fact which it disclosed would have suggested the possibility of its being susceptible of some practical application. It was not until the following year (1837), however, that the electro-deposition of metals, experimentally, seriously occupied the attention of persons devoted to research, the first of whom was Dr. Golding Bird, who decomposed solutions of the chlorides of ammonium, potassium, and sodium, and succeeded in depositing these metals upon a negative electrode of mercury,† whereby he obtained their amalgams. From the time when his interesting results became known, many persons repeated his experiments, while others turned their attention to electrolysis as a new subject of investigation, and pursued it with different objects, as will be shown hereafter.

Mr. G. R. Elkington, in 1836, obtained a patent for "Gilding copper, brass, and other metals" by immersing the articles in a boiling alkaline solution containing dissolved gold. This was followed, in 1837, by several other patents granted to Mr. H. Elkington for coating metals with gold and platinum, and for gilding and silvering articles. In 1838, Mr. G. R. Elkington, with Mr. O. W. Barratt, patented a process for coating articles of copper and brass with zinc, by means of an electric current generated by a piece of zinc attached to the articles by a wire, and immersing them in a boiling neutral solution of chloride of zinc. This was the first process in which a separate metal was employed in electro-deposition.

Announcement of Jacobi's Discovery.—About the period at which the above processes were being developed, it appears that several other persons were engaged in experiments of an entirely different character and of far greater importance, as will be seen by the results which followed their labours. In St. Petersburg, Professor Jacobi had been experimenting in the deposition of copper upon engraved copper-plates, a notice of which appeared in the *Athenæum*, May 4th, 1839. The paragraph ran as follows :—" *Galvanic Engraving in Relief.*—While M. Daguerre and Mr. Fox Talbot have been dipping their pencils in the solar spectrum,‡ and astonishing us with their

---

* *Philosophical Magazine*, 1836.

† "Philosophical Transactions of the Royal Society," 1837.

‡ It was about this period that the famous *Daguerreotype process* of portrait-taking was being developed in England.

inventions [photographic], it appears that Professor Jacobi, at St. Petersburg, has also made a discovery which promises to be of little less importance to the arts. He has found a method—if we understand our informant rightly—of converting any line, however fine, engraved on copper, into a relief by galvanic process. The Emperor of Russia has placed at the professor's disposal funds to enable him to complete his discovery."

Jordan's Process published.—Having seen a copy of the above paragraph in the *Mechanic's Magazine*, May 11th, 1839, Mr. J. C. Jordan, of London, eleven days afterwards sent a communication to the editor of that journal, in which he put in his claim—if not to priority, as far as Jacobi was concerned, at least to prove that he had been experimenting in electro-deposition some twelve months before the announcement of Jacobi's discovery was published in this country. Indeed, Jordan's communication did more, for it contained a *definite process*, and since this was undoubtedly the first publication of the kind which had appeared in England, the merit of originality—so far as *publication* goes—is clearly due to Jordan. As an important item in the history of electro-deposition, we give the subjoined extract from his letter from the *Mechanic's Magazine*, June 8th, 1839. The letter was headed "Engraving by Galvanism."

Jordan's Process.—"It is well known to experimentalists on the chemical action of voltaic electricity that solutions of several metallic salts are decomposed by its agency and the metal procured in a free state. Such results are very conspicuous with copper salts, which metal may be obtained from its sulphate (blue vitriol) by simply immersing the poles of a galvanic battery in its solution, the negative wire becoming gradually coated with copper. This phenomenon of metallic reduction is an essential feature in the action of *sustaining* batteries, the effect in this case taking place on more extensive surfaces. But the form of voltaic apparatus which exhibits this result in the most interesting manner, and relates more immediately to the subject of the present communication, may be thus described :—It consists of a glass tube closed at one extremity with a plug of plaster of Paris, and nearly filled with a solution of *sulphate of copper*. This tube and its contents are immersed in a solution of *common salt*. A plate of copper is placed in the first solution, and is connected by means of a wire and solder with a zinc plate, which dips into the latter. A slow electric action is thus established through the pores of the plaster which it is not necessary to mention here, the result of which is the precipitation of minutely-crystallised copper on the plate of that metal in a state of greater or less malleability, according to the slowness or rapidity with which it is deposited. In some experiments of this nature, on removing the copper thus formed, I remarked that the surface in contact with the plate equalled the latter in smoothness and polish, and mentioned this fact to some individuals of my acquaintance. It occurred to me therefore, that if the surface of the plate was *engraved*, an *impression* might be obtained. This was found to be the case, for, on detaching the precipitated metal, the more delicate and superficial markings, from the fine particles of powder used in polishing, to the deeper touches of a needle or graver, exhibited their corresponding impressions in *relief* with great fidelity. It is, therefore, evident that this principle will admit of improvement and that casts and moulds may be obtained from any form of copper.

" This rendered it probable that impressions might be obtained from those other metals having an electro-negative relation to the zinc plate of the battery. With this view a common printing type was substituted for the copperplate and treated in the same manner. This also was successful ; the reduced copper coated that portion of the type immersed in the solution. This, when removed, was found to be a perfect matrix, and might be employed for the purpose of casting when time is not an object.

" It appears, therefore, that this discovery may possibly be turned to some practical account. It may be taken advantage of in procuring casts from various metals as above alluded to : for instance, a copper die may be formed from a cast of a coin or medal, in silver, typemetal, lead, &c., which may be employed for striking impressions in soft metals. Casts may probably be *obtained* from a *plaster* surface surrounding a plate of copper ; tubes or any small vessel may also be made by precipitating the metal around a wire or any kind of surface to form the interior, which may be removed mechanically by the aid of an acid solvent, or by heat." [May 22nd, 1839.]

It is a remarkable fact that Jordan's letter, regardless of the valuable information it contained, commanded no attention at the time. Indeed, the subject of which it treated (as also did Jacobi's announced discovery), apparently passed away from public view, until a paper by Mr. Thomas Spencer, of Liverpool, was read before the Liverpool Philosophical Society on the 12th of September in the same year. Omitting the prefatory observations with which the paper commenced, its reproduction will form a necessary link in the chain of evidence respecting the origin of the electrotype process, and assist the reader in forming his own judgment as to whom the merit of the discovery is really due.

Spencer's Paper on the Electrotype Process.—" In September, 1837, I was induced to try some experiments in electro-chemistry with a single pair of plates, consisting of a small piece of zinc and an equal sized piece of copper, connected together with a piece of wire of the latter metal. It was intended that the action should be slow ; the fluids in which the metallic electrodes were immersed were in consequence separated by a thick disc of plaster of Paris.   In one of the cells was sulphate of copper solution, in the other a weak solution of common salt.   I need scarcely add that the copper electrode was placed in the cupreous solution, not because it is *directly* connected with what I have to lay before the society, but because, by a portion of its results, I was induced to come to the conclusion I have done in the following paper.   I was desirous that no action should take place on the wire by which the electrodes were held together.   To attain this object I varnished it with sealing-wax varnish ; but, in so doing, I dropped a portion of it on the copper that was attached.   I thought nothing of this circumstance at the moment, but put the experiment in action.

" The operation was conducted in a glass vessel ; I had, consequently, an opportunity of occasionally examining its progress. When, after the lapse of a few days, metallic crystals had covered the copper electrode, *with the exception of that portion* which had been spotted with the drops of varnish, I at once saw that I had it in my power to guide the metallic deposition in any shape or form I chose by a corresponding application of varnish or other non-metallic substance.

" I had been long aware of what every one who uses a sustaining galvanic battery with sulphate of copper in solution must know, that the copper plates acquire a coating of copper from the action of the battery : but I had never thought of applying it to a useful purpose before.   My first essay was with a piece of thin copper-plate, having about four inches of superfices, with an equal-sized piece of zinc, connected together with a piece of copper wire.   I gave the copper a coating of soft cement consisting of bees-wax, resin, and a red earth—Indian or Calcutta red.   The cement was compounded after the manner recommended by Dr. Faraday in his work on chemical manipulation, but with a larger proportion of wax.   The plate received its coating while hot.   On cooling, I scratched the initials of my own name rudely on the plate, taking special care that the cement was quite removed from the scratches, that the copper might be thoroughly exposed.   This was put into action in a cylindrical glass vessel about half filled with a saturated solution of sulphate of copper. I then took a common gas glass, similar to that used to envelop an argand burner, and filled one end of it with plaster of Paris to the depth of three-quarters of an inch.   In this I put some water, adding a few crystals of sulphate of soda to excite action, the plaster of Paris serving as a partition to separate the fluids, but sufficiently porous to allow the electro-chemical fluid to penetrate its substance.

" I now bent the wires in such a form that the zinc end of the arrangement should be in the saline solution, while the copper end should be in the cupreous one. The gas glass, with the wire, was then placed in the vessel containing the sulphate of copper.

"It was then suffered to remain, and in a few hours I perceived that action had commenced, and that the portion of the copper rendered bare by the scratches was coated with a pure bright deposited metal, whilst all the surrounding portions were not at all acted upon. I now saw my former observations realised ; but whether the deposition so formed would retain its hold on the plate, and whether it would be of sufficient solidity or strength to bear working if applied to a useful purpose, became questions which I now endeavoured to solve by experiment. It also became a question whether, should I be successful in these two points, I should be able to produce lines sufficiently in relief to print from. The latter appeared to depend entirely on the nature of the cement or etching ground I might use.

"This last I endeavoured to solve at once. And, I may state, this appeared to be the principal difficulty, as my own impression then was that little less than $\frac{1}{8}$th of an inch of relief would be requisite.

"I then took a piece of copper, and gave it a coating of a modification of the cement I have already mentioned, to about $\frac{1}{8}$th of an inch in thickness ; and, with a steel point, endeavoured to draw lines in the form of net-work, that should entirely penetrate the cement, and leave the surface of the copper exposed. But in this I experienced much difficulty, from the thickness I deemed it necessary to use ; more especially when I came to draw the cross lines of the net-work. When the cement was soft, the lines were pushed as it were into each other ; and when it was made of a harder texture, the intervening squares of net-work chipped off the surface of the metallic plate. However, those that remained perfect I put in action as before.

"In the progress of this experiment, I discovered that the solidity of the metallic deposition depended entirely on the weakness or intensity of the electro-chemical action, which I found I had in my power to regulate at pleasure, by the thickness of the intervening wall of plaster of Paris, and by the coarseness and fineness of the material. I made three similar experiments, altering the texture and thickness of the plaster each time, by which I ascertained that if the plaster partitions were *thin* and *coarse*, the metallic depositions proceeded with great *rapidity*, but the crystals were friable and easily separated ; on the other hand, if I made the partition thicker, and of a little finer material, the action was much slower, and the metallic deposition was as solid and ductile as copper formed by the usual methods, indeed, when the action was exceedingly slow, I have had a metallic deposition apparently much harder than common sheet copper but more brittle.

"There was one most important (and, to me, discouraging) circumstance attending these experiments, which was that when I heated the plates to get off the covering of cement, the meshes of copper net-work invariably *came off with it*. I at one time imagined this difficulty insuperable, as it appeared to me that I had cleared the cement entirely from the surface of the copper I meant to have exposed, but that there was a difference in the molecular arrangement of copper prepared by heat and that prepared by voltaic action which prevented their chemical combination. However, I then determined, should this prove so, to turn it to account in another manner, which I shall relate in a second portion of this paper. I then occupied myself for a considerable period in making experiments on this latter section of the subject.

"In one of them I found on examination a portion of the copper deposition, which I had been forming on the surface of a coin, adhered so strongly that I was quite unable to get it off; indeed, a chemical combination had apparently taken place. This was only in one or two spots on the prominent parts of the coin. I immediately recollected that on the day I put the experiment in action I had been using nitric acid for another purpose on the table I was operating on, and that in all probability the coin might have been laid down where a few drops of the acid had accidentally fallen. I then took a piece of copper, coated it with cement, made a few scratches on its surface until the copper appeared, and immersed it for a short time in dilute nitric acid, until I perceived, by an elimination of nitrous gas, that the exposed portions were acted upon sufficiently to be slightly corroded. I then washed the copper with water, and put it in action, as before described. In forty-eight hours I examined it, and found the lines were entirely filled with copper; I applied heat, and then spirit of turpentine, to get off the cement; and, to my satisfaction, I found that the voltaic copper had completely combined itself with the sheet on which it was deposited.

"I then gave a plate a coating of cement to a considerable thickness, and sent it to an engraver; but when it was returned, I found the lines were cleared out, so as to be wedge-shaped, or somewhat in form of a V, leaving a hair line of copper exposed at the bottom and broad space near the surface; and where the turn of the letters took place, the top edges of the lines were galled and rendered ragged by the action of the graver. This, of course, was an important objection, which I have since been able to remedy in some respects by alteration in the shape of the graver, which should be made of a shape more resembling a narrow parallelogram than those in common use; some of the engravers have many of their tools so made. I did not put this plate in action, as I saw that the lines, when in relief, would have been broad at the top and narrow at the bottom. I took another plate, gave it a coating of the wax, and had it written on with a mere point. I deposited copper on the lines and afterwards had it printed from.

" I now considered part of the difficulties removed ; the principal one that yet remained was to find a cement or etching-ground, the texture of which should be capable of being cut to the required depth, and without raising what is technically termed a burr, and at the same time of sufficient toughness to adhere to the plates where reduced to a small isolated point, which would necessarily occur in the operation which wood-engravers term cross-hatching.

" I tried a number of experiments with different combinations of wax, resin, varnishes and earths, and also metallic oxides, all with more or less success. The one combination that exceeded all others in its texture, having nearly every requisite (indeed, I was enabled to polish the surface nearly as smooth as a plate of glass), was principally composed of virgin wax, resin, and carbonate of lead—the white-lead of the shops. With this compound I had two plates, 5 inches by 7, coated over, and portions of maps cut on the cement, which I had intended should have been printed off and laid before the British Association at its meeting."

Effect of Spencer's Paper.—When Spencer's paper was published it at once commanded profound attention, and many persons practised the new art either for amusement or scientific research, while others turned their attention to it with a view to making it a source of commercial profit. It was not, however, until Mr. Robert Murray, in January, 1840, informed the members of the Royal Institution, London, that he had discovered a method of rendering non-conducting surfaces—such as wax, &c.—conductive of electricity by employing plumbago, or black lead, that the art became really popular in the fullest sense. This conducting medium was the one thing wanted to render the process facile and complete ; and soon after Mr. Murray's invaluable discovery had been made known, thousands of persons in every grade of life at once turned their attention to the electrotype process until it soon became the most popular scientific amusement that had ever engaged the mind, we may say, of a nation. The simplicity of the process, the trifling cost of the apparatus and materials, and the beautiful results which it was capable of yielding, without any preliminary knowledge of science, all combined to render the new art at once popular in every home. Every one practised it, including the youth of both sexes.

It is not to be wondered at that an art so fascinating should have produced more than an ephemeral effect upon the minds of some of those who pursued it. Indeed, it is within our own knowledge that many a youth whose first introduction to chemical manipulation was the electro-deposition of copper upon a sealing-wax impression of a signet-ring or other small object, acquired therefrom a taste for a more extended study of scientific matters, which eventually led up to his devoting himself to chemical pursuits for the remainder of his days. At the period we refer to there were but few institutions in this country for the encouragement of scientific study. One of the most accessible and useful of these, however, was that founded by Dr. Birkbeck, the well-known Literary and Scientific Institution at that time in Southampton Buildings, London.

Vindication of Jordan's Claim.—Although Jordan's letter was published, as we have shown, three months prior to the reading of Spencer's paper in Liverpool, that important communication was overlooked, not only by the editor of the journal in which it appeared, but also by the scientific men of the period. Even the late Alfred Smee, to whose memory we are indebted for the most delightful work on electro-metallurgy that has appeared in any language, failed to recognise the priority of Jordan's claim. Impelled by a strong sense of justice, however, the late Mr. Henry Dircks wrote a series of articles in the *Mechanic's Magazine* in 1844, in which he proved that whatever merit might have been due to Spencer and Jacobi, Jordan was unquestionably the first to *publish* a process of electrotyping. Indeed, he went further, for he proved that the electro-deposition of copper had been accomplished *practically* long before the publication of *any* process. Before entering into the merits of Jordan's priority. Mr. Dircks makes this interesting statement :—

Mr. Dircks on Jordan's Discovery.—"The earliest application of galvanic action to a useful and ornamental purpose that I am acquainted with was practised by Mr. Henry Bessemer, of Baxter House, Camden Town, who, above ten years ago [about 1832] employed galvanic apparatus to deposit a coating of copper on lead castings. The specimens I have seen are antique heads in relief, the whole occupying a space of 3 inches by 4 inches. They have lain as ornaments on his mantel-piece for many years, and have been seen by a great number of persons."

Appreciating—from its historic and scientific interest—the importance of the above statement, it occurred to the author that if the means adopted at so early a period in electro-metallurgical history could become known, this would form an important link in the chain of research respecting the deposition of metals by electrolysis. He, therefore, wrote to Sir Henry Bessemer, requesting him to furnish such particulars of the method adopted by him in depositing copper upon the objects referred to as lay in his power after so long a period of time. With kind courtesy, and a generous desire to comply with

the author's wishes, Sir Henry took the trouble to furnish the information conveyed in the following interesting communication, which
cannot fail to be read with much gratification by all who have
studied the art of electro-deposition, either from its scientific or practical aspect.   When we call to remembrance the numerous inventions
with which the active mind of Sir Henry Bessemer has been associated
during the greater portion of the present century, culminating in his
remarkably successful improvements in the manufacture of steel, it is
pleasing to read that at the youthful age of eighteen—when voltaic
electricity was but little understood, and Daniell's, Grove's, and
Smee's batteries unknown—he was engaged in experiments with
metals, which were evidently conducted with an amount of patience
and careful observation which would have been highly creditable in a
person of more advanced years.

Sir Henry Bessemer's Experiments.—Replying to the author's
inquiry of Sir Henry Bessemer (in January, 1885) as to the method he
had adopted in coating with copper the objects referred to above, Sir
Henry wrote as follows ; and the minuteness of the details, given after
so great a lapse of time, will doubtless strike the reader with some
astonishment :—

"I have much pleasure in replying to your note of inquiry in
reference to the deposition of copper from its solutions on white metal
castings.

"My first experiments began when I was about eighteen years of
age, say in 1831-2.   At that period, after much practice, I was most
successful in producing castings of natural objects in an alloy of tin,
bismuth, and antimony.   In this alloy I cast such things as beetles,
frogs, prawns, &c. ; also leaves of plants, flowers, moss-rose buds ;
and also medallions, and larger works in basso-relievo.   By my
system of casting in nearly red-hot metal, the metal was retained for
ten or fifteen minutes in a state of perfect fluidity in the mould,
and hence, by its pressure, forced itself into every minute portion of
the natural object, whatever it might be ; thus every minute thorn
on the stem of the rose was produced like so many fine projecting
needles.   I exhibited several of these castings, coated with copper, at
' Topliss's Museum of Arts and Manufactures,' at that time occupying the site of the present National Gallery, and which museum was
afterwards removed to a large building in Leicester Square, now the
Alhambra Theatre, where I also exhibited them.

"Beautiful as were the forms so produced, they had a common
lead-like appearance, which took much from their value and artistic
beauty, and as a remedy for this defect, it occurred to me that it was
possible to give them a thin coat of copper, deposited from its solution in dilute nitric acid.   This I made by putting a few pence [copper

coins were in currency in those days] into a basin with water and nitric acid. My early attempts were not very successful, for the deposited metal could be rubbed off, and was in other ways defective. I next tried sulphate of copper, both cold and boiling solutions. I found the sulphate much better adapted for the purpose than the nitrate solution. At first I relied on the property which iron has of throwing down copper from its solutions, and by combining iron, in comparatively large quantities, with antimony, and using this alloy with tin, bismuth, and lead, I succeeded in getting a very thin, but even, coating of copper; but it was not sufficiently solid, and easily rubbed off.

" In pursuing my experiments, I found that the result was much improved by using a metallic vessel for the bath instead of an earthenware one, such as a shallow iron, tin, or copper dish, as a slight galvanic action was set up, but the best results were obtained by using a zinc tray, on the bottom of which the object was laid, face upwards, and the solution then poured in. By this means a very firm and solid coating was obtained, which could be burnished with a steel burnisher without giving way. By adding to the copper solution a few crystals of distilled verdigris, I obtained some beautiful green bronze deposits, a colour far more suitable for *medallions* and *busts* than the bright copper coating obtained by the sulphate when used alone.

" I cast and coated with green copper a small bust of Shakespeare, which, with many other specimens, I sold to Mr. Campbell, the sculptor, who at that time was modelling a life-sized bust of Canning : he had arranged that I should cast it from the " lost-wax," and deposit green copper thereon. Unfortunately Campbell died before his model was completed. But for this incident I might possibly have carried the depositing process much further, but at that time my success in casting, in a very hard alloy, dies used for embossing cardboard and leather, offered a more direct and immediate commercial result, and thus the artistic branch was lost sight of. I remember showing some of these castings to my friend the late Dr. Andrew Ure, about the year 1835-6, with which he was much pleased. In referring to them several years later, in the second edition of his supplement to his ' Dictionary of Arts and Manufactures,' published in 1846, he mentions these castings as *lead castings*, at page 70, under the head of ' Electro-Metallurgy,' which commences in these words :—

" ' *Electro-Metallurgy.*—By this elegant art, perfectly exact copies of any object can be made in copper, silver, gold, and some other metals, through the agency of electricity. The earliest application of this kind seems to have been practised about *ten years ago*, by Mr. Bessemer, of Camden Town, London, who deposited a coating of copper upon lead castings so as to produce antique heads, in relief, about three or four inches in size. He contented himself with forming a few such ornaments for his mantel-piece, and though he made no secret of his purpose, he published nothing upon the subject. A letter of the 22nd of May, 1839, written by Mr. C. Jordan, which appeared in the *Mechanic's Magazine* for June 8th following, contains the first printed notice of the manipulation requisite for obtaining electro-metallic casts, and to this gentleman, therefore, the world is indebted for the first discovery of this new and important application of science to the uses of life.'

'' The first inception of the idea of coating works of art in metal with a deposited coating of another metal, if not resting solely with me, at least I certainly was within measurable distance of this great discovery some three or four years before it was brought forward by any other person, but I failed to see its true significance, and consequently lost a grand opportunity.

'' You are quite at liberty to make any use you like of this information.''

We will now return to Mr. Dircks' vindication of Jordan's claim.

Referring to Jordan's letter to the *Mechanic's Magazine,* Mr. Dircks says, '' In particular I would direct attention to the fact of the main incidents named by Mr. Jordan, published June 8th, 1839, agreeing with those published by Mr. Spencer, September 12th, 1839, and, curious enough, being called forth by the same vague announcement of Professor Jacobi's experiments which was then making the round of the periodicals. Both parties described Dr. Golding Bird's small galvanic apparatus ; one used a printer's type, the other a copper coin, and both recommend the application of heat to remove the precipitated copper.

'' I was aware of Mr. Jordan's letter at the time of its publication, and have frequently been surprised since, that his name has not transpired in any discussion I have heard upon the subject. Nothing can be clearer than his reasoning, the details of his experiments, and his several concluding observations.''

Dr. Golding Bird's Experiments.—There can be no doubt whatever that after Dr. Golding Bird published the results of his interesting experiments in 1837, and the means by which he obtained his important results, many scientific men devoted themselves to investigating the new application of electricity, amongst whom was Mr. Henry Dircks. '' It was particularly in September and October, 1837,'' wrote Mr. Dircks, '' that several parties attached to scientific pursuits in Liverpool, were engaged in repeating the experiments of Dr. Golding Bird, and of which he gave an account before the chemical section of the British Association at Liverpool, over which Dr. Faraday presided.

The apparatus used on that occasion by myself and others was precisely that recommended by Dr. Bird, consisting of simply any glass vessel capable of holding a solution of common salt, into which is inserted a gas lamp chimney, having its lower end plugged up by pouring into it plaster of Paris ; a solution of sulphate of copper is then poured into it, and the whole immersed into the contents of the glass, and tightened with pieces of cork. The result expected from this arrangement was the deposit of metallic veins of the copper within the plaster diaphragm, independent of any connection with the poles of the battery. Dr. Faraday, and every other electrician, expressed surprise and doubt at the results in this respect said to have been obtained by Dr. Bird ; and Dr. Faraday particularly urged the necessity and importance of caution in receiving as established a result so greatly at variance with all former experience, and proceeded to explain a variety of causes tending to lead to fallacious results in the curious and interesting experiments."

Up to this time, the possibility of obtaining electrical effects by means of a single metal, in the manner pursued by Dr. Bird, would have been considered theoretically impossible. It must not be wondered at, therefore, that even the greatest of our philosophers—Michael Faraday—should have been sceptical in the matter. It is clear now, however, that Dr. Golding Bird's results were based upon principles not then understood, and that to this gifted physician we are indebted for what is termed the " single-cell " voltaic arrangement—the first, and for some time after the only, apparatus employed in producing electrotypes.

Origin of the Porous Cell.—It appears that while Mr. Dircks was experimenting (in 1837) in obtaining crystals of copper by Dr. Bird's method, he was frequently in communication with Mr. John Dancer, a philosophical instrument maker in Liverpool, and in October of the following year (1838) that gentleman showed him a " ribbon of copper, thin, but very firm, granular on one side, while it was bright and smooth, all but some raised lines, on the other." This result, Mr. Dancer informed him, was obtained by galvanic action, observing that some specimens were as tenacious as rolled copper, while others were crystalline and brittle. Mr. Dancer attributed the superiority of the former to the following cause : " Having gone to the potteries to look out suitable jars for sustaining batteries, and having fixed on a lot which he was told would not answer as *they were not glazed*, and would not hold liquor," it occurred to him that such *unglazed* jars might be turned to account, and used instead of bladder, brown paper plaster of Paris, and other porous substances he had previously employed. Having obtained a sample for experiment, he subsequently found that he could obtain a more firm and compact deposit of copper

than in any previous experiment.   To the accidental circumstance
above referred to, we are undoubtedly indebted for that most important accessory to the single-cell apparatus and the two-fluid battery—
the porous cell.

In a letter to Mr. Dircks, relative to Spencer's claim to the discovery
of a means of obtaining " metallic casts " by electro-deposition, Mr.
Dancer says, " I met Mr. Spencer one morning in Berry Street,
Liverpool, and happened to have one of these precipitated copper
plates with me, which I showed to him.   When I told him how it had
been formed he would scarcely believe it, until I pointed out the impressions in relief of all the minute scratches that were on the plate
against which it had been deposited.   The surprise that Mr. Spencer
expressed very naturally led me to suppose that it was the first compact piece of precipitated copper he had seen."   At this early period
(1838) Mr. Dancer had not only deposited tough *reguline* copper, but
he went a step farther.   He attached to a copper plate, by means of
varnish, " a letter cut out from a printed bill.   The copper precipitated
on all parts of the plate, except where the letter was fixed ; when I
peeled the precipitated copper off, the letter came out, not having
connection with the outside edge.   I also obtained an impression by
stamping my name on a copper cylinder, the impression being the
reverse way. . . . . All this happened many months before I was
aware that Mr. Spencer had been engaged in anything of the kind,
except that he had Dr. Bird's experiments in action.   Some time after
this Mr. Spencer applied to me for one of my porous jars, and one day
at his house he told me for what purpose he wanted it."

It is perfectly evident that Mr. Dancer's results were obtained long
before the publication of Spencer's paper, and that both were indebted
to Dr. Golding Bird's simple but ingenious contrivance for prose-
cuting their first experiments ; and it is also clear that Dancer's
brilliant idea of substituting porous earthenware for the crude plaster
diaphragms greatly facilitated experimental researches in this
direction ; while at the same time it placed within our reach one of
the most valuable accessories of the two-fluid voltaic battery—the
porous cell.

Being desirous of placing Jordan's claim to priority—as the first
to make publicly known the process of electrotyping, or *electrography,*
as he termed it—Mr. Dircks followed up the subject in the *Mechanic's
Magazine,* in a series of papers, in which he not only traced Mr.
Spencer's experiments to their true origin, namely, Dr. Bird's
experiments published two years before, and the hints which he had
derived from Dancer, but he moreover showed that Spencer must
have been aware of Jordan's published process, for he says, in
summing up the evidence he had produced against Spencer's position

in the matter thus : "Lastly, therefore, that through the *Mechanic's Magazine* (which Mr. Spencer was regularly taking in) the experimental results obtained by Mr. Dancer, and the reports in April and May, 1839, in public papers, of Jacobi's experiments, all being broad hints, and abundant assistance to aid Mr. Spencer, that he is rather to be praised for his expression of what was already known, on a smaller and less perfect scale, than to be adjudged a discoverer, much less the father of electro-metallurgy, having a preference to every other claimant." Following the paper from which the foregoing extract is taken, is a footnote by the Editor of the *Mechanic's Magazine*, which is important as showing how strange it was that Jordan's communication not only escaped the attention of scientists, but even that of the conductor of the journal in which it appeared : "Mr. Dircks has proved beyond all doubt that we have made a great mistake in advocating so strenuously the claims of Mr. Spencer to the invention of electrography. No one, however, can suppose that we would intentionally exalt any one at the expense of our own journal, which we are now pleased to find was the honoured medium of the first distinct revelation of this important art to the public, by an old and esteemed correspondent of ours, Mr. Jordan. Whatever Mr. Bessemer, Mr. Dancer, Mr. Spencer, or others, may have previously said or done, it was in private—made no secret of, perhaps, but still not communicated to the public at large—not recorded in any printed work for general benefit. For anything previously done by any of them, they might have still remained in the profoundest obscurity. No public description of an earlier date than Mr. Jordan's can, we believe, be produced ; and when we look upon that description, it is really surprising to see with what fulness and precision the writer predicated of an art nearly all that has been since accomplished. In supporting, as we did, the claims of Mr. Spencer to be considered as the first discoverer, we had lost all recollection of Mr. Jordan's communication. We have no personal acquaintance with either of the gentlemen, and could have no motive for favouring one more than the other. We took up the cause of Mr. Spencer with spontaneous warmth because we thought him to be a person most unfairly and ungenerously used, as in truth he was so far as the intention went, by those who, having at the time none of those reasons we now have for questioning Mr. Spencer's pretensions, yet obstinately refused to acknowledge them. If it should seem to the reader more than usually surprising that Mr. Jordan's paper escaped the recollection of the editor, through whose hands it passed to the public, his surprise will be lessened, perhaps, when he observes how it appears to have escaped notice, or been passed over in silence, by every one else down to the present moment—even those, not a few, who have expressly occupied themselves in electrography. . . . . To us, the most surprising thing of any connected with the case is, that neither Mr. Jordan himself, nor any of his friends, should before now have thought it worth while to vindicate his claims to the promulgation of an art which justly entitles him to take a high place among the benefactors of his age and country.—*Ed. M. M.*"

While Mr. Dircks' "Contributions to the History of Electro-Metallurgy" were being published in the columns of the *Mechanic's Magazine*, the arguments and facts which he adduced created a deep impression in the minds of scientific men of the day, who had unfortunately accepted Spencer as the originator of electrotypy. Of all men, scientists are the most anxious to accord the merit of *discovery* to those who are really entitled to it. Devoting themselves to the investigation of natural laws, and their application to the useful purposes of man, they are naturally jealous of any attempt on the part of one to appropriate the honour—usually the only reward—due to another. It is not surprising, therefore, that when it became fully proved that to Jordan and not Spencer was due the credit of having been the first to publish a process for the practical deposition of copper by electrolysis, that such men should frankly acknowledge their mistake. Amongst those who came forward to do justice to Jordan's claim were the late Professor Faraday, Dr. Andrew Ure, and Professor Brande, then chemist to the Royal Mint. The latter eminent chemist and author of the best chemical manual in our language, sent the following letter to Mr. Dircks, which clearly acknowledges the error into which, in common with others, he had fallen in attributing to Spencer the merit of the electrotype process :—

" I am much obliged by your copy of the *Mechanic's Magazine* and the information it contains respecting Mr. Spencer's pretensions. I certainly always gave him credit for much more merit than he appears to have deserved."

When Spencer found that his position was so severely shaken by Mr. Dircks' powerful defence of Jordan's claim to priority, he wrote several letters in reply, which appeared in the columns of the above journal, with a view to refute his opponent's arguments, and shake his testimony ; but in this he was unsuccessful, for the facts which Mr. Dircks had made known were absolutely beyond refutation. It is not often that men of science enter into a controversy of this nature, but silence under such circumstances would have been an act of injustice to Jordan, by leaving the question still in doubt.

Amongst those who ascribed to Spencer the discovery of the electrotype process was Mr. George Shaw, of Birmingham, in the first edition of his " Manual of Electro-Metallurgy." In the second edition of his work, however, he made the *amende* to Jordan, by frankly acknowledging his mistake. The following letter from the late Dr. Andrew Ure to Mr. Dircks shows how fully he recognised that gentleman's advocacy of Jordan's claim : " I read with great interest your narrative of the discovery or invention of the electrotype art, and am much pleased to see justice done to modest retiring merit in the persons of Mr. Jordan and Mr. Dancer. The jay will feel a little awkward this cold weather, stripped of his peacock plumage."

The following letter from Faraday tends to show that the great philosopher, in common with most other persons, had, prior to Mr. Dircks' explanation of the facts, believed in Spencer being the origi-nator of electrotyping : " I am very much obliged by your kindness in sending me your account of the facts, &c., &c. It is very valuable as respects the fixing of dates, and has rather surprised me." *

It is a pity, but none the less true, that while Jordan's communica-tion received no attention whatever, although published in a well-read journal, Spencer's paper—which had merely been read before a local society in Liverpool, and afterwards printed for *private circulation only*—commanded the profoundest attention. In short, to use a common phrase, it "took the world by storm." The name of " Spencer, the discoverer of Electrotyping," was on every lip, and men of science of all nations regarded him as one who had made a great addition to the long roll of important discoveries which science had placed at the disposal of art. Henry Dircks' champion-ship of Jordan's just claim, however, eventually broke up Spencer's position, and to *the first publisher of the electrotype process*, Mr. C. J. Jordan, was at last accorded the *merit*—for he received no other recog-nition—of having published a process, if we may not say discovery, which was destined to prove of inestimable advantage to his fellows, not only in itself, but as being the means by which the minds of men were directed to the deposition of other metals by electrical agency. It would not be out of place to suggest that in commemoration of Jordan's *gift* to mankind of so useful and valuable a process, an appropriate testimonial should be set on foot—if not by the public, at least by those who have directly gained so much by his initiation of the art of electro-deposition.

The success which attended the electrotype process induced many persons to turn their attention to the deposition of gold and silver, by means of the direct current; but up to the year 1840 no really suc-cessful solution of either metal was available. In that year Mr. John Wright, a surgeon in Birmingham, and Mr. Alexander Parkes, in the employment of Messrs. Elkington, were engaged in making experiments in electro-deposition, when the former gentleman hap-

---

* The three foregoing letters, which we transcribed from the originals, are now, we believe, published for the first time.

pened to meet with a passage in Scheele's "Chemical Essays," in
which he found that cyanides of gold, silver, and copper, were
soluble in an excess of cyanide of potassium. It at once occurred to
him that solutions of gold and silver thus obtained might be employed
in electro-deposition, and he then formed a solution by dissolving
chloride of silver in a solution of ferro-cyanide of potassium, from
which he obtained, by electrolysis, a stout and firm deposit of silver,
a result which had never before been obtained. A few weeks after,
Mr. Wright prepared a solution with cyanide of potassium, instead of
the ferro-cyanide, and although various cyanide solutions of silver
and copper had already been employed in the simple immersion
process of depositing these metals, there is no doubt that it is to Mr.
Wright that we are really indebted for the practical application of
cyanide of potassium as a solvent for metallic oxides and other salts
used in electro-deposition. About this time (1840) Messrs. Elkington
were preparing to take out another patent, when Mr. Wright, having
submitted his results to them, agreed to include his process in their
patent, in consideration of which it was agreed that he should receive
a royalty of one shilling per ounce for all silver deposited under the
patent : on his decease, which took place soon afterwards, an annuity
was granted to his widow. This patent, with Wright's important
addition, namely the employment of alkaline cyanides, formed the
basis of the now great art of electro-gilding and plating : but it was
some time before the proper working strength of baths and the proportion of cyanide could be arrived at, the deposits being frequently
non-adherent, which caused them to strip or peel off the coated articles
in the process of burnishing. This was afterwards remedied to some
extent by dipping the articles (German silver chiefly) in a very dilute
solution of mercury. About this time, the author, in conjunction
with his brother, Mr. John Watt, introduced electro-gilt and silvered
steel pens, which were sold in considerable quantities.

In the same year, Mr. Murray discovered a means of rendering
non-conducting surfaces, as wax, &c., conductive, by coating them
with powdered plumbago, and this important suggestion proved of
inestimable advantage to those who desired to follow the art of
electrotyping commercially. Indeed, without the aid of this useful
substance, it is doubtful whether the important art would have greatly
exceeded the bounds of experiment. At this period, also, another
important improvement in the electrotype process was introduced by
Mr. Mason, which consisted in employing a separate battery as a substitute for the "single-cell" process up to that time adopted in
electrotyping. By the new arrangement, a copper plate was connected to the positive pole of a Daniell Battery, while the mould to be
coated with copper was attached to the negative pole. When these
were immersed in the electrotyping bath (a solution of sulphate of

copper), under the action of the current the copper-plate became dissolved as fast as pure copper was deposited upon the mould, whereby the strength of the solution was kept in an uniform condition. It is this method which is now almost universally adopted (when dynamo machines are not employed) in practising the art of electrotyping upon a large scale.

In 1841, Mr. Alfred Smee published his admirable work on Electrometallurgy, which at that period proved of the greatest service to all persons interested in the new art. In the year following, Mr. J. S. Woolrich introduced his magneto-electric machine, which for many years after occupied a useful position as a substitute for voltaic batteries, in several large plating works. In this year also, Dr. H. R. Leeson took out a patent for improvements in electro-depositing processes, in which he introduced the important elastic moulding material, "guiding wires," keeping articles in motion while in the bath, &c.

In 1843, Moses Poole obtained a patent for the use of a thermoelectric pile as a substitute for the voltaic battery; but the invention was not, however, successful. Many patents were taken out in the following years for various processes connected with electro-deposition ; but the next most important improvement was due to Mr. W. Milward, of Birmingham, who accidentally noticed that after waxmoulds, which had been covered with a film of phosphorus—by applying a solution of that substance in bisulphide of carbon to their surfaces —had been immersed in the cyanide of silver plating bath, the silver deposit upon other articles, such as spoons and forks, for example, which were afterwards coated in the same bath, presented an unusually bright appearance in parts, instead of the dull pearly lustre which generally characterises the silver deposit. This incident induced Mr. Milward to try the effect of adding bisulphide of carbon to the plating bath, which produced the desired result. For some time he kept the secret to himself; but finding that it eventually became known, he afterwards patented the process in conjunction with a Mr. Lyons, who had somehow possessed himself of the secret. From that time the addition of bisulphide of carbon to silver baths for the purposes of " bright " plating has been in constant use.

In the foregoing sketch of the origin and history of electro-deposition we have endeavoured to give such information as we hoped would be interesting to many who are engaged in the practice of the art, and also instructive to those who may be about to enter into a study of the subject, believing, as we do, that the present volume would be incomplete without some special reference to the interesting origin of so great and useful an art—an art which has many widespread applications of great commercial and decorative importance.

Further reference to subsequent inventions connected with electro-deposition and its developments in recent years, will be found in the later chapters of this volume.

CHAPTER IV.

ELECTRO-DEPOSITION OF COPPER.

Electrotyping by Single-cell Process.—Copying Coins and Medals.—Moulding Materials.—Gutta-percha.—Plastic Gutta-percha.—Gutta-percha
and Marine Glue.—Beeswax.—Sealing-wax.—Stearine.—Stearic Acid.—
Fusible Metal.—Elastic Moulding Material.—Plaster of Paris.

It may fairly be said that the discovery of the electrotype process formed the basis of the whole electrolytic industry ; and, in
its applications to various purposes of the arts and to literature, it
has proved of inestimable value. While, in its infancy, the electrotype process was a source of scientific recreation to thousands of
persons of all classes, many were those who saw in the new process a
wide field of research, from which much was expected and more has
been realised. While Faraday, Becquerel and others were investigating the process in its more scientific relations, practical men were
trying to apply it to various art purposes, until, in course of time,
electrotyping was added to our list of chemical arts.

The simplest form of arrangement for electrotyping small objects
is known as the "single-cell" process, which it will be well to consider before describing the more elaborate apparatus employed for
larger work.

Electrotyping by the Single-cell Process.—In its most simple
form, a small jar, Fig. 53, may be used as the outer
vessel, and in this is placed a small porous cell, made
of unglazed earthenware or biscuit porcelain, somewhat taller than the containing vessel. A strip of
stout sheet-zinc, with a piece of copper wire attached,
either by means of solder or by a proper binding
screw, is placed in the porous cell. A *saturated* solution of sulphate of copper (bluestone), made by
dissolving crystals of that substance in hot water,
and pouring the liquid, when cold, into the outer cell.

Fig. 53.

The porous cell is then filled to the same height as the copper
solution with a solution of sal-ammoniac or common salt. To keep
up the strength of the solution when in use a few crystals of sulphate
of copper are placed in a muslin bag, which is hooked on to the

edge of the vessel by means of a short copper hook, and the bag allowed to dip a little way into the liquid. The prepared mould is connected to the end of the wire (which is bent in this ∩ form) and gently lowered into the solution, when the whole arrangement is complete. In place of the porous cell the zinc may be wrapped in several folds of brown paper, enclosing a little common salt, but the porous cells are so readily obtained that it is never worth while to seek a substitute for them. This simple arrangement will easily be understood by referring to the cut.

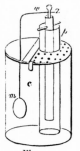

Fig. 54.

A more convenient single-cell apparatus is shown in Fig. 54, in which the containing vessel, or *cell*, is a glass or stoneware jar capable of holding about three pints. In this is placed a porous cell (*p*). A bar or plate of zinc (*z*), with binding screw attached, is deposited in the porous cell ; a short piece of copper wire (*w*), for suspending the *mould* (*m*) or object to be copied, has its shorter end inserted in the hole of the binding-screw. The outer vessel is about three parts filled with a *saturated* solution of sulphate of copper (*c*), and the porous cell is filled to the same height with a half-saturated solution of sal-ammoniac or common salt. If the zinc is *amalgamated*, however, dilute sulphuric acid is used instead of the latter solution in the porous cell, and a small quantity of oil of vitriol (from half an ounce to one ounce of acid to the quart of copper solution) added.

Amalgamating the Zinc.—Pour a little dilute sulphuric acid, or undiluted muriatic acid, into a dish, and, having tied a piece of flannel to the end of a stick, lay the zinc in the dish and proceed to brush the acid all over the plate ; now pour a little mercury (quicksilver) on the plate, and rub it over the zinc with the little mop, when it will readily spread all over the surface, giving the zinc a bright silvery lustre. It is important that the zinc should be *thoroughly* cleaned by the acid, otherwise the mercury will fail to amalgamate with the metal, and dark patches of unamalgamated zinc will appear. The perforated shelf, or tray, in the engraving is a receptacle for crystals of sulphate of copper, which, being placed upon it, gradually become dissolved while the deposit of copper is going on, and thus re-supply the solution as it becomes exhausted, whereby the operation progresses uniformly.

To prepare the copper solution for small experimental purposes, dissolve about 10 ounces of sulphate of copper in 1 quart of hot water and stir until the crystals are all dissolved : then set the vessel aside until cold, when the *clear* liquor is to be carefully poured into the depositing cell. When unamalgamated zinc is used in the single-cell arrangement the sulphate of copper should be simply a saturated solution of the salt without that addition of acid, though a *few drops only* may be added with advantage.

It is of great importance that the sulphate of copper should be *pure*. The crystals should be of a rich dark blue colour and absolutely free from greenish crystals (*sulphate of iron*), which not unfrequently get mixed with the copper salt by the carelessness of the shopkeepers' assistants.

Copying Coins and Medals.—Before explaining the various methods of obtaining moulds from different objects, for the purpose of producing fac-similes in copper, let us see how we may employ the above apparatus in a more direct way. Suppose we desire to obtain a copy, in reverse, of some medal or old coin, or even a bronze penny-piece, having decided which side of the coin it is intended to electrotype—say the *obverse* or "head" side—we must first render the surface clean and bright. This may be very readily done by means of rottenstone and a little olive oil, applied with a piece of chamois leather and briskly rubbed over the face of the coin. In two or three minutes the surface will be sufficiently bright, when the oil must be wiped off thoroughly either with cotton wool or blotting paper. A short piece of copper wire is next to be soldered to the back of the coin, and the polished side is then to be brushed-over with a soft plate-brush and plumbago, or blacklead, which will prevent the deposited copper from adhering to the medal. In order to prevent the copper from being deposited upon the back and rim these parts must be coated with some *non-conducting* material. For this purpose paraffin wax, applied by gently heating the medal and touching it with the wax, or red sealing-wax, dissolved in spirit of wine or wood spirit (pyroxylic spirit), brushed over the surfaces to be protected, will answer well; but if the latter is employed it must become thoroughly dry before being placed in the copper solution.

Being thus prepared, the end of the conducting wire is to be inserted in the binding screw attached to the zinc and securely fixed by turning the screw until it grips the wire firmly. The coin must be lowered into the solution steadily, with its face towards the porous cell, and if any air-bubbles appear upon its face they must be removed by means of a camel-hair brush, or, still better, by blowing upon them through a glass tube. It is a good plan to breathe upon the face of the coin before placing it in the solution, which, by covering it with a layer of moisture, effectually prevents the formation of air-bubbles.

In about twenty-four hours from the first immersion of the medal the deposit of copper will generally be sufficiently stout to bear re-

moving from the original, when the extraneous copper, which has
spread round the edge of the deposit, or electrotype, may be carefully
broken away by means of small pliers ; if the medal be gently heated
over a small lamp, the electrotype will readily become detached,
and will present, in *reverse*, a perfect copy of the original, in which
even the very finest lines will be accurately reproduced.   In its
present condition the electrotype is hard and brittle, and will, therefore, require careful handling.   To give it the toughness and flexibility of rolled copper it is only necessary to heat the electrotype to
dull redness, which may be conveniently done by placing it on a piece
of sheet-iron, and laying this on the clear part of a fire until red hot,
when it must be withdrawn and the " type " set aside to cool.   If
placed in a very weak solution of sulphuric acid for a few moments,
then rinsed and dried, and afterwards brushed over with a little rouge
or whiting, its surface may be readily brightened.

If we desire to obtain a copy *in relief* from our electrotype (also in
copper) we must now treat it as *the mould*, following the same routine
as before in all respects, by which we shall obtain a perfect fac-simile
of the original coin, which may be mounted and bronzed by any of
the processes hereafter given.

Having thus seen what results may be obtained with the most
simple application of the single-cell process, we will next turn our attention to the different methods of obtaining moulds from various
objects, but, before doing so, it will be necessary to consider the
nature of the several substances which are employed in *moulding* and
the methods of preparing them for use.

Moulding Materials.—The chief substances used in the electrotyping art for making moulds are gutta-percha, wax, and fusible
metal ; other materials, however, are employed in certain cases in
which the substances named would be inapplicable.   The various
materials will be considered under their separate heads, as follows :—

Gutta-percha.—This most useful moulding material is the concrete
juice of *Isonandra Gutta*, a tree growing only in the Malayan Archi-
pelago, and of other species of the same genus.   The stem of the
gutta-percha tree, which sometimes acquires the diameter of 5 or 6
feet, after being notched yields a milky juice which, when exposed to the air for some time, solidifies, and this constitutes the
gutta-percha of commerce.   As imported, it is in irregular blocks of
some pounds in weight, and commonly containing a large proportion
of impurities in the shape of bark, wood, stones, and earthy matter.
To purify the crude article it is first cut in thin slices, which are afterwards torn into shreds by machinery.   These are next softened by
hot water and afterwards kneaded in a *masticator*, by which the impurities become gradually washed away by the water.   After several

hours the gutta-percha is found to be kneaded into a perfectly homogeneous mass, which is rolled or drawn into sheets, bands, &c.

Gutta-percha becomes soft and plastic at the temperature of boiling water (212° Fahr.), when two pieces may be welded together. It is a non-conductor of electricity, and is indeed one of the best *insulating* materials known : it is impervious to moisture, and is scarcely at all affected by either acids or alkalies. Owing to its plasticity when soft, it is one of the most useful materials for making moulds, yielding impressions which are exquisitely sharp in the very finest lines. When used for making moulds from small objects, as coins, medallions, or sealing-wax impressions of seals, a piece of gutta-percha of the required size is placed in hot water (the temperature of which should be about 160° Fahr.), and, when sufficiently soft, it should be rolled while still wet in the palms of the hands until it assumes the form of a ball ; it should then again be soaked in the hot water for a short time, and be again rolled as before, care being taken to observe that the surface of the ball exhibits no seams or fissures. When larger objects have to be copied stout sheet gutta-percha is used, and a piece of the required size cut from the sheet, which is softened as before, then applied to the object, and the necessary pressure given to secure a faithful impression.

Plastic Gutta-percha.—When gutta-percha is steeped for a few hours in benzol or naphtha it becomes considerably swollen ; if afterwards soaked in hot water it is exceedingly plastic, and requires but moderate pressure to obtain most perfect copies from even such fragile objects as plaster of Paris models.

Gutta-percha and Marine Glue.—The following has also been recommended : gutta-percha 2 parts, Jeffrey's marine glue 1 part. Each of the materials is first to be cut up into thin strips ; they are then to be mixed, placed in a pipkin and heated gently, with continual stirring, until the substances have become well incorporated : the mixture is now ready for use, and should be rolled into the form of balls before being applied for taking impressions. A very useful mixture is made by melting thin strips of gutta-percha as before, and adding one-third part of lard, keeping the mixture well stirred. It is applied by pouring it over flat surfaces, as steel plates, &c.

Beeswax.—This is a very useful material for moulding, and may be applied either in the form of virgin or white wax, or the ordinary commercial article—yellow beeswax. Since this substance, however, is very commonly adulterated, it may be useful to know something of its natural characteristics. At the temperature of 32° Fahr. beeswax becomes brittle, at from 80° to 90° it becomes soft and plastic, and it melts at about 155° Fahr. Mr. B. S. Proctor says : "It becomes plastic or kneadable at about 85° Fahr., and its behaviour while worked between the finger and thumb is characteristic. A piece the size of a pea being worked in the hand till tough with the warmth, then placed upon the thumb and forcibly stroked down with the forefinger, curls up, following the finger, and is marked by it with longitudinal streaks." Its ordinary adulterants are resin, farina, mutton suet, and stearine, though more ponderous substances, such as plaster of Paris, have sometimes been detected. White wax is very commonly adulterated with spermaceti, sometimes to the extent of two-thirds of the latter to one of wax. These sophistications, although not necessarily fatal to the preparation of good moulds, are certainly objectionable, inasmuch as it not unfrequently happens that a wax mould splits or cracks, not alone from cooling too quickly, but owing to the presence of foreign substances which impair its toughness.

Sealing-wax.—This substance may be employed for taking impressions of seals or crests, and was, indeed, one of the first materials used in the earliest days of electrotyping. The material, however, should be of good quality, and only sufficient heat applied to melt, without inflaming it.

Stearine.   Stearic Acid.—The former substance is the solid constituent of tallow, and the latter (stearic acid) is the same substance separated from fats by chemical processes. Either may be used for making moulds instead of wax ; but the late C. V. Walker recommended the following mixture in preference to either :—

		ozs.
Spermaceti .	. . . . .	8
Wax .	. . . . . .	$1\frac{3}{4}$
Mutton Suet	. . . . .	$1\frac{3}{4}$

Another formula consists of :—

		ozs.
White Wax	. . . . .	8
Stearine	. . . . . .	3
Flake White or Litharge	. . . .	$\frac{1}{2}$

The whole ingredients are put into a pipkin and gently heated over a low fire, with continual stirring, for about half an hour, after which the mixture is allowed to rest until the excess of litharge (oxide of lead) has deposited. The clear residue is then to be poured into a shallow dish, and when cold is put aside until required for use.

Fusible Metal.—This alloy, which melts at the temperature of boiling water, and in some preparations very much below that point, is very useful for making moulds from metallic and some other objects : and since it can be used over and over again, and is capable of yielding exceedingly sharp impressions, it may be considered one of the most serviceable materials employed for such purposes. The following represent the principal formulæ for fusible metal, the last of which melts at the low temperature of 151° Fahr. or 61° below the boiling point of water :—

	ozs.			ozs.			ozs.
I. Bismuth .	8	II.	Bismuth .	8	III.	Bismuth	8
Lead .	4		Lead .	5		Lead .	4
Tin .	4		Tin .	4		Tin .	2
			Antimony	1		Cadmium	2
	—			—			—
	16			18			16

The metals are to be put into a crucible or clean iron ladle, and melted over a low fire ; when thoroughly fused, the alloy is poured out upon a cold surface in small buttons or drops, and these, when cold, are to be again melted and poured out as before, the operations to be repeated several times in order to ensure a perfect admixture of the metals. Another and better plan is to *granulate* the metal, or reduce it to small grains in the following way :—Fill a tall jar or other vessel with cold water, and on the surface of the water place a little chopped straw (about 3 inches in length). When the metal is melted, get an assistant to stir the water briskly in one direction, then pour in the metal, holding the ladle at some distance from the surface of the water ; by this means the metal will be diffused and separated into a considerable number of small grains. The water is then to be poured off, and the grains collected, dried, and re-melted, after which another melting and granulation may be effected, and the alloy finally melted and cast into a mould, or simply poured out upon a flat iron or other surface, when it will be ready for future use. By the repeated melting, the alloy loses a little by the oxidation of the metals ; but since the heat required to fuse it is less than that of boiling water, the loss is but trifling, as compared with the importance of obtaining a *perfect* alloy of the various metals. It should be the practice to remove the crucible or ladle from the fire the moment the alloy begins to melt, and to depend upon the heat of the vessel to complete the fusion.

Elastic Moulding Material.—For making moulds from objects which are much *under cut*, in which case neither of the foregoing substances would be available, an elastic material is employed which has the same composition as that from which printers' rollers are made, that is to say, a mixture of glue and treacle, the formula for which is :—

	ozs.
Glue of the best quality . . .	12
Treacle	3
	—
	15

The glue is first to be covered with *cold* water and allowed to stand for at least twelve hours, by which time it should be perfectly soft throughout. The excess of water is then to be poured off, and the vessel placed in a saucepan or other convenient utensil, containing a little water, and heat applied until the glue is completely melted, which may be aided by frequent stirring. When quite melted, pour in the treacle, and again stir until perfect incorporation of the ingredients is effected, when the composition may be set aside to cool . until required for use. To check evaporation and consequent drying of the surface, the vessel, when the material is quite cold, may be inverted over a piece of clean paper, by which, also, it will be protected from dust. The compound thus formed is exceedingly elastic, and may readily be separated from models even when severely undercut. Owing to the *solubility* of this composition, however, some care is necessary in using it, otherwise it will become partially dissolved in the copper solution or bath. This is more likely to occur, however, when the solutions are of less strength than *saturated*, by which term we understand that the water present holds as much sulphate of copper in solution as it is capable of doing. Various remedies for overcoming this disadvantage will be given when treating of the methods of obtaining moulds from the material.

Plaster of Paris.—This substance is also used for mould-making, either from metallic or natural objects; but the plaster should be of the finest quality, such as is used by Italian image makers for the *surface* of their work, and not the coarse material usually sold in the shops. The plaster should be *fresh* when purchased and preserved in a closely-covered jar until required for use.

Having thus far considered the materials used in making moulds for electrotype purposes, we will next explain the methods of applying them, confining our observations to the more simple examples in the initial stages of the process.

CHAPTER V.

ELECTRO-DEPOSITION OF COPPER (*continued*).

Moulding in Gutta-percha.—Plumbagoing the Mould.—Treatment of the Electrotype.—Bronzing the Electrotype.—Moulds of Sealing-wax.—Copying Plaster of Paris Medallions.—Preparing the Mould.—Plumbagoing.—Clearing the Mould.—Wax Moulds from Plaster Medallions.—Moulds from Fusible Metal.

Moulding in Gutta-percha.—In the former case, we explained how a copy of a coin could be obtained, in reverse, by making the original act as the mould. We will now turn our attention to obtaining fac-simile duplicates in relief, from impressions or moulds of similar objects, from such of the materials described in the last chapter as will best answer the purpose ; and since the application of these materials in the simple way we shall indicate will lead to an understanding of the general principles of mould-making, it is recommended that the student should endeavour to acquire adroitness in taking impressions which will be perfectly sharp and clear, before he attempts to obtain metallic deposits of copper upon them.

To obtain a copy of a medal, coin, or other similar object, the most convenient material to employ is gutta-percha. Take a small piece of this substance and place it in hot but not boiling water for a few minutes, or until it is perfectly soft ; while still wet, roll it between the palms of the hands until it assumes the form of a ball ; it should then be replaced in the water for a short time, and again rolled as before. The coin to be copied is now to be laid, face upward, upon a piece of plate-glass, slate, or polished wood. Now take the ball of gutta-percha and place it in the centre of the coin, and press it firmly all over it, *from the centre to its circumference*, so as to exclude the air, and in doing this it may be necessary to occasionally moisten the tips of the fingers with the tongue to prevent the gutta-percha from sticking to them. A flat piece of wood may now be laid over the gutta-percha, and if this be pressed forcibly by the hands this will ensure a perfect impression. After about a quarter of an hour or so, the gutta-percha mould may be readily removed from the coin, provided that the material has set hard.

Plumbagoing the Mould.—Having thus obtained a mould from a material which is a *non-conductor* of electricity, we next proceed to give it a conducting surface, without which it would be incapable of receiving the metallic deposit of copper which constitutes an electrotype. For this purpose, *plumbago*, or *graphite*,* is usually employed. To plumbago the surface of the gutta - percha mould proceed as follows :—Hold the mould between the fingers of the left hand, face upwards ; now dip a soft camel-hair brush in finely-powdered plumbago (which should be of good quality) and briskly brush it all over the surface, every now and then taking up a fresh supply of plumbago with the brush. Care must be taken to well brush the powder into every crevice of the impression, and it is better to work the brush *in circles*, rather than to and fro, by which a more perfect coating is obtained. When properly done, the face of the mould has a bright metallic lustre, resembling a well-polished (that is blackleaded) stove.

In order to prevent the deposit of copper from taking place on the upper edge (beyond the actual impression), the plumbago which has been accidentally brushed over this surface should be removed, which may be conveniently done by rubbing it off with a piece of damp rag placed over the forefinger. The mould is now to be attached to the conducting wire by gently heating its longer end in the flame of a candle or ignited match, and then placing it on the edge of the mould, as far as the circumference of the impression ; by giving it gentle pressure it will become sufficiently imbedded : the wire must not, however, be below the flat surface of the mould. If held steadily in the hand for a few moments, or until the wire and gutta-percha have cooled, the joint will *set*, and the mould may then be carefully laid aside until the point of junction has set firm. A little plumbago must now be brushed over the joint, so as to ensure a perfect *electrical connection* between the wire and the plumbagoed mould.

The mould being attached to the conducting wire, must now be connected to the zinc by its binding-screws as before (Fig. 54), and both should be immersed at the same time in their respective solutions, but this must be done with care, otherwise the mould may become separated from the wire. It may be well, in this place, to call attention to certain precautions which, if carefully followed, will prevent failure, and consequent disappointment, in electrotyping.

Precautions.—1. The solution of copper to be used in the single-cell apparatus must be kept as nearly as possible in a *saturated* condition, which is effected by keeping the shelf or tray constantly supplied with crystals of sulphate of copper. 2. The superficial surface of zinc immersed in the porous cell should not be much greater than that of the mould to be copied. 3. The solution should be stirred with a

---

* Commonly called *blacklead*, but in reality *carbon* in a crude state.

glass rod or strip of wood before immersing the mould, especially if it
has been previously used for electrotyping : if this is not done, the
deposit may become irregular in thickness. 4. The plumbagoed
mould should not be disturbed until its entire surface is covered with
copper. A few moments after immersion, a bright pinkish red
deposit of copper will be observed at the end of the wire, which in a
short time will radiate in the direction of the plumbagoed surface,
and this will gradually extend wherever this conducting medium has
been spread with the brush, provided the operation has been conducted with proper care, and an uniform coating obtained.

Treatment of the Electrotype.—A sufficiently stout deposit of
copper, upon a gutta-percha mould of a small coin, may generally be
obtained in about two days, or even in less time, under the most
favourable conditions : but it is not advisable to attempt to separate
the electrotype from the mould while the deposit is very thin, otherwise the former may become broken in the operation. Assuming the
deposit to be thick enough, the first thing to do is to cut the end of
the wire connected to the mould with a pair of cutting pliers or a file,
after which the superfluous copper may be removed from the outer
edge by breaking it away with the pliers, taking care not to injure
the " type " itself. The mould may then be placed in hot water for
a moment, when the electrotype will readily separate from the gutta-
percha. In order to give additional solidity to the electrotype, it should
be *backed up* with pewter solder, which may easily be done as follows :—
Put a small piece of zinc into about a teaspoonful of hydrochloric acid
(muriatic acid) : when the effervescence which takes place has ceased,
brush a little of the liquid, which is a solution of chloride of zinc,
over the back of the electrotype, and then apply solder by means of a
moderately-hot soldering iron, until the entire surface is *tinned*, as it
is called, when a further supply of solder should be run on to the
back to give the required solidity. When this is done, the rough
edge of the electrotype should be rendered smooth with a keen file.

Bronzing the Electrotype.—To impart an agreeable bronze appearance to the type, it should first be cleaned by brushing it with
a solution of carbonate of potash (about half a teaspoonful in an ounce
of water), and applying at the same time a little whiting. An
ordinary tooth-brush may be used for this purpose, and after brisk
rubbing the type must be well rinsed in clean water. The bronze
tint may be given by brushing over it a weak solution of chloride of
platinum (1 grain to an ounce of water) : when the desired tint is
obtained, the type is to be rinsed with hot water and allowed to dry.
The tone may be varied from a delicate olive-brown to deep black,
according to the proportion of platinum salt employed. A few drops
of sulphide of ammonium in water, or, still better, a few grains of

sulphide of barium dissolved in water, will give very pleasing bronze tints to the copper surface, the depth of which may be regulated at will by a longer or shorter exposure to the action of the bronzing material. If a solution of sulphide of barium be used, about 5 grains to the ounce of water will produce a pleasing tone in a few seconds. It is better to immerse the electrotype in the liquid (previously filtered) and to remove it the instant the desired tone is reached, and to place it at once in clean water.

Another method of bronzing electrotypes is by the application of plumbago, by which very pleasing effects may be obtained with a little care in the manipulation. The surface of the electrotype is to be first cleaned with rotten stone and oil ; the oil is then to be partially removed by a tuft of cotton wool, and the surface is next to be brushed lightly over with plumbago (a soft brush being used) until a perfectly uniform coating is given. It is next to be heated to a point that would singe the hair of the blacklead brush, and then set aside to cool, after which it must be brushed with considerable friction. The tint will depend upon the quantity of oil allowed to remain, this enabling the surface to retain more of the blacklead, consequently to appear of a darker colour. The effect is very fine, and gives high relief to the prominent parts, from their getting so much more polish than the hollows, thus obviating the disagreeable effect which all unbronzed bassi-relievi produce by reason of their metallic glare.— *Hockin.*

The beautiful red bronze tone which is seen on exhibition and other medals is produced by brushing over the medal a paste composed of peroxide of iron (jewellers' rouge) and plumbago, after which the article is moderatly heated, and when cold is well brushed until it acquires the necessary brightness and uniformity of surface. Equal parts of fine plumbago and jewellers' rouge are mixed up into an uniform paste with water, and the cleaned medal is then uniformly brushed over with the mixture, care being taken not to allow the fingers to come in contact with the face of the object. The medal is then placed on a stout plate of iron or copper, and this is heated until it acquires a dark colour ; it is then removed from the fire and allowed to become cold. It is next brushed for a long time, and in all directions, with a moderatly stiff brush, which is frequently passed over a block of yellow beeswax, and afterwards upon the paste of plumbago and rouge. The bronzing may also be produced by dipping the cleaned medal in a mixture composed of equal parts of perchloride and pernitrate of iron ; the medal is then to be heated until these salts are thoroughly dry. It is afterwards brushed as before with the waxed brush until a perfectly uniform and bright surface is obtained.

Bronzing may also be effected by dipping the medal in a solution of sulphide of ammonium, and when this has dried, the plumbago and rouge paste is to be applied as before, and the waxed brush again employed. If the object be heated after applying the sulphide of ammonium, a black bronze, called " smoky bronze," is produced, and if the high lights be lightly rubbed with a piece of chamois leather dipped in spirit of wine, a very pleasing effect of contrast is obtained.

Moulds of Sealing Wax.—This material is, as we have said, very useful for obtaining impressions of seals, signet rings, and other small objects. A simple way of taking an impression in sealing-wax is as follows : Hold a card over a small benzoline lamp, but not touching the flame ; now take a stick of the best red sealing-wax and allow it to touch the heated part of the paper, working it round and round until a sufficient quantity of the wax becomes melted upon the card. Now place the card upon the table, and having gently breathed upon the seal or signet ring, impress it in the usual way. Having secured an impression, cut away the superfluous portions of the card with a pair of scissors, and moisten the wax impression with a few drops of spirits of wine. When this has apparently dried, proceed to brush plumbago over the surface, using a camel-hair brush, and when perfectly coated, gently heat the end of the conducting-wire and apply it to the edge of the sealing-wax, allowing the point of the wire to approach the edge of the impression. Now brush a little plumbago on the point, and connect the short end of the wire to the binding-screw.

After having obtained several electrotypes successfully, and thereby become *au fait* to the manipulation of the single-cell apparatus, the student will naturally desire to extend operations to objects of a more important nature, such as medallions, busts, statuettes, and natural objects, as leaves, fishes, &c. But before attempting the more elaborate subjects it will be well to select, for our next operation, one of a simpler character, such as a plaster of Paris medallion, an admirable model to reproduce in metallic copper.

Copying Plaster of Paris Medallions.—These pleasing works of art, which may be obtained at small cost from the Italian image makers, are specially suited for the elementary study of the electrotype process, while a cabinet collection of such objects reproduced in copper forms an exceedingly interesting record of the manipulator's skill and perseverance. There are several materials from which moulds from plaster medallions may be obtained ; but we will first describe the method of preparing a mould with gutta-percha. To render the plaster more capable of bearing the treatment it will have to be subjected to, the face of the medallion should first be brushed over with boiled linseed oil, and this allowed to sink well into the plaster. After about two days the oil will have sufficiently dried and hardened upon the surface to render the plaster less liable to injury. The medallion thus prepared is next to be provided with a rim or collar of pasteboard or thin sheet tin, which must be tightly secured round its circumference either by means of thin copper wire, jeweller's "binding-wire," or strong twine. The rim should project about half an inch above the highest point on the face of the medallion, and must be on a level with its base ; it is then to be laid upon a perfectly smooth surface until the moulding material is ready. We recommend the student to practise upon small medallions at first ; say about two inches or two inches and a half in diameter.

Preparing the Mould.—A lump of gutta-percha is now to be taken of sufficient size to cover the medallion, fill the vacant space up to the top of the rim, and project above it. The gutta-percha is to be softened in hot water and rolled up into the form of a ball, as before directed, care being taken to obliterate all *seams* or cracks by repeatedly soaking in the hot water and rolling in the hands. It must on no account be applied until it is perfectly smooth, and as soft as hot water will make it. To give additional smoothness to the surface of the ball, it may be lightly rolled round and round, with one hand only, for an instant upon a polished table just before being used. Now take the ball in one hand and place it in the centre of the medallion ; then press it firmly from the centre towards the circumference, taking care not to shift it in the least degree. The gutta-percha must be pressed well into the cavity, and when this is done, a piece of flat wood may be placed on the mass and this pressed with both hands with as much force as possible for a few moments, when it may be left until the gutta-percha has set hard. If convenient, a weight may be placed upon the board after having pressed it with the hands. In about half an hour the board may be removed, and the mould allowed to rest until quite cold, when the rim may be removed and the mould separated by gently pulling it away from the medallion. As a precaution against breaking the plaster medallion, it may be well to suggest that its back should be examined, and if it be otherwise than perfectly flat, it may be advisable to gently rub it upon a sheet of glass-paper, which will readily remove all irregularities from the surface. It is also important that the surface upon which the medallion is laid, when applying the gutta-percha, should be quite level ; and it will be still better if several folds of blotting-paper are placed between the table and the medallion before the necessary pressure is given. These points being attended to, there is little fear of the medallion becoming broken.

Plumbagoing.—The gutta-percha mould is now to be well plumbagoed, for which purpose a soft brush, such as jewellers use for brushing plate and jewellery that has been rouged, may be used, and this being frequently dipped into the plumbago is to be lightly but briskly applied, special care being taken to well plumbago the *hollows*. When it is borne in mind that the most delicate line, even if imperceptible to the eye, will be reproduced in the metallic copy, the importance of not injuring the face of the mould will become at once apparent. It is also absolutely necessary that the gutta-percha should be of the best quality, and since the same material may be used over and over again, its first cost is of little consideration.

Clearing the Mould.— The mould being well coated with plumbago, all excess of this material which has become spread over the outer edges, beyond the impression itself, must be wiped away, and the more completely this is done the less trouble will there be afterwards in clearing away from the electrotype the crystalline deposit which, under any circumstances, forms around the circumference of the electrotype. Indeed, when the student has once or twice experienced the inconvenience of having to remove the superfluous copper from his electrotypes, he will not fail to exert his wits to diminish the labour which this involves as far as practicable, by every possible care before the mould goes into the copper bath. We therefore urge for his guidance, that the removal of the excess of plumbago should be deemed one of the important details of his manipulation, and that it should never be neglected. After wiping away the excess of blacklead, it will be found a good plan to place a piece of dry rag on the forefinger and to rub it on a common tallow candle, so as to make the part slightly greasy ; if now the edge of the mould (carefully avoiding the impression) be rubbed with the rag-covered finger, this will effectually prevent the deposit from taking place upon such part ; before doing this, however, the conducting wire should be gently heated and imbedded in the edge of the mould as before, taking care that the point of the wire touches the extreme edge of the impression, and a *perfect connection* between the wire and the latter must be secured by applying a little plumbago with a camel-hair brush or the tip of the finger. It is sometimes the practice to apply varnish of some kind to the edges of moulds, and also to the conducting wire as far as the joining, but until the student has thoroughly mastered the process of copying simple objects in the way we have indicated, we do not recommend him to employ varnishes : indeed not until dealing with objects of a larger and more elaborate kind.

The mould being now ready, is to be connected to the binding-screw by its wire, and since the material of which it is composed is much lighter than the copper solution, the wire must be sufficiently rigid, when bent at right angles, as in Fig. 54 to keep the mould well down in the bath. Being placed in the solution, it must be allowed to remain undisturbed until the entire surface of the impression is

I

covered.   In from two to three days the deposit should be of sufficient thickness to allow of its separation from the mould.

For copying small medallions of the size referred to, the single-cell apparatus shown in Fig. 54 may be used, but for larger sizes or for

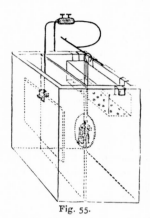

depositing upon several moulds at the same time, the arrangement shown in Fig. 55 will be most suitable.   This apparatus consists of a wood box well varnished in the interior, and divided into two cells or compartments by a partition of thin porous wood.   The larger cell is nearly filled with a saturated solution of sulphate of copper, and the smaller cell with a half-saturated solution of sal-ammoniac.   A perforated shelf is suspended in the larger compartment to contain a supply of crystals of the sulphate.   A plate of *pure* zinc, connected by a copper conducting wire, is suspended in the smaller cell, and the mould connected to the opposite end of the wire by suitable binding-screws.   In this ar-

Fig. 55.

rangement neither acid nor mercury are used, and although the action is not so rapid as in the former arrangements, it is very reliable for obtaining good results.

Wax Moulds from Plaster Medallions. — Beeswax is a very useful material for preparing moulds from plaster medallions, the following simple method being adopted :—The medallion, instead of being oiled as in the previous case, is simply soaked in hot water for a short time or until it has become completely saturated.   First put a sufficient quantity of wax into a pipkin and melt it by a slow fire ; when melted, place it on the hob until wanted.   Place the medallion face upwards in a plate or large saucer, into which pour boiling water until it reaches nearly half-way up its edge.   In a minute or two the face of the plaster will assume a moist appearance, when the excess of water is to be poured out of the plate.   A rim of card is now to be fastened round the edge of the medallion, which may be secured either by means of sealing-wax or a piece of twine.   As before, the rim should extend about half an inch above the most prominent point of the image.   The medallion being returned to the plate, the wax is now to be steadily poured on to the face of the object, the lip of the pipkin being placed near the pasteboard rim and nearly touching it, to prevent the formation of air-bubbles.   When the cavity is filled up to the top of the rim, if any air-bubbles appear they must be at once removed with a camel-hair brush kept for this purpose, or the feather end of a quill, or even a strip of paper may be used. The wax must now be allowed to cool as slowly as possible, and in order to favour this gradual cooling, a clean, dry jar may be inverted over the mould and there left until the wax is quite cold. This precaution will tend to prevent the wax from cracking, an event which sometimes, but not very frequently, occurs.

When quite cold, the wax mould will generally separate from the plaster by the application of moderate force to pull them asunder. If such is not the case, however, return the medallion to the plate and pour in a little boiling water. After a *few seconds'* immersion the mould will easily come away. If, however, owing to some irregularity in the face of the medallion, the mould still refuses to separate, plunge the whole into *cold* water, and, if necessary, use the edge of a knife as a lever between the two surfaces and force them asunder. If it be found that small portions of plaster adhere to the mould these may be carefully picked out with a fine-pointed piece of wood, and the mould afterwards very lightly brushed over with a soft plate brush. Should it be found that some particles still obstinately adhere to the wax, apply a little oil of vitriol with a thin strip of wood to the parts and set the mould aside for about twelve hours, by which time the acid, by attracting moisture from the air, will loosen the plaster, which may then be brushed away with a soft brush and water. The mould must then be put away to dry, or may be laid, face downward, upon a pad of blotting-paper or calico.

The mould is now to be plumbagoed with a very soft brush, but, owing to the yielding nature of the wax, the greatest care must be taken not to apply the brush too severely, only sufficient friction being used to coat the surface uniformly. It is a good plan to sprinkle a little plumbago over the face of the mould, and then to work the brush about in circles, by which means a well plumbagoed surface may readily be obtained. This operation being complete, the superfluous plumbago is to be brushed off, and, by blowing upon the face of the mould, any plumbago remaining in the crevices may be removed. The conducting wire is to be attached, as in the case of gutta-percha, by gently *warming* the end of the wire; but, if the mould be a tolerably large one (say, 3 inches in diameter) it will be well to bend the end of the wire so as to leave a length of about an inch or more to be embedded in the edge of the mould, by which means it will be more effectually supported than if the point of the wire only were attached. The joint must now be well plumbagoed, and the excess of this material which has been brushed over the edges may easily be removed by scraping it away with a pen-knife.

The same precautions must be observed with regard to wax-moulds as with those made from gutta-percha when immersing them in the bath, otherwise they will, from their exceeding lightness, be disposed to rise out of the solution. In the case of large moulds made from such light materials they require to be *weighted* in order to keep them beneath the surface of the copper solution, as we shall explain when treating of them.

The stearine composition may be employed instead of wax in the preceding operation, but we recommend the student to adopt the latter material for copying small medallions, since, with a little care, it will answer every purpose, and needs no preparation beyond melting it.

Moulds from Fusible Metal.—There are many ways of making moulds from fusible metal, but, for our present purpose, we will select the most simple. To obtain an impression of a coin or medal, melt a sufficient quantity of the alloy in a small ladle or iron spoon, then, holding the coin face downward between the forefinger and thumb of the right hand, pour the alloy into the rim of an *inverted* cup or basin, and, bringing the coin within a distance of about 2 inches from the molten alloy, allow it to fall *flat* upon the metal and there leave it until cold. If, when the metal is poured out, there is an appearance of dulness on the surface (arising from oxidation of the metals) a piece of card or strip of stiff paper should be drawn over it, which will at once leave the surface bright. As the metal soon cools, however, this may be more conveniently done by an assistant just before the coin is allowed to fall. If no other help is at hand a piece of card should be placed close to the cup, so that the moment the metal is poured out it may be applied as suggested, and the coin promptly dropped upon the cleaned surface of the alloy. A very little practice will render the student expert in obtaining moulds in this way, and, considering how very readily the material is re-melted, a few failures need not trouble him.

The fusible alloy may also be employed in the form of a paste, but, in this case, it is advisable to have the assistance of another pair of hands, since, in this condition, it soon becomes solid and therefore unusable. The coin should first have a temporary handle attached to it, which may readily be done by rolling a small lump of gutta-percha into the form of a ball : one part of this should now be held in the flame of a candle until the part fuses, when it is to be pressed upon the back of the coin and allowed to remain until cold. This gutta-percha knob will serve as a handle by which the coin may be held when the impression is about to be taken. The requisite quantity of the fusible alloy is now to be poured upon a piece of board and worked up into a stiff paste by means of a flat piece of wood—an operation that only occupies a few moments. *The instant* the alloy has assumed the pasty condition the coin, being held by its gutta-percha handle, is to be promptly and firmly pressed upon the mass until it is sufficiently imbedded in it. In the course of a minute or so the coin may be withdrawn, when the mould should present a perfect and delicate impression of the original—of course in reverse. Should any faults be visible, owing to want of dexterity on the part of the operator, the metal must be re-melted and the operation conducted again. A very little practice will enable the student to produce moulds in this alloy with perfect ease. The coin, in each of the above cases, should be perfectly cold before applying it to the alloy. Large medals are moulded by simply dropping them—a little sideways—into the metal when on the point of solidification.

Connecting the Mould to the Wire.—When a perfect mould is obtained the conducting wire is to be attached, which is done by first scraping the longer end of the wire so as to render it perfectly clean ; it is then to be held in the flame of a candle, but at a little distance from the clean end. The mould being now held in the left hand, is to be brought near, but not touching, the flame, and, when the wire is sufficiently hot, it is to be pressed against the *back* of the mould, when it will at once become imbedded in it, and in a few moments will be firmly set. A small portion of powdered resin applied to the spot will assist the union of the two metals. The back and upper edge of the mould must now be coated with sealing-wax varnish or some other quick-drying varnish, or, if carefully applied, paraffin wax (which melts at a very low heat) may be applied by first *gently* heating the mould and touching it with a small stick of the paraffin wax. It is well, also, to varnish that portion of the conducting wire *above* the joint which has to be immersed in the copper bath, in order to prevent it from receiving the copper deposit.

CHAPTER VI.

ELECTRO-DEPOSITION OF COPPER (*continued*).

Electrotyping by Separate Battery.—In employing the single-cell apparatus, we have seen that it is necessary to keep up the strength of the solution by a constant supply of crystals of sulphate of copper, otherwise the solution would soon become exhausted of its *metal*, and therefore useless. If we employ a *separate battery*, however, this method of sustaining the normal condition of the bath is unnecessary, as we will now endeavour to show ; but in doing so we must direct the reader's attention for the moment to the principles of electrolysis, explained in a former chapter. The practical application of those principles may be readily expressed in a few words : If, instead of making the mould, or object to be copied, the *negative element*, as in the single-cell apparatus, we take a separate battery composed of two elements—say, zinc and copper, as in Daniell's battery, we must then employ a *separate* copper solution or electrolytic bath, in which case the object to be deposited upon must be connected to the zinc element, as before, but the wire attached to the negative element of the battery (the free end of which is the *positive electrode*) must have attached to it a plate of sheet copper, which with the mould must be immersed in the solution of sulphate of copper. By this arrangement, while the copper is being deposited upon the mould, the sheet copper becomes dissolved by the sulphuric acid set free, forming sulphate of copper, which continued action re-supplies the bath with metal in the proportion (all things being equal) in which it is exhausted by deposition of copper upon the mould.

Arrangement of the Battery.—At Fig. 56 is shown a Daniell's battery, A, connected, by its negative conducting wire (proceeding from the zinc), to the mould, B, with its face turned towards the copper plate or anode, C. The depositing vessel, D, which may be of glass or stoneware, for small operations, is charged with an acid solution of sulphate of copper, which is composed as follows :—

Sulphate of Copper   .   .   .   .   .   1 lb.
Sulphuric Acid   .   .   .   .   .   1 ,,
Water (about) .   .   .   .   .   .   1 gallon.

The sulphate of copper, as before, is dissolved in a sufficient quantity of hot water, after which cold water is added to make up one gallon ; the sulphuric acid is then added and the solution is set aside until quite cold, when it is to be poured into the depositing bath, which should be quite clean. When first placing the mould to be copied in the bath, a small surface only of the copper plate should be immersed in the solution, and this

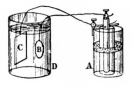

Fig. 56.

may be gradually increased (by lowering the copper plate) as the deposit extends over the surface of the mould.

In Fig. 54 is shown an arrangement in which several moulds are suspended by a brass rod laid across the bath B, the rod being connected to the zinc element of the battery, A, by the wire, x. Strips of sheet copper are suspended by a brass rod, c, which is connected by a binding-screw to the positive conducting wire, z, of the battery, which

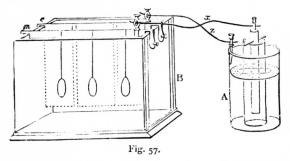

Fig. 57.

in the woodcut represents a Daniell cell. In this arrangement, the sheet copper, by becoming dissolved in the solution during the electrolytic action, keeps up the normal strength of the bath, which in the single-cell arrangement is attained by the supply of crystals of sulphate of copper. It may be well to mention that it is always preferable, besides being more economical of time, to deposit upon several moulds at a time in the bath, and this can be effected even with apparatus of small dimensions. The more extensive arrangements for depositing upon large objects by means of powerful battery currents will be considered in another chapter.

Copying Plaster Busts.—For this purpose, the elastic moulding material is used. Suppose we desire to obtain an electrotype from a small plaster bust, the object must first be well brushed over with boiled linseed oil, and then set aside for two or three days to allow the surface to harden. In applying the oil, care should be taken not to allow it to touch the lower surface surrounding the orifice at its base, over which a piece of stout paper must be pasted to prevent the elastic material from entering the cavity, but before doing this partly fill the cavity with sand, to increase its weight. The bust is next to be suspended, upside down, by means of twine or thin copper wire, inside a jar sufficiently wide and deep to leave at least half an inch all round and at the bottom. When thus placed in its proper position, the elastic composition (p. 105), having been previously melted, is poured in, and if any air-bubbles appear, these must be removed with the feather of a quill, when the vessel is allowed to rest until the composition is quite cold.

The vessel is now to be inverted, when the solidified mass and the imbedded bust will gradually slip out. To facilitate this by preventing the composition from sticking to the jar, it is a good plan to slightly oil the interior of the vessel in the first instance. Having removed the mould, it must now be separated from the plaster bust. This is done as follows :—First place the mould in an erect position, base downward, then, with a thin knife, make an incision from the top to the base of the mould, *at the back* of the bust. The mould may now be readily opened where the incision has been made, and while being held open, an assistant should be at hand to gently remove the bust, when the mould, owing to its elasticity, will readily close itself again. It must next be secured in its proper position by being carefully bound round with a bandage of tape. The mould is then to be inverted, and returned to the jar. A sufficient quantity of wax is now to be melted at the *lowest* temperature that will liquefy it, otherwise it will injure the mould ; it is then to be poured into the mould and allowed to rest until thoroughly cold. When cold, the elastic mould is to be again removed from the jar, and separated by untying the bandages from the wax-casting. This latter must now be well plumbagoed, a conducting wire attached, and the joint coated with plumbago as before directed. Since it will be difficult, however, to obtain an uniform deposit over such a comparatively large surface, it will be necessary to apply *guiding* wires, as they are called, and to which we must now direct special attention.

Guiding Wires.—The application of additional wires, to facilitate the deposition of copper in the cavities, or undercut surfaces, of moulds was first introduced by Dr. Leeson. A sufficient number of lengths of fine brass wire are twisted firmly round the main conducting wire, at a short distance from its junction with the mould, and these, one by one, are bent in such a way that their extreme points may rest, *lightly*, upon the hollow surfaces of the mould, whereby the current is diverted, to a certain extent, from the main

Fig. 58.

wire to the cavities or hollows, which are less favourably situated for receiving the metallic deposit than the plane surfaces. The application of guiding wires is more especially necessary when the object to be copied is of considerable dimensions ; the principle of their arrangement is shown in Fig. 58.

The mould, prepared as described, is to be put in connection with the battery, by suspending it from the negative conducting-rod, and then gently lowered into the coppering bath. In the present case only a moderately stout deposit, or " shell," of copper will be necessary, since, as we shall explain, this deposit will, in the next operation, act the part of a *mould*, in producing a fac-simile of the original. When a perfect coating is obtained, of sufficient thickness to bear handling, it is to be removed from the bath, rinsed, and allowed to drain. It must then be heated sufficiently to melt the wax, which is allowed to run into any convenient receptacle, and the interior of the electrotype (which now represents a mould) must be cleansed from all adhering wax, by continuing the heat until the last drop ceases to flow. It must then be treated with spirit of turpentine, with the application of moderate heat, to dissolve out the remaining wax, the operation being repeated so as to entirely remove all traces of the wax.

The next operation consists in depositing copper upon the *interior* of the copper mould, which may be readily done in the following way:—A small quantity of sweet oil is first to be poured into the mould, which must be moved about so that the oil may spread all over the surface ; it must then be tilted over a vessel to allow the oil to run out, and next placed upon several folds of blotting-paper before a fire, for several hours, until the oil ceases to flow. The mould must now be carefully examined, and if any " pin-holes," as they are called, are visible, these must be stopped by melted wax dropped upon each spot upon the outside of the mould.

The mould is now to be placed in a jar, in an inverted position, and held in its place by a padding of paper or rag, wedged around its base. The *negative* electrode (or wire connected to the zinc of the battery) is now to be connected to the mould, which may conveniently be done by soldering. A strip of stout sheet copper, attached to the *positive* electrode, is then to be suspended in the cavity of the copper mould, but not allowed to touch any part of it, and in this position it must be fixed securely, which may be conveniently done by a piece of wood laid across the orifice of the mould. The mould is now to be filled with the copper solution last mentioned, and the battery is then to be set in action. In order to obtain a good solid electrotype from the copper mould, it will be necessary to renew the copper plate, or anode, from time to time when it becomes worn away, unless it be of sufficient thickness to render such renewal unnecessary. The strength of the battery must also be well kept up by renewing the acid solution in the porous cell. When a deposit of sufficient thickness is obtained, the conducting wires may be disconnected, the copper solution poured out, and the interior rinsed with water.

The next operation is to remove the shell of copper constituting the mould, which is done by *breaking it away*—beginning at the base—

with a pair of pliers. When the first layer of metal has been lifted from the underlying deposit, the remainder may generally be *peeled off* with but little trouble, when the electrotype proper will be exhibited, and if successfully accomplished it will amply reward the operator for the trouble and care devoted to its production. The student should not, however, undertake the manipulation of the elastic moulding composition until he has acquired a skilful aptness in the simpler processes of electrotyping. It may be well to mention that the elastic composition may be re-used several times, provided it has been kept in a covered vessel, to exclude it from the action of either a moist or a very dry atmosphere.

Moulding in Plaster cf Paris.—This material, especially for copying natural objects, such as leaves, ferns, fishes, &c., is exceedingly useful, and we will, as in former instances, first give the more simple method of applying it, so that the student may have no difficulty in its manipulation. To obtain a plaster mould from a coin or medal, for example, first oil the face of the object slightly by applying a single drop of oil, with a tuft of cotton wool, and with a fresh piece of wool gently rub the coin all over, so as to leave but a trace of oil on the surface, the most trifling quantity being sufficient to prevent the adhesion of the plaster to the original. A rim of card is now to be fixed round the medal, to form a receptacle for the plaster. A little cold water is then to be poured in a cup, or other convenient vessel, and a small portion of *fine plaster* dropped into the water. The excess of water is now to be poured off and the plaster briskly stirred with a spoon. Now fill the spoon with the plaster (which should be about the consistency of cream) and pour it carefully over the face of the medal. If any air-bubbles appear, disperse them with a feather or camel-hair brush, which should be immediately after plunged into cold water, so that the plaster may easily be removed, and the brush thus left ready for future use. In about half-an-hour or so, the coin and mould may be detached, and the latter should then be placed in a moderately warm oven until dry. When perfectly dry, the face of the mould is to be well painted over with boiled linseed oil, repeating the operation several times ; or the mould may be saturated with wax, by pouring a little of this substance, in a melted state, over the face of the mould, and then placing it in the oven until the wax becomes absorbed by the plaster. When cold, the mould must be plumbagoed in the ordinary way, and a copper conducting wire attached by twisting the wire round its circumference, and forming a connection with the plumbagoed surface by means of a drop of melted wax, afterwards brushed over with plumbago. That portion of the wire which surrounds the mould should be coated with varnish to prevent the copper from being deposited upon it. The superfluous plumbago should, as in the former cases, be removed, by scraping it away with a knife, leaving the connection, of course, untouched. The mould is now ready for the depositing bath, into which it must be gently lowered, so as to avoid breaking the connection between the conducting wire and the plumbagoed surface, a precaution which must in all similar cases be strictly observed.

Copying Animal Substances.—Suppose we desire to obtain an electrotype of a small fish (the scaly roach being very suitable), for example. The object is first brushed over lightly with a little linseed oil; we next mix a sufficient quantity of plaster of Paris into a thinnish paste, and pour this in a shallow rim of metal or stout cardboard placed upon a piece of glass or sheet of paper, previously rubbed over with a little oil or grease ; before the plaster has time to *set*, the fish is to be held by its head and tail, and laid on its side upon the paste, using sufficient pressure to imbed one half of the fish. To assist this, the soft plaster may be worked up or guided to its proper places by means of a knife-blade, care being taken to avoid spreading the plaster beyond that part which is to form the *first half* of the mould. The plaster is now allowed to set hard, which occupies about half an hour. We next proceed to mould the second half of the fish. A small brush, say a painter's sash tool, is dipped in warm water, and then well rubbed over a lump of soap ; this is to be brushed all over the plaster, but avoiding the fish, and the soap and water applied several times to ensure a perfect coating. A rim of greater depth, say $\frac{3}{4}$ of an inch deeper, must be fixed round the mould, in place of the former rim, and a second quantity of plaster made into a thinnish paste, as before, which must then be carefully poured over the fish and upper surface of the mould, taking care not to let it flow over the rim. This second batch of plaster should be sufficient to form a thick half mould, as in the former case, otherwise it may break when being separated from the first half mould.

When the plaster has set quite hard, the two moulds may be separated by gently forcing them asunder, the soap and water having the effect of preventing the two plaster surfaces from adhering, while the oil applied to the fish also prevents the moulding material from sticking to it. When the two halves of the mould are separated, the fish is to be carefully removed, and the plaster moulds placed in a warm, but not very hot, oven, and allowed to become *perfectly* dry. They are then to be placed faced downwards in a plate or other shallow vessel, containing melted bees-wax, and allowed to remain until saturated with the material, especially on the faces of the moulds ; these are now allowed to become quite cold, when they are ready to receive a coating of plumbago, which must be well brushed into every part of the impression, until the entire surfaces present the bright metallic lustre of a well-polished fire-stove. The conducting wire must now be attached, which may be effected in this way : Bend a piece of stout copper wire in the form shown in Fig. 59, and pass the mould under the hook at *a*, and beneath the coil of the wire at *b* ; the shorter end of wire at *a* should just touch the edge of the impression, near the mouth or tail of the fish. The wire thus adjusted must be secured firmly in its place, by being bound to the mould with thin copper wire. Before placing the conducting wire in its position, as above described, it will be advisable to wipe away all superfluous plumbago from the face of the mould, carefully avoiding injury to the impression, and when the conducting wire is adjusted, it is a good plan to coat the wire at all parts but the extreme point at *a* with varnish, or melted paraffin wax, to prevent the copper from becoming deposited upon it. The end of the wire at *a* must be put in *metallic contact*, so to speak, with the plumbagoed impression, by brushing a little of that substance over the point of junction. Thus prepared, the long end of the conducting wire is to be connected to the negative pole of the battery, and the mould gently immersed in the bath, the copper anode previously being suspended from the positive electrode.

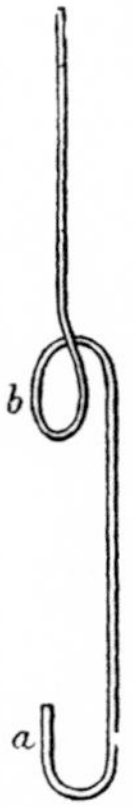

The second half mould may now be treated in same way as the above, and when two perfect electrotypes, or *shells*, are obtained, the superfluous copper should be removed by aid of a pair of pliers and a file ; when this is done the inner edges of each electrotype may be *tinned*, by first brushing a little chloride of zinc round the edge, and then passing a soldering iron, charged with pewter solder, over the surface. When the two halves of the fish are thus prepared, they may be brought together and held in position by means of thin iron " binding wire." The flame of a spirit-lamp or a blow-pipe flame may now be applied, which, by melting the solder, will soon complete the union, when a perfect representation of a fish will be obtained. This may afterwards be bronzed, gilt, or silvered by the processes described hereafter, and, if desired, mounted upon a suitable stand.

Fig. 59.

The elastic moulding material may also be used for copying animal substances ; in this case, one half of the fish must be imbedded in moulding sand ; a cylinder of thin sheet tin, bound together with fine copper wire, or by soldering, is then placed round the sand, so as to enclose it, and the sand is made as level as possible, by gently pressing it with any convenient instrument. The melted elastic material is now to be poured into the cylinder, which should be about two inches higher than the highest part of the object, until it nearly reaches the top : it is then allowed to rest for at least twelve hours, when the metal rim is to be removed and the mould withdrawn ; the object is next to be liberated from the mould, and the other half moulded in the same way.   The wax and stearine composition is to be poured into each half mould, and from the models thus obtained plaster moulds may be procured in the same way as from the natural object, but in this case the wax models must be well brushed over with plumbago before being embedded in the plaster.   Since electrotypes of fishes look exceedingly well as wall ornaments, it will be only necessary, for this purpose, to obtain an electrotype of one half of the fish, which may, after trimming and bronzing, be cemented to an oval board, stained black and polished, and, if desired, mounted in a suitable frame.

Electro-Coppering Flowers, Insects, &c.—Fragile objects, to which the ordinary methods of plumbagoing could not be applied, may be prepared to receive a deposit of copper in the sulphate bath by either of the following methods :—

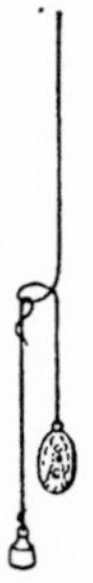

1. The object, say a rose-bud or a beetle, for instance, is first attached to a copper wire ; it is next dipped in a weak solution of nitrate of silver (about forty grains of the nitrate dissolved in one ounce of distilled water), and after being allowed to drain, but before it is dry, it is to be exposed to the vapour of phosphorus under a bell-glass.   To produce the vapour a small piece of phosphorus is dissolved in a little alcohol ; this is poured into a watch-glass (chemical " watch-glasses " are readily procurable), which is then placed in a plate containing hot sand.   The object being fixed by its wire in such a position that it cannot shift, the

Fig. 60.

bell-glass (an ordinary fern-glass will answer admirably) is to be placed over the whole, and allowed to remain undisturbed for about half an hour.   The sand should not be hot enough to endanger the bell-glass.   By this process, the silver of the nitrate is reduced to its metallic state, causing the object to become a conductor of electricity ; it is then ready for the coppering bath, in which it must be immersed with great care.   Since very light objects will not sink in the solution bath, it is a good plan to form a loop in the conducting wire, as shown in Fig. 60, to which a piece of strong silk thread or twine, having a small leaden weight connected to the opposite end, may be fastened, as in the sketch.   By this simple contrivance light objects and *floating moulds*, as those made of gutta-percha, wax, &c., may be easily sunk into the bath, and retained therein until sufficiently coated.

2. The most effective application of phosphorus for the above purpose consists in dipping the object in a solution of phosphorus in

bisulphide of carbon. This highly volatile and inflammable substance dissolves phosphorus very freely ; the solution, known as "Greek fire," is a most dangerous compound to handle, and if any of it drop upon the skin it may produce sores of a serious nature ; moreover, if it be incautiously allowed to drop upon the clothing, or upon the floor, it may afterwards ignite and do much mischief. In employing the solution of phosphorus, therefore, the greatest possible care must be observed. The object, being attached to a wire, is dipped into the solution, and after being allowed to rest for a few seconds, is next immersed in a weak solution of nitrate of silver, and afterwards allowed to dry in the light. If the object, after being dipped in the phosphorus solution, be allowed to remain in the air for more than a few seconds before being placed in the nitrate solution, it is very liable to become ignited. The solution of phosphorus is prepared by dissolving a small portion of the substance in bisulphide of carbon, about one part of phosphorus by weight being sufficient for the purpose in 20 of bisulphide of carbon.

3. A safer method of producing a conducting surface on these objects is to employ an alcoholic solution of nitrate of silver, made by adding an excess of powdered nitrate of silver to alcohol, and heating the mixture over a hot-water bath. The object is to be dipped in the warm solution for an instant, and then exposed to the air for a short time until the spirit has evaporated. If now submitted to the fumes of phosphorus, as before described, the film of nitrate of silver soon becomes reduced to the metallic state, when the object is ready for the coppering bath.

To render non-metallic substances conductive, Mr. Alexander Parkes introduced the subjoined ingenious processes.

1. A mixture is made from the following ingredients :—

Wax or tallow	1 ounce
India-rubber	1 drachm
Asphalte	1 ounce
Spirit of turpentine	1½ fl. ounce

The india-rubber and asphalte are to be dissolved in the turpentine, the wax is then to be melted, and the former added to it and incorporated by stirring. To this is added one ounce of a solution of phosphorus in bisulphide of carbon, in the proportion of one part of the former to fifteen parts of the latter. The articles, being attached to a wire, are dipped in this mixture : they are next dipped in a weak solution of nitrate of silver, and when the black appearance of the silver is fully developed, the article is washed in water : it is afterwards dipped in a weak solution of chloride of gold, and again washed. Being now coated with a film of gold, it is ready for immersion in the copper bath.

2. In this process, the solution of phosphorus is introduced into the materials used for making the mould, thus :—

Wax and deer's fat, of each   .   .   .   . ¼ pound

Melt together and then add :—

Phosphorus   .   .   .   . 10 grains
Dissolved in bisulphide of carbon   .   .   . 150   „

The wax mixture must be allowed to become nearly cool, when the phosphorus solution is to be added very carefully, through a tube dipping under the surface of the mixture ; the whole are then to be well incorporated by stirring.   Moulds prepared from this composition are rendered conductive by being first dipped in a solution of nitrate of silver, then rinsed, and afterwards dipped in a weak solution of chloride of gold, and again washed, when they are ready for the coppering solution.

Copying Vegetable Substances.—The leaves of plants, seaweeds, ferns, &c., may be reproduced in electrotype, and form very pleasing objects of ornament when successfully produced.   If we wish to copy a vine-leaf, for example, the leaf should be laid face downwards upon a level surface, and its back then covered with several layers of thin plaster of Paris until a tolerably stout coating is given ; the leaf is then to be inverted and embedded in a paste of plaster, care being taken not to allow the material to spread over the face of the leaf.   When the plaster has become hard, finely powdered plumbago is to be dusted over the entire surface from a muslin bag.   A rim of pasteboard, slightly greased on one side, is now to be fixed round the outer edge of the plaster, and secured by a piece of twine. To render this more easy, the plaster may be pared away with a knife, so as to leave a broad flat edge for the card rim to rest against. Melted wax is now to be poured into the pasteboard cylinder thus formed, in sufficient quantity to make a tolerably stout mould.   When thoroughly cold, the rim is to be removed and the mould liberated carefully.   It is then to be plumbagoed, connected to the negative electrode of the battery, and immersed in the copper bath.   The elastic material may also be employed in making moulds from vegetable objects.

Depositing Copper upon Glass, Porcelain, &c.—The article should first be brushed over with a tough varnish, such as copal, or with a solution of gutta-percha in benzol ; when dry it is to be well plumbagoed.   In some cases it may be necessary to render the surface of the glass rough, which is effected by submitting it to the fumes of hydrofluoric acid ; this is only necessary, however, when the vessel is of such a form that the deposited copper might slip away from the glass. Porcelain capsules, or evaporating dishes, may receive a coating of copper at the outside, by varnishing this surface, extending the coating to the upper rim of the vessel, then applying the plumbago and depositing a coating of copper of sufficient thickness. Another and more effectual way of obtaining an adhesive deposit upon glass or porcelain is to send the article to a glass or porcelain gilder, and have gold burnt into its surface, and then depositing upon the gold coating in the usual manner. MM. Noualhier and Prevost patented a process for producing a conducting surface upon glass or vitreous substances, which consists in first coating the object with varnish or gold size, and then covering it with leaf copper. By another method they triturated bronze powder with mercury and common salt, and then dissolved out the salt with hot water, leaving the bronze powder to settle. When dry, this powder is to be applied to the varnished object in the same way as plumbago. For this purpose, however, Bessemer bronzes, which are exquisitely impalpable, and produce a very good conducting surface, may be employed with or without being mixed with plumbago.

Coppering Cloth.—In 1843, Mr. J. Schottlaender obtained a patent for depositing either plain or figured copper upon felted fabrics. The cloth is passed under either a plain or engraved copper roller, immersed horizontally in a sulphate of copper bath, containing but little free acid. The deposit takes place upon the roller as it slowly revolves; the meshes of the cloth are thus filled with metal, and the design of the roller copied upon it. The coppered cloth is slowly rolled off and passes through a second vessel filled with clean water. The roller is previously prepared for a non-adhesive deposit.

CHAPTER VII.

ELECTRO-DEPOSITION OF COPPER (continued).

Of all the purposes to which the art of electrotyping is applied, none is of greater importance than its application to letterpress printing and the copying of wood engravings to be printed from instead of from the wooden blocks themselves. Although this latter branch of the art is very extensively adopted in this country, in the reproduction of large and small engraved blocks for illustrated works and periodicals, newspaper titles, &c., the application of electrotyping as a substitute for stereotyping in letterpress printing has not, as yet, attained the dignity of an art in England. In America, however, the art of reproducing set-up type in electrotype copper has not only acquired a high state of development as a thoroughly practical branch of electro-deposition, but it has almost entirely superseded the process of stereotyping. There are several reasons why this art has been more fully developed in the States than here. In the first place our transatlantic kindred are more prompt in recognising and adopting real improvements ; they are less mindful of cost for machinery when the object to be attained is an important one ; they are not so much under the influence of so-called " practical men " as to ignore scientific help : finally, they do not wait until all their competitors have adopted a process before they run the risk of trying it for themselves.

During the past few years we have been much impressed by the extreme beauty of the American printing, and the exquisite brilliancy of their engravings. Being printed from copper surfaces, the ink delivers more freely than from stereotype metal, while, we believe, a smaller amount of ink is required. Again, the Americans extensively employ wood pulp in the manufacture of their paper, and this material being less absorbant than cotton-pulp, causes the ink to remain *on the* *surface* rather than to sink into the substance of the paper—a fact which was established by the author's father, the late Mr. Charles Watt (the inventor of the wood-paper process), when it was first exhibited in London in the year 1853,* in the presence of the present Earl of Derby and many scientific men and representatives of the press.

With a full belief that the American system of electrotyping, as applied to letterpress printing, will eventually be adopted in this country—at first by the more enterprising members of the printing community—we propose to explain as concisely as the subject will admit the method which has been practically adopted in the United States, and we have to thank the distinguished firm of R. Hoe and Co., of New York, the well-known manufacturers of printing and electrotyping machinery, for much of the information we desire to convey, as also for their courtesy in furnishing us, at our request, with electrotypes of their machinery for the purposes of illustration. We are also indebted to Mr. Wahl † for additional information on this subject.

"As applied to letterpress printing, electrotyping is strictly an American art." This is the claim put forward by the firm referred to, and we freely acknowledge the fact. We gave our cousins the art of electrotyping, and in exchange they show us how we may apply it to one of the most useful of all purposes—the production of good printing from a more durable metal than either ordinary type or stereotype metal.

Electrotyping Printers' Set-up Type.—In pursuing the art of electrotyping, as applied to letterpress printing, the *compositor*, *electrotyper*, and *mounter* must work with one common object, each having a knowledge of what the other requires to perform his part of the work properly. In carrying out the operation on an extensive scale, the depositing room should be on the ground floor, owing to the weight of the vats, and the flooring should be cemented and well drained. The apartment should be well lighted, and provided with an ample supply of water. The depositing vats may be of wood, lined with pitch ; and where a magneto or dynamo-electric machine is employed, this should be fixed at such a distance from the vats as not to be in the way, but at the same time to be as near to them as possible without inconvenience.

---

* Manufacturers in this country refused to adopt this process. It was, however, "taken up" in America in the same year, where it has been worked ever since. It is now used in this country to some extent, as also in many other parts of the world.

† "Galvanoplastic Manipulations." By W. H. Wahl.

Preparing the Formes.—When the formes, or pages of set-up type, have to be electrotyped, it is necessary that great care should be exercised in selecting the types, rules, &c., in *justifying* the same, and in locking-up the forme. When the art of electrotyping comes to be a recognised substitute for stereotyping, it is probable that some modifications in the structure of printers' type may be made to suit more fully the requirements of the electrotyper than the ordinary type. The following suggestions are given relative to the *composition* of the type for reproduction in electrotype, and these should be well understood by those who may hereafter be called upon to produce electrotypes from printers' formes.

Composition.—Every *quadrat, space, lead-slug, reglet,* or piece of *furniture* should be *high.* Some leads have one or both edges bevelled ; but even though the bevel is small it is sufficient to cause considerable trouble, and such leads should not be used in moulding, as the wax is sure to be forced into the space of the bevel, to be broken off, and to require extra labour in distributing the type, besides making it necessary to scrape the wax from the leads before they can be used again. So far as possible, use thick rules and those having a bevel on each side of the face.   Thin rules make so small an opening in the wax that there is great difficulty in blackleading the mould, and in the bath the copper may *bridge* across a small opening, leaving the face and sides of the rule uncovered, or at most with but a thin, imperfect deposit that is useless.   For this reason, type having considerable bevel, is best for electrotyping.   English type has more bevel than American.   Bevelled rules also make impressions in which the hairs of the blackleading brush can penetrate more deeply.   Type-high bearers, or *guards*, about ¼ of an inch thick, should be put around each page, and scattered through blank spaces, to prevent the wax from spreading while the forme is pressed in it, and also to facilitate the operation of "backing."   If there are several pages in a forme, separate them by two guards ; one guard does not give sufficient room to saw between the pages and leave enough of the bearer to protect the edges of the plate in "shaving."   When the matter occupies but a portion of a page, or the lines are shorter than the full width of the page, as in poetry, an *em* dash or a letter should be placed bottom up in each corner of the page, as a guide to the finisher in trimming the the plate.   When the folio is at one corner, that will answer for one of the guides.   All large blanks, chapter heads, and lines unprotected by other matter, should have type-high bearers so placed as to guard the exposed parts from injury.

Locking-up.—The formes must be locked much tighter than for printing, for, in order that the mould shall be perfect, the wax must enter and fill solidly all the interstices of the forme.   This requires great pressure, and the movement of the wax caused by the entering of the type in taking the impression, or mould, is very likely to displace any portions of the forme that may be loose. A proof should always be taken after the forme is locked up for the foundry, and both should be examined to make sure that no part has shifted in driving the *quoins*. Sometimes the matter is set with high spaces but low leads, or *vice versâ*, or low spaces but no leads; frequently copperfaced and white-faced type are used in the same forme. None of these combinations should be allowed, but the whole forme should be either high spaces and high leads or low spaces and low leads. In offices having no high quads, &c., low material must be used; but greater care is necessary in preparing the forme, more labour required of the electrotyper, and the plate is much less satisfactory than when high material is used. Woodcuts which are locked up with the type must be perfectly cleaned with naphtha or benzine, and dried thoroughly before the forme is blackleaded, and great care must be taken not to clog the fine lines of the engraving.

Moulds should not be taken from electrotype cuts, since much better ones can be obtained direct from the woodcut.

Correcting the Matter.—When necessary to make alterations in electrotype plates, the matter for corrections should be set up and electrotyped, but the compositor should separate each correction by a space about a pica, in order that there may be room to saw between them. If the alteration is but a single letter or short word, it is usual to solder the type to the plate. By setting up corrections in their regular order, the labour and cost of plate alterations may frequently be much reduced.

The above technical hints will aid the electrotyper into whose hands a printers' forme may be placed for reproduction in electrotype copper.

Plumbagoing the Forme.—The forme of type must first be cleansed from printing-ink, if very dirty, either with potash ley or benzine; or, if not very dirty, with water distributed from a rubber pipe with rose sprinkler, after which it must be dried. The forme is next to be well brushed over with plumbago, to prevent the wax from sticking. This is applied with a soft hand-brush, the plumbago being made to penetrate every crevice. In doing this, great care must be taken not to fill up the fine lines of the forme with the plumbago.

Preparation of the Mould.—For this purpose a *moulding case* (Fig. 61) is employed, which is a flat brass pan about three-sixteenths of an inch in depth, with two flanges, which fit into the clamps of the *moulding press*. This is fitted with an "electric connection gripper." The *moulding composition* consists of the best pure yellow beeswax, to which is added from five to twenty per cent. of virgin turpentine, to prevent it from cracking. If the temperature of the apartment is from 90° to 95° Fahr., the wax may not require any addition. The composition should be melted by steam heat.*

Filling the Case.—The moulding case having been slightly warmed, on the *steam-heating* table, *a*, Fig. 62, is placed on the *case-filling table*, *b*, truly levelled, and the melted wax, contained in the small jacketed pan, is poured into it with a clean iron or copper ladle, great care being taken to run the wax entirely over the case while it is hot, so that it may not, by cooling too quickly in any part, cause irregularities.

Fig. 61.—Moulding Case.

The air-bubbles which rise to the surface must be touched with the heated *building-iron*, Fig. 64, when they will disappear. If, on cooling, the wax shrinks away from the edges of the case, it can be re-melted there by running the point of the heated building-iron over it, so as to close up any fissure. When cool, the wax should present a smooth, even surface; if this be not

Fig. 62.—Case-filling and Steam-heating Tables.

the case it is useless, and must be put back into the pot and re-melted. The whole surface is now to be carefully and thoroughly rubbed over with plumbago, and polished with soft hand-brush; when this is effected, the wax is ready to receive the impression.

Taking the Impression, or Moulding.—For this purpose considerable and steady pressure is necessary, and this is given either by

---

* Gutta-percha is seldom used in America for making moulds.

means of a hydraulic press, or by the "toggle" press, one form of which, as manufactured by Hoe & Co., is shown in Fig. 63. This form of press consists of a massive frame, having a planed bed, over which is a fixed head. There is a projecting table, on which the forme and case may be arranged before sliding them to receive the pressure, which is put upon them by raising the bed by means of the hand wheel and screw, and the two toggles. In this way enormous pressure is obtained with but little manual exertion.

Fig. 63.—Toggle Press for Electrotype Mould.

The Cloth.—Where low spaces are used, it is customary to make a preliminary impression with a thin sheet of gum cloth interposed; this is then removed and the pressure put on again. Where the cloth is not used, it is necessary to shave off, with a wide, thin knife, the projecting wax ridges.

Removing the Forme.—In case the forme should stick to the wax, it may be relieved by touching the chase gently in two or three places with a long screw driver, taking care not to break the face of the wax. The case is now to be placed upon a table, ready for the process of *building*.

Building.—The mould is now taken in hand by a workman who, with the wide, thin bladed knife, shaves off the projecting wax ridges forced up about the edges and low parts of the mould by the press, and which, if not removed, would impede the separation of the "shell" from the face of the mould, when removed from the depositing tank. The operation of " building " is thus performed : the workman takes an implement such as is shown in Fig. 64 (called a " building-iron "),

several of which are laid on a rack in a small oven heated by gas, and applies to it from time to time a thin strip of wax, allowing the melted wax to run from the point of the tool on to the open spaces or *blanks* of the mould. The operation requires a skilful and steady hand of a practised workman. Upon this point Wahl says, " It is essential, in order to avoid the chiseling (routing or deepening of the open spaces) that would otherwise be necessary to perform upon the finished electrotype, for, unless these open spaces are considerably lower than the spaces between the fine lines of the subject, they are apt to *smut* in the printing process. To cut these out with the chisel, or routing machine, from the finished electrotype would be a difficult and dangerous operation, difficult because of the comparative hardness of the copper surface, and dangerous because the breaking of the continuity of the copper surface will be liable to curl up on the edge of the cut, and to gradually destroy its attachment to the stereotype metal with which it is backed up." To avoid the necessity of chiseling, with the risks which it entails, a ridge of wax is built up on those parts of the mould which require to be depressed in the finished electrotype, but great care is necessary to prevent the wax from running where it is not wanted. The wax used for the above purpose is cut into strips of six or eight inches in length, and about half an inch in thickness.

Fig. 64.

Plumbagoing the Mould.—The wax mould being prepared as above, is next coated with plumbago, the material used in America being obtained from Ceylon graphite. The plumbagoing is generally performed by a machine, the most approved form of which is represented in Fig. 65, its cover being removed to show its construction. The machine has a travelling carriage, holding one or more forms, which passes to and fro under a laterally-vibrating brush. An apron is placed below to receive the loose plumbago, which is used over and over again. As soon as the mould is sufficiently plumbagoed, it is removed from the press, and the surplus material is either dislodged by a hand-brush or with broad-nosed bellows. It is essential that all excess of plumbago be removed, otherwise a coarse and faulty electrotype will be obtained.

Owing to the unavoidable *dust* created by the dry plumbagoing machine, by the impalpable graphite powder, some electrotypists prefer to adopt the *wet* process invented by Mr. Silas P. Knight, of

Fig. 65.—Plumbagoing Machine.

the electrotyping department of Messrs. Harper Brothers, New York. This process is said to work more speedily and delicately than the former, the moulds being thinly and uniformly coated, neither omitting the dot of an *i*, nor allowing the *bridging* over of fine lines.

Knight's Plumbagoing Process.—By this method, the moulds are placed upon a shelf, in a suitable receptacle, and a rotary pump forces an emulsion of plumbago and water over their faces, through a travelling fine-rose nozzle. This process is said to be "rapid, efficient, neat, and economical."

Wiring.—When the plumbagoing is complete, the workman takes one or more lengths of stout copper wire, the ends of which are first cleaned, and then gently heated ; the wires are then embedded in the wax composition on the *side* of the mould, and the joints are then plumbagoed with the finger so as to ensure a perfect *electrical connection* between the wire and the plumbagoed surface. In order to prevent the copper deposit from taking place upon such surfaces beyond the face of the mould as may have become coated with the graphite, the workman takes his hot building-iron, and passes it over these outlying parts of the mould so as to destroy the conductibility of the superfluous plumbago ; this is termed *stopping*.

When moulds of large size have to be treated, it is necessary to place a series of copper wires on the edges of the mould, by which means the deposit commences uniformly at the several points of junction ; these wires are then brought in contact with the slinging wires by which the mould is suspended, and thus receive the current from the conducting rod connected to the dynamo-electric machine or batteries.

Hoe's Electric Connection Gripper.—A very practical arrangement for conducting the current to several points, or parts of the mould, is effected by the "electric connection gripper" of Messrs. R. Hoe and Co., which is represented in Fig. 61, as connected to the moulding case. "This arrangement is designed to hold and sustain the moulding case, and at the same time to make an electric connection with the prepared conducting face of the mould itself, consequently leaving the metal case itself entirely out of the current (circuit), so that no copper can be deposited on it."

Metallising the Moulds.—Plumbago being but a moderately good conductor, many attempts have been made both to improve its conductibility, and to provide a substitute for it altogether. With the former object, we have mixed moderate proportions of Bessemer bronze powder with advantage, as also copper reduced from the sulphate by metallic zinc, and afterwards triturated with honey, an impalpable powder, or *bronze*, being obtained by washing away the honey with boiling water, and afterwards collecting the *finest* particles of the reduced metal by the process of *elutriation ;* that is, after allowing the agitated mixture of water and metallic powder to repose for a few seconds, the liquid, holding the finest particles in suspension, is poured off and allowed to settle, when an exceedingly fine deposit of metallic copper is obtained. The process of coppering the mould, devised by Mr. Silas Knight, is generally adopted in America. By this method, a thin film of copper is deposited on the mould in a few seconds, the operation being conducted as follows : "After stopping out those portions of the mould that are not to receive the deposit, it is laid in a shallow trough, and a stream of water turned upon it from a rose jet, to remove any particles of blacklead that may remain in the lines or letters. The workman then ladles out of a conveniently placed vessel some sulphate of copper solution, pours it upon the face of the mould, then dusts upon it from a pepper box some impalpably fine iron filings, and brushes the mixture over the whole surface, which thus becomes coated with a thin, bright, adherent coat of copper. Should any portion of the surface, after such treatment, remain uncoppered, the operation is repeated. The excess of copper is washed off, and the mould is then ready for the bath." The washing of the mould is effected by means of a stream of water applied from a rubber hose and pipe, and the mould must be placed in the bath directly after the washing is complete.

Adams' Process of Metallising Moulds.—This process, which was patented in America in 1870, is said to give a perfect conducting surface to wax moulds with greater certainty and rapidity than any other, and will accomplish in a few minutes that which plumbago alone would require from two to four hours. The process is conducted as follows : While the mould is still warm in the moulding case, apply freely powdered tin (tin bronze powder, or white bronze powder) with a soft brush until the surface presents a bright, metallic appearance ; then brush off the superfluous powder. The forme of type or wood-cut is then plumbagoed, and an impression or mould taken in the wax as before described, the mould being built up and connected as before. The tin powder is now to be brushed over it either by hand or machine, and the superfluous tin blown away by the bellows, after which the building-iron is applied for stopping all parts upon which copper is not to be deposited. The mould is then to be immersed in alcohol, then washed with water "to remove the air from the surface," when it is ready to be immersed in a solution prepared as follows : Fill a depositing tank nearly full of water, keeping account of the number of gallons poured in ; hang a bag of crystals of sulphate of copper until the water is saturated ; for every gallon of water used add from half a pint to three gills of sulphuric acid, and mix the whole thoroughly. In this solution hang a sheet of copper, connected to the positive pole of the battery, and when the solution becomes cool and settled, immerse the mould and connect it with the negative electrode, when the surface of the mould will be quickly covered with thin copper. Then remove for completion to another and larger depositing vat, containing a solution made in the proportion of one pound of sulphate of copper and one gill of sulphuric acid to each gallon of water. If crystals of sulphate of copper form on the copper plate in the first depositing vat, disconnect it and dissolve them off, substituting for it a clean plate.

Since, in the above process, the tin powder becomes dissolved and enters into the solution, when this liquid becomes saturated with tin, after being long in use, it must be cast aside and replaced by fresh solution. The tin powder may be employed, as a substitute for plumbago, without changing from one bath to another, thus : After the mould has received the desired impression, it is taken to the plumbago table, and held face downward with one end resting on the table, while the other is supported by the hand.   It is then struck on the back several times to loosen the blacklead that is pressed on the wax while moulding, and all the fine dust that may cling to the mould must be blown away.   After building up and making all connections, it is to be placed in the hand-case or plumbagoing machine, and the tin powder applied in the same way as plumbago.   Both the machine and hand-case should be kept free from plumbago, the tin powder only being used to metallise the surface of the mould.   If the machine be used, place the mould, or moulds, on the carriage, cover well over with tin powder, close the door, and run once forward and backward under the vibrating brush ; then turn the moulds round, put on more tin powder, and run through again.   It takes three minutes for the whole operation.   The tin powder is to be beaten out on the table used for this powder as before, and then thoroughly well blown out.   Instead of using the building-iron for stopping off, any suitable varnish, or an alcoholic solution of sealing-wax, may be used.

Quicking.—To prevent the copper deposit from being broken over lines of set-up type, the lines may be wetted with a dilute solution of nitrate of mercury, or with the cyanide quicking solution used in preparing work for plating.   A further deposit is then given in the sulphate of copper bath.

The Depositing Bath.—The solution employed is a saturated solution of sulphate of copper, acidulated with sulphuric acid, and large copper anodes are suspended in the bath, between which the cases containing the prepared moulds are suspended, *back to back*, so that the faces of the moulds may be directly opposite the anodes.   The time occupied in obtaining the electro deposit of copper depends upon the power of the current employed and the thickness of metal desired. For ordinary book or job work, the shell of copper should be about the thickness of good book paper, and this should be obtained in from three to five hours.   Electros for newspaper, titles, and such blocks as are subjected to much use, should receive a stouter deposit.

Batteries.—Several modifications of the Smee battery have been extensively adopted in the United States, including copper plates, electro-silvered, and platinised ; but the most generally accepted improvement consists in employing *platinised platinum* plates for the negative element instead of platinised silver of the Smee battery.   The battery plates, instead of nearly touching the bottom of the cell, as in the ordinary Smee battery, whereby, after being in use some time, they become immersed in a saturated solution of sulphate of zinc, causing great diminution of the current, only extend to about one-third of the depth of the battery cell.   By this arrangement, which was devised by Mr. Adams, of America, in 1841, an equal action of the battery is kept up for a much longer period than would be possible with a Smee battery of ordinary construction. Wahl says that a Smee battery of twenty-six pairs, each 12 by 12 inches, will deposit from six to six-and-half square feet of copper upon prepared moulds in four hours. Batteries, however, are not now much used in the States, having been greatly superseded by the dynamo-electric machine, whereby the electrotyping and electro-depositing arts in general have become enormously increased.

Treatment of the Electrotype.—When the mould has received the requisite deposit, it is to be removed from the bath, and is next to be separated from the wax composition. This is done by placing the mould in an inclined position, and passing a stream of hot water over the copper surface, which, by softening the wax, enables the copper shell to be stripped off, by raising it from one corner while the hot water is passing over the mould. The shell should be removed with care and must not be allowed to *bend* in the least degree. The thin film of wax which adheres to the face of the electro is removed by placing it upon a wire rack, resting on a vessel containing a solution of caustic potash, which is poured over the electro by means of a ladle, the liquor returning to the vessel beneath. The potash has the effect of dissolving the wax in a short time, after which the electro is well rinsed in cold water.

Tinning and Backing the Electrotype.—The first of these operations, *tinning*, is necessary in order to ensure a perfect union between the "backing-up metal" (stereotype metal) and the electrotype. The back of the electro is first brushed over with a solution of *chloride of zinc*, made by dissolving zinc in muriatic acid, and diluting it with about one-third of water, to which, sometimes, a little sal ammoniac is added. The electrotype is now laid, face downwards, upon an iron soldering plate, floated on a bath of melted stereotype metal, and when sufficiently hot, melted solder, composed of equal parts lead and tin, is poured over the back, by which it acquires a clean bright coating of solder. Another method is the following : The shell being placed face downward, in the *backing-pan*, is brushed

Fig. 66.

over with the "tinning liquid" as before, and *alloyed tin foil* is spread over it, and the pan again floated on the hot backing-up metal until the foil melts and covers the whole back of the electrotype. When the foil is melted, the backing-pan is swung on to a levelling stand, and the melted backing metal is carefully poured on the back of the shell from

an iron ladle, commencing at one of the corners and gradually running over the surface until it is covered with a backing of sufficient
thickness. A convenient form of backing-pan and stand is shown in
Fig. 66. The thickness of the backing is about one-eighth of an inch,
or sufficient to enable the electro, when trimmed and mounted, to withstand the pressure of the printing press. The backing-up alloy is
variously composed, but the following is a good practical formula :—

Tin	4 parts
Antimony	· 5 „
Lead	· 91 „
	100

Finishing.—As they pass from the hands of the " backer,"
the plates present a rough and uneven surface on the back, and the
blanks are higher than they should be for mounting. It is the finisher's

Fig. 67.—Saw Table, with Squaring Table.

duty to remedy all such defects. If the backed electrotype consists of
several pages, it is first taken to the saw table, Fig. 67, where it is
roughly sawn apart by a circular saw, the eyes of the workman being
protected from the particles of flying metal by a *square plate of glass*, as
shown in the figure. Each plate is then trimmed all round to remove
rough edges, and if there are any projections which would prevent it
from lying flat, these are carefully cut down with a small chisel. The
plate is next *shaved* to remove the roughness from the back and make

it of uniform thickness in all parts. This is effected in small establishments by the hand shaving machine, Fig. 68, but since this opera-
is the most laborious part of the finishing process, it is far preferable
to employ a power machine for this purpose. The plate being now
brought to nearly its proper thickness, and almost true, is next tested
with a straight-edge, and all unevennesses beaten down with a light
hammer and planer, preparatory to the final shaving: the plate is
then passed through the hand shaving machine, accurately adjusted,
and two or three light cuts are taken off. The face is then tested by
rubbing with a flat piece of willow charcoal, which, by not blackening
the low parts, or hollows, enables the workman to see if any such
exist, in which case he puts
a corresponding mark to
indicate these places on the
back with a suitable tool.
The plate is then laid, face
downward, and the marked
places are struck with a ball-
faced hammer which forces
up the printing surface beneath to its proper level.
The plate is next subjected
to the hand shaving machine
(Fig. 68), by which the back
becomes shaved down to its
proper thickness and ren-
dered perfectly level and
smooth. The edges are next
planed square and to the
proper size, after which they
are transferred to the car-
penter, who mounts them,

Fig. 68.—Hand Shaving Machine.

type-high, on blocks of wood, which may be either of cherry, mahogany,
or other suitable wood, which is cut perfectly true and square in every
direction. The plates, when mounted, are ready for the printer.

Bookwork is usually not mounted on wood. the plates being left un-
mounted, and finished with bevelled edges, by which they are secured
on suitable plate blocks of wood or iron, supplied with gripping pieces
which hold them firmly at the proper height, and enable them to be
properly locked up.

Fig. 69 represents Messrs. R. Hoe & Co.'s power planing and
sawing machine, which is intended for roughing off plates before
sending them to the shaving machine, and is said to be very simple,
quick, and efficient in operation. A circular saw runs in an elevating

table at one corner, for squaring up, and an outside cutter, with
sliding table, is attached for squaring up metal bodies, &c.

Electrotyping from Plaster Moulds.—Plaster of Paris may be employed for making electrotype moulds instead of wax, in which case
the plaster mould is first soaked in wax ; it is then coated with a mixture composed as follows : nitrate of silver 1 gramme, dissolved in
water, 2 grammes ; to this is added 2½ grammes of ammonia, and then
3 grammes of absolute alcohol.   The mould is then to be exposed to

Fig. 69.—Power Planing and Sawing Machine.

sulphuretted hydrogen gas—made by pouring dilute sulphuric acid on
powdered sulphide of iron.

Electrotyping Wood Engravings, &c.—One of the most useful
and extensive applications of electrotyping is in the copying of wood
engravings in electrolytic copper, to form metallic printing surfaces in
lieu of printing from the less durable material, wood.   The value and
importance of electrotype blocks to the proprietors of illustrated
publications—many of which have an enormous sale—will be at once
recognised when we state that the electrotype heading of *The Times*
newspaper is reputed to have produced no less than twenty millions of
copies or impressions before it required renewal.   It would be difficult
to estimate how many wood blocks would have been required to

furnish so large a number of impressions, equally perfect. Indeed, if we take the trouble to examine some of the illustrations of our periodical literature which have been produced direct from wood blocks, we cannot fail to notice the gradual depreciation of the original engraved blocks.

In copying an engraved wood block, it is first well brushed over with plumbago, or simply moistened with water ; it is then placed upon a level bench, and a metal frame somewhat higher than the block is fastened round it. A lump of softened gutta-percha is then placed in the centre of the engraving, and forcibly spread outward (towards the frame), by which air becomes excluded. A plate of cold iron is now placed over the gutta-percha, with gentle pressure, which is afterwards gradually increased, by means of a press, as the gutta-percha becomes harder. When the mould has cooled, it is carefully separated from the block, and well plumbagoed, after which the connecting wire and " guiding wires " are attached ; it is then ready for the depositing bath, where it is allowed to remain until a shell of sufficient thickness is obtained, which will depend upon the size of the mould and the strength of the current employed. Under favourable conditions, a shell of copper, say, of about one square foot of surface, will be obtained in about eight or ten hours, or even less ; it is commonly the practice to put a series of moulds in the bath towards the evening, and to leave them in the bath all night ; on the following morning the deposit is found to be ready to separate from the mould. In electrotype works where magneto or dynamo machines are employed (as in some of our larger printing establishments), a good shell is obtained in from three to five hours,* according to the dimensions of the mould. After removing the mould from the bath, it is rinsed in water, and the shell carefully detached, and the electrotype is next backed-up with solder or a mixture of type metal and tin, the back of the electrotype being first brushed over with a solution of chloride of zinc. The edges of the electrotype are next trimmed with a circular saw, and are afterwards submitted to the planing machine, by which the backing metal is planed perfectly level and flat ; the edges are then bevelled by a bevelling machine, when the plate is ready for mounting on a block of cedar or mahogany, which is effected by

---

* An American electrotypist, on a visit to London, told the author, about five years ago, that, having adopted the Weston dynamo machine in place of voltaic batteries, he could deposit a shell of copper upon fifteen moulds, each having about two square feet of surface, in about two and a half hours ; that is to say, by the time the fifteenth or last mould was put into the bath the one which had first been immersed was sufficiently coated for backing up.

L

means of small iron pins driven into the bevel edges of the backing metal. When complete, the block, with its mounted electrotype, should be exactly *type high*. Respecting electrotypes from wood engravings, or " electros," as they are commonly called in the printing trade, we may mention that many of our larger illustrations are produced from electrotypes. Engraved steel plates are copied in the same way as above, and their reproduction in copper by the electrotype process is extensively practised.

Tin Powder for Electrotyping.—Grain tin may be reduced to an impalpable powder by either of the following methods:—1. Melt the grain tin in an iron crucible or ladle, and pour it into an earthenware mortar, heated a little above its melting point, and triturate briskly as the metal cools. Put the product in a muslin sieve and sift out the finer particles, and repeat the trituration with the coarser particles retained in the sieve. To obtain a still finer product place the fine powder in a vessel of clean water and stir briskly; after a few seconds' repose, pour off the liquor in which the finer particles are suspended, and allow them to subside, when the water is to be again poured off and a fresh quantity of the powder treated as before. The impalpable powder is finally to be drained and dried, and should be kept in a wide-mouthed stoppered bottle for use. 2. Melt grain tin in a graphite crucible, and when in the act of cooling, stir with a clean rod of iron until the metal is reduced to a powder. The powder should then either be passed through a fine sieve or *elutriated* as above described, which is by far the best method of obtaining an absolutely impalpable product. In using this powder for electrotyping purposes in the manner previously described, it must not be forgotten that the tin becomes dissolved in the copper bath; it should therefore only be employed in a bath kept specially for the purpose, and not be suffered to enter the ordinary electrotyping vat.

CHAPTER VIII.

ELECTRO-DEPOSITION OF COPPER (*continued*).

Deposition of Copper by Dynamo-electricity.—Within the past few years, owing to the great advance made in the production of powerful and reliable magneto and dynamo-electric machines, the reduction of copper by electrolysis in the various branches of electrodeposition has assumed proportions of great magnitude ; and while nickel-deposition—which fifteen years ago was a comparatively undeveloped art—has quietly settled down into its legitimate position as an important addition to the great electrolytic industry, the electrodeposition of copper, and its extraction from crude metal, have progressed with marvellous rapidity, both at home and abroad, but more especially so within the past five or six years, and we may safely predict from our knowledge of the vast number of magneto and dynamo machines which are now being constructed, under special contracts, that in a very short time the electrolytic reduction of copper will reach a scale of magnitude which will place it amongst the foremost of our scientific industries in many parts of the world. Before describing the processes of coppering large metallic objects, we must turn our attention to the production of electrotypes of larger dimensions than those previously considered. At a very early period of the electrotype art, Russia, under the guidance of the famous Professor Jacobi, produced colossal statues in electrolytic copper, which at the time created profound astonishment and admiration. About the same period our own countrymen directed their attention to this application of electrotypy, and at subsequent periods electrotypes of considerable dimensions were produced not only in this country but on the Continent. Some exceedingly fine specimens have been produced by Messrs. Elkington & Co., one of the most notable of which is that of the Earl of Eglinton, 13½ feet high and weighing two tons, while some other equally good specimens of life-size busts and bas-reliefs are to be seen in Wellington College, the House of Lords, &c. The well-known Paris firm, Messrs. Christofle & Co., have also produced colossal electrotype statues, one of which is 29 feet 6 inches in height, and weighs nearly three tons and a half ; the completion of the deposit occupied about ten weeks.

Copying Statues, &c.—When very large objects have to be reproduced in electrotype, the method adopted is usually as follows : The original, formed of plaster of Paris, produced by the modeller or sculptor, is first brushed all over with boiled linseed oil, until the surface is completely saturated with the drying oil. After standing for two or three days, according to the temperature and condition of the atmosphere, the object, which is thus rendered impervious to moisture, and readily receives a coating of plumbago, is thoroughly well brushed over with blacklead until the entire surface is perfectly coated with the conducting material. The model is next connected to conducting wires, assisted by guiding wires, and placed in a sulphate of copper bath, where it receives a deposit of about one-sixteenth of an inch in thickness, or a shell sufficiently stout to enable it to retain its form after the inner plaster figure has been removed, which is effected in this way : the electrotype, with its enclosed model, being taken out of the bath, is first thoroughly well rinsed, the copper shell is then cut through with a sharp tool at suitable places, according to the form of the original figure, by which these various parts, with their guiding wires attached, become separated ; the plaster figure is then carefully broken away, and all parts of it removed. After rinsing in hot water, the outer surface of the copper " formes " are well varnished over to prevent them from receiving the copper deposit in the next operation. The formes are next exposed to the fumes of sulphide of hydrogen, or dipped in a weak solution of sulphide of potassium (liver of sulphur), to prevent the adhesion of the copper deposit. These " formes," or parts of the electrotype shell, constitute the moulds upon which the final deposit, or electrotype proper, is to be formed, and these are returned to the depositing tank and filled with the solution of sulphate of copper, anodes of pure electrolytic copper being suspended in each portion. Deposition is then allowed to take place until the interior parts or moulds receive a coating of from one-eighth to one-third of an inch in thickness. The various pieces are then removed from the bath, and after well rinsing in water, the outer shell, or mould, is carefully stripped off, and the respective parts of the electrotype figure are afterwards fixed together when the operation is complete.

Lenoir's Process.—A very ingenious method of electrotyping large figures was devised by M. Lenoir, which consists in first taking impressions in gutta-percha of the object in several pieces, which may afterwards be put together to form a perfect figure ; the inner surfaces of these impressions, or parts of the mould, are then well coated with plumbago. A "dummy" of the form of the interior of the mould, but of smaller dimensions, is now formed of platinum wire, to act as an anode, and the several parts of the plumbagoed gutta-percha mould are put together to form a complete mould all round it. The mould, with its platinum wire core (the anode)—which is insulated from metallic contact with the mould by a covering of india-rubber thread—is then placed vertically in the bath, weights being attached to allow the mould to sink into the solution. The platinum anode and the plumbagoed mould are then put in circuit and deposition allowed to progress. To keep up the strength of the copper solution within the mould, in the absence of a soluble anode, a continual flow of fresh copper solution is allowed to enter the mould, from a hole at the top of the head, which makes its escape through holes in the feet of the mould. When a sufficiently stout deposit is obtained, the flexible wire anode is withdrawn through the aperture in the head, after which the various portions of the gutta-percha mould are removed, and the seams at the junctions of the electrotype are cleared away by appropriate tools.

Deposition of Copper on Iron.—Since iron receives the copper deposit from acid solutions without the aid of a separate current, and the deposit under these conditions is non-adherent, it is the practice to give a preliminary coating of copper to iron objects in an *alkaline* bath, ordinary cyanide solutions being most generally adopted for this purpose. Many other solutions have, however, been recommended, some of which may deserve consideration. In any case, the iron article is first steeped in a hot potash bath, when the presence of greasy matter is suspected, and after rinsing, is immersed in a pickle of dilute sulphuric acid, $\frac{1}{2}$ lb. of acid to each gallon of water. After well rinsing, the article is scoured with coarse sand and water, applied with a hard brush, and after again rinsing, is immersed in the alkaline bath until perfectly coated with a film of copper. It is then again rinsed, and at once placed in a sulphate of copper bath, where it is allowed to remain until a sufficiently stout coating of copper is obtained. In some cases, where the object is of considerable proportions, it is kept in motion while in the solution, by various mechanical contrivances, as in Wilde's process, to be referred to shortly.

Coppering Printing Rollers.—Many attempts have been made, during the past thirty years or so, to substitute for the costly solid copper rollers used in calico-printing, iron rollers coated with a layer of copper by electrolysis. The early efforts were conducted with the ordinary voltaic batteries, but the cost of the electricity thus obtained

was far too great to admit of the process being practically successful,
while at the same time the operation was exceedingly slow. A method
which was partially successful consisted in depositing, in the form of
a flat plate, an electrotype bearing the design, which was afterwards
coiled up in a tubular form, and connected to an iron cylinder or roller
by means of solder, the seam being afterwards touched up by the
engraver. A far better system, however, is now adopted, which is in
every way perfectly successful ; and printing rollers are produced in
large quantities by electro-deposition at about one-half the cost of
the solid copper article. Before describing the methods by which
cast-iron rollers are faced with copper at the present time, it may be
instructive to consider briefly some of the means that have been
adopted to deposit a sufficient thickness of copper upon a cast-iron
core to withstand the cutting action of the engraver's tools.

Schlumberger's Process.—This consists in depositing copper upon
previously well-cleaned cast-iron cylinders by means of the "single-
cell" process. The solution bath consists of a mixture of two solutions composed of (1) Sulphate of copper, 1 part; sulphate of soda,
2 parts ; carbonate of soda, 4 parts ; water, 16 parts. (2) Cyanide of
potassium, 3 ; water, 12 parts. The interior of the bath is surrounded
by porous cells containing amalgamated zinc bars with copper wires
attached, and dilute sulphuric acid. The solution is worked at a
temperature of from 59° to 65° Fahr., and the iron cylinder, being
put in contact with the zinc elements, remains in the bath for twenty-
four hours, at the expiration of which time it is removed, well washed,
rubbed with pumice-powder, again washed in a solution of sulphate
of copper having a specific gravity of 1·161, containing $\frac{1}{300}$th part of its
volume of sulphuric acid ; scraps of copper are kept in the bath, to supply the loss of copper, and prevent the liquid becoming too acid. The
cylinder is then returned to the bath, or placed in a mixture composed
of the following two solutions : (1) Acetate of copper, 2 ; sulphate of
soda, 2 ; carbonate of soda, 4 ; water, 16 parts. (2) Cyanide of potassium, 3 ; liquid ammonia, 3 ; water, 10 parts. The cylinders are
to be turned round once a day, in order to render the deposit uniform,
and the action is continued during three or four weeks, or until the
deposit is $\frac{1}{25}$th of an inch thick.

Another method consists in first coppering the well-cleaned cast-
iron cylinder in an ordinary alkaline coppering bath, and then transferring it to an acid bath of sulphate of copper, the cylinder in each
case being surrounded by a hollow cylinder of copper for the anode ;
the process is allowed to proceed slowly, in order to obtain a good
reguline coating, and when this is obtained of sufficient thickness to
bear engraving upon, the surface is rendered smooth by turning at
the lathe.

Producing Printing Rollers by Magneto-electricity. — *Wilde's Process.*—It is obvious that the electrical power obtained from magneto and dynamo-electric machines is more capable of depositing economically the requisite thickness of copper upon cast-iron cylinders to form printing rollers than could be expected from voltaic electricity, which necessarily involves the solution of an equivalent of zinc and the consumption of sulphuric acid to deposit a given weight of copper. It is well known that deposition takes place more freely upon the lower surfaces of the cathode, and consequently, when the deposited metal is of any considerable thickness, the irregular surface thus produced is often a source of great trouble to the electro-depositor ; in the case of printing rollers, however, in which a perfectly uniform thickness of the deposit is absolutely indispensable, some means must be adopted to render the deposit as uniform as possible from end to end of the cylinder. To accomplish this, Mr. Henry Wilde, of Manchester, effected an arrangement for which he obtained a patent in 1875, which consists in "giving to the electrolyte or depositing liquid in which the roller to be coated is immersed, or the positive and negative electrodes themselves, a rapid motion of rotation, in order that fresh particles of the electrolyte may be brought successively in contact with the metallic surfaces. By this," says the patentee, "powerful currents of electricity may be brought to bear upon small surfaces of metal without detriment to the quality of the copper deposited, while the rate of the deposit is greatly accelerated.

" Motion may be communicated to the electrolyte, either by the rotation of the electrodes themselves, or when the latter are stationary, by paddles revolving in an annular space between them. The iron roller to be coated with copper is mounted on an axis, the lower end of which is insulated, to prevent its receiving the deposit of copper at the same time as the roller. The roller, after having received a film deposit of copper from an alkaline solution in a manner well understood, is immersed in a vertical position in a sulphate solution of copper, and a motion of rotation is given to the roller or rollers by means of suitable gearing. The positive electrodes are copper rollers or cylinders, of about the same length and diameter as the roller to be coated, and are placed parallel with it in the sulphate solution. The electrical contacts are made near the upper and lower extremities of the electrodes respectively, for the purpose of securing uniformity in the thickness of the deposit. The sulphate solution may be maintained at an uniform density, from the top to the bottom of the bath, by rotating a small screw propeller, enclosed in a tube communicating with the liquid, and driven by the same gearing that imparts motion to the roller."

The electric current employed for depositing copper by the above method may be obtained from Wilde's magneto-electric machine, which has been very extensively adopted for this purpose, or from any dynamo electric machine capable of yielding an adequate current. Mr. Wilde says, in the specification above quoted, "Although I have only mentioned cast-iron as the metal upon which the copper is deposited, the process is applicable to rollers made of zinc or other metals, and their alloys. The method of accelerating the rate of deposit, by giving to the electrolyte, or to the electrodes, a motion of rotation, may be applied to the electrolytic method of refining copper described in Mr. J. B. Elkington's patent." Mr. Wilde's system of coppering cast-iron rollers was established in 1878, but he subsequently disposed of his patent rights to the Broughton Copper Company, who carry on the process successfully, and have extended it to the coating of hydraulic rams, &c.

Coppering Cast Iron.—The great progress which has been made in the production of artistic castings in iron during the past thirty years or so, not only in this country, but on the Continent, has always created a desire that some economical and reliable method of coating such work with copper could be devised, not alone to preserve the iron from atmospheric influence, but also to enhance the beauty of the work by facing it with a superior metal. To deposit a protective coating of copper upon large pieces of cast iron, however, has generally been a matter of considerable difficulty, owing to the almost inevitable presence of sand-holes and other flaws which, even when not of large size, are often of sufficient depth to retain particles of silicious or other matter which cannot readily be dislodged by the ordinary methods of pickling and scouring ; and since these defective spots do not receive the deposit of copper, the underlying metal must always be liable to corrosion at such parts, when subjected to the effects of moisture. These observations are chiefly directed to the coppering of cast-iron work destined to be exposed to the vicissitudes of the weather, as street lamp-posts, for example; and though we have not yet devoted much attention to this branch of industry in this country, it has received a good deal of attention in France, but more especially in Paris. To overcome the difficulties above mentioned, copper is not deposited direct upon the iron, as will be seen below, but upon a coating of varnish, rendered conductive by the application of plumbago. The system adopted by M. Oudrey, at his works at Auteuil, may be thus briefly described :—The cast-iron object is first coated all over with a varnish composed of resinous matters dissolved in benzol, to which is added a sufficient quantity of red or white lead, the varnish being then allowed to dry. The surface thus prepared is next brushed over with plumbago, and the article then coated with copper in the ordinary sulphate bath by the "single-cell " method, for which very large porous cells are employed. In about four or five days a sufficiently thick coating of copper is obtained, when, after rinsing and drying, a bronzed appearance is given to the work by the application of a solution of ammonio-acetate of copper. With respect to this process M. Fontaine observes : " It is evident that a coating of copper so deposited can be possessed of no other solidity than its own, and the latter is entirely dependent on the thickness and tenacity of the deposit M. Oudry was accordingly led to effect depositions having one-half a millimetre on ordinary objects and one millimetre or more on fine works. If to that thickness is added those of a layer of plumbago and three layers of insulating coating material, it will be readily conceived that such a system of coppering is only suitable in the case of very large objects. In the case of small objects—such as a bust, for example—the nicety of the details would be irretrievably spoiled by these five layers, and it would amount to sacrificing to too great an extent the artistic worth of the object for the purpose of attaining its preservation. It is, nevertheless, certain that this process has really become a branch of industry, and that it is the first one which has been applied on a large scale. All the lamp-posts of the city of Paris, the beautiful fountains of the Place de la Concorde and of the Place Louvois, and a considerable number of statues and bas-reliefs, have been coppered at Auteuil, in the inventor's factory." It appears that M. Oudry's son subsequently modified the above process by substituting for the coatings of paint and plumbago an immersion of the cast-iron objects in a thick paint composed of hot oil and copper-dust suspended in the liquid. The objects, when removed from this bath, are first dried in an oven and then rubbed with a wire-brush and copper dust. They are afterwards immersed in a sulphate of copper bath.

It is obvious that in either of the above processes a quantity of copper far in excess of what would be required as a protective coating for iron—provided it could be deposited *direct* upon the metal—must from necessity be deposited upon the plumbagoed, or copper-dusted surface ; and it is also clear that since the copper represents merely a thin shell upon the surface, that a very moderate amount of rough usage, such as the Parisian *gamin* or London street Arab could inflict on very easy terms, by the simple process of climbing the lamp-posts with metal-tipped boots, would quickly break this " shell " and expose the underlying layer of plumbagoed varnish. When all these objections are taken into consideration—the partial obliteration of the finer details of the object, the labour, cost of material, the length of time required to complete a single article before it is ready to be placed in position, and add to this the constant liability to damage from accident or mischief, this method of coppering iron does not appear to have much to recommend it.

Cold Coppering Solution.—One of the chief reasons why alkaline coppering solutions seldom work vigorously when used at the ordinary temperature, is that they are too frequently prepared with cheap commercial cyanide, containing but a small percentage of real cyanide, and consequently overloaded with carbonate of potash, a salt which has no solvent action on the anode, and is of little or no service in the coppering bath. While experimenting in this direction some time since, we found that a good coppering solution, to be worked in the cold, could be prepared from the following formula; but it is essential that the cyanide be of good quality. For each gallon of solution required, 3 ounces of chloride of copper are to be dissolved in about a pint and a-half of cold water; 12 ounces of soda crystals are next dissolved in about a quart of water. The latter solution is then to be gradually added to the chloride of copper solution, with gentle stirring after each addition, until the whole of the alkaline liquid has been added, when the resulting carbonate of copper is allowed to settle. After an hour or so the supernatant liquor is poured off and fresh water added to wash the precipitate, which is again allowed to subside as before, the washings being repeated several times, and the precipitate then dissolved in a solution of cyanide of potassium composed of six ounces of the cyanide dissolved in about a quart of water, the whole being well stirred until the copper salt has become dissolved. The solution thus formed is now to be set aside for several hours and the clear liquor then carefully decanted from any sediment that may be present; water is then added to make up one gallon of bath. This solution will coat cast or wrought iron very readily with a current from two to three Daniells, in series, and may be used to give a preliminary coating to iron work which is to be afterwards thickly coppered in an ordinary sulphate bath. The anode used in this, and all other alkaline coppering baths, should be of pure electrolytic copper.

Coppering Steel Wire for Telegraphic Purposes.—It had always been held that if iron wire could be successfully and economically coated with copper, it would be of incalculable service in telegraphy; and, indeed, many attempts to accomplish this were made at a period when magneto and dynamo-electric machines were unknown. It soon became apparent, however, that, independent of other difficulties, the object could never be practically attained by means of the voltaic battery. Now that we are enabled to obtain electricity simply at the cost of motive power, that which was impossible thirty years ago has been to some extent accomplished, and the coppering of steel wire for telegraph purposes forms an extensive branch of manufacture in connection with one of the telegraph systems of America. The manufacture of "compound wire," as it is called, has been carried out on a very large scale at Ansonia, Connecticut, by the Postal Telegraph

Company, who, Professor Silliman, of Yale College, U.S.A., states, have acquired "the largest electro-plating establishment in the world; yet its capacity is soon to be trebled. The works are employed in coppering steel wire used in the company's system of telegraphy, and now deposit two tons of pure copper per day. The steel core of the wire gives the required tensile strength, while the copper coating gives extraordinary conducting power, reducing the electrical resistance enormously. The compound wire consists of a steel wire core weighing 200 lbs. to the mile, and having a tensile strength of 1650 lbs., upon which copper is deposited, by dynamo-electricity, of any required thickness. Twenty-five large dynamo machines are employed, which deposit collectively 10,000 lbs. of copper per day, representing 20 miles of 'compound wire,' carrying 500 lbs. of copper to the mile. When the works are completed, three 300 horse-power engines will drive dynamo machines for supplying the current to deposit copper upon 30 miles of wire per day. In the process of deposition the wire is drawn slowly over spiral coils, through the depositing vats, until the desired thickness is obtained." The advantages of coppered steel wire over ordinary galvanised iron wire for telegraph purposes cannot well be over-estimated, and if the process prove as successful as it is stated to be, it will undoubtedly be a great electrolytic achievement.

Coppering Solutions.—In preparing alkaline coppering solutions, for depositing a preliminary coating of copper upon iron, and for other purposes of electro-coppering, either of the formulæ for brassing solutions may be used,[*] by omitting the zinc salt and doubling the quantity of copper salt; or either of the following formulæ may be adopted. As a rule, copper solutions should be worked hot, say at a temperature of about 130° Fahr., with an energetic current, especially for cast-iron work, since even with the best solution deposition is but slow when these solutions are worked cold. It is important to bear in mind in making up copper solutions—and the same observation applies with at least equal force to brassing solutions—that commercial cyanide of potassium is largely adulterated with an excess of carbonate of potash, and unless a cyanide of known good quality be employed, the solution will be not only a poor conductor of the current, but the anodes will fail to become freely dissolved, whereby the solution will soon become exhausted of a greater portion of its metal in the process of deposition. The cyanide to be used for making up such solutions should contain at least 75 per cent. of *real* cyanide.

Solution 1.—Dissolve 8 ounces of sulphate of copper in about 1 quart of hot water; when cold, add liquid ammonia of the specific gravity of ·880 gradually, stirring with a glass rod or strip of wood after each addition, until the precipitate which at first forms becomes

---

[*] See pp. 374 *et seq.*

re-dissolved; dilute the solution by adding 1 quart of cold water. Now prepare a solution of cyanide of potassium by dissolving about 1¼ lb. of the salt in 2 quarts of water, and add this gradually to the copper solution, with stirring, until the blue colour of the ammonio-sulphate entirely disappears; finally add the remainder of the cyanide solution, and allow the mixture to rest for a few hours, when the clear liquor may be decanted into the depositing vessel or tank, and is then ready for use. This solution may be used cold, with a strong current, but it is preferable to work it at about 110° to 130° Fahr.

Solution 2.—The acetate or chloride of copper may be used instead of the sulphate in making up a coppering bath, the latter salt being preferable.

Solution 3.—A solution prepared as follows has been recommended: Dissolve cyanide of copper in a solution of cyanide of potassium, consisting of 2 pounds of cyanide to 1 gallon of water, then adding about 4 ounces more of the salt as free cyanide; the solution is then ready, and should be worked at a temperature of about 150° Fahr. Cyanide of copper is not freely soluble in a solution of cyanide of potassium. and the liquid does not readily dissolve the anodes, nor is it a good conductor. It has also a tendency to evolve hydrogen at the cathode; this, however, may be lessened or wholly prevented by avoiding the use of free cyanide, employing a weaker current, and adding liquid ammonia and oxide of copper. From our own experience, the addition of liquid ammonia to copper solutions, if not applied in the first instance, becomes a necessity afterwards.

᷿ *Solution* 4.—Roseleur gives the following formula for a coppering solution: 20 parts of crystallised acetate of copper are reduced to a powder, and formed into a paste with water; to this is added 20 parts of soda crystals, dissolved in 200 parts of water, the mixture being well stirred. To the green precipitate thus formed, 20 parts of bisulphite of sodium, dissolved in 200 parts of water, are added, by which the precipitate assumes a dirty yellow colour. 20 parts of *pure* cyanide of potassium, dissolved in 600 parts of water, are finally added, and the whole well stirred together. If the solution does not become colourless, an addition of cyanide must be given. It is said that this solution may be worked either hot or cold, with a moderately strong current.

Dr. Elsner's Solution.—In the preparation of this solution, 1 part of powdered bitartrate of potassium is boiled in 10 parts of water, and as much recently prepared and wet hydrated carbonate of copper, which has been washed with *cold* water, stirred with it as the above solution will dissolve. The dark blue liquid thus formed is next filtered, and afterwards rendered still more alkaline by adding a small quantity of carbonate of potash. This solution is stated to be applicable to coating iron, tin, and zinc articles.*

Walenn's Coppering Solution.—This solution, to be employed for coppering iron, consists in dissolving cyanide of copper in a solution composed of equal parts of cyanide of potassium and tartrate of ammonia. Oxide of copper and ammoniuret of copper are added in sufficient quantity to prevent the evolution of hydrogen at the surface of the work during deposition. The solution is worked at about 180° Fahr. The current from one Smee cell may be used with this solution. It has been found that 1½ ounce of copper per square foot will protect iron from rust.

Another process of Mr. Walenn's is as follows :—

The first part of this invention "relates to electro-depositing copper upon iron, or upon similar metals, so that the coating may be soft and adherent. This consists in using the solution at a boiling heat, or near thereto, namely, from 150° Fahr. to the boiling point of the solution. The second part is to prevent the evaporation of a solution which is heated during deposition. A cover, with a long condensing worm tube, is used in the depositing bath ; the other end of the tube opens into a box containing materials to condense or appropriate the gases that escape. The liquids flow back down the tube into the tank. The third part of the invention consists in working electro-depositing solutions in a closed vessel under known pressure, being applied by heating the solution or otherwise. The closed vessel may be used for solutions in which there is free ammonia, or where other conditions arise in which it is necessary to enclose the solution, although neither appreciable increase of pressure arises nor is heat applied. If there be much gas coming off, the condensing tube, opening into a box of the second part of the improvements, may be employed." The fourth part of the invention consists in adding to the charged, and fully made, copper, brassing, or bronzing solution, cupric ammonide in the cold, until the solution is slightly green.

Gulensohn's Process.—A bath is made by first obtaining a solution of chloride of copper, the metal from which is precipitated in the form of phosphate, by means of pyrophosphate of soda. The precipitate is then thoroughly washed until all traces of the chloride of soda formed have been removed ; the phosphate of copper is next dissolved in a solution of caustic soda, and, if necessary, a small quantity of liquid ammonia is added to assist the solution of the phosphate, and to render the deposit brighter and more solid. The strength of the solution must be regulated according to the strength of the current

* *The Chemist*, vol. vii. p. 124.

employed in the deposition. The bath may be used for depositing upon iron or other metals.

Weil's Coppering Processes.—(1.) For coating large objects, as cast-iron fountains, lamp-posts, &c. M. Weil's patent gives the following process : Dissolve in 1,000 parts of water, 150 of sodio-potassic tartrate (Rochelle salt), 80 of caustic soda, containing from 50 to 60 per cent. of free soda, and 35 of sulphate of copper. Iron and steel, and the metals whose oxides are insoluble in alkalies, are not corroded in this solution. The iron or steel articles are cleaned with dilute sulphuric acid, of specific gravity 1·014, by immersing them in that liquid from five to twenty minutes, then washing with water, and finally with water made alkaline by soda. They are next cleaned with the scratch-brush, again washed, and then immersed in the cupreous bath, in contact with a piece of zinc or lead, or suspended by means of zinc wires ; the latter is the most economical way. The articles must not be in contact with each other. They thus receive a strongly-adherent coating of copper, which increases in thickness (within certain limits) with the duration of immersion. Pure tin does not become coppered by contact with zinc in this solution ; it oxidises, and its oxide decomposes the solution, and precipitates red sub-oxide of copper, and by prolonged action, all the copper is thus removed from the liquid. The iron articles require to be immersed from three to seventy-two hours according to the colour, quality, and thickness of the required deposit. The copper solution is then run out of the vat, and the coated articles washed in water, then cleaned with a scratch-brush, washed, dried in hot sawdust, and lastly in a stove. To keep the bath of uniform strength, the liquid is renewed from below, and flows away in a small stream at the top. After much use, the exhausted liquid is renewed by precipitating the zinc by means of sulphide of sodium (not in excess), and re-charging the solution with cupric sulphate. Weil also supplies to the bath hydrated oxide of copper.

(2.) A coppering bath is prepared as follows : 35 parts of crystallised sulphate, or an equivalent of any other salt of copper, are precipitated as hydrated oxide by means of caustic soda or potash. The oxide of copper is to be added to a solution of 150 parts of Rochelle salt, and dissolved in 1,000 parts of water. To this, 60 parts of caustic soda, of about 70 per cent., is to be added, when a clear solution of copper will be obtained. Other alkaline tartrates may be substituted for the Rochelle salt above mentioned, or even tartaric acid may be employed ; but in the case of tartaric acid, or acid tartrates, a small additional quantity of caustic alkali must be added, sufficient to saturate the tartaric acid or acid tartrate. Oxide of copper may also be employed, precipitated by means of a hypochlorite, but in all cases the proportions between the copper and tartaric acid should be maintained as above, and it is advantageous not to increase to any notable extent the proportion of the caustic soda.

The object to be coppered is to be cleaned with a scratch-brush and then placed in the bath, when it will become rapidly coated with an adherent film of metallic copper. As the bath gradually loses its copper, oxide of copper as above prepared should be added to maintain it in a condition of activity, but the quantity of copper introduced should never exceed that above prescribed, as compared with the quantity of tartaric acid the bath may contain. If the copper notably exceeds this proportion, certain metallic iridescences are produced on the surface of the object. These effects may be employed for ornamental and artistic purposes. According to the time of the immersion, the strength of the current, and the proportion of copper to the tartaric acid, these iridescences may be produced of different shades and tints, which may be varied or intermingled by shielding certain parts of the object by a coating of paraffin or varnish, the iridescent effect being produced on the parts left exposed. All colours, from that of brass to bronze, scarlet, blue, and green, may be thus produced at will.

Electro-Etching.—When we bear in mind the fact that, with few exceptions, the anodes employed in electrolytic processes become dissolved in the bath during electro-deposition, it is evident that if certain portions of an anode were protected, by means of a suitable varnish, from the solvent action of the solution, that such parts, after the plate had been subjected to electro-chemical action in the bath, would, on removal of the varnish, appear in relief, owing to the exposed surfaces having been reduced in substance by being partially dissolved in the solution. Suppose a smooth and bright plate of copper, for instance, were to have a design sketched upon it with a suitable varnish, and the plate then connected to the positive electrode of a voltaic battery and immersed in a solution of sulphate of copper, a cathode of the same metal being suspended from the negative electrode ; if, after a few hours' immersion, the plate be taken from the bath, and the varnish removed, the design will appear in bright relief, while the unvarnished parts will have been eaten away, or dissolved, leaving hollows of a comparatively dull appearance ; the design now forms a printing surface, from which copies may be impressed upon paper in the usual way.

The process of voltaic etching is performed in various ways, but the following will explain the general principle upon which the art is conducted. A copper wire is first soldered to the plate, and the back is then coated with a tough varnish ; when this is dry, the face of the plate is coated with engraver's " etching-ground," a composition of beeswax 5 parts, linseed oil 1 part, melted together ; it is sometimes the practice to *smoke* the surface, before applying the etching needle, in order to render its tracings more visible. The design is then drawn upon the face of the plate, cutting through to the clean surface of the copper. When the etching is complete, the plate is made the anode in a sulphate of copper bath, while a plate of copper is immersed as the cathode. The electric current, passing out of the engraved lines, causes the copper to be dissolved from them, whereby they become *etched*, much in the same way, and with the same effect, as when acid is used in the ordinary etching process. The required gradations of light and shade are produced by suspending cathodes of different forms and sizes opposite the plate to be etched, in various positions, and at different distances from it, thus causing the plate to be acted upon in unequal depths in different parts, the deepest action being always at those portions of the electrodes which are nearest to each other.

Instead of using wax, or other etching-ground, as an insulating material, the plate may be coated with a film of some metal which will not be dissolved in the bath. For example, the plate may be first strongly gilt by electro-deposition, and the design then produced by means of a graver, the tool cutting just sufficiently deep to expose the copper; if now the plate be used as an anode, the copper will become dissolved, as before, leaving the gilt surface unacted upon, since the sulphuric acid set free during the voltaic action has no effect upon gold.

Again, the design may be made with lithographic ink or varnish, and the exposed parts of the plate then strongly gilt; if, thereafter, the varnish, or other insulating material be cleaned off the plate, the voltaic etching will follow the ungilt portions, causing them to become hollowed out as before.

The baths used for etching by electrolysis should be composed of the same metal as that to be etched; thus, a sulphate of copper bath is employed for etching copper plates, sulphate of zinc for zinc plates, and gold or silver solutions when their metals are to be treated in the same way. Copper and zinc plates, however, may be etched by means of the voltaic battery, in dilute solutions of nitric, sulphuric, hydrochloric, or acetic acid, a process which is said to be coming very much into practice.

Glyphography.—This process was invented by Mr. E. Palmer, and consists in first staining copper plate *black* on one side, over which a very thin layer of a *white* opaque composition, resembling white wax, is spread. The plate is then drawn upon with various etching needles in the usual way, which remove portions of the white composition, by which the blackened surface becomes exposed, forming a strong contrast to the surrounding white ground. When the drawing is complete, it is carefully inspected, and then passes into a third person's hands, " by whom it is brought in contact with a substance having a chemical affinity for the remaining portions of the composition, by whom they are heightened, *ad libitum*. Thus, by careful manipulation, the *lights* of the drawing become thickened all over the plate equally. . . . The depths of these non-printing parts of the block must be in some degree proportionate to their width ; consequently the larger breadths of *lights* require to be thickened on the plate to a much greater extent. It is indispensably necessary that the printing surfaces of the block prepared for the press should project in such relief from the block itself as shall prevent the inking roller touching the interstices ; this is accomplished in wood engraving by cutting out these intervening parts, which form the lights of the print, to a sufficient depth ; but in glyphography the depth of these parts is formed by the remaining portions of the white composition on the plate, analogous to the thickness or height of which must be the depth on the block, seeing that the latter is in fact a *cast* or *reverse* of the former.'' The plate, thus prepared, is well plumbagoed all over, and is then placed in a sulphate of copper bath, and a deposit of sufficient thickness obtained, which, on being separated, will be found to be a perfect cast of the drawing which formed the *cliché*. The metallic plate thus obtained is afterwards backed up with solder and mounted in the same way as a stereotype plate, and is then ready for the printing press.

Making Copper Moulds by Electrolysis.—A drawing is made upon a varnished copper plate, as before described ; the plate is then dipped into a weak " quicking " solution, and then laid upon a flat and level surface. The mercury attacks the surfaces exposed by the graver or etching needle, and takes the *meniscus*, or curved form, that is, the relief is greater as the etching lines are larger ; the drawing, therefore, is reproduced in relief by the mercury. The plate is next covered with a thin paste of plaster of Paris, and when this has set, the two moulds are to be separated. A counter mould may now be taken from this, or it may be prepared in the usual way, and, after being well plumbagoed, receive a deposit of copper. By the following plan a mould is produced, which is at once ready for the bath. A copper plate is varnished and etched as before. A neutral solution of chloride of zinc is then poured upon the plate, and after this a quantity of fusible metal, which melts at from 175° to 212° Fahr. The flowing of the fusible metal over the surface of the plate is aided by the application of a spirit-lamp held beneath the plate, or by spreading the metal over the surface with a hot iron rod. The mould thus obtained may then be reproduced by the ordinary electrotype process.

Making Electrotype Plates from Drawings.—This invention relates to an improved process of forming matrices of designs for the

M

production of electrotype plates directly by the hand of the artist or designer, in which the design is produced by means of a pointed tool upon a thin sheet of soft metal supported upon a peculiar backing of semi-plastic inelastic material of sufficient body or consistence to support the metal without pressure, but sufficiently yielding to give to the slightest touch of the artist, and allow the material to be depressed under the tool for the formation of the lines of design. In carrying out this invention a mixture is made of plaster of Paris 1 lb., chromate of potassa $\frac{1}{4}$ oz., and common salt, 1 oz., which forms a compound that will give the most delicate touch of the artist, and will allow the finest lines to be produced upon the metal by the tool. These ingredients may be mixed in various proportions, which will depend somewhat upon the boldness or delicacy of the design to be produced. The mixture may be brought to a semi-plastic state by the addition of about 1 pint of water, or sufficient to bring it to the proper consistence, and the plasticity of the compound may be modified to suit various requirements by using more or less water. The semi-plastic composition is moulded or otherwise formed into a flat tablet of suitable size, and a sheet of soft metal is carefully secured on the upper face of same, projecting edges being left, which are afterwards turned down over the sides of the tablet. The metal is then ready for the artist, who, with a pointed tool or tools, produces the required design by indenting the lines thereon. Wherever touched by the tool the metal will be depressed into the backing, which has just sufficient body to support the untouched parts, but yields to the slightest pressure of the tool. When the design is finished, the metal is carefully removed from the backing, having the design in relief on one side and in intaglio on the other, and is ready for the production of the electrotype plate in the ordinary way, which may be taken from either side, as circumstances require.

Coppering Steel Shot.—The electro-deposition of copper is being extensively applied by the Nickel Plating Company, Greek Street, Soho, London, to the coating of large and small steel shot with copper for the Nordenfelt gun.

Coppering Notes.—1. In preparing cast-iron work for electro-coppering, after the pieces have been pickled and scoured, they should be carefully examined for sand-holes, and if any such cavities appear upon the work, they must be well cleared from black or dirty matter, which may have escaped the brushing, by means of a steel point. It must always be borne in mind that copper, and indeed all other metals, refuse to deposit upon dirt. After having cleared out the objectionable matter from the sand-holes, and again well brushed the article with sand and water, it is a good plan to give the piece a slight coating of copper in the alkaline bath, and then to examine it again, when if any cavities show signs of being foul, they must be cleared with the steel point as before. The article should then have a final brushing with moist sand, and after well rinsing be placed in the alkaline coppering bath and allowed to remain, with an occasional shifting of position, until sufficiently coated. If the piece of work is required to have a stout coating of copper, it should receive only a moderate deposit in the cyanide bath, and after being well rinsed suspended in the sulphate of copper, or *acid* bath, as it is sometimes termed, and allowed to remain therein until the desired coating is obtained. To secure an uniform deposit, however, the object should be occasionally shifted while in the bath, except when mechanical motion is applied, as in coppering iron rollers and other similar work.

2. Respecting the working of copper solutions, Gore makes the following observations : " If the current is too great in relation to the amount of receiving surface, the metal is set free as a brown or nearly black metallic powder, and hydrogen gas may even be deposited with it and evolved. In the sulphate solution, if the liquid is too dense, streaks are apt to be formed upon the receiving surface, and the article (especially if a tall one) will receive a thick deposit at its lower part, and a thin one at the upper portion, or even have the deposit on the upper end redissolved. If there is too little water, crystals of sulphate of copper form upon the anode, and sometimes even upon the cathode, at its lower part, and also at the bottom of the vessel. If there is too much acid the anode is corroded whilst the current is not passing. The presence of a trace of bisulphide of carbon in the sulphate solution will make the deposit brittle, and this continues for some time, although the solution is continually depositing copper ; in the presence of this substance the anode becomes black, but if there is also a great excess of acid, it becomes extremely bright. Solutions of cupric sulphate, containing sulphate of potassium, and the bisulphide of carbon applied to them, are sometimes employed for depositing copper in a bright condition. The copper obtained from the usual double cyanide of copper and potassium solution, by a weak current, is of a dull aspect, but with a strong current it is bright." For depositing copper from alkaline solutions, we prefer the Bunsen battery to all others.

3. The anodes used in electrotyping, as also those employed for depositing copper generally, should consist of pure electrolytic copper, in preference to the ordinary sheet metal, which invariably contains small traces of arsenic and other metals, which are known to diminish its conductivity considerably. Clippings and other fragments of copper from electrotypes may be used up as anodes, either by suspending them in a platinum-wire cradle or in a canvas bag, the fragments being put in connection with the positive electrode of the battery by

means of a stout rod or strip of copper. These make-shift anodes, however, should be used for thickening the deposit (if an electrotype) after the mould is completely coated with copper, and not in the earlier stage of the process.

4. When it is desired to obtain an electrotype of considerable thickness, this may be hastened in the following way : After the complete shell is obtained, clean copper filings are to be sifted over the surface, and deposition allowed to proceed as usual, when the newly deposited metal will unite with the copper filings and the original shell, and thus increase the thickness of the electrotype. By repeated additions of copper filings, followed by further deposition of copper, the back of the electrotype may be strengthened to any desired extent.

5. For coating with copper non-conducting substances, such as china or porcelain, the following process has been adopted in France : Sulphur is dissolved in oil of spike lavender to a sirupy consistence, to which is added either chloride of gold or chloride of platinum, dissolved in ether, the two liquids being mixed under gentle heat. The compound is next evaporated until it is of the consistency of ordinary paint, in which condition it is applied with a brush to such parts of a china or porcelain article as it is desired to coat with copper : the article is afterwards baked in the usual way, after which it is immersed and coated with copper in the ordinary sulphate bath.

CHAPTER IX.

DEPOSITION OF GOLD BY SIMPLE IMMERSION.

Prepartion of Chloride of Gold.—Water Gilding.—Gilding by Immersion in a Solution of Chloride of Gold.—Gilding by Immersion in an Ethereal Solution of Gold.—Solution for Gilding Brass and Copper.—Solution for Gilding Silver.—Solution for Gilding Bronze.—French Gilding for Cheap Jewellery.—Colouring Gilt Work. Gilding Silver by Dipping, or Simple Immersion.—Preparation of the Work for Gilding.—Gilding by Contact with Zinc, Steele's Process.—Gilding with the Rag.

Preparation of Chloride of Gold.—Since for all gilding purposes by the *wet way*, as we may term it in contradistinction to the process of mercury gilding, this metal requires to be brought to the state of *solution*, it will be well to explain the method of preparing the salt of gold commonly known as the *chloride of gold*, but which is, strictly speaking, a *terchloride* of the metal, since it contains three equivalents of chlorine. The most convenient way of dissolving the precious metal is to carefully place the required quantity in a glass flask, such as is shown in Fig. 70, and to pour upon it a mixture consisting of about 2 parts of hydrochloric acid and 1 part nitric acid *by measure*. This mixture of acids was called *aqua regia* by the ancients because it had the power of dissolving the king of metals—gold. To dissolve 1 ounce of gold (troy weight) about 4 ounces of *aqua regia* will be required, but this will depend upon the strength of the commercial acids. Soon after the mixed acids have been poured on the gold, gas is evolved, and the chemical action may be accelerated by placing the flask upon a sand-bath moderately heated. It is always advisable, when dissolving this or other metal, in order to avoid excess of acid, to apply less of the solvent than the maximum quantity in the first instance, and, when the chemical action has ceased, to pour off the dissolved metal and then add a further portion of the solvent to the remainder of the undissolved metal, and so on

Fig. 70.

until the entire quantity is dissolved without any appreciable excess of acid, after which the various solutions are to be mixed together.

The solution of chloride of gold is to be carefully poured into a porcelain evaporating dish * (Fig. 71), and this, placed on a sand-bath or otherwise, gently heated until nearly all the acid is expelled, when the solution will assume a reddish hue. At this period the author prefers to move the evaporating dish round and round gently so as to spread the solution over a large surface of the interior of the vessel : in this way the evaporation of the acid is hastened considerably. When the solution assumes a blood-red colour the dish should be gently, but repeatedly, moved about as before until the semi-fluid mass—which gradually becomes deeper in colour and more dense in substance—*ceases to flow*. Towards the end of the operation the last remaining fluid portion flows torpidly, like molten metal, until it finally ceases altogether, at which moment the dish should be removed from the sand-bath and allowed to cool. It is necessary to mention that if too much heat be applied when the solution has acquired the blood-red colour the gold will quickly become reduced to the metallic state. If such an accident should occur the reduced metal, after dissolving out the chloride with distilled water, must be treated with a little *aqua regia*, which will again dissolve it.

Fig. 71.

The red mass resulting from the above operation (if properly conducted) is next to be dissolved in distilled water, in which it is readily soluble, and should form a perfectly clear and bright solution of a brownish-yellow colour. If, on the other hand, the evaporation has not been carried to an extent sufficient to expel all the acid the solution will be of a pure yellow colour. It invariably happens, after the chloride of gold is dissolved in water, that a white deposit remains at the bottom of the evaporating dish—this is *chloride of silver*, resulting from a trace of that metal having been present in the gold.

Water-Gilding.—Previous to the discovery of the electrotype process and the kindred arts of electro-gilding and silvering to which it gave rise, a process was patented by Mr. G. R. Elkington for gilding metals by the process of simple immersion or "dipping," and this process, which acquired the name of *water-gilding*, was carried on by Messrs. Elkington at Birmingham for a considerable time with success for a certain class of cheap jewellery. The solution was prepared as follows: A strong solution of chloride of gold was first obtained, to which *acid* carbonate of potash was added in the proportion of 1 part of gold, in the form of chloride, to 31 parts of the acid carbonate ; to this mixture was added 30 parts more of the latter salt previously dis-

---

* Evaporating dishes made from Berlin porcelain are the best for this purpose, since they are not liable to crack when heated.

solved in 200 parts of water. The mixture was then boiled for two hours, during which period the solution, at first yellow, assumed a green colour, when it was complete. To apply the above solution the metal articles, of brass or copper, are first well cleaned and then immersed in the solution, which must be hot, for about half a minute. Articles of silver or German-silver to be gilt in this solution must be placed in contact with either a copper or zinc wire.

Gilding by Immersion in a Solution of the Chloride of Gold. —Articles of steel, silver, copper, and some other of the baser metals, may be gilt by simply immersing them in a weak solution of the chloride of gold ; this is, however, more interesting as a fact than of any practical value.

Gilding by Immersion in an Ethereal Solution of Gold.— Chloride of gold is soluble in alcohol and in ether. The latter solution may be obtained by agitating a solution of gold with ether, after which the mixture separates into two portions ; the upper stratum, which is of a yellow colour, is an ethereal solution of chloride of gold, while the lower stratum is merely water and a little hydrochloric acid. Steel articles dipped in the ethereal solution become instantly covered with gold, and, at one time, this method of gilding steel was much employed for delicate surgical instruments, as also for the ornamentation of other articles of steel. After being applied, the ether speedily evaporates, leaving a film of gold upon the object. If the ethereal solution be applied with a camel-hair brush or quill pen, initials or other designs in gold may be traced upon plain steel surfaces. Or, if certain portions of a steel object be protected by wax or varnish, leaving the bare metal in the form of a design, the ethereal solution may then be applied to the exposed surfaces, which will appear in gold when the wax or varnish is dissolved or otherwise cleared away. Various ways of applying this solution for the ornamentation of steel will naturally occur to those who may be desirous of utilising it.

Solution for Gilding Brass and Copper.—The following formula has been adopted for " water-gilding " as it is termed :—

Fine gold . . . . . 6¼ dwts.

Convert the gold into chloride, as before, and dissolve it in 1 quart of distilled water, then add

Bicarbonate of potassa . . . . 1 lb.

and boil the mixture for two hours. Immerse the articles to be gilt in the warm solution for a few seconds up to one minute according to the activity of the bath.

Solution for Gilding Silver.—Dissolve equal parts, by weight, of bichloride of mercury (corrosive sublimate) and chloride of ammonium (sal-ammoniac), in nitric acid ; now add some grain gold to the mixture and evaporate the liquid to half its bulk ; apply it, whilst hot, to the surface of the silver article.

Solution for Gilding Bronze, &c.—A preparatory film of gold may be given to large bronze articles that are to be fully gilt by either of the processes hereafter described, or small articles of "cheap" work may be gilt by immersing them in the following solution, which must be used at nearly boiling heat :—

Caustic potash	180 parts
Carbonate of potash . . .	20 „
Cyanide of potassium . . .	9 „
Water	1,000 „

Rather more than $1\frac{1}{2}$ part of chloride of gold is to be dissolved in the water, when the other substances are to be added and the whole boiled together. The solution requires to be strengthened from time to time by the addition of chloride of gold, and also, after being worked four or five times, by additions of the other salts in the proportions given. This bath is recommended chiefly for gilding, economically, small articles of cheap jewellery, and for giving a preliminary coating of gold to large articles, such as bronzes, which are to receive a stronger coating in the pyrophosphate bath described further on, or in cyanide solutions by aid of the battery. In this bath articles readily receive a light coating of gold, and it will continue to work for a very long period by simply adding, from time to time as required, the proper proportions of gold and the other substances comprised in the formula. By keeping the bath in proper order a very large number of small articles may be gilt in it at the expense of a very small proportion of gold.

Another method of gilding by simple immersion, applicable to brass and copper articles, is to first dip them in a solution of protonitrate of mercury (made by dissolving quicksilver in nitric acid and diluting with water), and then dipping them into the gilding liquid—this plan being sometimes adopted for large articles. It is said that copper may be gilded so perfectly by this method as to resist for some time the corrosive action of strong acids. During the action which takes place, the film of mercury, which is electropositive to the gold, dissolves in the auriferous solution, and a film of gold is deposited in its place.

French Gilding for Cheap Jewellery.—The bath for gilding by dipping, recommended by Roseleur, is composed of—

Pyrophosphate of soda or potassa . .	800 grammes
Hydrocyanic acid of $\frac{1}{8}$ (prussic acid) .	8 „
Gold in the form of chloride (crystallised)	20 „
Distilled water	10 litres.

The pyrophosphate of soda is generally employed, and this may be prepared by melting, at a white heat, ordinary crystallised phosphate of soda in a crucible. The quantity of gold given in the above formula represents the grammes of the pure metal dissolved by aqua regia. In making up the bath, 9 litres of water are put into a porcelain or enamelled-iron vessel, and the pyrophosphate added, with stirring, a little at a time, moderate heat being applied until all the salt is dissolved. The solution is then to be filtered and allowed to cool. The chloride of gold must not be evaporated to dryness, as previously described, but allowed to crystallise ; the crystals are to be dissolved in a little distilled water, and the solution filtered to keep back any chloride of silver that may be present in the dissolving flask, derived from the gold. The filter is next to be washed with the remainder of the distilled (or rain) water. The chloride solution is now to be added to the cold solution of pyrophosphate of soda, and well mixed by stirring with a glass rod. The hydrocyanic acid is then to be added, with stirring, and the whole heated to near the boiling point, when the solution is ready for use. If the pyrophosphate solution is tepid, or indeed in any case, Roseleur thinks it best to add the prussic acid before the solution of chloride of gold is poured in. The employment of prussic acid in the above solution is not absolutely necessary, indeed many persons dispense with it, but the solution is apt to deposit the gold too rapidly upon articles immersed in it, a defect which might be overcome by employing a weaker solution. If the solutions are cold when mixed, the liquor is of a yellowish colour, but it should become colourless when heated. It sometimes happens that the solution assumes a wine-red colour, which indicates that too little prussic acid has been used ; in this case the acid must be added, drop by drop, until the solution becomes colourless. An excess of prussic acid must be avoided, since it has the effect of retarding the gold deposit upon articles immersed in the solution. The proper condition of the bath may be regulated by adding chloride of gold when prussic acid is in excess, or this acid when chloride of gold predominates. In this way the bath may be rendered capable of gilding without difficulty, and of the proper colour.

Respecting the working of this solution, Roseleur says, "The bath will produce very fine gilding upon well-cleaned articles, which must also have been passed through a very diluted solution of nitrate of mercury, without which the deposit of gold is red and irregular, and will not cover the soldered portions. The articles to be gilded must be constantly agitated in the bath, and supported by a hook, or placed in a stoneware ladle perforated with holes, or in baskets of brass gauze, according to their shape or size."

In gilding by dipping, it is usual to have three separate baths placed in succession, and close to each other, all being heated upon the same furnace by gas or otherwise. The first bath consists of an old and nearly exhausted solution in which the articles are first dipped to free them from any trace of acid which may remain upon them after being dipped in aqua fortis. The second bath, somewhat richer in gold than the former, is used for the next dipping, and the articles then receive their final treatment in the third bath. By thus working the baths *in rounds*, " the fresh bath of to-day becomes the second of to-morrow, and the second takes the place of the first, and so on. This method of operating allows of much more gilding with a given quantity of gold than with one bath alone," and consequently is advantageous both on the score of economy and convenience. The gilding is effected in a few seconds, when the articles are rinsed in clear water and dried by means of hot sawdust, preferably from white woods : they are afterwards burnished if necessary. Roseleur does not approve of boxwood sawdust for this purpose, since it is liable to clog the wet pieces of work, besides being less absorbent than the sawdust of poplar, linden, or fir. The sawdust should neither be too fine nor too coarse, and kept in a box with two partitions, with a lining of zinc at the bottom. The box is supported upon a frame of sheet-iron or brickwork, which admits, at its lower part, of a stove filled with bakers' charcoal, which imparts a gentle and uniform heat, and keeps the sawdust constantly dry. After drying very small articles in sawdust, they are shaken in sieves of various degrees of fineness, or the sawdust may be removed by winnowing.

The above process of gilding by dipping, or "pot gilding," as it was formerly called, is applied to articles of cheap jewellery, as bracelets, brooches, lockets, &c., made from copper or its alloys, and has been extensively adopted in France for gilding the pretty but spurious articles known as French jewellery.

Colouring Gilt Work.—In working gold solutions employed in the dipping process, it may sometimes occur that the colour of the deposit is faulty and patchy instead of being of the desired rich gold colour. To overcome this, certain " colouring salts " are employed, the composition of which is as follows :—

Nitrate of potash	. .
Sulphate of zinc	. .
Sulphate of iron	. .
Alum	. . . .

Of each equal parts.

These substances are placed in an earthenware pipkin, and melted at about the temperature of boiling water. When fused, the mixture is ready for use. The articles are to be brushed over with the composition, and are then placed in a charcoal furnace in which the fuel burns between the sides and a vertical and cylindrical grate, as shown in Figs. 72—73).

The work is placed in the hollow central portion where the heat radiates. A vertical section of the furnace is shown in Fig. 73. When put into the furnace, the salts upon the articles first begin to dry, after which they fuse, and acquire a dull, yellowish-red colour. On applying the moistened tip of the finger to one of the pieces, if a slight hissing sound is heard, this indicates that the heat has been sufficient, when the articles are at once removed and thrown briskly into a very weak sulphuric acid pickle, which in a short time dissolves the salts, leaving the work clear and bright, and of a fine gold colour. It must be borne in mind that this " colouring "

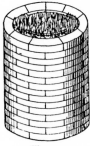

Fig. 72.

process has a rather severe action upon gilt work, and should the gilding be a mere film, or the articles only gilt in parts, the fused salts will inevitably act upon the copper of which the articles are made, and strip the greater portion of the gold from the surface ; as it would be a great risk to submit a large number of indifferently gilt articles to the colouring process unless it was known that sufficient gold had been deposited upon them, although of inferior colour, it would be better to operate upon one or two samples first, when, if the result prove satisfactory, the bulk of them may then be treated as above. Some operators, when the

Fig. 73.

" dipping " has not been satisfactory as to *colour*, give the articles a *momentary* gilding with the battery in the usual way.

When it is desired to gild articles strongly by the dipping process, they are gilt several different times, being passed through a solution of nitrate of mercury previous to each immersion : the film of mercury thus deposited on the work becomes dissolved in the pyrophosphate bath, being replaced by the subsequent layer of gold. In this way articles may be made to receive a substantial coating of gold. In France, large articles, such as clocks, ornamental bronzes, &c., are gilt in this manner, by which they acquire the beautiful colour for which French clocks and goods of a similar character are so justly famed. Roseleur states that he has succeeded in gilding copper by this method sufficiently strong to resist the action of nitric acid for several hours. When articles are strongly gilt by the dipping process, they may be scratch-brushed, or subjected to the process called *or-moulding* described in another place.

Gilding Silver by Dipping, or Simple Immersion.—The articles are first cleaned and scratch-brushed, after which they are boiled for about half an hour in the pyrophosphate gilding bath, to which a few extra drops of prussic acid or sulphurous acid have been added. The former acid dissolves a small portion of silver from the articles, which is replaced by an *equivalent* proportion of gold, while the sulphurous acid acts as a reducing agent in the gold solution, and causes the metal to deposit upon the silver from the affinity existing between the two metals, especially when one of them is in the nascent state, that is, just disengaged from a combination. This gilding is very fine, but without firmness. The deposit is rendered more rapid and thicker when the articles of silver are continually stirred with a rod of copper, zinc, or brass.—*Roseleur.* The deposition by contact of other metals, is, however, due to voltaic action set up by the pyrophosphate solution, and is altogether different to the action which takes place during the simple dipping process, in which a portion of the metal of which the article is composed is dissolved by the solution, and replaced by an *equivalent* proportion of gold.

Preparation of the Work for Gilding—As a rule, the articles should first be placed in a hot solution of caustic potash for a short time, to remove greasy matter, then well rinsed, and afterwards either scratch-brushed, or dipped in aqua fortis or "dipping acid" for an instant, and then thoroughly well rinsed. If the articles merely require to be brightened by scratch-brushing, after being gilt, it is only necessary to put them through the same process before gilding, which imparts to the work a surface which is highly favourable to the reception of the deposit, and which readily acquires the necessary brightness at the scratch-brush lathe as a finish. Articles which are to be left with a dead or frosted surface, must be dipped in dipping acid and rinsed before being placed in the gilding bath. It is commonly the practice to "quick" the articles, after dipping in acid, by immersing them in a solution of nitrate of mercury until they become white ; after this dip, they are rinsed, and at once put into the bath.

Gilding by Contact with Zinc—Steele's Process.—In this process, a solution is made by adding chloride of gold to a solution of cyanide of potassium : in this the articles to be gilt are placed, in contact with a piece of zinc, which sets up electro-chemical action, by which the gold becomes deposited upon the articles ; but since the metal also becomes reduced upon the zinc, the process would not be one to recommend on the score of economy. In some cases, however, in which it is necessary to deposit a film of gold upon some portion of an article which has *stripped* in the burnishing, a cyanide solution of gold may be dropped on the spot, and this touched by a zinc wire, when it will receive a slight coating of gold, and thus save the necessity of re-gilding the whole article. This system of " doctoring " is sometimes necessary, but should be avoided if possible, as it is undoubtedly a fraud upon the customer, since the doctored spot must, sooner or later, yield up its film of gold and lay bare the metal beneath.

Gilding with the Rag.—This old-fashioned process, which was at one time much used for gilding the insides of snuff-boxes, bowls of mustard and salt spoons, &c., is conducted as below. Instead of forming the chloride of gold in the ordinary way, the following in-gredients are taken :—

Nitric acid	5 parts.
Sal-ammoniac (chloride of ammonium)	2 „
Saltpetre (nitrate of potassa)	1 „

A quantity of finely rolled gold is placed in a glass flask, and the other substances are then introduced ; the flask is next heated over a sand-bath. During the action which takes place, the nitric acid de-composes the chloride of ammonium, liberating hydrochloric acid, which combines with the nitric acid, forming aqua regia, which dis-solves the gold, forming chloride ; the nitrate of potash remains mixed with the chloride of gold. The flask is then set aside to cool : when cold, the contents of the flask are poured into a flat-bottomed dish, and pieces of linen rag, cut into convenient squares, are laid one above another in the solution, being pressed with a glass rod, so that they may become thoroughly impregnated with the liquid. The squares of rag are next taken up, one by one, and carefully drained, after which they are hung up in a dark closet to dry. When nearly dry, each piece of rag, supported upon glass rods, is placed over a charcoal fire until it becomes ignited and burnt to tinder, which is promoted by the nitrate of potash ; the burning rag is laid upon a marble slab until the combustion is complete, when the ashes are to be rubbed with a muller, which reduces them to a fine powder. The powder is now collected and placed between pieces of parchment, round which a wet cloth is to be folded ; it is thus left for about a week, being stirred each day, however, to ensure an equal damping of the powder by the moisture which permeates the parchment.

To apply the powder, a certain quantity is placed on a slab and made into a paste with water ; the workman then takes up a small portion with his thumb, which he rubs upon the cleaned surface of the part to be gilt ; the crevices, fillets, or grooves are rubbed with pieces of cork cut to the shape required for the purpose, and the corners, or sharp angles, are rubbed with a stick of soft wood ; such as willow or poplar. When the articles have been gilt in this way, they are finished by burnishing in the usual manner. When a red-coloured gold is required, a small portion of copper is added to the other ingredients when preparing the salt of gold as above described.

CHAPTER X.

ELECTRO-DEPOSITION OF GOLD.

Gilding by Direct Current, or Electro-Gilding.—Preparation of Gilding Solutions. — Gilding Solutions: Becquerel's. — Fizeau's.—Wood's.—M. de Briant's. — French Gilding Solutions. — Gilding Solutions made by the Battery Process.—De Ruolz's.—Cold Electro-Gilding Solutions. —Observations on Gilding in Cold Baths.—Ferrocyanide Gilding Solution.—Watt's Gilding Solution.—Record's Gilding Bath.

Gilding by Direct Current, or Electro-gilding.—In gilding by dipping, or simple immersion, it is obvious that, as a rule, only a limited amount of gold can be deposited upon the work, and that the application of this method of gilding, therefore, must be confined to cheap classes of work, or to articles which will not be subjected to much friction in use. In gilding by the separate current, on the other hand, we are enabled to deposit the precious metal not only of any required thickness, but also upon many articles which it would be practically impossible to gild properly by simple immersion in a solution of gold.

Electro-gilding is performed either with hot or cold solutions; but for most practical purposes hot solutions are employed. When gold is deposited from *cold* solutions, the colour of the deposited metal is usually of a yellow colour, and not of the rich orange-yellow tint which is the natural characteristic of fine gold. the deposit, moreover, is more crystalline, and consequently more porous in cold than hot solutions, and is therefore not so good a protective coating to the underlying metal. The gold deposited from hot solutions is not only of a superior colour and of closer texture, but it is also obtained with much greater rapidity; indeed, from the moment the articles are immersed in the gilding bath, all things being equal, the colour, thickness, and rapidity of the deposit are greatly under the control of the operator. In a few seconds of time an article may be gilded of the finest gold colour, with scarcely an appreciable quantity of the precious metal, while in the course of a very few minutes a coating of sufficient thickness may be obtained to resist a considerable amount of wear.

The superior conductivity of hot gilding solutions enables the operator to gild many metallic surfaces, as tin, lead, Britannia metal, and steel, for example, which he could not accomplish satisfactorily with cold solutions ; moreover, hot gilding solutions readily dissolve any trace of greasy matter, or film of oxide which may be present on the surface of the work, through careless treatment, and thus clean the surface of the work for the reception of the gold deposit.

Since *cold* gilding solutions are occasionally used in electro-deposition, these will be treated separately, as also the special purposes to which they are applied.

Preparation of Gilding Solutions.—In making up gilding baths from either of the following formulæ, except in such cases as will be specified, the gold is first to be converted into chloride, as before directed : but the actual weight of the pure metal required for each specified quantity of solution will be given in each case.

Of all the solutions of gold ordinarily employed in the operations of electro-gilding by the direct current, the *double cyanide of gold and potassium*, when prepared from pure materials, is undoubtedly the best, and has been far more extensively employed than any other. It is very important, however, in making up gold solutions, to employ the purest cyanide that can be obtained. A very good article, commonly known as " gold cyanide," if obtained from an establishment of known respectability, is well suited to the purpose of preparing these solutions. The following formulæ are those which have been most extensively adopted in practice ; but it may be well to state that some persons employ a larger proportion of gold per gallon of solution than that given, a modification which may be followed according to the taste of the operator ; but we may say that excellent results have been obtained by ourselves when employing solutions containing much less metal than some extensive firms have been known to adopt.

Gilding Solutions.—I. To make one quart of solution, convert 1½ dwt. of fine gold into chloride as before, then dissolve the mass in about half a pint of distilled water, and allow the solution to rest so that any trace of *chloride of silver* present may deposit. Pour the clear liquor, which is of a yellow colour, into a glass vessel of convenient size, and then dissolve about half an ounce of cyanide in four ounces of cold water, and add this solution, gradually, to the chloride of gold, stirring with a glass rod. On the first addition of the cyanide, the yellow colour of the chloride solution will disappear, and on fresh additions of the cyanide being made, a brownish precipitate will be formed, when the cyanide solution must be added, gradually, until no further precipitation takes place. Since the precipitate is freely soluble in cyanide of potassium, great care must be exercised not to add more of this solution than is necessary to throw down the metal in the form of *cyanide of gold*. To determine the right point at which to stop, the precipitate should now and then be allowed to fall, so that the clear supernatant liquor may be tested with a drop of the cyanide solution, delivered from one end of the glass rod; or a portion of the clear liquor may be poured into a test tube, or other glass vessel, and then tested with the cyanide. If cyanide has been accidentally added in excess, a little more chloride of gold must be added to neutralise it. The precipitate must be allowed to settle, when the supernatant liquor is to be poured off, and the precipitate washed several times with distilled water. Lastly, a little distilled water is to be added to the precipitate, and a sufficient quantity of cyanide solution poured in to dissolve it, after which a little excess of cyanide solution must be added, and the solution then made up to one quart with distilled water. Before adding the final quantity of water, however, it is a good plan, when convenient to do so, to pour the concentrated solution into an evaporating dish, and to evaporate it to dryness, which may be most conveniently done by means of a sand-bath, after which the resulting mass is to be dissolved in one quart of hot distilled water, and, should the solution work slowly in gilding, a little more cyanide must be added. The solution should be filtered before using, and must be worked hot, that is at about 130° Fahr.

II. Take the same quantity of gold, and form into chloride as before, and dissolve in half a pint of distilled water ; precipitate the gold with ammonia, being careful not to add this in excess. The precipitate is to be washed as before, but must not be allowed to become *dry*, since it will explode with the slightest friction when it is in that state. A strong solution of cyanide is next added until the precipitate is dissolved. The concentrated solution is now to be filtered, and finally, distilled water added to make one quart. Of course it will be understood that the quantity of solution given in this and other formulæ merely represents the basis upon which larger quantities may be prepared. This solution must *not* be evaporated to dryness.

III. *Becquerel's Solution.*—This is composed of—

Chloride of gold	. . . .	1 part
Ferrocyanide of potassium	. .	10 parts
Water	. . . . . .	100 „

The above salts are first to be dissolved in the water ; the liquid is then to be filtered ; 100 parts of a saturated solution of ferrocyanide of potassium are now to be added, and the mixture diluted with once or twice its volume of water. " In general, the tone of the gilding varies according as this solution is more or less diluted ; the colour is most beautiful when the liquid is most dilute, and most free from iron [from the ferrocyanide]. To make the surface appear bright, it is sufficient to wash the article in water acidulated with sulphuric acid, rubbing it gently with a piece of cloth."

IV. *Fizeau's Solutions.* — (1.) 1 part of dry chloride of gold is dissolved in 160 parts of distilled water ; to this is added, gradually, solution of a carbonated alkali in distilled water, until the liquid becomes cloudy. This solution may be used immediately. (2.) 1 gramme of chloride of gold and 4 grammes of hyposulphite of soda are dissolved in 1 litre of distilled water.

V. *Wood's Solution.* — 4 ounces (troy) of cyanide of potassium and 1 ounce of cyanide of gold are dissolved in 1 gallon of distilled water, and the solution is used at a temperature of about 90° Fahr., with a current of at least two cells.

VI. *M. de Briant's Solution.* — The preparation of this solution is thus described : " Dissolve 34 grammes of gold in aqua regia, and evaporate the solution until it becomes neutral chloride of gold ; then dissolve the chloride in 4 kilogrammes of warm water, and add to it 200 grammes of magnesia : the gold is precipitated. Filter, and wash with pure water ; digest the precipitate in 40 parts of water mixed with 3 parts of nitric acid, to remove magnesia, then wash the remaining [resulting] oxide of gold, with water, until the wash-water exhibits no acid reaction with test-paper [litmus-paper]. Next dissolve 400 grammes of ferrocyanide of potassium [yellow prussiate of potash] and 100 grammes of caustic potash in 4 litres of water, add the oxide of gold, and boil the solution about twenty minutes. When the gold is dissolved, there remains a small amount of iron precipitated which may be removed by filtration, and the liquid, of a fine gold colour, is ready for use ; it may be employed either hot or cold."

VII. *French Gilding Solutions.* — The following solutions are recommended by Roseleur as those which he constantly adopted in practice — a sufficient recommendation of their usefulness. In the first of these both phosphate and bisulphite of soda are employed, with a small percentage of cyanide. The first formula is composed of—

Phosphate of soda (crystallised)	.	60 parts
Bisulphite of soda . . .	.	10 „
Cyanide of potassium (pure) .	.	1 part
Gold (converted into chloride)	.	1 „
Distilled or rain water .	.	1,000 parts.

The second formula consists of—

Phosphate of soda . . .	.	50 parts
Bisulphite of soda . :	.	12½ „
Cyanide of potassium (pure) .	.	½ part
Gold	.	1 „
Distilled water . . .	.	1,000 parts.

In making up either of the above baths, the phosphate of soda is first dissolved in 800 parts of hot water : when thoroughly dissolved, the solution should be filtered, if not quite clear, and allowed to cool.

N

The gold having been converted into solid chloride, is next to be dissolved in 100 parts of water, and the bisulphite of soda and cyanide in the remaining 100 parts. The solution of gold is now to be poured slowly, with stirring, into the phosphate of soda solution, which acquires a greenish-yellow tint. The solution of bisulphite of soda and cyanide is next to be added, promptly, when the solution becomes colourless and is ready for use. If the solution of phosphate of soda is not allowed to become cold before the chloride of gold is added, a portion of this metal is apt to become reduced to the metallic state. Roseleur considers it of great importance to add the various solutions in the direct order specified.

The first-named bath is recommended for the rapid gilding of articles made from silver, bronze, copper, and German silver, or other alloys of copper. The second bath is modified so as to be suitable for gilding steel, as also cast and wrought iron *directly* ; that is, without being previously coated with copper. The solutions are worked at a temperature of from 122° to 176° Fahr. In working the first bath, Roseleur says, " Small articles, such as brooches, bracelets, and jewellery-ware in general, are kept in the right hand with the conducting wire, and plunged, and constantly agitated in the bath. The left hand holds the anode of platinum wire, which is immersed more or less in the liquor according to the surface of the articles to be gilt. Large pieces are suspended by one or more brass rods, and, as with the platinum anode, are moved about. The shade of the gold deposit is modified by dipping the platinum anode more or less in the liquor, the paler tints being obtained when a small surface is exposed, and the darker shades with a larger surface. Gilders of small articles generally nearly exhaust their baths, and as soon as they cease to give satisfactory results, make a new one, and keep the old bath for coloured golds, or for beginning the gilding of articles, which are then scratch-brushed and finished in a fresh bath. Those who gild large pieces maintain the strength of their baths by successive additions of chloride of gold, or, what is better, of equal parts of ammoniuret of gold and cyanide of potassium." Articles of copper or its alloys, after being properly cleaned, are sometimes passed through a very weak solution of nitrate of mercury before being immersed in the gilding bath.

The above system of working *without* a gold anode is certainly economical for cheap jewellery, or such fancy articles as merely require the *colour* of gold upon their surface ; but it will be readily understood that solutions worked with a platinum anode would be useless for depositing a durable coating of gold upon any metallic surface, unless the addition of chloride of gold were constantly made.

VIII. *Gilding Solutions Made by the Battery Process.*—The system of forming gold solutions by electrolysis has much to recommend it ; the process is simple in itself : it requires but little manipulation, and in inexperienced hands is less liable to involve waste of gold than the ordinary chemical methods of preparing gilding solutions. A gold bath made by the battery process, moreover, if the cyanide be of good quality, is the purest form of solution obtainable. To prepare the solution, dissolve about 1 pound of good cyanide in 1 gallon of hot distilled water. When all is dissolved, nearly fill a perfectly clean and *new* porous cell with the cyanide solution, and stand it upright in the vessel containing the bulk of the solution, taking care that the liquid stands at the same height in each vessel. Next attach a clean block of carbon or strip of clean sheet copper to the negative pole of a voltaic battery, and immerse this in the porous cell. A gold anode attached to the positive pole is next to be placed in the bath, and the voltaic action kept up until about 1 ounce of gold has been dissolved into the solution, which is easily determined by weighing the gold both before and after immersion. The solution should be maintained at a temperature of 130° to 150° Fahr. while it is under the action of the current.

Another method of preparing gold solutions by the battery process, is to attach a large plate of gold to the positive, and a similar plate of gold or block of carbon to the negative electrode, both being immersed in the hot cyanide solution as above, and a current from 2 Daniell cells passed through the liquid. The negative electrode should be replaced by a clean cathode of sheet German silver for a few moments occasionally, to ascertain whether the solution is rich enough in metal to yield a deposit, and when the solution is in a condition to gild German silver promptly, with an anode surface of about the same extent, the bath may be considered ready for use. The proportion of gold, per gallon of solution, may be greatly varied, from ½ an ounce, or even less, to 2 ounces of gold per gallon of solution being employed, but larger quantities of cyanide must be used in proportion. While the gold is dissolving into the solution, the liquid should be occasionally stirred. The bath should be worked at from 130° to 150° Fahr., the lower temperature being preferable. In making solutions by the battery process, the position of the anode should be shifted from time to time, otherwise it is liable to be cut through at the part nearest the surface of the solution (the *waterline*) where the electro-chemical action is strongest. A good way to prevent this is to punch a hole at each corner of the gold anode, and also a hole midway between each of the corner holes, through which the supporting hook may be successively passed ; this arrangement will admit of **eight** shiftings of the anode. Another plan is to connect a stout platinum wire or a bundle of fine wires to the anode by means of gold solder, and to immerse *the whole* of the anode in the cyanide solution ; this is a very good plan for dissolving the gold uniformly, since the platinum is not acted upon by the cyanide. Sometimes gold wires are used to suspend the anode, in which case the wire should be protected from the action of the cyanide by slipping a glass tube or piece of vulcanised india-rubber tubing over it.

IX. *De Ruolz's Solution.*—10 parts of cyanide are dissolved in 100 parts of distilled water, and the solution then filtered ; 1 part of cyanide of gold, carefully prepared and well washed, and dried out of the influence of light, is now added to the filtered solution of cyanide. It is recommended that the solution be kept in a closed vessel at a temperature of 60° to 77° Fahr. for two or three days, with frequent stirring, and away from the presence of light.

X. *Cold Electro-gilding Solutions.*—The cold gilding bath is some-times used for very large objects, as clocks, chandeliers, &c., to avoid the necessity of heating great volumes of liquid. As in the case of hot solutions, the proportions of gold and cyanide may be modified considerably. Any double cyanide of gold solution may be used cold, provided it be rich both in metal and its solvent, cyanide of potassium, and a sufficient surface of anode immersed in the bath during electro-deposition. For most practical purposes of cold gilding, the following formulæ are recommended by Roseleur :—

Fine gold	10 parts
Cyanide of potassium of 70 per cent. . .	30 „
Liquid ammonia	50 „
Distilled water	1,000 „

The gold is converted into chloride and crystallised, and is then dis-solved in a small quantity of water ; the liquid ammonia is now to be added, and the mixture stirred. The precipitate, of a yellowish brown colour, is *aurate of ammonia, ammoniuret of gold,* or *fulminating gold,* and is a highly explosive substance, which must not on any account be allowed to become dry, since in that state it would detonate with the slightest friction, or an accidental blow from the glass stirrer. Allow the precipitate to subside, then pour off the supernatant liquor and wash the precipitate several times ; since the washing waters will retain a little gold, these should be set aside in order that the metal may be recovered at a future time. The same rule should apply to all washing waters, either from gold or silver precipitates. The aurate of gold is next to be poured on a filter of *bibulous paper,* that is filtering paper specially sold for such purposes. The cyanide should, in the interim, have been dissolved in the remainder of the water. The cyanide solution is now to be added to the precipitate, which it readily dissolves, and this may be conveniently done, if a large filter is used, by pouring it on to the wet precipitate while in the filter, a portion at a time, until the aurate of ammonia has disappeared, and the whole of the cyanide solution has passed through the filter. This will be a safer plan than removing the precipitate from the filter ; or the filter may be suspended in the cyanide solution until the aurate is all dissolved. The solution is finally to be boiled for about an hour, to drive off excess of ammonia.

After this solution has been worked for some time it is apt to become weaker in metal, in which case it must be strengthened by additions of aurate of ammonia. For this purpose, a concentrated solution of the gold salt in cyanide of potassium is kept always at hand, and small quantities added to the bath from time to time when necessary. It is preferable to employ good ordinary cyanide in making up the bath, and *pure* cyanide for the concentrated solution.

2. This solution is composed of—

Fine gold	10 parts
Pure cyanide of potassium . .	20 ,,
Or commercial cyanide . . .	30 to 40 parts
Distilled water	1,000 parts.

The gold is to be formed into chloride and crystallised, as before, and dissolved in about 200 parts of the water ; the cyanide is next to be dissolved in the remainder of the water, and, if necessary, filtered. The solutions are now to be mixed and boiled for a short time. When the solution becomes weakened by use, its strength is to be augmented by adding a strong solution of cyanide of gold, prepared by adding a solution made from 1 part of solid chloride dissolved in a little water, and from 1 to 1½ parts of pure cyanide of potassium, also dissolved in distilled water, the two solutions being then mixed together.

3. This solution consists of—

Ferrocyanide of potassium (yellow prussiate of potash)	20 parts
Pure carbonate of potash	30 ,,
Sal-ammoniac	3 ,,
Gold	15 ,,
Water	1,000 ,,

All the salts, excepting the chloride of gold, are to be added to the water, and the mixture boiled, and afterwards filtered. The chloride of gold is next to be dissolved in a little distilled water and added to the filtered liquor. Some persons prefer employing the aurate of ammonia in place of the chloride of gold, and sometimes small quantities of prussic acid are added to the bath, to improve the bright-ness of the deposit ; but this acid makes the bath act more slowly.

The deposit of gold from cold solutions varies greatly as to colour. When the bath is in its best working condition, and a brisk current of electricity employed, the gold should be of a pure yellow colour : sometimes, however, it is several shades lighter, being of a pale yellow ; it sometimes happens that the gold will be deposited of an earthy grey colour, in which case the articles require to be cautiously scratch-brushed, and afterwards coloured by the or-moulu process to be described hereafter. The proportion of cyanide in these baths should be about twice that of the chloride of gold ; but since the cyanide is of variable quality, it may often be necessary to employ an excess, which is determined by the colour of the deposit : if the gold is in excess the deposit may be of a blackish or dark red colour : or if, on the contrary, cyanide preponderates, the operation is slow and the gold of a dull grey colour, and not unfrequently, when the bath is in this condition, the gold becomes re-dissolved from the work in solution, either entirely or in patches.

When the bath is not in good working order, the gold anode must be withdrawn from the solution, otherwise it will become dissolved by the cyanide. It is a "remarkable phenomenon," says Roseleur, "that solutions of cyanides, even without the action of the electric current, rapidly dissolve in the cold, or at a moderate temperature, all the metals, except platinum, and that at the boiling point they have scarcely any action upon the metals."

Observations on Gilding in Cold Baths.—When a pure yellow colour is desired, a newly-prepared double cyanide of gold solution, in which a moderate excess only of cyanide is present, and containing from 1 to 2 ounces of gold per gallon, will yield excellent results with the current from a single Wollaston or Daniell battery : but suffi-cient anode must be exposed *in the solution* to admit of the deposit taking place *almost* immediately after the article is immersed in the bath. The anode may then be partially raised out of the solution, and the deposition allowed to take place without further interference than an occasional shifting of the object to coat the spot where the slinging wire touches. After the article has been in the bath a minute or so, the operator may assure himself that deposition is pro-gressing satisfactorily by dipping a piece of clean silvered copper wire in the bath and allowing it to touch the object being gilt, when, if the end of the wire becomes coated with gold, he may rest assured that deposition is proceeding favourably. Care must be taken, how-ever, that the deposit is not taking place too rapidly, for it is abso-lutely necessary that the action should be gradual, otherwise the gold may strip off under the operation of the scratch-brush. If any portions of the work appear patchy or spotted, the pieces must be removed from the bath, rinsed, and well scratch-brushed. As in hot gilding, the plater will find the scratch-brush his best friend when the work presents an irregular appearance.

It is not advisable to employ a current of high intensity in cold gilding; the Wollaston or Daniell batteries, therefore, are most suitable, and when a series of cells are required to gild large surfaces or a considerable number of objects, the poles of the batteries should be connected in parallel, that is all the positive electrodes should be connected to the anodes, and all the negative electrodes put in communication with the conducting-rod supporting the work in the bath. After deposition has taken place to some extent, an extra cell may be connected, followed by another, if necessary, and so on; but while only a thin coating of gold is upon the work, the strength of the current should be kept low; deposition takes place more slowly upon gold than upon copper or its alloys, therefore an increase of battery power becomes a necessity after a certain thickness of gold has been deposited. If the current be too weak, on the other hand, the deposit is apt to occur only at the prominent points of the article, and upon those portions which are nearest the anode. It sometimes happens, with newly made baths, that when the articles are shifted to expose fresh surfaces to the anode, the gold already deposited upon the work becomes dissolved off: when such is the case, it generally indicates that there is too great an excess of cyanide in the solution, although the same result may occur if there be too little gold or the current too feeble.

When the gold deposited in a cold bath is of an inferior colour, the article may be dipped in a weak solution of nitrate of mercury until it is entirely white: it is then to be heated to expel the mercury, and afterwards scratch-brushed. Or the article may be brushed over with the "green colour," described in another chapter, and treated in the same way as bad-coloured gilding from hot solutions.

XI. *Ferrocyanide Gilding Solution.*—To avoid the use of large quantities of cyanide of potassium in gilding solutions, the following process has been proposed: In a vessel, capable of holding 4 litres, are dissolved in distilled water 300 grammes of ferrocyanide of potassium, and 50 grammes of sal-ammoniac: 100 grammes of gold, dissolved in aqua regia and evaporated to expel the acid as usual, are dissolved in 1 litre of distilled water. Of this solution, 200 cubic centimetres are added, little by little, to the ferrocyanide solution, when oxide of iron (from the ferrocyanide) is precipitated. The liquid is allowed to cool, and is then filtered and made up to 5 litres, when the bath is ready for use. Since it is not a good conductor, however, and deposits oxide of iron upon the anode, a small quantity of cyanide is added, but not sufficient to evolve hydrocyanic acid on boiling. The bath should be worked at from 100° to 150° Fahr. When the bath ceases to yield a good deposit, 200 c.c. of the gold solution must be added gradually, as before ; if it is desired to increase this proportion of gold, one-tenth of the quantity of the other salts must also be added to the bath.

XII. *Watt's Gilding Solution.*—A gilding solution which the author has used very extensively, and which he first adopted about the year 1838, is formed as follows : 1½ pennyweight of fine gold is converted into chloride, as before described, and afterwards dissolved in about ½ pint of distilled water. Sulphide of ammonium is now added gradually with stirring, until all the gold is thrown down in the form of a brown precipitate. After repose the supernatant liquor is poured off, and the precipitate washed several times with distilled water ; it is then dissolved in a strong solution of cyanide of potassium, a moderate excess being added as free cyanide, and the solution thus formed is diluted with distilled water to make up one quart. Before using this solution for gilding it should be maintained at the boiling point for about half an hour, and the loss by evaporation made up by addition of distilled water. This bath yields a fine gold colour, and if strengthened from time to time by a moderate addition of cyanide, will continue to work well for a very considerable period : it should be worked at about 130° Fahr. The above solution gives very good results with a Daniell battery, and the articles to be gilt do not require *quicking*, as the deposit is very adherent.

XIII. *Record's Gilding Bath.*—This solution, for which a patent was obtained in 1884, is formed by combining nickel and gold solutions, by which, the patentee avers, a considerable saving of gold is effected. To make this solution, he dissolves 5 ounces of nickel salts in about 2 gallons of water, to which 12 ounces of cyanide of potassium is added, " so that the nickel salts may be taken up quite clear." The solution is then boiled until the ammonia contained in the nickel salts is entirely evaporated. This solution is then added to the ordinary gold solution containing 1 ounce of gold. The proportions given are preferred, but may be varied at will.

CHAPTER XI.

ELECTRO-DEPOSITION OF GOLD (*continued*).

General Manipulations of Electro-gilding.—Preparation of the Work.—Dead Gilding.—Causes which affect the Colour of the Deposit.—Gilding Gold Articles.—Gilding Insides of Vessels.—Gilding Silver Filigree Work.—Gilding Army Accoutrement Work.—Gilding German Silver.—Gilding Steel.—Gilding Watch Movements.

General Manipulations of Electro-Gilding.—In small gilding operations, the apparatus and arrangements are of an exceedingly simple character, and need not involve more than a trifling outlay. A 12-inch Daniell cell, or a small battery (say a half-gallon cell), constructed as follows, will answer well for gilding such small work as Albert chains, watch cases, pins, rings, and other work of small dimensions. This battery consists of a stone jar, within which is placed a cylinder of thin sheet-copper, having a binding screw attached. Within this cylinder is placed a porous cell, furnished with a plate or bar of amalgamated zinc, to the upper end of which a binding screw is connected. A dilute solution of sulphuric acid is poured into the porous cell, and a nearly saturated solution of sulphate of copper, moderately acidified with sulphuric acid, is poured into the outer cell. This simple battery costs very little, is very constant in action, and may readily be constructed by the amateur or small operator. The gilding bath may consist of one quart of solution, prepared from any of the formulæ given; a square piece of rolled gold, about 2 by 2 inches, weighing about five pennyweights, or even less, will serve for the anode; and an enamelled iron saucepan may be used to contain the solution. Since gilding baths require to be used hot (about 130 Fahr.), except for special purposes, the solution may be heated by means of a small 4-burner oil lamp, such as is shown in Fig. 74, the gilding vessel being supported upon an iron tripod or ordinary meat stand.

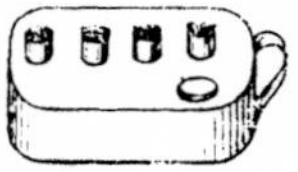

Fig. 74.

With this simple arrangement, it is quite possible to gild such articles as we have named, besides smaller articles, such as brooches, lockets, and scarf-pins; and provided the gold anode be replaced, as it becomes "worn away" in use, and the solution kept up to its normal height by additions of distilled water to make up for loss by evaporation, the same bath will be capable of gilding a good amount of small work. The bath will, however, require small additions of cyanide every now and then, that is when it shows signs of working slowly, or yields a deposit of an indifferent colour: the battery, also, will need proper attention by renewal of the dilute acid occasionally. In working on the small scale referred to, in the absence of a proper scratch-brush lathe, the hand "scratch-brush," Fig. 76, may be resorted to : this consists simply of a single scratch-brush, cut open at one end, and spread out before using, by well brushing it against some hard metal substance : to mollify the extreme harshness of the newly cut brass wire, of which the brush is composed, it may advantageously be rubbed to and fro upon a hard flagstone, after which it should be rinsed before using. To apply the hand scratch-brush, prepare a little warm soap and water, into which the brush must be dipped frequently while being used. In brushing Albert chains or similar work, the swivel may be hooked on to a brass pin, driven into the corner of a bench or table, while the other end of the chain is held in the hand ; while thus stretched out, the moistened brush is dipped in the suds, and *lightly* passed to and fro from end to end, and the position of the article must be reversed to do the opposite side ; to brush those parts of the links which cannot be reached while the chain is outstretched, the chain is held in the hand, and one part at a time passed over the first finger, by which means the unbrightened parts of the links may be readily scratch-brushed. It is important, in scratch-brushing, to keep the brush *constantly* and freely wetted as above.

Gilding on a somewhat larger scale— say with one or two gallons of gold solution—may be pursued without any very great outlay, and yet enable the gilder to do a considerable amount of work of various kinds and dimensions in the course of an ordinary working day. The arrangement we would suggest may be thus briefly explained : for the battery, a one-gallon Bunsen, or Smee, or an 18-inch Daniell cell ; for the anode, two or more ounces of fine gold rolled to about 6 by 3 inches, to which a stout piece of platinum wire, about 4 inches in length, should be attached by means of gold solder. A small binding screw may be employed to connect the platinum wire with the positive electrode of the battery. The object of using platinum wire is to enable *the whole* of the anode to be immersed in the solution when a large surface is necessary, and which could not be properly done if copper wire were used, since this metal (unlike platinum, which is not affected by the solution) would become dissolved by the bath, and affect the colour of the deposit. A simple method of heating the gilding solution and keeping it hot while in use will be seen in the accompanying engraving, Fig. 75. The gilding bath rests upon a short-legged iron tripod, beneath which is a perforated gas burner, supplied with gas by means of flexible india-rubber tubing connected to an ordinary gas-burner. Perforated burners are readily procurable, and are of trifling cost. For brightening small articles the hand scratch-brush referred to (Fig. 76) may be used, but, for the convenience of handling, it should be

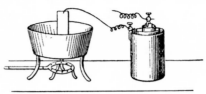

Fig. 75.

tied to a stick, to prevent it from bending in the hand. The brush is to be dipped in soap-suds or stale beer frequently while being applied to the work.

In gilding upon the above moderate scale, however, the lathe scratch-brush, described further on, will be as necessary as in still larger operations: an ordinary foot lathe, such as is used in silver plating (which see), is the machine generally used for this purpose, and is of very simple construction. Such lathes, or their chief parts, may often be procured second hand for a very moderate sum. As in scratch-brushing electro-silvered or plated work, stale beer is employed to keep the brushes constantly wet while the lathe is being used, and the work is pressed very lightly against the revolving brushes. It is important, however, when the scratch-brushes are new, that they should have some hard metallic surface pressed against them while in brisk motion for a few minutes, to spread them well out or make them *brushy*, and to reduce the extreme harshness of the newly-cut brass wire; if this precaution be not followed, the gold, if the coating be

Fig. 76.

thin, may become partially removed from the gilt article, rendering its surface irregular and of an indifferent colour, necessitating regilding and scratch-brushing.

Preparation of the Work.—In electro-gilding watch chains of various kinds, brooches, lockets, scarf-pins, and other small articles of jewellery, it is generally sufficient to well scratch-brush and rinse them, after which they are at once put into the bath. A preparatory dip in a hot potash bath, however, may be resorted to. After scratch-brushing, a short length of copper "slinging" wire is attached to the article, and the free end is connected to the negative electrode of the battery by simply coiling it around the stouter wire several times : the ends of both wires, however, should previously be cleaned by means of a piece of emery cloth. When the articles are first dipped into the solution, they should be gently moved about, so that the deposit may be regular. Chains should be shifted from their position occasionally, so that those portions which are in contact with each other may become gilt; this may generally be done by giving the chain a brisk shake from time to time, and also by slipping the chain through the loop of the slinging wire. If brooches and other similar articles are slung by a loose loop of wire, gentle shaking is all that is necessary to shift their position on the slinging wire.

Some operators, when gilding metal chains or other work manufactured from copper or its alloys—brass, gilding metal, and German silver—prefer to *quick* them after steeping in the potash bath and scratch-brushing. In this case it will be necessary to have a quicking bath or " mercury dip " always at hand. The mercury dip consists of a very dilute solution of nitrate or cyanide of mercury, and after the articles have been *whitened* in this bath, they must be well rinsed in clean water before being immersed in the gilding bath. The object of mercury dipping is to ensure a perfect adhesion of the gold deposit. The author has never, either in electro-gilding or silvering, found it necessary to apply the quicking process, but the solutions both of gold and silver were not prepared in the same way as those ordinarily adopted by the trade. The solutions which the author worked for a great number years without the aid of the mercurial coating are mentioned in the chapters describing the preparation of gilding and silvering baths.

Dead Gilding.—There are several methods of preparing the work so that the deposit instead of being more or less bright when removed from the gilding-bath, may present a *dead* or *frosted* appearance, which is not only exceedingly beautiful in the rich dulness of its lustre, but is absolutely necessary for certain classes of work, portions of which are relieved by burnishing. To obtain a deposit of a somewhat dead lustre, copper and brass articles are dipped for a moment in a mixture of equal parts of oil of vitrol and nitric acid, to which is added a small quantity of common salt. The articles are slung on a stout wire, coiled into a loop, and dipped in the nitro-sulphuric acid " dip " for an instant, and immediately rinsed in clean water, kept in a vessel close to the dipping acid : if not sufficiently acted upon during the first dip, they must be again steeped for a moment, then rinsed in several successive waters, and at once put into the gilding bath. There should be as little delay a possible in transferring the articles to the gold bath, after dipping and rinsing, since copper and its alloys, after being cleaned by the acid and rinsed, are very susceptible of oxidation, even a very few moments being sufficient to tarnish them. If the mercury dip is employed, the work must be dipped in the quicking bath immediately after they have been rinsed from the acid dip.

The surface of articles may be rendered still more *dead*, or frosted, by slightly brushing them over with finely powdered pumice, or, still better, ordinary bath brick reduced to a powder. By this means the extreme point of dulness, or deadness, may be reached with very little trouble. Work which requires to be burnished after gilding should first be steeped in the potash bath, and after rinsing be well scratch-brushed, or scoured with silver sand, soap, and water, when, after again rinsing in hot water, it is ready for the bath. In scouring the work with sand and soap, it is necessary to use warm water freely ; the soap may be conveniently applied by fixing a large piece of this material—say ½ lb. of yellow soap—to the scouring-board by means of four upright wooden pegs or skewers, forming a square about 2½ inches each way, within which the soap may be secured firmly, and will retain its position until nearly used up. By this simple plan the soap, being a fixture, may be rubbed with the scouring-brush, as occasion may require, without occupying a second hand for the purpose.

Causes which Affect the Colour of the Deposit.—In the operation of gilding, the colour of the deposit may be influenced almost momentarily in several ways. Assuming that the current of electricity is neither too strong nor too weak, and the bath in perfect order, if too small a surface of anode is immersed in the bath, the gold deposit will be of a pale yellow colour. Or, on the other hand, if too large a surface of anode is exposed *in solution*, the deposit may be of a dark brown or "foxy" colour, whereas the mean between these two extremes will cause the deposit to assume the rich orange-yellow colour of fine or pure gold. Again, the colour of the deposit is greatly affected by the motion of articles while in the bath : for example, if the gilding be of a dark colour, by briskly moving the articles about in the bath, they will quickly assume the proper colour. The temperature of the solution also affects the colour of the deposit, the tone being deeper as the solution becomes hotter, and *vice versâ*. The colour of the gilding is likewise much affected by the nature of the current employed. A weak current from a Wollaston or Daniell battery may cause the deposit to be of a paler colour than is desired, whereas a Smee, Grove, or Bunsen (but more especially the latter) will produce a deposit of a far richer tone. The presence of other metals in the solution, but copper and silver more particularly, will alter the colour of the deposit, and therefore it is of the greatest importance to keep these metals out of the ordinary gilding solution by careful means. When gilding in various colours is needed, recourse must be had to the solutions described elsewhere, but on no account should the gilding bath used for ordinary work be allowed to become impregnated with even small quantities of any other metal. When we state that trifling causes will sometimes interfere with the natural beauty of the pure gold deposit, the importance of preserving the baths from the introduction of foreign matters will be at once apparent. Another thing that affects the colour of the gilding is the accumulation of *organic matter*, that is, vegetable or animal matter, which is introduced into the bath by the articles immersed in it : thus, greasy matter from polished work, and beer from the scratch-brush, will sometimes lodge in the interstices of hollow work, and escape into the bath even after the articles have been rinsed : each in their turn convey organic matter to the gold solution, by which it acquires a darkened colour ; indeed, we have known solutions acquire quite a brown colour from these causes. In our experience, however, the presence of a *small* amount of such foreign matter, in moderation, has often proved of advantage, especially in the gilding of insides of vessels, when a rich and deep-toned gilding is required : a solution in this condition we should prefer, for insides of cream ewers, sugar-bowls, and goblets, to a newly-prepared gold solution ; indeed, when a bath works a little *foxy*, it is, to our mind, in the best condition for these purposes, since the former is apt to yield a deposit which is too yellow for such surfaces. There is an extreme, however, which must be avoided, that is when the bath yields a *brown-yellow* deposit, which is very unsightly, though not uncommonly to be seen in our shop windows.

When the gilding upon chains or articles of that class is of a deep brownish-yellow colour when removed from the bath, it will, when scratch-brushed, exhibit a fine gold appearance, specially suited to this class of work, and more like jewellers' " wet colour work " than electro-gilding, which will render it more acceptable to those who are judges of gold colour. Indeed, when the electro-gilding process was first introduced, it was a general complaint amongst shopkeepers that electro-gilding was too yellow, and that electro-gilt work could easily be distinguished from *coloured* gold in consequence, which was admitted to be a serious defect, since a person wearing a gilt article would naturally wish it to be assumed by others to be of gold. In gilding such articles, therefore, the aim of the gilder should be to imitate as closely as possible the colour of gold jewellery, whether it be *dry* or *wet* coloured work. In the latter there is a peculiar depth and softness of tone which is exceedingly pleasing; in dry coloured work a rich dead surface is produced which it is not so difficult to imitate in electro-gilding. The processes of " colouring " articles of gold will be given in another chapter, since a knowledge of these processes is not only useful but often necessary to an electro-gilder, in whose hands such work may sometimes be placed for restoration or recolouring.

Gilding Gold Articles.—Although "painting the lily" would not be a very profitable or successful operation, articles made from inferior gold alloys are frequently sent to the electro-gilder to be "coloured," that is, to receive a slight film of pure gold, to make them look like gold of a superior quality, like *coloured gold*, in fact. Although such an imposition is a positive fraud upon the purchaser, the electro-gilder has little choice in the matter ; if his natural scruples would tempt him to refuse such unfair work, as it may be called, he knows full well that others will readily do the work and "ask no questions ; " he must therefore undertake it or lose a customer—perhaps an important one. Albert chains, rings, pins, brooches, and a host of other articles manufactured from gold alloys of very low standard, are frequently "coloured" by electro-deposition, simply because the process of colouring by means of the "colouring salts " would *rot* them, if not dissolve them entirely.

Gilding Insides of Vessels.—Silver or electro-plated cream ewers, sugar-basins, mugs, &c., are electro-gilt inside in the following way : The inside of the vessel is first well scratch-brushed, for which purpose a special scratch-brush, called an *end-brush*, is used. Or this surface may be scoured with soap and water with a piece of stout flannel ; the vessel, after well rinsing, is then placed upon a level table or bench ; a gold anode, turned up in the form of a hollow cylinder, is now to be connected to the positive electrode of a battery, and lowered into the vessel, and supported in this position, care being taken that it does not touch the vessel at any point.

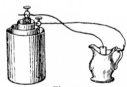

Fig. 77.

The negative electrode is to be placed in contact with the vessel (Fig. 77), and hot gold solution then carefully poured in, up to its extreme inner edge, below the mount, if it have one. A few moments after pouring in the gold solution, the anode should be *gently* moved to and fro, without coming in contact with the vessel itself, so as to render the deposit more uniform ; it may then be allowed to remain without interruption for a minute or so, when the gentle movement of the anode may be renewed for a few moments, these alternations of motion and repose being kept up for about five or six minutes—or perhaps a little longer—by which time a sufficiently stout coating is generally obtained. Moving the anode occasionally has the effect of rendering the deposit more regular, while it also exposes fresh surfaces of the solution to the metal surfaces under treatment ; great care, how-

ever, is necessary to avoid driving the solution over the ornamental mount on the rim of the vessel. The *lips* of cream ewers,
which the gold solution cannot reach when the vessel is filled with
solution, are gilt by conducting the solution to such parts as follows :
A small gold anode, with a short piece of copper wire attached,
is enclosed in a piece of rag or chamois leather ; the end of this wire
is then connected to the positive electrode (the article itself being in
direct contact with the negative), and the pad, or "doctor," as it is
sometimes called, is dipped in the gold solution and applied to the part
to be gilt ; in this way, by repeatedly dipping the pad in the solution
and conducting it over the surface, this part in a short time becomes
sufficiently gilt ; since the lip of a cream ewer, however, is the most
important part of the gilt surface, the application of the pad should be
continued until a proper coating is obtained, and care must be taken
that the point of junction between the two deposits of gold is not
visible when the gilding is complete. We should prefer to gild the
lip of such vessels first, and after well scratch-brushing, or scouring the interior, and especially the line where the two gildings will
meet, then to gild the interior of the body of the vessel, and finally to
scratch-brush the whole surface. In gilding the insides of vessels, it
is important that the outsides and mounts, or mouldings, should be ·
perfectly dry, otherwise the gold solution may, by *capillary attraction*,
pass beyond its proper boundary and gold become deposited where
it is not required, thus entailing the trouble and annoyance of removing it.

Gilding Silver Filigree Work.—A *dead* surface of silver is very
apt to receive the gold deposit ununiformly, and this is specially so in
the case of silver filigree work, the interstices of which cannot fully
be reached by the scratch-brush ; the surfaces brightened by the
scratch-brush readily receive the deposit, while those portions of the
article which escape the action of the wire brush will sometimes fail to
"take" the gold. When this is found to be the case, a large surface
of anode should be immersed in the bath, and the article briskly moved
about until the whole surface is coated, when the anode may be partially withdrawn, and a sufficient surface only exposed in the bath to
complete the article as usual. In gilding work of this description it is
necessary that a fair amount of *free* cyanide should be in the bath, but
the excess must not be too great, or the deposit will be *foxy*—a colour
which must be strenuously avoided, since the brown tint will be visible
more or less upon those interstices (especially the soldered parts)
which the scratch-brush cannot reach. As a rule, filigree work
should not be risked in an old gold solution in which organic matter or
other impurities may be present. It is a good plan, after giving the
article a quick coating in the way indicated, to rinse and "scratch" it

again, and then to re-immerse it in the bath. A solution for gilding
filigree work should also be tolerably rich in gold—about 5 pennyweights to the quart of solution being a good proportion, though some
gilders use a still larger proportion of metal. In gilding filigree work
a rather intense current is necessary ; a Bunsen battery, therefore,
should be employed, or two Daniell cells arranged for intensity.

Gilding Army Accoutrement Work.—In the early days of
electro-gilding, great difficulty was experienced by electro-gilders in
imparting to sword-mounts, the threaded ornamentation of scabbards,
and other army accoutrements the rich dead lustre, as the French term
it, which the mercury gilders produced with so much perfection, and
for a long period electro-gilders, in their anxiety to obtain contracts
for gilding this class of work, made many unsuccessful attempts and
suffered much disappointment from the repeated rejection of their
work by the government authorities. At the period referred to, there
was a great desire, if possible, to render the pernicious art of mercury
gilding unnecessary, since it was too well known that those engaged
in the art suffered severely from the effects of mercurial poisoning, by
which their existence was rendered a misery to them, and their lives
abbreviated to a remarkable degree. It may be stated, however, that
the operations of gilding with an amalgam of gold and mercury were
frequently conducted with little or no regard to the dangerous nature
of the fluid metal which the workpeople were constantly handling, and
the volatilised fumes of which they were as constantly inhaling. It
was a happy epoch in the gilding art when deposition of gold by electricity rendered so baneful a process, incautiously practised, comparatively unnecessary. We say comparatively, because amalgam or
mercury gilding is still adopted, though with a little better regard to
the health of the workmen, for certain classes of work, for which, even
up to the present period, electro-gilding is not recognised as a perfect
substitute.

To gild army accoutrement work, so as to resemble, as closely as
possible, mercury gilding, the *colour* and general appearance of the
matted or *dead* parts must be imitated very closely indeed. There are
no articles of gilt work that look more beautiful by contrast than
those in which dead surfaces are relieved by the raised parts and
surrounding edges being brightened by burnishing, and this effect is
charmingly illustrated in the mountings of the regulation sword of the
British officer. Indeed this class of work, when properly finished,
may be considered the perfection of beauty in gilding.

To give the necessary *matted* surface to the chased portions of sword
mounts, and work of a similar description, these parts should be
brushed over with finely-powdered pumice, or bath-brick reduced to a
powder and sifted, the latter substance answers the purpose very

o

well. The application of either of these materials should be confined, as far as is practicable, to the chased parts of the article, so as to avoid rendering the surfaces to be afterwards burnished rough by the action of the pumice powder. The plain surfaces of the article may then be scoured with silver-sand, soap and water, or scratch-brushed ; but great care must be taken not to allow the scratch-brush to touch the surfaces that are to be left dead. Sometimes it is the practice to add a little aurate of ammonia to the gilding solution to produce a dead lustre in gilt work. When it is preferred to adopt the *quicking* process, in gilding this class of work, the articles, after being quicked in the usual way, are placed in the bath until they have nearly received a sufficient deposit, when they are removed, rinsed, and the chased parts quickly brushed with pumice, as before, after which they are returned to the bath for a short time, or until the proper colour and matted appearance are imparted to the work.

Gilding German Silver.—This alloy of copper, as also brass, will receive a deposit of gold in strong and warm cyanide solutions of gold without the aid of the battery ; this being the case, in order to prevent the deposit from taking place too rapidly, when electro-gilding articles made from these alloys, the temperature of the solution should be kept rather low—that is not beyond 120° Fahr.—and only sufficient surface of anode immersed in the solution to enable the article to become gilt with moderate speed when *first* placed in the bath. It is also advisable that the gold solution should be weaker, both in gold and cyanide, than solutions which are used for gilding silver or copper work. If, however, *quicking* be adopted, these precautions are not so necessary, since the film of mercury checks the rapidity of the gilding. Either method may be adopted according to the fancy of the gilder ; but for our own part, we would not suffer a particle of mercury to enter the gilding-room (except upon the amalgamated plates of a battery) under any circumstances.

Gilding Steel.—The rapidity with which this metal receives a deposit of gold, even with a very weak battery current, in ordinary cyanide solutions, renders it imperative that a separate solution should be prepared and kept specially for steel articles. We have obtained excellent results by employing a bath composed of

> Ordinary double cyanide of gold solution . 　. 1 part.
> Water . . . . 4 to 6 parts.

To this weakened solution a small quantity of cyanide of potassium may be added, and the current employed should be of low tension—a Wollaston or Daniell battery being preferable. The temperature of the bath should be warm, but not hot. The surface of anode in solution must be just so much as will enable the gold to deposit soon after the article is placed in the bath, but not immediately after its immersion. In other words, if the gold is allowed to *jump* on, it will most assuredly as quickly jump off when the scratch-brush is applied.

In preparing steel articles for gilding, the author has found that by scratch-brushing the work with vinegar, or very dilute hydrochloric acid, instead of sour beer, a very fine coating of copper (derived from the brass wire of the brushes) has been imparted to the articles, to which the gold deposit, from a weak bath, adhered with great firmness.

A very successful method of gilding steel is to first copper or brass the articles in the alkaline solutions of these metals, as recommended for silvering steel and iron.[*] The brass or copper solutions should be used warm, and be in good working order, so as to yield bright deposits of good colour. Before electro-brassing the articles, however, they must be thoroughly cleansed by scouring with silver-sand, soap and water, or scratch-brushed. Bright steel articles which are not required to have a durable coating of gold, but merely a slight film or "colouring" of the precious metal, generally need no preparation whatever, but may receive a momentary dip in the gilding bath, then rinsed in hot water, and at once placed in hot boxwood sawdust. In doing this cheap class of work, however, it is better to use a copper or platinum anode in place of the gold anode, and to make small additions of chloride of gold when the solution shows signs of becoming exhausted. It must be remembered, however, that the very dilute gold solution we have recommended for gilding steel contains in reality but very little gold, therefore, as it becomes further exhausted by working without a gold anode, additions of the chloride, in very small quantities, will require to be made so soon as the bath exhibits inactivity.

For gilding polished steel, a nearly neutral solution of chloride of gold is mixed with sulphuric ether, and well shaken ; the ether will take up the gold, and the ethereal solution float above the denser acid. If the ethereal solution be applied by means of a camel-hair brush to brightly-polished steel or iron, the ether evaporates, and gold, which adheres more or less firmly, becomes reduced to the metallic state on the steel, and may be either polished or burnished.

In gilding upon an extensive scale, where large objects, such as time-pieces, chalices, patens, and other work of large dimensions, have to be gilt, the depositing tanks are generally enamelled iron jacketed pans, heated by steam. These vessels are placed in rows near the wall of the gilding-room, in a good light, and suitable iron piping conveys the steam to the various tanks, each of which is provided with a suitable stopcock to admit or shut off the steam as required ; an exit pipe at the bottom of each " jacket " allows the water from the con-

---

[*] See Chaps. XVI., XXIV., and XXV.

densed steam to escape into a drain beneath. Each of these tanks is provided with the usual conducting rods, and the current, which is sometimes derived from a magneto or dynamo machine in large establishments, is conveyed by suitable leading wires or rods, attached to the wall at a short distance from the series of depositing vessels. In gilding large quantities of small articles, as steel pens, for example, a considerable number of gilding tanks, of an oblong form, are placed in a row, at a moderate distance apart, and the pens or other small objects are introduced into these as the gilders receive the work prepared for them.

Gilding Watch Movements—Continental Method.—The remarkable beauty of the Swiss watch movements has always been the subject of much admiration, and for a long period this pleasing industry was solely confined to Switzerland; France, however, eventually got possession of the method, and the art has been extensively practised in that country, but more especially at Besançon and Morez, in Jura, and in Paris. M. Pinaire, a gilder at Besançon, generously communicated the process to the late M. Roseleur, to whom we are indebted for the process.

Pinaire's Method of Gilding Watch Movements.—In gilding watch parts, and other small articles for watchmakers, gold is seldom applied directly upon the copper. In the majority of cases there is a preliminary operation, called graining, by which a vary agreeable grained and slightly dead appearance is given to the articles. If we examine carefully the inside of a watch we may see the peculiar pointed dead lustre of the parts.

This peculiar bright dead lustre, if it may be so expressed, is totally different from that ordinarily obtained. For instance, it does not resemble the dead lustre obtained by slow and quick electro-deposition of gold, silver, or copper, which is coarser and duller than that of watch parts. Neither does it resemble the dead lustre obtained with the compound acids, which is the result of a multitude of small holes formed by the juxtaposition, upon a previously even surface, of a quantity of more or less large grains, *always in relief.*

The graining may be produced by different methods, and upon gold, platinum, and silver; and since the latter metal is that preferred we shall describe the process applied to it.

This kind of gilding requires the following successive operations :—

1. *Preparation of the Watch Parts.* — Coming from the hands of the watchmaker, they preserve the marks of the file, which are obliterated by rubbing upon a wet stone, and lastly upon an oilstone.

2. The oil or grease which soils them is removed by boiling the

watch parts for a few minutes in an alkaline solution made of 100 parts of water and 10 of caustic soda or potassa, and rinsing them in clean water, which should wet them thoroughly if all the oil has been removed. The articles are threaded upon a brass wire.

3. A few gilders then cleanse them rapidly by the compound acids for a bright lustre ; others simply dry them carefully in sawdust from white wood.

4. *Holding the Parts.*—The parts thus prepared are fastened by means of brass pins with flat heads upon the even side of a block of cork.

5. The parts thus held upon the cork are thoroughly rubbed over with a brush quite free from greasy matters, and charged with a paste of the finest pumice-stone powder and water. The brush is made to move in circles in order not to abrade one side more than the other. The whole is thoroughly rinsed in clean water, and no particle of pumice dust should remain upon the pieces of the cork.

6. Afterwards we plunge the cork and all into a mercurial solution, which very slightly whitens the copper, and is composed of—

Water	10 litres.
Nitrate of binoxide of mercury .	2 grammes.
Sulphuric acid	4 „

The pieces are simply passed through the solution, and then rinsed. This operation, which too many gilders neglect, gives strength to the graining, which without it possesses no adherence, especially when the watch parts are made of white German silver, dignified by the name of nickel by watchmakers, or when the baths contain tin in their composition.

7. *Graining.*—In this state the parts are ready for the graining—that is to say, a silvering done in a particular manner.

Nothing is more variable than the composition of the graining powders ; and it may be said that each gilder has his own formula, according to the fineness of the grain desired.

The following formulæ are used in the works of M. Pinaire :—

Silver in impalpable powder. . . .	30 grammes.
Bitartrate of potassa (cream of tartar) finely pulverised and passed through a silk sieve	300 „
Chloride of sodium (common salt) pulverised and sifted as above	1 kilogramme.

It is stated that the majority of operators, instead of preparing their graining-silver, prefer buying the Nuremburg powder, which is produced by grinding a mixture of honey and silver-foil with a muller upon a ground-glass plate, until the proper fineness is obtained. The silver is separated by dissolving the honey in boiling water, and washing the deposited metal in a filter until there is no remaining trace of honey. The silver is then carefully dried at a gentle heat. This silver, like bronze powder, is sold in small packages :—

Silver powder	30	grammes.
Cream of tartar	120 to 150	,,
Common salt (white and clean) . . .	100	,,

Or—

Silver powder	30	,,
Cream of tartar	100	,,
Common salt	1	kilogramme.

All these substances should be as pure as possible, and perfectly dry. Cream of tartar is generally dry, but common salt often needs, before or after it has been pulverised, a thorough drying in a porcelain or silver dish, in which it is kept stirred with a glass rod or a silver spoon.

The mixture of the three substances must be thorough, and effected at a moderate and protracted heat.

The graining is the coarser as there is more common salt in the mixture ; and conversely, it is the finer and more condensed as the proportion of cream of tartar is greater ; but it is then more difficult to scratch-brush.

8. *The Graining Proper.*—This operation is effected as follows : A thin paste of one of the above mixtures with water is spread by means of a spatula upon the watch parts held upon the cork. The cork itself is fixed upon an earthenware dish, in which a movement of rotation is imparted by the left hand. An oval brush with close bristles is held in the right hand, and rubs the watch parts in every direction, but always with a rotary motion. A new quantity of the paste is added two or three times, and rubbed in the manner indicated. The more we turn the brush and the cork, the rounder becomes the grain, which is a good quality ; and the more paste we add, the larger the grain.

The watchmakers generally require a fine grain, circular at its base, pointed at its apex, and close—that is to say, a multitude of juxtaposed small cones. A larger grain may, however, have a better appearance, but this depends on the nature and the size of the articles grained.

9. When the desired grain is obtained, the watch parts are washed and then scratch-brushed. The wire brushes employed also come from Nuremburg, and are made of brass wires as fine as hair. As these wires are very stiff and springy, they will, when cut, bend and

turn in every direction, and no work can be done with them. It is, therefore, absolutely necessary to anneal them more or less upon an even fire. An intelligent worker has always three scratch-brushes annealed to different degrees : one which is *half soft*, or half annealed, for the first operation of uncovering the grain ; one *harder*, or little annealed, for bringing up lustre ; and one *very soft*, or fully annealed, used before gilding, for removing the erasures which may have been made by the preceding tool, and for scratch-brushing after the gilding. Of course the scratch-brushing operation, like the graining proper, must be done by striking circles, and giving a rotary motion between the fingers to the tool. The cork is now and then made to revolve. After a good scratch-brushing, the grain, seen through a magnifier, should be regular, homogeneous, and with a lustre all over. Decoctions of liquorice, saponaire (soapwort), or Panama wood are employed in this operation.

CHAPTER XII.

VARIOUS GILDING OPERATIONS.

Electro-gilding Zinc Articles.—Gilding Metals with Gold Leaf.—Cold Gilding.—Gilding Silk, Cotton, &c.—Pyro-gilding.—Colour of Electro-deposited Gold.—Gilding in various Colours.—Colouring Processes.—Re-colouring Gold Articles-—Wet-Colour Process.—French Wet-Colouring.—London Process of Wet-Colouring.

Electro-gilding Zinc Articles.—About thirty years ago a very important industry was introduced into France, which at once commanded universal admiration, and a rapid sale for the beautiful products which were abundantly sent into the market. We allude to the so-called *electro-bronzes*. These exquisite works of art, many of which would would bear comparison with the finest of real bronzes, were in fact zinc castings or copies from original works of high merit, coated with brass, or, as it was then called, *electro-bronze*, and artificially coloured, so as to imitate as closely as possible the characteristic tone of real bronze. At the time we speak of, articles of every conceivable form, from the stag beetle, mounted upon a leaf, electrotyped from nature, and reproduced in the form of a zinc casting, each object being electro-bronzed, to a highly-finished statuette or massive candelabrum, appeared in our shop windows and show-rooms, and presented a really beautiful and marvellously varied and cheap addition to our rather meagre display of art metal work. It was soon discovered by those who had the taste for possessing bronzes, but not the means to satisfy it, that the imitation bronzes lacked nothing of the beauty of the originals, while they presented the advantage of being remarkably cheap, and thus within the reach of many. The process by which the electro-bronzing upon zinc castings is conducted is considered in another place, and we will now explain how articles of this description, that is zinc castings, may be electro-gilt, and either a bright or dead surface imparted to the work according to the artistic requirements of the article to be treated.

Preparation of Zinc Castings for Gilding.—In order to obtain the best possible results, the zinc casting—presuming it to be a work of art which deserves the utmost care to turn it out creditably—should first be examined for air or sand-holes, and these, if present, must be stopped or plugged with *easy-running* pewter solder, and the spot afterwards touched up so as to resemble the surrounding surface, whether it be smooth or chased. When the whole article has been carefully examined and treated in this way, it is to be immersed for a few minutes only in a moderately strong potash bath, after which it must be well rinsed. It is next to be placed in a weak sulphuric acid pickle, consisting of—

Sulphuric acid	. . . .	10 parts
Water	. . . . . .	100 „

but should not remain in the acid liquor more than a minute or two, after which it is to be thoroughly well rinsed in clean water. The article, having a copper wire attached, is now to be placed in either a cold or warm brassing solution or alkaline coppering bath for a short time, or until it is covered with a thin deposit of either metal.* If on removing it from the brassing bath it is found that the soldered spots have not received the deposit, and present a blackish appearance, the article must be well scratch-brushed all over, and again placed in the bath, which, by the way, will deposit more readily upon the solder if the bath be warm, a brisk current employed, and gentle motion given to the article when first placed in the bath. The object, when placed in the bath a second time, should be allowed to remain therein for about half an hour or somewhat longer, by which time, if the solution be in good order, and the current sufficiently active, it will yield a deposit sufficiently thick either for bronzing or gilding. It is a common practice to deposit a slight coating of brass or copper upon zinc-work in a warm solution in the first instance, and then to complete the operation in a cold bath.

When the object is to be left bright, that is merely scratch-brushed, after being coated with copper or brass as above, it is simply gilt in an ordinary cyanide gilding bath, and is then treated in the same way as ordinary brass or copper work. If, however, the article is to be left dead, the following method may be adopted: After being well rinsed, the object is to be immersed in a silvering bath in which it is allowed to remain until it assumes the characteristic white and dead lustre of electro-deposited silver. When the desired effect is produced, the article must be well rinsed in warm (not hot) water and *immediately* placed in a gilding bath which is in a good condition for yielding a deposit of the best possible colour.

Another method, which has been much practised on the Continent, is thus described by Roseleur: "Add to the necessary quantity of water one-tenth of its volume of sulphuric acid, and dissolve in this acid liquor as much sulphate of copper as it will take up at the ordinary temperature. This solution will mark from 20° to 24° Baumé

---

* See Chaps. XXIV. and XXV.

(about 1·1600); now add water to reduce its specific gravity to
16° or 18° B. (about 1·1260). This galvano-plastic * bath is
generally contained in large vessels of stoneware, slate, wood, or
gutta-percha, and porous cells are immersed in it, which are filled
with a weak solution of sulphuric acid and amalgamating salts. Plates
or cylinders of zinc are put into these cells, and are connected with
one or more brass rods, which rest upon the sides of the vat, and
support the articles which are to receive the dead lustre."

The articles of zinc, previously coated with copper or brass in an
alkaline solution, are suspended in the above bath until they have
acquired the necessary dead lustre, after which they are treated as
follows: After being thoroughly well rinsed, they are immersed for
a moment in a bath composed of—

Nitrate of mercury	. . .	1 part
Sulphuric acid	. . . .	2 parts
Water	.	1,000 „

After again rinsing, the articles are steeped in the following solution :—

Cyanide of potassium	. . .	40 parts
Nitrate of silver .	. . . .	10 „
Water	.	1,000 „

The articles are well rinsed after removal from this bath, and are
then ready for gilding, the solution recommended for which is composed of—

Phosphate of soda	. . . .	60 parts
Bisulphite of soda	. . .	10 „
Cyanide of potassium	. . .	1 to 2 parts
Neutral chloride of gold .	. .	2 parts
Water	.	1,000 „

This bath is used at nearly the boiling point, with an intense voltaic
current. The anode consists of platinum wire, which at first is dipped
deeply into the solution, and afterwards gradually raised out of the
bath, as the article becomes coated with gold, until, towards the end
of the operation, but a small surface of the wire remains in the bath.
It is said that the colour of the gilding by this method is remarkable
for its "freshness of tone." Some operators first gild the article by
the *dipping* process before described, and then deposit the requisite
quantity of gold to produce a dead surface by the electro process in a
bath specially suited to the purpose. Other gilders first half-gild the

---

* This term, though never a correct one, is still generally used on the continent to designate the art of electrotyping, or the deposition of copper from
its sulphate.

article with the battery, then dip it in the mercury bath, and after well rinsing, finish the operation by a second deposit of gold. In either case, the article is finally well rinsed in warm water, and afterwards dried in hot sawdust or a warm stove. Great care is taken to avoid handling the article so as not to stain it with the fingers, or to scratch it in any way, since the delicate frosted surface is very readily injured. It is also very important that the rinsing, after each operation, should be perfectly carried out, and that the final drying is complete ; for if any of the gold solution remain upon any part of the work, voltaic action will be set up between the zinc and copper at the spot, and the article disfigured by the formation of verdigris. The foregoing process is specially applied to articles of zinc, such as clock-cases, &c., which are generally kept under glass, but may be applied to smaller ornamental articles which are not liable to friction in use.

In our own practice we have found when gilding zinc, that the best results were obtained when *all* the various stages of the process, from the first pickling to the drying, were conducted with *rapidity*, the greatest possible attention being devoted to the various rinsing operations. If all baths are in proper order, the various *dips* and electro-deposits should each only occupy from a few seconds to a few minutes, while the drying should be effected with the greatest possible despatch, so that the object, being but thinly coated with metals which are electro-negative to itself, may not be subjected to electro-chemical action in parts owing to the presence of moisture or traces of the gilding solution.

Gilding Metals with Gold Leaf.—Articles of steel are heated until they acquire a bluish colour, and iron or copper are heated to the same degree. The first coating of gold leaf is now applied, which must be gently pressed down with a burnisher, and again exposed to gentle heat ; the second leaf is then applied in the same way, followed by a third, and so on ; or two leaves may be applied instead of one, but the last leaf should be burnished down while the article is cold.

Cold Gilding.—A very simple way of applying this process is to dissolve half a pennyweight of standard gold in aqua regia ; now steep several small pieces of rag in the solution until it is all absorbed ; dry the pieces of rag, and then burn them to tinder. To apply the ashes thus left, rub them to a powder, mix with a little water and common salt, then dip a cork into the paste thus formed, and rub it over the article to be gilt.

Gilding Silk, Cotton, &c.—There are several methods by which textile fabrics may be either gilt or silvered. One method is to stretch the fabric tightly upon a frame, after which it is immersed in a solution of acetate of silver, to which ammonia is added until the precipitate at first formed becomes dissolved, and a clear solution obtained.

After immersion in this solution for an hour or two, the thread or fabric is first dried, and then submitted to a current of hydrogen gas, by which the silver becomes reduced and the surface metallised. In this condition it is a conductor of electricity, and may be either gilt or silvered in any ordinary cyanide solution. By another method, the piece of white silk is dipped in an aqueous solution of chloride of gold ; it is then exposed to the fumes of sulphurous acid gas, produced by burning sulphur in a closed box, when in a very short time the entire piece will be coated with the reduced metal.

Pyro-gilding.—This process, which is recommended for coating iron and steel, is conducted upon the same principle as pyro-plating,* except that the precious metal is deposited in several layers, instead of, as in the former case, depositing the required coating in one operation. The steel article being prepared as recommended for pyro-plating, first receives a coating of gold in the gilding-bath ; it is next heated until the film of gold disappears ; it is then again gilt, and heated as before, these operations being repeated until the last layer remains fully on the surface.

Colour of Electro-deposited Gold.—It might readily be imagined that gold, when deposited from its solution upon another metal, would necessarily assume its natural colour, that is, a rich orange yellow. That such is not the case is well known to all who have practised the art of gilding, and the fact may easily be demonstrated by first gilding a piece of German silver in a *cold* cyanide solution of gold, and then raising the temperature of the solution to about 130° Fahr. If now a similar piece of metal be gilt in the warm solution, and the two gilt surfaces compared, it will be found that while the deposit from the cold solution is of a pale yellow colour, that obtained by the warm solution is of a deeper and richer hue. The colour of the deposit may also be influenced by the nature of the current, the same solution being used. For example, the gold deposited by the current from a Bunsen battery is generally of a finer and deeper colour than that obtained by the Wollaston battery. In the former case, the superior intensity of the current seems to favour the colour of the deposit. This difference, however, is not so strongly marked in the case of some other gold solutions, as that prepared by precipitating gold with sulphide of ammonium, and redissolving the precipitate with cyanide, for example, which yields an exceedingly good coloured deposit with copper and zinc elements (Wollaston or Daniell). Since the colour of the gold deposit is often of much importance to the electro-gilder, we purpose giving below the various means adopted for varying the colour of the deposit to suit the requirements of what we may term *fancy gilding*.

Gilding in Various Colours.—A very deep coloured deposit of

---

* See p. 274.

gold may be obtained in an old gold solution, in which organic matter has accumulated from imperfect rinsing of the work after scratch - brushing, and in which there is a good proportion of free cyanide, by employing a strong current and exposing a large surface of anode. In this case the deposit is of a *foxy* colour, as it is termed, and when scratch-brushed exhibits a depth of tone which, while being unsuited for most purposes, may be useful as a variety in some kinds of fancy gilding where a strong contrast of colours is a requisite. The colour of the deposit is also much influenced, as before observed, by the extent of *anode surface* exposed in the bath during the operation of gilding ; if a larger surface be exposed than is proportionate to the *cathode surface* (or work being gilt) the colour is dark, whereas when the anode surface exposed is *below* the proper proportion, the deposit will be of a pale colour. *Motion* also affects the colour of the gold deposit—sometimes in a very remarkable degree —the colour being *lighter* when the article is moved about in the solution, and *darker* when allowed to rest. These differences are more marked, however, with old and dark coloured solutions than with recently prepared solutions, or such as have been kept scrupulously free from the introduction of organic impurities.

For ornamental gilding, as in cases where chased or engraved silver or plated work is required to present different shades of colour on its various surfaces, solutions of gold may be prepared from which gold of various tints may be obtained by electro-deposition. These solutions are formed by adding to ordinary cyanide gilding baths varying proportions of silver or copper solution, or both, as also solutions of other metals ; but in order to insure uniformity of results, the solutions should be worked with anodes formed from an alloy of the same character ; or at least, if an alloy of silver and gold, for example, is to be deposited, an anode of gold and one of silver should be employed in order to keep up the condition of the compound solution.

Green Gold.—This is obtained by adding to a solution of double cyanide of gold and potassium a small proportion of cyanide of silver solution, until the desired tint is obtained. The solution should be worked cold, or nearly so.

Red Gold.—To a solution of cyanide of gold add a small quantity of cyanide of copper solution, and employ a moderately strong current. It is best, in making these additions, to *begin low*, by adding a very small proportion of the copper solution at first, and to increase the quantity gradually until the required tone is obtained, since an excess of the copper solution would produce a deposit of too *coppery* a hue. The tint generally required would be that of the old-fashioned gold and copper alloy with which the seals and watch cases of the last, and earlier part of the present, century were made.

Pink Gold.—This may be obtained by first gilding the article in the usual way, then depositing a slight coating in the preceding bath, and afterwards depositing a mere pellicle of silver in the silvering bath. The operation requires great care to obtain the desired pink tint. The article is afterwards burnished; but since the silver readily becomes oxidised (unless protected by a colourless varnish) the effect will not be of a permanent character.

Pale Straw-coloured Gold.—Add to an ordinary cyanide solution a small quantity of silver solution, and work the compound solution *cold*, with a small surface of anode and a weak current.

Colouring Processes.—When the gilding is of an inferior colour it is sometimes necessary to have at command some method by which the colour may be improved. There are several processes by which this may be effected, but in all cases there must be a sufficient coating of gold upon the article to withstand the action of the materials employed. This condition being fulfilled, the artificial colouring processes may be applied with advantage, and gold surfaces of great beauty obtained. Of the processes given below, the first formula will be found exceedingly useful, since it may be applied to work which, though fairly well gilt, need not be so stoutly coated as is necessary when employing the second formula. It is specially useful for bringing up a good colour upon brooches, albert chains, and small articles generally. It is technically known by the name "green colour," and is composed as follows :—

	ozs.	dwts.	grs.
I. Sulphate of copper . . .	0	2	0
French verdigris . . .	0	4	12
Sal ammoniac	0	4	0
Nitre	0	4	0
Acetic acid (about) . . .	1	0	0

The sulphate of copper, sal ammoniac, and nitre are first to be pulverised in a mortar, when the verdigris is to be added and well mixed with the other ingredients. The acetic acid is then to be poured in, a little at a time, and the whole well worked up together, when a thin mass of a bluish green colour will result. The article to be coloured is to be dipped in the mixture and then placed on a clean piece of sheet copper, which is next to be heated over a clear fire, until the compound assumes a dull black colour ; it is now allowed to cool, and is then plunged into a tolerably strong sulphuric acid pickle, which soon dissolves the colouring salts, leaving the article of a fine gold colour. It is generally advisable to well scratch-brush the article before colouring, when it will come out of the pickle perfectly bright. When removed from the pickle, the article must be well rinsed in hot water, to which a small quantity of carbonate of potash should be added ; it should next be brushed with warm soap and water, a soft brush being employed, and again rinsed in hot water, after which it may be placed in warm box sawdust, being finally brushed with a long-haired brush.

II. When the work is strongly gilt, but of an indifferent colour, the following mixture may be used :—

Powdered alum	3 ounces
„ nitre	6 „
„ sulphate of zinc . .	3 „
„ common salt . . .	3 „

These ingredients are to be worked up into a thickish paste, and the articles brushed over with it ; they are then to be placed on a piece of sheet iron, and heated over a clear charcoal or coke fire until they become nearly black ; when cool they are to be plunged into dilute muriatic or sulphuric acid pickle.

Recolouring Gold Articles.—It not unfrequently happens that an electro-gilder is required by his customers to renovate articles of gold jewellery, so as to restore them to the original condition in which they left the manufacturers. Although it has been the common practice, with some electro-gilders, to depend upon their baths to give the desired effect to what is called " coloured " jewellery, in some cases it would be better to apply the methods adopted by goldsmiths and jewellers for this purpose, by which the *exact* effect required can be more certainly obtained. There are two methods of colouring gold articles ; namely, " dry colouring," which is applied to articles made from 18-carat gold and upwards, and " wet-colouring," which is adopted for alloys of gold below that standard, but seldom lower than 12-carat.

The mixture for *dry-colouring* is composed of

Nitre	8 ounces
Alum	4 „
Common salt	4 „
	16 „

Or the following :—

Sal ammoniac	4 ounces
Saltpetre	4 „
Borax	4 „
	12 „

The ingredients must first be reduced to a powder, and then put into an earthen pipkin, which is to be placed over a slow fire to allow the salts to fuse *gradually* ; to assist this, the mixture should be stirred with an iron rod. When the fused salts begin to *rise* in the vessel, the pieces of work, suspended by a fine silver or platinum wire, should be at once immersed, and kept moved about until the liquid begins to sink in the colouring-pot, when the work must be removed, and plunged into clean muriatic acid pickle, which will dissolve the adhering salts. The colouring mixture will again rise in the pot, after the withdrawal of the work, when it may be reimmersed (when dry) for a short time, and then pickled as before ; it is then to be rinsed in a weak solution of carbonate of soda or potash, and afterwards well washed in hot soda and water, next in clean boiling water, and finally put into warm box sawdust to dry. Previous to colouring the work, it should be highly polished or burnished, although the latter operation may be performed *after* the work has been coloured ; the former method is, however, the best, and produces the most pleasing effect.

Wet-colouring Process.—This is applied to gold articles made from alloys *below* 18-carat, and though there are many formulæ adopted for colouring gold of various qualities below this standard, we must limit our reference to one or two only, and for ample information upon this subject direct the reader's attention to Mr. Gee's admirable *Goldsmith's Handbook.** The ordinary "wet colour," as the jewellers term it, consists chiefly in adding a little water to the ingredients formerly given, the proportions of the salts being generally about the same ; that is, nitre 8 ounces, alum and common salt of each 4 ounces. These ingredients being reduced to a fine powder and mixed together, are worked up into a thick paste with a little hot water in a good-sized pipkin or crucible, which is placed over a slow fire and heated gradually, the mixture being stirred with a wooden spoon until it boils up. The work is now to be introduced as before and allowed to remain for several minutes, when it must be withdrawn and plunged into boiling water, which will dissolve the colouring salts and show how far the colouring has progressed. When the mixture exhibits a tendency to boil dry, an occasional spoonful of hot water must be added to thin it, but never while the work is in the pot. When the work is first put into the colour it becomes nearly black, but assumes a lighter tone after each immersion until the characteristic colour of fine gold is obtained. When the operation is complete, the work will bear a uniform appearance, though somewhat dead, and may be brightened by burnishing or scratch-brushing. After each dipping the work must be well rinsed in clean boiling water. It must be finally plunged into hot water, and, after well shaking, be put into warm boxwood sawdust.

---

* "The Goldsmith's Handbook," by George E. Gee. London : Crosby Lockwood and Son.

French Wet Colouring.—The formula for this is :—

Saltpetre	8 ounces
Common salt	4 „
Alum	4 „

The ingredients must be finely pulverised, as before, and intimately mixed ; they are then to be put into a good-sized pipkin or crucible, and sufficient hot water added to form the whole into a thick paste. The mixture should be slowly heated, and stirred with a wooden spoon, when it will soon boil up. The work is then to be immersed for several minutes, then withdrawn and plunged into boiling water, which, dissolving the salts, will allow the work to be examined, when, if not of a sufficiently good colour, it must be reimmersed for a short time. As the mixture thickens by evaporation small quantities of boiling water must be added occasionally, but only after the work has been withdrawn. On the first immersion the work assumes a blackish colour, but at each successive immersion it becomes lighter, as the baser metals become removed from the surface of the work, until it finally assumes the characteristic colour of fine gold. This process should be applied to gold of less than 16 carats.

London Process of Wet Colouring.—For gold of not less than 15 carats the following mixture is used :—

Nitre	15 ounces
Common salt	7 „
Alum	7 „
Muriatic acid	1 „
	30

The salts are to be powdered, as before ; into a crucible about 8 inches high and 7 inches in diameter, put about two spoonfuls of water, then add the salts, place the crucible on the fire, and heat gradually until fusion takes place, keeping the mixture well stirred with a wooden spoon. The article, which should first be boiled in nitric acid pickle, is then to be suspended by a platinum wire, and immersed in the fused mixture for about five minutes, then withdrawn and steeped in boiling water. The muriatic acid is now to be added to the mixture, and when it again boils up the article is to be immersed for about five minutes, then again rinsed in boiling water. A spoonful of water is now to be added to the mixture, and the work again put in for about three minutes, and again rinsed ; now add two spoonfuls of water to the mixture, boil up, and immerse the work for two minutes, and rinse again. Finally, add about three spoonfuls of water, and, after boiling up, put in the work for one minute, then rinse in abundance of clean boiling water, when the work will present a beautiful colour. The work should then be rinsed in a very dilute hot solution of potash, and again in clean boiling water, after which it should be placed in clean, warm boxwood sawdust.

P

CHAPTER XIII.

MERCURY GILDING.

Preparation of the Amalgam.—The Mercurial Solution.—Applying the
Amalgam—Evaporation of the Mercury.—Colouring.—Bright and Dead
Gilding in Parts—Gilding Bronzes with Amalgam.—Ormoulu Colour.
-Red-Gold Colour.—Ormoulu. — Red Ormoulu.—Yellow Ormoulu.—
Dead Ormoulu.—Gilders' Wax.—Notes on Gilding.

ALTHOUGH the process of gilding metals with an amalgam of gold and
mercury, or quicksilver, is not, strictly speaking, an electro-chemical
art, it is important that this system of gilding should be known to
the electro-gilder for several reasons : it is the chief process by which
metals were coated with gold before the art of electro-gilding was
introduced ; it is still employed for certain purposes, and many articles of silver which have been mercury gilt occasionally come into the
electro-depositor's hands for regilding, and which are sometimes
specially required to be subjected to the same process, when the voltaic
method is objected to.

Mercury-gilding, formerly called wash-gilding, water-gilding, or
amalgam-gilding, essentially consists in brushing over the surface of
silver, copper, bronze, or brass, an amalgam of gold and quicksilver,
and afterwards volatilising the mercury by heat. By repeated applications of the amalgam and evaporation of the mercury, a coating of
gold of any desired thickness may be obtained, and when properly
carried out the gilding by this method is of a far more durable
character than that obtained by any other means. As we have before
observed, the process, unless conducted with great care, is a very
unhealthy one, owing to the deleterious nature of the fumes of mercury to which the workmen are exposed, if these are not properly
carried off by the flue of a suitable furnace.

Preparation of the Amalgam.—Mercury, as is well known, has
the peculiar property of alloying or amalgamating itself with gold,
silver, and some other metals and alloys, with or without the aid of
heat. To prepare the amalgam of gold for the purpose of mercury
gilding, a weighed quantity of fine or standard gold is first put into a
crucible and heated to dull redness. The requisite proportion of
mercury—8 parts to 1 part of gold—is now added, and the mixture is

stirred with a slightly crooked iron rod, the heat being kept up until the gold is entirely dissolved by the mercury. The amalgam is now to be poured into a small dish about three parts filled with water, in which it is worked about with the fingers under the water, to squeeze out as much of the excess of mercury as possible. To facilitate this, the dish is slightly inclined to allow the superfluous mercury to flow from the mass, which soon acquires a pasty condition capable of receiving the impression of the fingers. The amalgam is afterwards to be squeezed in a chamois leather bag, by which a further quantity of mercury is liberated ; the amalgam which remains after this final treatment consists of about 33 parts of mercury and 67 of gold in 100 parts. The mercury which is pressed through the bag retains a good deal of gold, and is employed in preparing fresh batches of amalgam. It is very important that the mercury employed for this purpose be pure. The gold employed may be either fine or standard, but water-gilders generally use the metal alloyed either with silver or copper ; if to be subjected to the after process of *colouring*, standard alloys should be employed, since the beauty of the colouring process depends upon the removal, chemically, of the inferior metals, silver, copper, or both, from the alloy of gold, leaving the pure metal only upon the surface. The amalgam is crystalline, and produces a peculiar crackling sound when pressed between the fingers close to the ear.

It is usual to keep a moderate supply of gold amalgam in hand when mercury-gilding forms part of the gilder's ordinary business, and the compound is divided into a series of small balls, which are kept under water ; it is not advisable, however, to allow the amalgam to remain for a long period before being employed, since a peculiar phenomenon known as *liquation* takes place, by which the amalgam loses its uniformity of composition, the gold being more dense in some parts than in others.

The Mercurial Solution.—To apply the amalgam, a solution of nitrate of mercury is employed, which is prepared by dissolving, in a glass flask, 100 parts of mercury in 110 parts of nitric acid of the specific gravity 1·33, gentle heat being applied to assist the chemical action. The red fumes which are given off during the decomposition must be allowed to escape into the chimney, since they are highly deleterious when inhaled. When the mercury is all dissolved, the solution is to be diluted with about 25 times its weight of distilled water, and bottled for use.

Applying the Amalgam.—The pasty amalgam is spread with the blade of a knife, upon a hard and flat stone called *the gilding stone,* and the article, after being well cleaned and scratch-brushed, is treated in the following way : the gilder takes a small scratch-brush, formed of stout brass wire, which he first dips in the solution of nitrate of mercury, and then next draws it over the amalgam, by which it takes up a small quantity of the composition ; he then passes the brush carefully over the surface to be gilt, repeatedly dipping the brush in the mercurial solution and drawing it over the amalgam until the entire surface is uniformly and sufficiently coated. The article is afterwards well rinsed and dried, when it is ready for the next operation.

Evaporation of the Mercury.—For this purpose a charcoal fire, resting upon a cast-iron plate, has been generally adopted, a simple hood of sheet iron being the only means of partially protecting the workmen from the injurious effects of the mercurial vapours. M. D'Arcet, of Paris, invented a furnace, or forge, with an arrangement by which the workman could watch the progress of his work through glass, and thus escape the injurious effects of the mercury vapours. The difficulty of seeing the process clearly, however, during the more important stages of the operation (owing doubtless to the condensation of the mercurial vapour upon the glass), caused the arrangement to be disapproved by those for whose well-being it was specially designed, and the simple hood, regardless of its fatal inadequacy, is still preferred by many mercury gilders. When the amalgamated article is rinsed and dried, the gilder exposes it to the glowing charcoal, turning it about, and heating it by degrees to the proper point ; he then withdraws it from the fire by means of long pincers or tongs, and takes it in his left hand, which is protected with a leather or padded glove, and turns it over the fire in every direction, and while the mercury is volatilising, he strikes the work with a long-haired brush, to equalise the amalgam coating, and to force it upon such parts as may appear to require it.

When the mercury has become entirely volatilised, the gilding has a dull greenish-yellow colour, and the workman examines it to ascertain if the coating is uniform ; if any bare places are apparent, these are touched up with amalgam, and the article again submitted to the fire, care being taken to expel the mercury gradually.

Colouring.—The article is next well scratch-brushed, when it assumes a pale greenish colour ; it is afterwards subjected to another heating to expel any remaining mercury, when, if sufficient amalgam has been applied, it acquires the characteristic orange-yellow colour of fine gold. It is next submitted to the process of *colouring*. If required to be bright, the piece of work is burnished in the ordinary way, or, according to the nature of the article, is subjected to the ormoulu process described further on. When the surface is required to be dead, or frosted, the article is treated somewhat in the same way as " dry coloured " gold jewellery work, that is, it is brushed over with a hot paste composed of common salt, nitre, and alum, fused in

the water of crystallisation of the latter, after which it is heated upon
a brisk charcoal fire, without draft, and moved about until the salts
become first dried and then fused ; the article is then plunged into a
vessel containing a large quantity of cold water, in which the colouring salts are dissolved, and the dead or matted appearance of the
work becomes at once visible.   When applying the amalgam for dead
gilding, great care must be exercised to insure a sufficiently stout
coating of gold upon the work, otherwise the colouring salts will
surely attack the underlying metal.   When about to colour the work
as above, the operator binds the article by means of iron wire to a
short rod of the same metal : he then either dips the article in the
colouring paste or applies it with a brush, and after gently drying it,
holds the piece over the fire until the perfect fusion of the composition
has taken place, when it is at once dipped in water.   The coloured
marks left by the wire are removed by a weak solution of nitric acid.

Bright and Dead Gilding in Parts.—When it is desired to have
some parts of an article burnished and other parts left dead, the
former are protected by a mixture of Spanish white (pure white chalk),
bruised sugar candy, and either gum or glue, dissolved in water.   The
mixture of alum, nitre, and common salt is then applied to the parts
to be left dead, the article afterwards dried, and heated over the charcoal fire as before until the dried salts have been fused, when it is at
once plunged into cold water, and subsequently in dilute nitric acid,
being finally well rinsed and dried.   The protected parts are then
subjected to the operation of burnishing, when the article is complete.

Another method adopted in France, in which electro-gilding takes
a part, is described as follows : Those parts which are intended for
a dead lustre are first gilt with the amalgam ; the article is then
heated, scratch-brushed, and re-heated to the orange-yellow colour.
Then, with the battery, a sufficiently strong gold deposit is given to
the whole, without regard to the parts already mercury-gilt.   All the
surfaces are next carefully scratch-brushed, and the electro-gilt portions are brushed over, first with a thin mixture of water, glue, and
Spanish white, and afterwards with a thick paste of yellow clay.
After drying, the mercury-gilt portions are covered with the paste
for dead-gilding (alum, nitre, &c.), and the article heated until the
salts fuse, when it is plunged into water and treated as above.

Roseleur, however, considers this method open to several objections,
among which is, red spots are apt to be produced upon such places as
may have been too much heated, or where the gold has not been sufficiently thick.   He recommends the following by preference : " Gild
with amalgam, and bring up the dead lustre upon those portions
which are to receive it, and preserve [protect] them entirely with a
stopping-off varnish.   After thorough drying, cleanse the object by

dipping it into acids in the usual manner, and gild in the electro-bath. The varnish withstands all these acids and solutions. When the desired shade is obtained, dissolve the varnish with gazoline or benzine, which, unless there has been friction applied, do not injure in any way either the shade or velvety appearance of the dead lustre. Wash in a hot solution of cyanide of potassium, then in boiling water, and allow to dry naturally. . . . Gilding with dead lustre, whatever process be employed, suits only those objects which will never be subjected to friction ; even the contact of the fingers injures it.''

Gilding Bronzes with Amalgam.—The article is first annealed very carefully, as follows : The gilder sets the piece upon burning charcoal, or peat, which yields a more lively and equal flame, covering it up so that it may be oxidised as little as possible, and taking care that the thinner parts do not receive an undue amount of heat. This operation is performed in a dark room, so that the workman may see when the desired cherry-red heat is reached. He then lifts the piece from the fire, and sets it aside to cool in the air gradually. When cold, the article is steeped in a weak sulphuric acid pickle, which removes or loosens the coating of oxide. To aid this he rubs it with a stiff and hard brush. When the article has been thus rendered bright, though it may appear uniform, it is dipped in nitric acid and rinsed, and again rubbed with a long-haired brush. After washing in clean water, it is dried in hot sawdust or bran. This treatment somewhat reduces the brightness of the surface, which is favourable to the adhesion of the gold. The amalgam is next applied with the scratch-brush, as before, and the object then heated to expel the mercury. If required to be dead, it is treated with the colouring-salts, as before described.

Ormoulu Colour.—To obtain this fine colour upon bronze or other work, the gilt object is first lightly scratch-brushed, and then made to *come back again*, as it is termed, by heating it more strongly than if it were to be left dead, and then allowed to cool a little. The ormoulu colouring is a mixture of hematite (peroxide of iron), alum, and sea-salt, made into a thin paste with vinegar, and applied with a brush until the whole of the gilded surface is covered, except such parts as are required to be burnished. The object is then heated until it begins to blacken, the proper heat being known by water sprinkled over it producing a hissing noise. It is next removed from the fire, plunged into cold water, and washed, and afterwards rubbed with a brush dipped in vinegar if the object be smooth, but if it be chased, dilute nitric acid is employed for this purpose. The article is finally washed in clean water, and dried at a gentle heat.

Red Gold Colour.—To produce this colour, the composition known as *gilders' wax* is used. The article, after being coated with amalgam, is heated, and while still hot is suspended by an iron wire, and coated with gilders' wax, a composition of beeswax, red ochre, verdigris, and alum. It is then strongly heated over the flame of a wood fire; sometimes small quantities of the gilders' wax are thrown into the fire to promote the burning of the fuel. The object is turned about in every direction, so as to render the action of the heat uniform. As soon as all the wax has become burnt off, the flame is put out, and the article plunged into cold water, well washed, and brushed over with a scratch-brush and pure vinegar. Should the colour not be uniform or sufficiently good, the article must be coated with verdigris dissolved in vinegar, dried over a gentle fire, then plunged into cold water and brushed over with vinegar; and if the colour is of too deep a tone, dilute nitric acid may be substituted for the vinegar. After well washing, the article is burnished, then again washed, and finally wiped with soft linen rag, and lastly dried at a gentle heat.

Ormoulu.—The beautiful surface noticeable on French clocks and other ornamental work is produced by the process called *ormoulu*. The article is first gilt, and afterwards scratch-brushed. It is then coated with the thin paste of saltpetre, alum, and oxide of iron before mentioned, the ingredients being reduced to a fine powder, and worked up into a paste with a solution of saffron, annatto, or other colouring matter, according to the tint required, whether red or yellow. When the gilding is strong, the article is heated until the coating of the above mixture curls over by being touched with a wet finger. But when the gilding is only a slight film of gold, the mixture is merely allowed to remain upon the article for a few minutes. In both cases, the article is quickly washed with warm water containing in suspension a certain quantity of the materials referred to. The article must not be dried without washing. Such parts as may have acquired too deep a colour are afterwards struck with a brush made with long bristles. By a series of vertical strokes with the brush the uniformity of surface is produced. If the first operation has not been successful, the colouring is removed by dipping the article in dilute sulphuric acid, and after well rinsing, the operation is repeated until the desired effect is obtained.

Red Ormoulu is produced by employing a mixture composed of alum and nitre, of each 30 parts; sulphate of zinc, 8 parts; common salt, 3 parts; red ochre, 28 parts; and sulphate of iron, 1 part. To this may be added a small quantity of annatto, madder, or other colouring matter, ground in water.

Yellow Ormoulu is produced by the following: red ochre, 17; potash alum, 50; sulphate of zinc, 10; common salt, 3; and saltpetre 20 parts, made up into a paste as before.

Dead Ormoulu, for clocks, is composed of saltpetre, 37 ; alum, 42 ; common salt, 12 ; powdered glass and sulphate of lime, 4 ; and water, 5 parts. The whole of these substances are to be well ground and mixed with water.

Gilders' Wax, for producing a rich colour upon gilt work, is made from oil and yellow wax, of each 25 parts; acetate of copper, 13 parts; and red ochre, 37 parts. The oil and wax are to be united by melting, and the substances, after being well pulverised, added gradually.

Notes on Gilding.—When gilding single small articles, it is a good plan to hold the anode by its conducting wire in the left hand, so as to be able to control the amount of surface to be immersed in the bath, which must be considerably less (with hot solutions especially) than that of the article to be gilt. The object being slung by thin copper wire, the free end of the wire is to be twisted round the negative electrode (the wire issuing from the zinc of the battery), and the article then dipped into the bath. The article should *gradually* become coated, that is, in a few seconds, but not *immediately* after it is immersed. *Gentle motion* will secure an uniform deposit. After the article has become gilt all over, the anode may be lowered a little deeper into the bath, and the gentle motion of the article kept up for a short time, say from three to five minutes, or until it appears to be fairly coated. The length of time the article is to remain in the bath must be regulated by the price to be paid for the gilding. If a really good gilding is required, it may be necessary, after about five minutes' immersion, to rescratch-brush the article, dip it in the mercurial solution for a moment, or until it is white, and then, after well rinsing, give it a second coating. Ordinary gilding, however, is generally accomplished in a single immersion.

1. *Gilding Jewellery Articles.*—Chains, brooches, rings, pins, and other small articles of silver or metal jewellery should first be slung upon thin copper wire, then dipped for a few moments only in a warm potash bath. The articles are then to be rinsed in warm water and scratch-brushed, after which they are again rinsed, and at once immersed in the gold bath. When sufficiently gilt, the work should be rinsed in a vessel kept specially for the first rinsing, which should be saved, and afterwards in clean water. It is then to be properly scratch-brushed, and plunged into hot water ; next shaken about to remove as much water as possible, and finally put into warm boxwood sawdust. After moving it about in the boxdust for a few moments, the article requires to be shaken or knocked against the palm of the hand, to dislodge the sawdust. It is now ready to be wrapped up for the customer, pink tissue paper being preferable for gilt work, and blue or white tissue paper for silver or plated work.

2. *Treatment of Gilding Solutions.*—When the gilding bath has been

heated for a few hours, it will have lost a considerable proportion of its water, which must be made up by adding an equivalent of hot water. If this is not done, the bath, being stronger than it was originally, will probably yield a non-adhering deposit, and the gold may strip off the work under the scratch-brush. The solution should be kept up to its standard height in the gilding vessel by frequent additions of hot water during the whole time it is subjected to evaporation by the gas-burner, or other heating medium. The solution-line of the bath should be marked upon the inside of the vessel when the liquid is first poured in.

4. *Gilding different Metals.*—Silver and metal articles should not be slung upon the same wire and immersed in the bath at the same time, since brass, gilding metal, and copper receive the deposit more readily than silver. The latter metal should first receive a coating, after which, if time is an object, the metal articles may be placed in the bath with the partly-gilt silver articles.

5. *Employment of Impure Gold.*—When it is desired to make up a gold solution from impure material, as from " old gold," for instance, the alloy should first be treated as follows : To 1 ounce of the alloyed gold, if of good quality—say 18-carat gold, for example—add 2 ounces of silver, which should not be below *standard* ; melt them in a crucible with a little borax, as a flux. When the alloy is thoroughly melted it is to be poured into a deep vessel containing cold water, which must be briskly stirred *in one direction*, while the molten alloy is being poured in. This operation, termed *granulation*, causes the metal to assume the form of small lumps, or *grains*, as they are called. The water is now to be poured off and the grains of alloy collected and placed in a flask, such as is shown in Fig. 70. To remove the silver and copper from the granulated metal, a mixture of two parts water and one part strong nitric acid is poured into the flask, which is then placed on a sand-bath, moderately heated, until the red fumes which at first appear have ceased to be visible in the bulb of the vessel. The clear liquid is now to be carefully poured off into a suitable vessel—a

Fig. 78.

glass " beaker," such as is shown in Fig. 78, being a convenient vessel for the purpose. A small quantity of the dilute acid should then be poured into the flask, and heat again applied, in order to remove any remaining copper or silver. If, on the addition of the fresh acid, red fumes do not appear in the flask the operation is complete, and the grains of metal will have assumed a dark brown colour. The acid must now be poured off, and the grains well washed, while in the flask, with distilled water. The residuum is pure gold and may be at once dissolved in aqua regia, and treated in the same way as recommended for ordinary grain gold.   The silver may readily be recovered from the decanted liquor, which, owing to the presence of copper removed from the original alloy, will be of a green colour, by immersing in it a strip of stout sheet copper, which in the course of a few hours will reduce the silver to the metallic state, in the form of a grey, spongy mass.   When all the silver is thus thrown down, the green liquor is to be poured off and the silver deposit well washed with hot water.   Being now pure silver, it may be used for making up solution, or fused with dried carbonate of potash into a button.

6. *Gilding Filigree Work.*—Silver filigree work which has been annealed and pickled assumes a dead-white surface, which does not readily " take " the gilding unless the bath is rich in gold and free cyanide, and the current strong.   If such parts of the article as can be reached by the scratch-brush are brightened by this means, the interstices which have escaped the action of the brush will sometimes be troublesome to gild, while the brightened parts will readily receive the deposit.   In this case, if the bath is wanting in free cyanide, an addition of this substance must be made, and the article must be kept rather briskly moved about in the solution, and a good surface of anode immersed until the dead-white portions of the article are gilt. The anode may then be raised a little, and the piece of work allowed to rest in the bath, without movement, until the desired colour and thickness of coating are obtained.   Some persons prefer dipping this kind of work in the mercury solution before gilding, by which a more uniform deposit is obtained.   This plan is useful when the gold bath has been recently prepared.   It must not be forgotten, however, that in gilding filigree work the battery current must be brisk.

7. *Gilding Insides of Vessels.*—It sometimes happens, when gilding the interior of silver or electro-plated tankards, mugs, &c., which have been highly embossed or chased, that the gold, while depositing freely upon the prominent parts, refuses to deposit in the hollows. To overcome this, and to render the deposit uniform, the solution should be well charged with free cyanide ; the current must be of high tension (a Bunsen, for example), and the anode should be kept in motion during the first few moments.   In this way very little trouble will be experienced from the causes referred to.   It is important, however, that insides of such embossed work should be very thoroughly scratch-brushed in the first instance ; indeed, as a mechanical assistant, the scratch-brush lathe is the gilder's best friend.

8. *Old Solutions.*—When a gold solution has been much used it acquires a dark colour, from being contaminated by impurities as beer from the scratch-brush lathe, &c., and in this condition is likely to yield a deposit of a dull red-brown colour, which,

while being favourable to certain classes of work which can be readily *got at* by the scratch-brush, is very objectionable to articles of jewellery which are required to present a clear orange-yellow colour in all parts, including the interstices and soldered joints which cannot be reached by the lathe-brush. When the solution is in this condition we have found it advantageous to evaporate it to dryness, then to re-dissolve it in hot distilled water, filter the solution when cold, and add a small proportion of free cyanide, finally making up the bath to about three-fourths of its original volume. The solution thus treated yields a very rich colour in gilding. It is necessary to mention, however, that gold solutions which have been prepared by precipitating the gold from its chloride with *ammonia* should not be evaporated to dryness, since the explosive *fulminate of gold* may be present to some extent, which would render the operation hazardous.

9. *Management of Gold Baths.*—The colour of the gilding may be varied from a pale straw or lemon colour to a dark orange-red at the will of the operator ; thus, when the solution is cold, a pale lemon-coloured deposit will be obtained. If the bath be warm, a very small surface of anode exposed in the solution, and the article kept in brisk motion, the deposit will also be of a pale colour. If, on the other hand, there be a large excess of cyanide in the bath, a considerable surface of anode immersed and a strong current, the gilding will be of a dark red colour, approaching a brown tone, and the article, when scratch-brushed, will assume a rich orange-yellow colour, specially suited to certain classes of work, as the insides of cream-ewers, goblets, &c., and chains of various kinds. In order to obtain uniform results in any desired shade, when gilding a large number of articles of the same class, care must be exercised to keep the temperature of the bath uniform ; the anode surface immersed in the solution the same for each batch of work, consisting of an equal number of pieces of the same dimensions ; the battery current as uniform as possible, and, lastly, fresh additions of warm distilled water must be added frequently to the bath to make up for loss by evaporation. If these points be observed there will be no trouble in obtaining uniform results. It is scarcely necessary to state that a large bulk of gilding solution will keep in an uniform condition for a longer period than a smaller quantity, since the effect of evaporation is less marked than in the latter case.

10. *Worn Anodes.*—It is not advisable to employ anodes which have become ragged at the edges for gilding the insides of vessels, since particles of the metal are liable to be dislodged during the gilding process, and, falling to the bottom of the vessel, protect those parts upon which they drop from receiving the deposit ; indeed, the smaller fragments will sometimes become *electro-soldered* to the bottom of the vessel, causing some trouble to remove them. When the edges of an

anode are very ragged it is well to trim them with shears or a pair of sharp scissors before using the anode for gilding insides. The anode should always be formed into a cylinder, and not used as a flat plate for these purposes, otherwise the deposition will be irregular, and the hollow surfaces of chased or embossed work may not receive the deposit at all.

11. *Defects in Gilding.*—When the gold becomes partially dissolved off portions of an article while in the gilding-bath, it generally indicates that there is too great an excess of cyanide in the solution. The same defect, however, may be caused by the current being too weak, the liquid poor in gold, too small a surface of anode in the solution, or by keeping the articles too briskly in motion in a bath containing a large excess of cyanide. Before attempting an alteration of the solution, the battery should be looked to, and, if necessary, its exciting liquids renewed. The solution should then be well stirred and tried again ; if the same defect is observed an addition of chloride of gold should be made to the bath to overcome the excess of cyanide. If the deposit is of a very dark red colour, and of a dull appearance, this may be caused by employing too strong a current, by excess of cyanide, or too great a quantity of gold in the bath. If from the latter causes, the solution must be diluted ; if from the former, the articles should be suspended by a very thin slinging wire, or the positive element of the battery partially raised out of the battery-cell.

12. *Gilding Pewter Solder.*—Common jewellery is frequently repaired with pewter solder, which does not so readily take the gilding as the other parts. A good plan to overcome this is first to well scratch-brush the articles, after which the solder may be treated as follows : Make a weak acid solution of sulphate of copper, dip a camel-hair brush into the solution and apply it to the soldered joint, and at the same time touch the spot with a steel point ; in a few seconds the solder will become coated with a bright deposit of copper. Now rinse the article, and proceed to the gilding as usual, when it will be found that the soldered part upon which the film of copper has been deposited will readily receive a coating of gold, more readily, in fact, than the body of the article itself. The article, when gilt, is then scratch-brushed and treated as usual. The copper solution for the above purpose may be prepared by dissolving about $\frac{1}{2}$ oz. of sulphate of copper in $\frac{1}{2}$ pint of water, and adding to the solution about $\frac{1}{2}$ oz. of oil of vitriol.

13. *Gilding Cheap Jewellery.*—This class of work, whether of French or Birmingham manufacture, seldom requires more than a mere *dip* to meet the requirements of the customer ; indeed, the prices obtainable for gilding articles of this character will not admit of *gilding* in the proper sense of the term. In France it is usual to employ a platinum

anode, and to renew the gilding solution as it becomes exhausted of its metal by fresh additions of gold salt. The author has found it a very economical plan to use a *copper* anode for gilding work of this description, and by making small additions of chloride of gold when the bath exhibited signs of weakness, he has been able to gild a very large number of articles, of a very fine colour, with an infinitesimal amount of the precious metal. The only preparation such work received was a good scratch-brushing before gilding, and a very slight scratch-brushing after. In his experience, although the prices were very low, the result was exceedingly profitable. Against the employment of a copper anode, it has been argued that the solution must of necessity become highly impregnated with copper. To which we may reply that we did not find such to be the case in practice.

14. *Gilding German Silver.*—Since this alloy of copper, &c., will generally receive a coating of gold in ordinary cyanide solution, without the aid of the battery, the solution should be somewhat weaker, and the battery current also, otherwise the gold will not firmly adhere. The temperature of the solution should also be lower than is required for gilding articles of silver or electro-plate. When German silver articles are first placed in the gilding-bath a small surface of anode only should be immersed, and the deposit allowed to take place *gradually*. If these precautions be not observed the operator may suffer the annoyance of finding the work strip when the scratch-brush is applied, or at all events under the operation of burnishing.

15. *Stripping Gold from Silver.*—This may be done by making the article the anode in a strong solution of cyanide of potassium, or in an old gold solution containing a moderate amount of free cyanide. A quicker process, however, consists in immersing the articles in strong nitric acid, to which a little dry common salt is added. Care must be taken not to allow the article to remain in the stripping solution one moment after the gold has been removed, and the articles should be moved about in the liquid, especially towards the close of the operation, to facilitate the solution of the gold from the surface. The gold may afterwards be recovered from the exhausted acid bath by immersing in it several stout pieces of sheet zinc or iron, which will precipitate the gold in the metallic state, and this may be collected, dried, and fused with a little dried carbonate of potash. Or the exhausted stripping solution may be evaporated to dryness, and the residuum fused with dried carbonate of potash or soda, a little nitre being added towards the end of the operation, to refine it more completely.

16. *Spurious Gold—"Mystery Gold."*—Many attempts have been made from time to time to form an alloy which, having somewhat the colour

of gold, would also withstand the action of the usual test for gold—nitric acid. The introduction of the electro-gilding art greatly favoured such unscrupulous persons as desired to prey upon the public by selling as gold electro-gilt articles, which had not a fraction of the precious metal in their composition. An alloy of this kind entered the market many years ago, in the form of watch-chains and other articles of jewellery, the composition of which was, copper 16 parts, platinum 7 parts, and zinc 1 part. This alloy, when carefully prepared, bears a close resemblance to 16-carat gold, and when electro-gilt would readily pass for the genuine article. The manufacture of this variety of spurious gold seems to have received a check for a certain period ; but somewhat recently, in a modified formula, it has reappeared, not only in the form of articles of jewellery, but actually as current coin, and from its highly deceptive character, being able to resist the usual test, it has acquired the name of " Mystery Gold." It appears that, when converted into jewellery, the chief aim of the " manufacturers " is to defraud pawnbrokers, to whom the articles are offered in pledge ; and, since they readily withstand the nitric acid test, the " transactions " are often successful. According to Mr. W. F. Love, in a communication to the *Chemical News*, a bracelet made from an alloy of this character had been sold to a gentleman in Liverpool, and when the gilding was removed the alloy presented the colour of 9-carat gold. The *qualitative* analysis proved it to be composed of platinum, copper, and a little silver. A *quantitative* analysis yielded the following result :—

Silver	.	.	.	.	.	.	.	.	2·48
Platinum	.	.	.	.	.	.	.	32·02
Copper (by difference)	.	.	.	.	.	.	65·50

It was found that strong boiling nitric acid had apparently no effect upon it, even when kept in the acid for some time.

17. *Gilding Watch Dials.*—To prepare a silver watch dial (for example) for gilding, it should be laid, face upward, upon a flat piece of cork, and the face then gently rubbed all over with powdered pumice, sifted through a piece of fine muslin, and slightly moistened with water, using the end of the third finger of the right hand for the purpose. The finger being dipped in the pumice paste, should be worked with a rotary motion over the surface of the dial, so as to produce a perfectly uniform and soft dulness. When this is done, a piece of copper slinging wire is passed through the centre hole of the dial, and formed into a loop ; the dial is then to be rinsed, and placed in the bath, care being taken not to touch the face of the article either before or after gilding, except in the way indicated. The dial must afterwards be repainted.

18. Gilding solutions which have been worked with but a small excess of cyanide are apt to deposit more gold than is dissolved from the anode, by which the action of the bath becomes lessened, while the colour of the gilding is indifferent. It must always be borne in mind that in all cyanide solutions, but more especially such as are worked hot, the cyanide of potassium gradually becomes converted into carbonate of potassium by the action of the atmosphere, and therefore loses its solvent power.

19. For producing a dead or matted surface upon brass articles of jewellery, as brooches, lockets, &c., they are first dipped for an instant in a mixture composed of equal parts of sulphuric and nitric acids, to which a small quantity of common salt is added, and immediately plunged into cold water. After being rinsed in one or two other waters, they are promptly immersed in the gilding-bath, in which, after a moment's immersion, they acquire the desired colour of gold. After rinsing in hot water, they are finally dried in hot boxwood sawdust. In treating this class of work, care should be taken to avoid handling the pieces ; after they have been removed from the sawdust they should be at once wrapped up ready for delivery.

20. When it is desired to give a stout coating of gold to an article, it should be occasionally removed from the gilding-bath, then scratch-brushed, rinsed, and returned to the bath. If the article were allowed to remain in the bath undisturbed until a thick coating was deposited, the surface would probably be rough and crystalline, and moreover liable to strip when being scratch-brushed. It is sometimes the practice to dip the article for an instant in the quicking solution after each gilding, by which the respective layers of gold are less apt to separate in scratch-brushing or burnishing.

21. The wires used for slinging articles in the gilding-bath should never be reversed, but one end only employed for suspending the articles, the other being used for connection with the negative electrode of the battery. By adopting this system, the gold deposited upon the ends of the slinging wires is less liable to become wasted than when both ends of the wires are used indiscriminately. After the slinging wires have been used a few times, and before the gold upon them begins to chip or peel off, they should be carefully laid aside, with all the gilt ends in one direction, so that the gold may be removed, by stripping, at any convenient time. After stripping off the gold, the wires should be annealed, then pulled out straight, and placed in bundles. Before being again used, each end of the bundle of wires should be dipped for a few moments in an old dipping-bath, and then rinsed, when they will be ready for future use. It is better to treat slinging-wires thus carefully than to suffer them—which is commonly the case—to be scattered about.

22. *Gilding Lead, Britannia Metal, &c.*—When articles composed of lead, tin, Britannia metal, iron, or steel are required to be gilt, it is best to give them a preliminary coating of copper in an alkaline bath, or to electro-brass them, after which they may be gilt with perfect ease, and with but little liability to strip when scratch-brushed. The softer metals, however, will require to be burnished with great care, owing to their yielding nature under the pressure of the burnishing tools. The same observation also applies, inversely, to silvered or gilt steel work, in which case the superior metals, being softer than steel, become expanded under the influence of the burnishing tools, and are consequently liable to become separated, in *blisters*, from the underlying harder metal.

23. *Excess of Cyanide Injurious.*—When a newly prepared gilding solution is first used (hot), the deposit is usually of a rich, fine gold colour, if a sufficient quantity of free cyanide has been employed in its preparation, a proper surface of anode immersed in the solution, and the current brisk. If, on the other hand, the colour is pale—that is yellow, without the characteristic orange tint—this indicates that one or other of the above conditions is wanting. Before venturing to increase the amount of free cyanide, the condition of the battery should be examined, the temperature of the solution raised a little, and a larger anode surface immersed, when, if the solution still yields a light-coloured deposit, an addition of strong cyanide solution must be given gradually until the gilding assumes the proper orange-yellow colour. The addition of cyanide must always be made with caution, for if too great an excess be applied, the solution is apt to yield brown deposits, quite unsuitable for many articles of jewellery ware. This quality of gilding, however, is frequently taken advantage of for articles which are required to have a deep gold colour after scratch-brushing, as the insides of tankards, &c., and also Albert chains and work of a similar description. If a gold solution is in really good working condition, and the current sufficiently brisk, a copper or brass article should gild readily, in hot solutions, with an anode surface considerably less than that of the cathode, or article being gilt, especially if no motion be given to either electrode. A silver article, however, would require, in the same bath, a much larger surface of anode, but more especially if the surface were frosted, as in filigree work. In gilding articles of this description it is better to expose a large anode surface and keep the article in gentle motion when first put into the bath, after which a portion of the anode may be withdrawn, and the object allowed to rest undisturbed until the coating is sufficiently thick.

24. In working small baths, additions of hot distilled water must be given frequently to make up for the loss by evaporation ; but where large quantities of solution are employed, this addition need not be made more than once a day, or at the close of the operation. With this exception, a good gilding solution will continue to give uniformly good results for many days—especially if large in bulk—without alteration. When it begins to work tardily, however, which may readily be seen by the extra anode surface required to gild the articles promptly, moderate additions of cyanide must be given until the bath acquires its normal activity.

25. After working gilding-baths for a lengthened period, they generally assume a brown colour, and the gilding is, under such circumstances, of an indifferent colour. The chief causes of this discoloration have been already explained, and can be to a great extent avoided, by thoroughly rinsing the articles before putting them in the bath. When a solution is in this condition the best remedy is to evaporate it as before, and then redissolve the dried mass in distilled water, using about one-third less water than the original bulk. A little fresh cyanide must also be added, and the solution filtered, after which it will generally yield a deposit of excellent colour. Old solutions which give deposits of a greenish-black colour may be improved by evaporation, but the heating of the dried product should be carried somewhat farther than in the former cases. It is better, however, to abandon such a solution altogether and to make a fresh one. The gold from the waste solution may afterwards be recovered by the processes given in another chapter. When gilding solutions, after being worked some time, yield a pale straw-coloured gilding, this is attributed by some to the gradual accumulation of silver from the gold anodes (which always contain a trace of silver) ; we are, however, disinclined to accept this view, owing to the exceedingly small quantity of silver present in fine gold ; moreover, since silver deposits first, if present in a gold solution, we doubt its liability to accumulate in the bath. We would rather attribute the paleness of deposit referred to, to one or both of the following causes :—1. To the presence of a large excess of carbonate of potash in the bath from using an inferior cyanide ; 2. To the presence of tin derived from pewter soldered articles, imperfectly prepared for gilding.

26. The Bunsen battery is most generally used for gilding, and indeed the current from this source produces a gold deposit of very fine colour. It must be used with caution, however, when gilding articles at a low price, since it deposits the metal very freely from hot solutions, and would soon yield a coating of gold of greater thickness than would pay for ordinary cheap work. In gilding with this battery, the regulation of the anode surface in solution should be strictly observed, only a sufficient surface being exposed to enable the article to become gilt *almost* immediately after immersion ; the anode

Q

may be gradually lowered a little as the deposition progresses. Articles that only require to receive a mere colour of gold upon them (as in cheap jewellery) should be first scratch-brushed, then well rinsed in hot water ; dipped for a moment in the gold bath, then rinsed, and lightly scratch-brushed again, and after again rinsing receive a momentary dip in the gilding bath ; they are to be finally rinsed in boiling water, then shaken well, and placed in hot boxwood sawdust, from which they are afterwards removed and well shaken to cleanse them from this material.

CHAPTER XIV.

ELECTRO-DEPOSITION OF SILVER.

Preparation of Nitrate of Silver.—Observations on Commercial Cyanide.—
Preparation of Silver Solutions.—Bright Plating.—Deposition by Simple
Immersion.—Whitening Articles by Simple Immersion.—Whitening
Brass Clock Dials, &c.

THE process of " electro-plating " may be considered the most important branch of the great art of electro-deposition. Not only is it
invaluable in giving to articles manufactured from German silver,
Britannia metal, and other metallic surfaces, a beautifully white
coating of the precious metal even superior in brilliancy to that of
standard silver, but old plated and electro-silvered articles, from which
the silver has worn off, may be resilvered by this process and made
to look nearly equal to new, which there was no practical means of
doing before the introduction of electro-plating. This term, by-the-
bye, though generally used, is erroneous, since the process of *plating*
consists in attaching two plates, or ingots of metal, and rolling them
into sheets, from which, as in the old manufacture of Sheffield plate,
various articles of utility are, or rather were, made.

Preparation of Nitrate of Silver.—Since the silvering or "plating" solutions—with one exception—are prepared from the *nitrate of
silver*, it will be necessary to consider its preparation previous to
explaining the various ways in which silver baths are made up from
this salt of silver. To prepare nitrate of silver, the required quantity
of *grain silver* is carefully put into a glass flask * or evaporating dish,
the former by preference, since during the chemical action which
ensues while the solution of the metal is taking place, a portion of the
metal may be lost by the *spitting* of the solution when the chemical
action is at its height. In dissolving silver, take, say—

Grain silver	. . . . . .	2 ounces.
Pure nitric acid	. . . . .	3¼ ,,
Distilled or rain water	. . . .	1½ ,,

---

* When dissolving large quantities of silver, a stoneware vessel may be
employed.

Put the silver carefully into the flask, then add the water, and lastly the acid. In a few moments vigorous ebullition takes place, with the disengagement of red fumes of nitrous gas, which should be allowed to escape through the chimney. When the action begins to quiet down a little, the flask must be placed on a warm sand-bath. For small operations, or where a proper sand-bath is not provided, an ordinary frying-pan nearly half-filled with silver-sand will answer the purpose well. The flask should remain upon the sand-bath until the red fumes cease to appear in the bulb, at which period the chemical action is at an end. It may be well to mention that, in dissolving silver, it is advisable in the first instance to use rather less of the acid than is necessary to dissolve the whole of the silver, and to treat the undissolved portion separately, by which means excess of acid is avoided. The nitrate of silver solution must now be decanted into an evaporating dish and placed in the sand-bath, where it is allowed to remain until a film or *pellicle* forms on the surface of the liquor, when the vessel must be set aside to cool. A few hours after, crystals of nitrate of silver will have deposited, from which the remaining liquor is to be poured off, and this again evaporated as before. Instead of crystallising the nitrate, it may simply be evaporated to nearly dry-ness, by which the free acid will become expelled.

Should the nitric acid used in dissolving silver contain even a slight portion of *hydrochloric acid*, an insoluble white precipitate will be found at the bottom of the flask, which is *chloride of silver*. This, however, will not be injurious to the plating solution. Sometimes, also, a slight deposit of a brownish-black colour is found at the bottom of the vessel in which silver is dissolved ; this is *gold*, left in the grain silver through imperfect *parting* in the refining process. We have occasionally discovered more than a mere trace of gold at the bottom of the dissolving flask ; indeed in several instances an appre-ciable quantity.

When dissolving the crystals of nitrate of silver, for the preparation of either of the following plating solutions, distilled or rain water only should be used, since river water always contains traces of sub-stances which form a white precipitate in the presence of nitrate of silver.

In describing the silver solutions, the proportion of silver in the *metallic state* will be given, and it will be understood that in each case the weighed metal must be first converted into *nitrate*. We may also state that the proportion of silver to each gallon of solution may be varied according to the practice of the plater, some persons preferring solutions in which there is a moderate percentage of the metal, while others employ much greater quantities. The proportion of silver per gallon of solution ranges from $\frac{1}{2}$ an ounce to 5 or 6 ounces, and even more ; but for most practical purposes from 1½ to 2 ounces will be quite sufficient ; indeed, some of our best results have been obtained with 1 ounce of silver per gallon.

In most of the formulæ given, 1 gallon of solution will be taken as the basis for making up any required quantity of silvering bath ; and it will be readily understood that when larger proportions of silver to the gallon are preferred, a proportionate increase of cyanide must be used, not only to dissolve the precipitated metal, but also to play the part of *free cyanide* in the solution. It must be remarked here, that unless the silvering-bath contains an excess of cyanide of potassium, the anode, or *dissolving plate*, whose function it is to resupply the solution with silver in the proportion in which it is deposited upon the articles, will not keep up the *metallic strength* of the bath, and consequently it will deposit the metal slowly. A large excess of cyanide, on the other hand, is not only unnecessary, but is liable to cause the deposited silver to *blister* and *strip*, or peel off the work under the pressure of the burnishing tools ; and when very greatly in excess, the coating will be so non-adherent that it may even yield to the scratch-brush, and separate from the underlying metal.

Observations on Commercial Cyanide of Potassium.—Since the cyanide of potassium is one of the most important and useful substances that come under the command of the electro-depositor, while the success of his operations greatly depends upon its *active* quality, it is advisable to state that ordinary commercial cyanide varies considerably in this property ; so much so, indeed, as to render it absolutely necessary that the user should be put on his guard, lest in purchasing a cheap and worthless article, he should commit an error which may cost him much trouble and annoyance, as also pecuniary loss. Before making up *any solution* in large quantity, in which cyanide of potassium is the solvent, we advise him first to obtain samples of the commercial article, and to test them by either of the processes given in another chapter.[*] We may state that some of the cheap cyanides contain a large excess of carbonate of potash. This substance, while being a necessary ingredient in the manufacture, is also frequently used greatly in excess to produce a cheap article, and may be called its natural adulterant. This salt (carbonate of potash), however, unless specifically recommended in the preparation of certain depositing solutions, is not only useless, but when greatly in excess reduces the conductivity of both silvering and gilding baths.

Preparation of Silver Solutions.—*Solution I.* The solutions of silver most generally used for electro-plating are those commonly called "cyanide solutions," the foremost of which is the *double cyanide of silver and potassium*, which is prepared as follows : 1 ounce of silver is converted into *nitrate*, as previously described, and the crystals dis-

---

* See p. 431, et seq.

solved, with stirring, in about 2 quarts of distilled or rain water, which may, in the case of small quantities, be effected in a glass vessel or glazed earthenware pan. For large quantities a stoneware vessel should be used. When the crystals are all dissolved, a strong solution of cyanide (about $\frac{1}{2}$ a pound, dissolved in 1 quart of water) is added, a little at a time, when a precipitate of *cyanide of silver* will be formed, which will increase in bulk upon each addition of the cyanide. Each time, after adding the cyanide solution, the mixture must be well stirred with a glass rod or strip of wood free from resin. When it is found that the addition of cyanide produces but little effect, it must be added *very cautiously*, since an excess will redissolve the precipitate, and cause waste in the after process of washing this deposit. To avoid adding too much cyanide, the precipitate should be allowed to fall down an inch or so, when a glass rod may be dipped in the cyanide solution, and the clear liquor touched with this, when if a milkiness is produced, a little more cyanide must be added, and the stirring resumed. After a short repose, the same test may be applied, and so on, until a drop of cyanide solution produces no effect. Great care must be taken not to add more cyanide than is absolutely necessary to throw down the silver. As an additional precaution, when nearly the whole of the silver is precipitated, the vessel may be allowed to rest for an hour or so, and the clear liquor then poured off and treated separately, by which means the bulk of the precipitate will be saved from the risk of coming in contact with an excess of cyanide. If, through accident or faulty manipulation, too much cyanide has been added, more nitrate of silver solution must be poured in, which, combining with the surplus cyanide, will again produce the characteristic milkiness ; and if the additions of nitrate are made with care, the clear liquor will be perfectly free from silver, and after allowing the cyanide of silver to deposit, may be poured off and thrown away.

Washing the Precipitate.—In all such cases the precipitate should be allowed fully to settle ; the *supernatant liquor*, or "mother liquor," is next to be poured off slowly, so as not to disturb the solid matter (cyanide of silver) ; a large quantity of fresh water—which for this purpose may be common drinking water—is now to be poured on to the precipitate with brisk stirring, and the vessel again left to rest, after which the clear liquor is to be poured off as before, these *washings* being repeated three or four times.

Dissolving the Precipitate.—To convert the cyanide of silver into the *double cyanide of silver and potassium*, the strong solution of cyanide must be added in moderate portions at a time, constantly stirring as before, until the precipitate appears nearly all dissolved, at which period the additions of cyanide must be made with more caution. In this case, as in the former, it is a good plan, when nearly the whole of the precipitate is dissolved, to allow the vessel to stand for a short time, then to pour off the clear liquor—which is now a solution of the double cyanide of silver and potassium—and to treat the remainder of the precipitate with cyanide solution ; by this means too great an excess of the solvent is avoided.　When all the precipitated silver is redissolved, add about one-fourth more cyanide solution than that originally used, and pass the solution through a filter into the plating vat or depositing vessel, which may be conveniently done by means of a piece of unbleached calico (previously washed in lukewarm water to remove the " dressing ") stretched over three strips of wood bound together in the form of a triangle either with copper wire or string, as in Fig. 79.　When all the solution has passed through the filter, this may be washed by pouring a little water over it while resting over the bath.　The solution is finally to be made up to the full quantity by adding the necessary proportion of water, when its preparation is complete ; it will be better, however, to allow it to rest for twenty-four hours before using it for electro-plating.

Fig. 79.

　Free Cyanide.—This term is applied, as we have before hinted, to a moderate excess of cyanide of potassium which it is always necessary to have in the bath, to dissolve the insoluble cyanide of silver which forms on the anode, but since the ordinary commercial article is of very variable quality, the addition of this substance must to a great extent depend upon the judgment of the plater, subject to the precautions we have previously given ; from one-fourth to one-half the quantity of cyanide used in dissolving the precipitate of cyanide of silver may be added to the solution as free cyanide, and water then added to make up 1 gallon.　If the cyanide, in the first instance, be dissolved in a definite *measured* quantity of water, say at the rate of half a pound to a quart of water (40 ounces), the proportion of cyanide used in each of the former cases can be readily ascertained by simply measuring the balance of the solution and deducting its proportion of cyanide from the original weight taken.　A fair quality of ordinary commercial cyanide should not contain less than 50 per cent. of pure cyanide, but we have frequently met with an article which contains a very much lower percentage, which should never be used for making up plating solutions, but may be employed in the less important process of *dipping* in the preparation of work for nickel-plating.　Of course it will be understood that when cyanide containing a high percentage of the pure substance is obtained, a proportionately smaller quantity must be used.

　Solution II.—One ounce or more of silver is converted into nitrate as before, and the crystals dissolved in from three to four pints of distilled water. A solution of carbonate of potash (salt of tartar), consisting of about six or eight ounces of the salt to a pint of water, is to be gradually added to the solution of nitrate of silver, with constant stirring, until no further precipitation takes place. After settling, the clear liquor is poured off, and the precipitate (*carbonate of silver*) washed with water several times, as before directed. A strong solution of cyanide is then to be added until all the precipitate is thoroughly dissolved, when a moderate excess is to be added as free cyanide. The solution should now be filtered and water added to make up one gallon, or such quantity in the same proportion of materials as may be required. In adding excess of cyanide to this and other solutions, it is always preferable to add it moderately at first, otherwise the work is very liable to strip. After working the bath for some time, an addition of cyanide may be made, but so long as the anode keeps perfectly clean while the work is being plated, the less free cyanide there is in the bath the better. A solution which has been worked for a considerable time acquires a good deal of organic matter, becoming dark in colour in consequence, and is then capable of bearing, without injury to the work, a larger proportion of free cyanide than newly prepared solutions.

Solution III.—Mr. Alexander Parkes, in 1871, patented a solution for depositing solid articles. One ounce of silver is first converted into nitrate, from which the silver is thrown down in the form of *oxide* of silver, by means of a solution of caustic potash gradually added, until no further precipitation takes place. After washing the oxide, it is dissolved in 2 gallons of water containing 16 ounces of cyanide of potassium.

Solution IV.—The best solution for depositing silver upon German silver without recourse to the process of "quicking," is one which the author employed upon an extensive scale for many years with great success; it is composed as follows, and although it is rather more expensive in its preparation than many other solutions, it is, so far as he is aware, the best solution in which German silver work may be plated without being previously coated with mercury, as in the "quicking" process hereafter referred to. To prepare the solution, 1 ounce of silver is converted into nitrate and dissolved in two or three pints of distilled water as before. About three ounces of *iodide of potassium* are next to be dissolved in about half a pint of distilled water. The iodide solution is to be gradually added to the nitrate solution, the operation being performed in a dark corner of the room, or preferably by feeble gaslight, when a bright lemon yellow precipitate will be formed. The liquid must be briskly stirred upon each addition of the iodide, and care must be taken not to add the latter salt on any account in excess, otherwise it will redissolve the precipitate. When the precipitation is nearly complete, it is better to allow the vessel to rest, and to put a little of the clear liquor in a test tube, Fig. 80, and to add a drop or two of the iodide solution, when if a cloudiness is produced, a moderate quantity of the iodide is to be added to the bulk and well stirred in. The clear liquor should be repeatedly tested in this way, until a single drop of the iodide solution produces no further effect upon it. In case of an accidental addition of too much iodide, nitrate of silver solution must be added to neutralise it. When the precipitation is complete, the vessel must be set aside—in a dark place, since the iodide of silver is affected by light—for an hour or so, after which the clear liquor must be poured off, and the precipitate repeatedly Fig. 80. washed with cold water. Lastly, the iodide of silver is to be dissolved in a solution of cyanide of potassium, and a moderate excess added as before recommended. In working this solution, whenever the anode becomes coated with a greenish film, an addition of cyanide must be made to the bath.

Since the system of working the above solution differs in several respects from that adopted with other solutions, it may be well to describe our own practice in the treatment of German silver work. The articles are first placed in a warm solution of caustic potash, to remove greasy matter, after which they are well rinsed. Each article is then well scoured all over with powdered pumice and water, or finely powdered bath brick—an excellent substitute for the former—and water. As each article is brushed, it is to be well rinsed in clean water, and is then ready for the plating bath, in which it should be suspended without delay.

Solution V.—Mr. Tuck obtained a patent, in 1842, for "improvements in depositing silver upon German silver," in which he recommends, for plating inferior qualities of German silver, a liquid composed of *sulphate of silver* dissolved in a solution of carbonate of ammonia. Sulphate of silver is formed by adding a solution of sulphate of soda (Glauber's salt) to a solution of nitrate of silver, or by boiling silver with its weight of sulphuric acid. For plating the better qualities of German silver, cyanide of silver is dissolved in a solution of carbonate of ammonia. The proportions used are :—

Sulphate of silver	156 parts.
Carbonate of ammonia (dissolved in distilled water) .	70 ,,

Or,

Cyanide of silver	134 ,,
Carbonate of ammonia	70 ,,

The silver salt in each case is boiled with the solution of the carbonate of ammonia until it is dissolved. For coating common German silver, he adds half an ounce of sulphate of silver to a solution containing 107 grains of bicarbonate of ammonia.

Solution VI.—For producing very white deposits of silver, the following may be used :—One ounce of silver is dissolved and treated as before, and the crystals of nitrate redissolved in about half a gallon of distilled water. A moderately strong solution of common salt is then prepared, and this is gradually added to the former, when a copious white precipitate of *chloride of silver* is formed, which must be well washed with cold water. After pouring off the last wash water, a strong solution of cyanide is to be added to the precipitate until it is all dissolved, when a moderate excess is to be added, and the solution carefully filtered through filtering paper, the first runnings to be passed at least twice through the same filter. Lastly, add water to make up 1 gallon. The solution may be used immediately, but will work better after a few hours' rest. This solution is very useful for obtaining delicately white deposits, but is not suitable for ordinary plating purposes, since the deposited silver is liable to strip under the action of the burnisher. If used somewhat weaker than in the above proportions, with a moderate current and small anode surface, the deposit will adhere to most metals with tolerable firmness ; it is, however, most suitable for coating articles which are either to be merely scratch-brushed or left a dead white. Chased figures and cast metal work receive a brilliant white coating in this solution, but to retain their beauty they must be kept beneath a glass case, since the fine silver surface soon discolours in the atmosphere.

Solution VII.—This plating solution, which is one of the best for most purposes, is prepared by dissolving silver in a solution of cyanide,

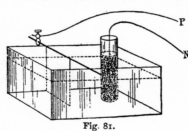

by aid of the electric current. Suppose we wish the solution to contain 1 ounce of silver per gallon. The required quantity of water is first put into the bath, and for each gallon of liquid about 3 ounces of good cyanide is added, and allowed to dissolve. A porous cell is now to be nearly filled with this

Fig. 81.

solution, and stood upright in the vessel containing the bulk of cyanide solution, the liquid being at the same height in both vessels. A strip of sheet copper is next to be connected to the negative electrode of a strong voltaic battery, and placed in the porous cell. A large sheet of silver is next to be connected to the positive electrode, and immersed in the larger vessel. The arrangement is shown in Fig. 81. The weight of the sheet silver being ascertained beforehand, the battery is allowed to remain in action for several hours, when the silver plate may be weighed to determine how much of it has been dissolved in the solution, and the action is to be kept up until the proper proportion has been dissolved by the cyanide in the outer vessel. When this point has been reached the porous cell is to be removed, and its contents may be thrown away.

Another electrolytic method of preparing a silver bath is the following :—To make a bath containing, say, one ounce of silver per gallon, the cyanide should be of known strength. Assuming the commercial article to contain 50 per cent. of real cyanide, about 3 ounces are to be dissolved in each gallon of water; a large silver anode connected to the positive pole of a battery is to be suspended in the solution, and a smaller plate of silver as a cathode. A strong current should be used, and the anode weighed from time to time until the desired proportion of silver has become dissolved into the solution. The condition of the bath may then be tested by suspending a clean plate of brass from the negative rod, at the same time immersing about an equal surface of anode, using a moderate current, when if the solution be in good working order, the cathode will at once receive a bright deposit of silver. During the action some caustic potassa is formed in the liquid, which may be converted into cyanide, by adding a moderate quantity of hydrocyanic acid, which must be done, however, with exceeding care, owing to the deadly nature of the acid, the vapour of which must not on any account be inhaled. Respecting the use of this acid, however. we must strongly recommend that its employment should never, under any circumstances whatever, be placed in the hands of persons unacquainted with its highly poisonous character.

Besides the foregoing, many other processes for preparing silver solutions have been proposed ; but since they are comparatively of little or no practical value, they deserve but a passing reference. By one of these processes chloride of silver is dissolved in hyposulphite of soda. The salt of silver thus formed (*hyposulphite of silver*) is readily acted upon by light, and the silver, being thus converted into an insoluble *sulphide*, gradually becomes deposited at the bottom of the bath. Solutions have also been prepared by employing ferrocyanide of potassium (yellow prussiate of potash) as the solvent of cyanide of silver. Again, the silver has been precipitated from its nitrate solution by lime-water, forming oxide of silver ; as also by ammonia, soda, magnesia, &c. ; the various precipitates being subsequently dissolved in a solution of cyanide of potassium. The ordinary double cyanide of silver and potassium solution, however, will be found most useful for the general purposes of electro-plating.

Bright Plating.—The silver deposited from ordinary cyanide solutions is of a pearly white or dull white, according to the condition and nature of the silver solution, and the strength of the current; and it is necessary to brighten the work by scratch-brushing before it is subjected to the operations of burnishing or polishing. It is possible, however, by adding to the plating-bath a small quantity of *bisulphide of carbon*, to obtain deposits of silver which are *bright* instead of dull, for the discovery of which important improvement we are indebted to Mr. W. Milward, of Birmingham, who made the discovery in the following way:—He had observed that when wax moulds for electro-typing, which had been coated with a film of phosphorus by applying a solution of that substance in bisulphide of carbon, were put into the cyanide plating solution to receive a deposit of silver, that other articles—as spoons and forks, for example—silvered in the same solution, assumed a brightness more or less uniform, sometimes extending all over the articles, and at others occurring in streaks and patches. This led him to try the effect of adding bisulphide of carbon alone to the plating solution, which produced very satisfactory results. The improvement was worked as a secret for some time, but eventually leaked out, in consequence of which Mr. Milward and a person named Lyons (who had become acquainted with the secret) took out a patent in March, 1847, for bright silver deposition by adding "compounds of sulphur and carbon," bisulphide of carbon being preferred. From that time the bisulphide of carbon has been constantly employed for producing bright deposits of silver.

To make up the solution for "bright plating," the following methods are adopted: 1. "6 ounces of bisulphide of carbon are put into a stoppered bottle, and 1 gallon of the usual plating solution added to it; the mixture is first to be well shaken, and then set aside for 24 hours. 2 ounces of the resulting solution are then added to every 20 gallons of ordinary plating solution in the vat, and the whole stirred together; this proportion must be added every day, on account of the loss by evaporation, but where the mixture has been made several days, less than this may be used at a time. This proportion gives a bright deposit, but by adding a larger amount a dead surface may be obtained, very different to the ordinary dead surface."

Another method of preparing a solution for bright plating is the following:—Put 1 quart of ordinary silver plating solution into a large stoppered bottle; now add 1 pint of strong solution of cyanide and shake well; 4 ounces of bisulphide of carbon are then to be added, as also 2 or 3 ounces of liquid ammonia, and the bottle again well shaken, which operation must be repeated every two or three hours,

The solution is then set aside to rest for about 24 hours, when it will be ready for use. About 2 ounces of the clear liquid may be added to every 20 gallons of ordinary plating solution, and well mixed by stirring. A small quantity of the brightening solution, or "bright," as it is termed in the plating-room, may be added to the solution every day, and the liquid then gently stirred. In course of time the bisulphide solution acquires a black colour, to modify which a quantity of strong cyanide solution, equal to the brightening liquor which has been removed from the bottle, should be added each time. In adding the bisulphide solution to the plating bath an excess must be avoided, otherwise the latter will be spoilt. Small doses repeated at intervals is the safer procedure, and less risky than the application of larger quantities, which may ruin the bath·

A very simple way to prepare the brightening solution is to put from 2 to 3 ounces of bisulphide of carbon in an ordinary "Winchester" bottle, which holds rather more than half a gallon. Now add to this about 3 pints of old silver solution, and shake the bottle well for a minute or so : lastly, nearly fill the bottle with a strong solution of cyanide, shake well as before, and set aside for at least 24 hours. Add about 2 ounces (not more) of the "bright" liquor, without shaking the bottle, to each 20 gallons of solution in the plating vat. Even at the risk of a little loss from evaporation, it is best to add the brightening liquor to the bath the last thing in the evening, when the solution should be well stirred so as to thoroughly diffuse the added liquor. The night's repose will leave the bath in good working order for the following morning.

Other substances besides the bisulphide of carbon have been used, or rather recommended, for producing a bright deposit of silver, but up to the present no really successful substitute has been practically adopted. Amongst other compounds which have been suggested, are a solution of iodine and gutta-percha in chloroform, which is said to have a more permanent effect than the bisulphide ; carbonate, and acid carbonate of potassium, $1\frac{1}{2}$ ounce of each, added once every nine or ten days to a plating solution containing 12 ounces of cyanide and $3\frac{1}{2}$ ounces of silver per gallon. According to Planté, bright silver deposits may be obtained by adding a little sulphide of silver to the plating solution.

Although the solution for "bright plating" is useful for some purposes, it is not adopted as a substitute for the ordinary cyanide of silver bath for the general purposes of electro-plating. For articles which have deep hollows and interstices which cannot be burnished without considerable difficulty, and for the insides of tea and coffee-pots, and articles of a similar description, which are required to be bright, but which cannot be rendered so by mechanical methods, a slight coating of bright silver is an advantage. A bath of brightening solution is usually kept for this purpose, in which the articles, after being plated in the ordinary way, are immersed for a short time, by which they receive a superficial coating of the bright deposit. In the " bright solution " the articles first become bright at their lower surface, the effect gradually spreading upward, until in a short time they become bright all over, when they are removed from the bath and immediately immersed in boiling water, otherwise the silver would quickly assume a dark colour.

Deposition by Simple Immersion.—Articles of brass and copper readily become coated with a film of silver, without the aid of the battery current, in tolerably strong solutions of the double cyanide of silver and potassium ; the deposit, however, is not of such a degree of whiteness as to be of any practical use. Other solutions of silver are therefore employed when it is desired to give a slight coating of this metal to small brass or copper work which will present the necessary brilliant white colour. Silver may also be deposited upon these surfaces by means of a paste of chloride of silver, to which common salt or cream of tartar is added. The following processes are those most generally adopted :—

Whitening Articles by Simple Immersion.—For small brass and copper articles, as buttons, hooks and eyes, coffin-nails, &c., silvering by simple immersion is employed ; and in order to produce the best possible effect, the solution bath should not only be prepared with care, but kept as free as possible from contamination by other substances. To prepare a bath for this purpose, a given quantity of fine grain silver is dissolved in nitric acid. The solution of nitrate of silver thus formed is added to a large quantity of water, a strong solution of common salt is then poured in, which precipitates chloride of silver in the form of a dense white precipitate. When the whole of the silver is thrown down (which may be ascertained by adding a drop or two of hydrochloric acid or solution of salt to the clear liquor) the precipitate is allowed to subside, after which the supernatant liquor is to be poured off, and the precipitate washed several times. The last rinsing water should be tested with litmus paper, when, if there be the least trace of acid, further washing must be given. The precipitate, which is very readily acted upon by sunlight, should be prepared in a dull light, or by gaslight, and, if not required for immediate use, it should be bottled and kept in a dark cupboard. The chloride of silver is to be mixed with at least an equal weight of bi-tartrate of potassa (cream of tartar), and only sufficient water added to form a pasty mass of the consistency of cream. In this mixture the articles, having been previously cleaned or *dipped*, are immersed, and stirred about until they are sufficiently white, when they are to be rinsed in hot water, and shaken up with boxwood sawdust. These preparations are also used for silvering clock-faces, thermometer and barometer scales, and other brass and copper articles, by being rubbed over the surface to be whitened with a cork.

Another chloride of silver paste, for whitening articles of brass or copper, may be made by taking chloride of silver and prepared chalk, of each one part, common salt $1\frac{1}{4}$ part, and pearlash 3 parts, made into a paste, as before. A third mixture is prepared by taking chloride of silver 1 part, cream of tartar at least 80 parts, to which is sometimes added about 80 parts of common salt. The whole are dissolved in boiling water. This solution acquires a greenish tint, from the presence of copper, which takes the place of the silver in the liquid. In using this solution the articles are introduced by means of a perforated basket, which is briskly stirred about in the hot liquid until uniformly white. Some operators modify the above solution by adding common salt, Glauber's salt, corrosive sublimate, caustic lime, &c., but it is doubtful whether any advantage is derived therefrom.

A process, which is to be applied cold, was proposed by Roseleur, and seems to have worked exceedingly well in his experienced hands. A solution of the double sulphite of silver is formed in the following way : Four parts of soda crystals are dissolved in five parts of distilled water, sulphurous acid gas (prepared by heating in a glass retort strong oil of vitriol with some small pieces of copper wire) is then passed through the liquid, by allowing it to bubble through mercury at the bottom of the vessel, to prevent the exit tube from becoming clogged with crystals ; the gas is allowed to pass until the fluid re-dissolves the crystals of bicarbonate, and slightly reddens blue litmus paper. It is then allowed to repose for twenty-four hours, so that some of the bisulphite of sodium formed may crystallise. The liquid portion is then to be poured off, and stirred briskly to expel the carbonic acid. If alkaline to test paper, more sulphuric acid gas must be added ; or if acid, a little more carbonate of soda. After well stirring, the solution should only turn blue litmus paper violet, or at most slightly red. A solution of nitrate of silver is now to be added to the above liquid, with stirring, until the precipitate at first produced begins to dissolve slowly, when the bath is ready for use. The solution thus prepared is said to be always ready for work, and " produces, quite instantaneously, a magnificent silvering upon copper, bronze, or brass articles which have been thoroughly cleansed, and passed through a weak solution of nitrate of binoxide of mercury, although this is not absolutely necessary." To keep up the strength of the bath fresh additions of nitrate of silver must be made from time to time, and after awhile some bisulphite of soda must also be added. In working this bath the solution is placed in a copper vessel, which also receives a deposit of silver. Roseleur states that he used this bath for five years, during which period he daily silvered " as many articles as a man could conveniently carry." He also states that, without the aid of a separate current, the deposit from this solution may become nearly as thick as desired, and in direct ratio to the time of immersion.

Whitening Brass Clock-Dials, &c., with the Paste.—For this purpose chloride of silver and cream of tartar, with or without the addition of common salt, is made into a paste, as before described, and this should be well triturated in a mortar until it is impalpable to the touch. The paste is then spread, a little at a time, upon the brass surface—which may be a clock-face or thermometer-scale, for instance —and rubbed upon the metal surface with a piece of soft cork, or " velvet " cork as it is called. By thus working the silver paste over the metal it soon becomes silvered, and a coating of sufficient thickness for its purpose obtained in a very short time, according to the size of the object. When the silvering is complete the article is to be rinsed and dried in the hot sawdust. Although a very slight film of silver only is obtained in this way, its somewhat dull tone is specially applicable to barometer and galvanometer scales, clock-dials, and objects of a similar nature, and, as far as its non-liability to tarnish is concerned, it may be considered superior to all other methods of silvering.

CHAPTER XV.

ELECTRO-DEPOSITION OF SILVER (continued).

Preparation of New Work for the Bath. —Quicking Solutions, or Mercury
Dips.—Potash Bath.—Acid Dips.—Dipping.—Spoon and Fork Work.—
Wiring the Work.—Arrangement of the Plating Bath.—Plating Battery.
—Motion given to Articles while in the Bath.—Cruet Stands, &c.—Tea
and Coffee Services.—Scratch-brushing.

Preparation of New Work for the Bath.—In order to insure
a perfect adhesion of the silver deposit to the surface of the article
coated, or *plated*, as it is erroneously termed, with this or any other
metal, the most important consideration is *absolute cleanliness*. By this
term we do not mean that the article should be merely *clean* in the
ordinary sense, but that it must be what is termed *chemically clean*,
that is, perfectly and absolutely free from any substance which would
prevent the silver from attaching itself firmly to the metal to be coated.
As evidence of the extreme delicacy which it is necessary to observe in
this respect, we may mention that if, after an article (say a German
silver spoon, for example) has been well scoured with powdered
pumice and water, it be exposed to the atmosphere even for a few
seconds, it becomes coated with a slight film of oxide—owing to the
rapidity with which copper (a constituent part of German silver)
attracts oxygen from the air ; this effect is still more marked in the
case of articles made from copper and brass. Now, this slight and
almost imperceptible film is quite sufficient to prevent perfect contact
between the deposited metal and that of which the article is composed.
This fact, in the early days of electro-plating, created a great deal of
trouble, for it was found that the work, after being silvered, was very
liable to strip under the pressure of the burnisher. To overcome the
difficulty, and to secure a perfect adhesion of the two metals, a third
metal—mercury or quicksilver—which has the power of alloying itself
with silver, gold, German silver, copper, and brass, was employed, and
though the author for many years obtained most successful and perfectly adherent deposits of silver without its aid, the process of
quicking was, and still is, practised by the whole of the electro-plating
trade. The silver solution which the author employed, however, and
which is described in the foregoing chapter, was differently prepared

R

to those ordinarily adopted by the trade. Since the process of "quicking" is generally adopted, it will be necessary to describe it in detail.

Quicking Solutions, or Mercury Dips.—This term is applied, as before hinted, to coating articles made of brass, copper, or German silver—the metals most usually subjected to the process of *electro-plating*—with a thin film of quicksilver, which may be effected by either of the following solutions :—*Nitrate of Mercury Dip.*—Put an ounce of mercury into a glass flask, and pour in an ounce of pure nitric acid diluted with three times its bulk of distilled water ; if, when the chemical action ceases, a small amount of undissolved mercury remains, add a little more acid, applying gentle heat, until the whole is dissolved. The solution is then to be poured into about 1 gallon of water, and well mixed by stirring. It is then ready for use, and is termed the *quicking solution*, or *mercury dip*. Articles of brass, copper, or German silver dipped into this solution at once become coated with a thin bright film of mercury.

Cyanide of Mercury Dip.—Dissolve one ounce of mercury as before, and dilute the solution with about 1 quart of distilled water. Now take a solution of cyanide of potassium, and add this gradually, stirring after each addition, until the whole of the mercury is precipitated, which may be determined by dipping a glass rod in the cyanide solution, and applying it to the clear liquor after the precipitate has subsided a little, when, if no further effect is produced, the precipitation is complete. The liquor is next to be separated by filtration. When all the liquor has passed through the filter, a little water is to be poured on to the mass, and when this has thoroughly drained off, the precipitate is to be placed in a glass or stoneware vessel, and strong solution of cyanide added, with constant stirring, until it is all dissolved, when a small excess of the cyanide solution is to be added, as also sufficient water to make up one gallon of solution.

Another mercury dip is made by dissolving *red precipitate* (red oxide of mercury) in a solution of cyanide, afterwards diluted with water.

Pernitrate of Mercury Solution.—This solution is composed of—

Pernitrate of mercury	1 part.
Sulphuric acid	2 parts.
Water 1000	„

A very good mercury dip may be made by simply dissolving two ounces of mercury in two ounces of nitric acid, without the aid of heat ; the solution thus formed is to be diluted with about three gallons of water.

The quicking bath should contain just so much mercury in solution as will render a clean copper surface *white* almost immediately after immersion ; if the solution be too strong, or too acid (when the nitrate of mercury solution is used), or if the solution has become nearly exhausted by use, when copper is dipped it may turn *black* or dark coloured instead of white, in which case the quicking bath must be rectified, otherwise it will be impossible to obtain an adherent coating of silver upon the article treated in it.   As a rule, the articles merely require to become perfectly and uniformly white from the coating of mercury, but the practice is to give a stronger film to work which is required to receive a stout deposit of silver, or gold, as the case may be. When the mercury dip becomes nearly exhausted and the mercurial coating, in consequence, becomes dark coloured, the liquor should be thrown away and replaced by a new solution, which is considered better than strengthening the old liquor : indeed, the small amount of mercury which remains in the bath after having been freely used is of so little consideration, that the liquor may be cast aside without sacrifice the moment it gives evidence of weakness by the dark appearance of the work instead of the characteristic brightness of metallic mercury.

It is a good plan to keep a quantity of concentrated mercurial solution always in stock, so that when a bath becomes exhausted it may be renewed in a few minutes by simply throwing away the old liquor and adding the due proportions of strong solution and water to make up a fresh "dip."

Potash Bath.—To remove greasy matter communicated to the work by the polishing process, all articles to be plated must first be steeped for a short time in a hot solution of caustic potash, for which purpose about half a pound of American potash is dissolved in each gallon of water required to make up a bath, and as this solution becomes exhausted by use it must receive an addition of the caustic alkali.   The workman may readily determine when the solution has lost its active property by simply dipping the tip of his finger in the solution and applying it to the tip of the tongue, when, if it fails to tingle or " bite " the tongue, the solution has lost its caustic property, and may either be thrown away or strengthened by the addition of more caustic potash.   When the bath has been once or twice revived in this way it is better to discard it altogether, when inactive, than to revive it. Indeed, when we consider that the object of the caustic alkali is to convert the greasy matters on the work into soap, by which they become soluble and easily removed by brushing, it will be apparent that the bath can only be effective so long as the *causticity* of the alkali remains.   Many persons, from ignorance of this matter, have frequently used their potash baths long after they have lost their activity, and as a natural consequence the work has come out of such baths nearly in the same state as they entered it, greasy and dirty.

Those who cannot conveniently obtain caustic potash (American potash, for example) may readily prepare it as follows: Obtain a few lumps of fresh lime and slake them by pouring water over them, and then covering them with a cloth ; soon after, the lime will fall into a powder, which must be made into *milk of lime*, as it is called, by mixing it with water to the consistence of milk or cream. A solution of pearlash is then made in boiling water, to which is added the cream of lime, and the mixture is to be boiled for at least an hour, in an iron vessel. About half or three-quarters of a pound of pearlash to each gallon should be employed, and about one-fourth less lime than potash. If the solution is thoroughly *causticised*, no effervescence will occur in the liquor if a drop or two of hydrochloric acid are added ; if, on the contrary, effervescence takes place on the addition of the acid, the boiling must be continued.

Acid Dips.—In the preparation of certain kinds of work, acid solutions or mixtures are employed which may be advantageously mentioned in this place. It is well to state, however, that after dipping the work in acid solutions it should be *thoroughly* rinsed in clean water, since the addition of even small quantities of acid to the alkaline plating or gilding baths would seriously injure these solutions. Indeed, careless and imperfect rinsing must always be avoided in all depositing operations, otherwise the baths will soon become deteriorated ; the rinsing waters should be frequently changed, and the workman taught that in this item of his labour his motto should be "water no object."

Fig. 82.

Nitric Acid Dip.—This is frequently used for dipping copper, brass, and German silver work, and is the ordinary aquafortis of commerce, or *fuming* nitric acid (*nitrous* acid). Stoneware jugs of the form shown in Fig. 82 are used for conveying strong acids. A dipping acid, composed as follows, is also much used for producing a bright and clean surface upon certain classes of work :—

Nitric acid, commercial (by measure) . . .	1 part.
Sulphuric acid	2 parts.
Water	2 „

To this mixture some persons add a little hydrochloric acid, and others a small quantity of nitrate of potassa (nitre).

Dip for Bright Lustre.—To give a bright appearance to copper, &c., the following mixture may be employed :—Old aquafortis, or nitric acid dip which has been much used, 1 part ; water, 2 parts ; muriatic acid, 6 parts. The articles are immersed in this solution for a few minutes, when they are to be briskly shaken in clean cold water, and if not sufficiently bright must be dipped again. If they become covered with a dirty deposit, the articles should be scoured with pumice and water, then immersed in the dip for a short time and again rinsed. Another method is to first dip the articles in a weak *pickle*, formed by diluting old and nearly exhausted nitric acid dip with water for a few minutes, after which they are to be dipped in the same old acid dip in its undiluted condition, and finally in strong aquafortis for a moment ; they are next to be well rinsed in several waters.

Dip for Dead Lustre.—To produce a *dead* or *matted* surface upon copper, brass, or German silver work, the following mixture is used :—

Brown, or fuming aquafortis (by measure)   .   . 2 parts.
Oil of vitriol   .   .   .   . 1 part.

To the above mixture a small quantity of common salt is added. The articles are allowed to remain for some time in the dip, after which they are withdrawn and promptly dipped in the preceding liquid and immediately well rinsed.

Respecting old aquafortis dips, Gore says these may be " revived to a certain extent by addition of oil of vitriol and common salt ; the sulphuric acid decomposes the nitrate of copper in it, and also the common salt, and sets free nitric and hydrochloric acids, and crystals of sulphate of copper form at the bottom of the liquid. All the nitric acid may be utilised in this manner." This is perfectly true, but as a rule acid "dips" which have become exhausted seldom produce the required brilliancy or tone of colour (when that is an object), even if strengthened by fresh additions of the concentrated acids with which they were first prepared. Zinc, tin, and lead, as also organic matter, generally find their way into these dips, and more or less interfere with the direct action of the nitric acid.

Dipping.—The article to be dipped should be suspended by a wire of the same metal, or by a wire covered with gutta-percha or india-rubber tubing, and after a moment's immersion in the acid solution, promptly plunged into clean cold water ; if the desired

Fig. 83.

effect—a bright or a dead lustre—is not fully produced by the first dip, the article must be again dipped for a moment and again rinsed. In order to remove the acid effectually, several washing vessels should be at hand, into each of which the article is plunged consecutively, but the last rinsing water, more especially, should be renewed frequently. When a number of small articles require to be dipped, they may be suspended from a wire, looped up as in Fig. 83, or they may be placed in a perforated stoneware basket (Fig. 84), provided with a handle of the same material. These perforated baskets are specially manufactured at the potteries for acid dipping and other purposes, and if carefully treated will last for an indefinite period. The basket containing the articles to be dipped is plunged into the dipping acid, and moved briskly about, so as to expose every surface of the metal to the action of the acid ; as the vessel is raised the liquid escapes through the perforations, and after a brisk shaking the basket and its contents

Fig. 84.   Fig. 85.

are plunged into the first washing water, in which it is again vigourously shaken, to wash away the acid as far as possible ; it is then treated in the same way in at least two more rinsing waters. The dipped articles are then to be thrown into a weak solution of crude bitartrate of potash, called *argol*, to prevent them from becoming oxidised or tarnished. From this liquid they are removed as required, and again rinsed before being quicked and plated. For dipping purposes, stoneware and gutta-percha bowls (Fig. 85) are also used, and sometimes platinum wire trays, supported by a hook, as in Fig. 87, are employed for very small articles. Hooks of the same kind, but in

Fig. 86.

various forms, are likewise used for supporting various pieces of work during the dipping operations. One of these is shown in Fig. 86.

Spoon and Fork Work.—In large establishments this class of work may be said to hold the leading position, since, as articles of domestic utility, the spoon and fork are things of almost universal requirement. As in all other kinds of electro-plated ware—and we may add everything else under the sun—the silvering, or plating, is accomplished according to the requirement of the customer and the price to be paid for the work when done. In other words, the actual deposit of silver which each article receives depends upon whether it is intended *to wear well*, or merely required *to sell*. In the former

case, it is usual for the customer to weigh the spoons and forks before
he sends them to the plater, and again on their return, and he pays so
much per ounce for the silver deposited, allowing a moderate discount
from the original weight to cover any loss which may be sustained in
preparing the work for the plating bath. When the goods, however,

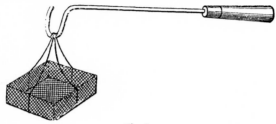

Fig. 87.

are merely required to look "marketable," the amount of silver
deposited upon such a class of work often ranges from little or nothing
to less, if possible.

Wiring the Work.—Spoons and forks are first *wired*, as it is
termed. For this purpose copper wire is cut into lengths of about
12 inches. A length of the wire is coiled once round the shank or narrow
part of the article, and secured by twisting it several times ; the loop
thus formed should be quite loose, so that the position of the spoon or
fork may be easily reversed or shifted while in the plating vat, to
equalise the deposit, and to allow the parts where the wire has been
in contact to become coated with silver. The copper wire used for
" slinging " is usually about No. 20 B.W.G. (Birmingham Wire
Gauge), which is the gauge most generally adopted in this country.
The spoons, &c., are next placed in the hot potash bath, where they
are allowed to remain for a short time, when
they are removed, a few at a time, and
rinsed in cold water. They are next to be
brushed or scoured all over with fine
pumice-powder moistened with water, and
then thrown into clean water, where they
remain until a sufficient number have

Fig. 88.

been scoured, when these are taken out by their wires and immersed in the quicking solution, which, for spoon work, may conveniently be in a shallow oval pan of the form shown in Fig. 88.
After remaining in the quicking bath a short time, they are
examined, and if sufficiently *quicked,* and uniformly bright, like
quicksilver, they are rinsed in water and at once suspended in the

plating tank, as close together as possible without touching.   When
the bath is filled with work, the spoons or forks should be turned
upside down by slipping the shank through the loop; and the workman who does this must be very careful to handle them as little as
possible, and only to grip them with the fingers by the *edges*, which
an experienced plater will do with great smartness, and with very
trifling contact with his fingers.   The objects of thus changing the
position of the work are twofold, namely, to allow the *wire mark* to
become coated with silver, and to equalise the deposit, which always
takes place more energetically at the *lower* end of the article while in
the bath.   This system of shifting should be repeatedly effected until
the required deposit is obtained.   When shifting the spoons, &c., all
that is necessary is to raise the straight portion of the suspending
wire which is above the solution with one hand, which brings, say, the
handle of the spoon, out of the solution; if this be now gripped
between the finger and thumb of the other hand at its edges, and
raised until the bowl end touches the loop, by simply turning the
spoon round its bowl will be uppermost, in which position the article
is carefully but quickly lowered into the bath again.

Another method of suspending spoons and forks in the plating bath
is the following : Copper wire, about the thickness of ordinary bell
wire, is cut up into suitable lengths, and which will depend
upon the distance between the negative conducting-rod and
the surface of the silver solution.   These wires are next to
be bent into the form of a hook at one end, and at the other
end is formed a loop, as in Fig. 89, leaving an opening
through which the shank of a spoon or fork may pass into
the ring or loop and be supported by it.   To prevent the
silver from being deposited upon the vertical portion of
the wire, where it would be useless and unnecessary, this
portion of the wire should be protected by means of glass,
gutta-percha, or vulcanised india-rubber tubing, which is
slipped over the wire before the upper hook is formed.

Fig. 89.

After being some time in use, the lower ring becomes thickly coated
with a crystalline deposit of pure silver, when these wires must be
replaced by new ones, and the insulating tubes may be again applied
after removal from the old wires.

Ordinary slinging wires, as those previously described, should never
be used more than once, and for this reason : when a certain amount
of silver or other metal is deposited upon wire—except under certain
conditions—it is invariably more or less brittle, and in attempting to
twist it round an article it is very liable to break, often causing the
article to fall from the hand—perhaps into the bath—and rendering
the silver-covered fragments of wire liable to be wasted by being

swept away with the dirt of the floor. It is more economical to employ fresh wires for each batch of work, and to *strip* the silver or other metal from the wires by either of the processes hereafter given, by which not only may all the metal be recovered, but by *annealing*, cleaning, and straightening the wires, they may be used again and again. Moreover, a wire that has been twisted once becomes hardened at that part, and cannot with safety be twisted again without being *annealed*.

Arrangement of the Plating Bath.—The size and form of depositing tanks for silver plating vary in different establishments, as also does the material of which they are constructed. For small quantities of silver solution, say from ten up to thirty gallons, oval stoneware pans may be used, and with ordinary care will last a great number of years. Wooden tubs, if absolutely clean, may also be employed for small operations, but since that material absorbs the silver solution, such vessels should be well soaked with hot water before pouring in the solution. Tanks made from slate, with india-rubber joints, have also been much used in silver-plating. Very good plating tanks may be made in the same way as directed for nickel - plating baths, that is, an outer vessel of wood, secured by screwed bolts, lined with sheet lead, and re-lined with matched boarding. Wrought-iron tanks, lined with wood, are, however, greatly preferred, and when properly constructed and lined, form the most durable of all vessels for solutions of this description. Depositing tanks for large operations are usually about six feet in length, three feet in width, and about two feet six inches in depth, and hold from two to three hundred gallons of solution; tanks of greater length are, however, sometimes employed. An ordinary wrought-iron plating tank is shown in Fig. 90, in which also the arrangement of the silver anodes and sundry articles in solution is seen. The upper rim of the tank is furnished with a flange of wood, firmly fixed in its position, upon which rest two rectangles of brass tubing or stout copper rod. The outer rectangle frequently consists of brass tubing about an inch in diameter, at one corner of which a binding screw is attached, by means of solder, for connecting it with the positive pole of the battery

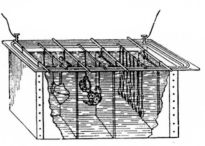

Fig. 90.

or other generator of electricity. The inner rectangle should be of stout copper rod, or wire—usually about one-half the thickness of the former, and is also provided with a binding screw at one corner, to connect it with the negative pole of the battery. A series of brass rods, from half to one inch in diameter, and each about the length of the tank's width, are laid across the outer rectangle, and from these are suspended the silver anodes ; similar but shorter rods of brass are placed between each pair of anodes, and rest upon the inner rectangle ; from these rods the articles to be plated are suspended, as shown in the engraving. Before the respective conducting rods are placed in position, they must be thoroughly well cleaned with emery cloth, as also must be the rectangular conductors and wire holes of binding screws which are to receive the positive and negative conducting wires, the ends of which must likewise be cleaned with emery cloth each time before making connection with the battery. It may be well here to remark that *all* the points of connection between the various rods, wires, and binding screws must be kept perfectly clean, otherwise the electric current will be obstructed in its passage. When the conducting rods become foul by being splashed with the cyanide solution, they should be well cleaned with emery cloth, and the operation of cleaning these rods should always be performed each morning before the first batch of work is placed in the solution ; the emery cloth should only be applied when the conducting rods are perfectly dry. It is always a characteristic of a really good plater that all his conducting rods are kept bright and clean, and every appliance in its proper place.

Plating Battery.—The most useful form of battery for depositing silver, either upon a large or small scale, is a modification of the

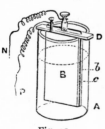

Fig. 91.

Wollaston battery shown in Fig. 91. For depositing upon a large scale, a stone jar capable of holding about ten gallons forms the battery cell. A bar of wood, D, having a groove cut in it, so as to allow a stout plate of zinc to pass freely through it, rests across the battery jar, A. Two sheets of copper, B b, connected by strips of the same metal soldered to the upper corners, are placed over the wooden bar, and a binding screw connected to one of the copper plates, either by means of solder or by a side screw. The copper plates should nearly reach to the bottom of the jar. A suitable binding screw is attached to the zinc plate, c, which must be well amalgamated. The exciting fluid consists of dilute sulphuric acid, in the proportion of one part of the latter to fifteen parts of cold water. To prevent the zinc from coming in contact with the copper plates, a small block of wood, having a tolerably deep groove of the same width as the thickness of the sheet of copper, may be fixed on to each edge of the pair of plates about midway between the top and the bottom. In order to regulate the amount of current in working these batteries, it is commonly the practice to drill a hole in the centre of the upper part of the zinc plate, to which a strong cord is attached, and allowed to pass over a pulley, the other end of the cord being connected to a counterweight. A windlass arrangement, as in Fig. 92, is also used for this purpose, by which the zinc plate can be raised and lowered by simply turning a handle connected to a revolving spindle, supported by uprights of wood, round which the cord becomes wound or unwound according to the motion given to the handle.

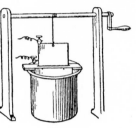

Fig. 92.

When the bath is about to be filled with work, the zinc plate should only be lowered a short distance into the acid solution, and the surface is to be increased as the filling of the bath progresses ; if this precaution is not observed, the deposition will take place too rapidly upon the work, and the deposited metal will assume a grey colour instead of the characteristic white, besides which the silver will be liable to strip or separate from the underlying metal in the subsequent processes of scratch-brushing and burnishing, or even under the less severe process of polishing. A very safe way to check the too rapid deposit, is to suspend an anode from the negative conducting rod as a *cathode* when the first batch of articles is being placed in the bath. When very powerful batteries or dynamo-machines are used, the resistance coil (*vide* Nickel-plating) must be employed.

Motion given to Articles while in the Bath.—In order to insure uniformity of deposit while employing strong electric power from magneto or dynamo machines, it has been found that by keeping the articles slowly in motion while deposition is taking place, this desirable end can be effectually attained. There are several ingenious devices adopted for this purpose, to several of which we may now direct attention. It is a fact that deposition takes place first at the *extreme end* of the article in solution—that is the point *farthest from* the source of electricity ; * and this being so, we may be sure that the deposition

* Except in *electrotyping*, in which case the surface which receives the deposit (plumbago) is but an indifferent conductor of electricity.

progresses in the same ratio during the whole time the articles are receiving the deposited metal provided the solution and the work remain undisturbed. Indeed, in the case of table forks, if we suffer them to remain, with their prongs *downward*, undisturbed for a considerable time, we shall find, on removing them from the bath, that the prongs, from the extreme tips upward, will be coated with a crystalline or granular deposit, while the extreme upper portion of the article will be but poorly coated. In no case is the fact of the deposit taking place from the lowest part of an article upward more practically illustrated than in the process of " stripping " (to which we shall refer hereafter) or dissolving the silver from the surface of plated articles, when, after they have been in the stripping solution for some time, we find that the *last* particles of silver which will yield to the

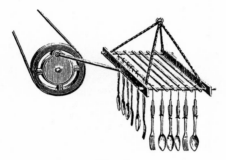

Fig. 93.

chemical action of the liquid are the points of the prongs of a fork, the lowest part of the bowl of a spoon, as also (if the articles have been duly shifted during the plating) the extreme ends of the handles of either article.

To keep the articles in gentle motion while in the bath, one method is to connect the suspending rods to a frame of iron, having four wheels about three inches in diameter connected to it, which slowly travel to and fro to the extent of three or four inches upon inclined rails attached to the upper edges of the tank, the motion, which is both horizontal and vertical, being given by means of an eccentric wheel driven by steam power. By another arrangement, the articles are suspended from a frame (as in Fig. 93), and the motion given by the eccentric wheel as shown in the engraving. The simplicity of the former arrangement, however, will be at once apparent.

Cruet Stands, &c.—Before being submitted to the cleansing operations, *quicking*, &c., before described, the " wires " of cruet and liqueur stands must be separated from the bottoms, to which they are generally connected by small nuts, and these latter should be slung upon a wire and laid aside until the other parts of the article are ready for plating. A wire is then to be connected to each part of the cruet frame, and these are then to be immersed in the hot potash liquor, being left therein sufficiently long to dissolve or loosen any greasy matter which may attach to them. After being rinsed, they are to be well brushed with powdered pumice and water. The brushes used for this and similar purposes are made from hog hair, and are supplied with one or more rows, to suit the various purposes for which they are required ; for example, a one-rowed brush is very useful for cleaning the joints connecting the rings with the framework of cruet stands, as also for all crevices which cannot be reached by a wider tool ; a two-rowed brush is useful for crevices of greater extent and for hollows ; and three, four, five, and six-rowed brushes for flat surfaces, embossed work, and so on. One of these useful tools is shown in Fig. 94.

Fig. 94.

After scouring and rinsing, the parts of the cruet stand or liqueur stand are to be immersed in the quicking solution until uniformly white in every part, after which they must be well rinsed and immediately put into the plating bath ; after a short immersion, the pieces should be gently shaken, so as to shift the slinging wire from its point of contact, and thus enable that spot to become coated with silver ; it is always advisable to repeatedly change the position of the wire so as to avoid the formation of what is termed a *wire mark*, and which is of course due to the deposit not taking place at the spot where the wire touches the article, thereby leaving a depression when the article is fully plated. The flat base of the cruet stand should be suspended by two wires, each being passed through one of the holes at the corner, and it should be slung sideways and not lengthwise ; its position in the bath should be reversed occasionally, so as to render the deposit as uniform as possible ; the same observation applies to the " wire " part of the cruet stand. When *mounts* are sent with the cruet stand, not separate, but cemented to the cruets, which is often the case, it will be well, if it can be conveniently done, to remove the pin which connects the top or cover with the rim of the mustard mount, so as to plate these parts separately, otherwise the cover will require shifting repeatedly in order to allow those parts of the joint which are protected from receiving the deposit when the cover is *open*, to become duly coated.

Tea and Coffee Services.—Like the foregoing articles, these are of very variable design, and are either plain, chased and embossed, or simply engraved. Unless sent direct from the manufacturer in the proper condition for plating—that is, with their handles and covers unfixed—it will be better to remove the pins connecting these parts with the bodies of tea and coffee pots before doing anything else to them, unless, as is sometimes the case, they are so well riveted as to render their severance a matter of difficulty. The disadvantages attending the plating of these vessels with their handles and lids on are that the solution is apt to get inside the sockets of the handles, and to ooze out at the joints when the article is finished, while the joint which unites the lid with the body can only be properly plated when the lid is shut, at which time the interior of the lid can receive no deposit. When sent to the plater by the manufacturer, the various parts are usually either separate, or merely held together by long pins, which may readily be withdrawn by a pair of pliers, and the parts again put together in the same way when the articles are plated and finished—that is burnished or polished, as the case may be.

In plating work of this description, the articles are potashed, scoured and quicked as before, and when ready for the plating bath, the tea and coffee pots are generally *wired* by passing the slinging wire through the rivet-holes of the joints ; but in order to equalise the deposit as far as possible, it is a good plan, after the article has received a certain amount of deposit, to make a loop at one end of a copper wire, and to pass it under one of the feet of the teapot, then to raise the vessel somewhat, and connect the other end of the wire with the conducting rod ; care must be taken, however, not to let the wire touch the body of the vessel, or if it does so, to shift it frequently.

Since deposition always takes place more fully at the *points* and projections of an article, it will be readily understood that the interiors of vessels—being also *out of electrical sight*, so to speak, of the anodes—will receive little if any deposit of silver. This being the case, if we wish to do the work thoroughly well in every part, it will be necessary to deposit a coating of silver upon the inside either before or after the exterior has been plated. To do this, the vessel being well cleaned inside, is placed upright on a level bench, and a wire connected to the negative pole of the battery is slipped through the joint as before. A small silver anode, being either a strip of the metal or a narrow cylinder, is to be attached to the positive pole, and the anode lowered into the hollow of the vessel, care being taken that it does not touch in any part. The vessel is then to be filled to the top with silver solution dipped out of the bath with a jug, and the whole allowed to rest for half an hour or so, at the end of which time the interior will generally have received a sufficient coating of silver.

Scratch brushing.—One of the most important mechanical operations connected with silver-plating is that of scratch-brushing. For this purpose skeins of thin brass wire, bound round with stout brass or copper wire (Fig. 97), are used. When the plated articles are removed from the bath, they present a pearly white appearance not unlike very fine porcelain ware, but still more closely resembling standard silver that has been heated and pickled in dilute sulphuric acid, as in the process of *whitening* watch dials. The dead white lustre of electro-deposited silver is due to the metal being deposited in a crystalline form, and the dulness is of so fugitive a nature that even scratching the surface with the finger nail will render the part more or less bright by burnishing the soft and delicate crystalline texture of the deposit. The object of scratch-brushing is to obliterate the white " burr," as it is called, before the work is placed in the hands of the burnisher or polisher, otherwise it would be apt to show in such parts of the finished article as could not be reached by the tools employed in those operations. As

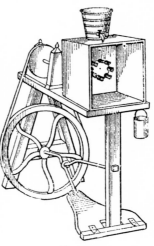

Fig. 95.

in the case of gilding, the revolving scratch-brushes are kept constantly wetted by a thin stream of stale beer, or half beer and water, supplied, by means of a tap, from a small vessel (which may conveniently be a wooden bucket) placed on the top of the scratch-brush box. A tin can, or other light vessel, stands upon the floor, beneath the box, to catch the beer runnings, which escape through a pipe let into a hole in the bottom of the box. A still more handy plan is to have a small hook fixed below the right-hand corner of the scratch-brush box, for supporting a tin can or other vessel : and by giving the box a slight inclination forwards, and towards the right-hand corner, the liquor will flow out through a hole at the corner, in

which a short piece of lead pipe should be inserted. By this arrangement (Fig. 95), the workman can empty the can into the vessel above,
whenever the beer liquor ceases to drip upon the scratch-brushes,
without allowing the driving wheel to stop. Much time may be saved
in this way, especially when the liquid happens to run short, at which
time the can requires to be emptied
frequently. To prevent the beer
runnings from overflowing, and
thus making a mess on the floor,
while wasting the liquor, no more
liquor should be put into the
cistern above than the vessel below will contain. A quart or

Fig. 96.

three-pint can full will be quite sufficient for ordinary work, and a
vessel of this latter capacity will be quite as large as the workman
can manipulate readily without stopping the lathe.

The lathe scratch-brush consists of a series of six or eight scratch-
brushes (according to the number of grooves in the " chuck ") bound
to the chuck by strong cord, as in Fig. 96.   Previous to
fixing the brushes, the skein of fine brass wire forming a
single scratch-brush, Fig. 97, is to be cut with a pair of
shears or strong scissors.   Before applying the compound
brush—which is connected to the lathe-head by means of its
screwed socket—to the plated work, the brushes should be
opened, or spread, by pressing rather hard upon them, while
revolving, with a piece of stout metal, or the handle of one
of the cleaning brushes ; this will spread the bundles of wire
into a brush-like form suitable for the purpose to which they
are to be applied.   It may be well to state that the revolving
scratch-brush should on no account be applied to the work in
a dry state, but only when the beer liquor is running sufficiently free to keep the brushes *wet*.

In working the scratch-brush, it must be allowed to revolve to the right of the operator, otherwise the " chuck " will
be liable to come unscrewed ; moreover, this is the most convenient motion for enabling the workman to guide the articles
without risk of their being jerked out of his hand—an accident that might readily occur if he inadvertently turned the
wheel the wrong way. In scratch-brushing spoons and forks,
a very moderate pressure is all that is necessary to render
the surface bright ; a little more pressure, however, is required
for the edges of salvers, dishes, handles and feet of cruet stands, and
other work in which hollows of some depth form a necessary feature
of the ornamental mounts.

Fig. 97.

Plating by Dynamo-Electricity.—In the larger electro-plating establishments, magneto or dynamo-electric machines are employed, and the current from these powerful machines is conveyed by stout leading wires to the various baths, the force of the current entering the baths being regulated by resistances. In works of moderate dimensions, a good machine, either of the magneto or dynamo-electric type, will supply sufficient electricity to work a large bath of each of the following solutions : nickel, silver, brass and copper, as also a good-sized gold bath. In working with these machines, it is of the greatest importance that they should be driven at an uniform speed ; and though some machines require to be driven at a higher rate of speed than others, the maximum allowed by the respective makers should never be exceeded, or the machine may become considerably heated and seriously injured. When starting the machine, the number of its revolutions should be ascertained by means of the *speed indicator* referred to elsewhere, and as far as practicable the normal speed should be maintained without sensible variation while the current is passing into the vats. Although this uniformity of speed is more certainly obtained, we believe, with gas engines than with steam power, if proper care and attention are given, and frequent examination of the speed of the dynamo-armature made by the plater, tolerable regularity may be attained from the latter source of power. It must always be remembered by the plater, that when the engine which drives the dynamo is also employed for driving polishing lathes, emery wheels, &c., when very heavy pieces are being treated in the polishing shop the speed of the dynamo may be greatly influenced ; indeed we have frequently known the belt to be suddenly thrown off the pulley of a dynamo from this cause, and the machine, of course, brought to a full stop.

CHAPTER XVI.

ELECTRO-DEPOSITION OF SILVER (*continued*).

Plating Britannia Metal, &c.—Plating Zinc, Iron, &c.—Replating Old Work. --Preparation of Old Plated Ware.—Stripping Silver from Old Plated Articles.—Stripping Gold from Old Plated Articles.—Hand Polishing. —Resilvering Electro-plate.—Characteristics of Electro-plate.—Depositing Silver by Weight.—Roseleur's Argyrometric Scale.—Solid Silver Deposits.—On the Thickness of Electro-deposited Silver.—Pyro-plating. —Whitening Electro-plated Articles.—Whitening Silver Work.

Plating Britannia Metal, &c.—It was formerly the practice to give a coating of copper or brass to articles made from Britannia metal, tin, lead, or pewter, since it was found difficult otherwise to plate such metals and alloys successfully, that is without being liable to strip. It is usual now, however, to immerse the articles first in the hot potash solution, and to place them, with or without previous rinsing, in the depositing-bath. Since the potash bath dissolves a small quantity of metal from the surface of articles made from these metals, a favourable surface is left for the reception of the silver deposit, to which the metal adheres tolerably well—indeed sufficiently so to bear the pressure of the burnishing tools. Since Britannia metal, pewter, &c., are not such good conductors of electricity as German silver, copper, or brass, an energetic current must be applied when the articles are first immersed in the bath, and when the whole surface of each article is perfectly coated with silver, the amount of current may be somewhat diminished for a time, and again augmented as the deposit becomes stouter ; care being taken not to employ too strong a current, however, in any stage of the plating process. It may be mentioned that articles made from Britannia metal—which are generally sold at a very low price—are seldom honoured with more than a mere film of silver, in fact just so much as will render them marketable, and no more ; still, however, a very extensive trade is done in work of this description, much of which presents an exceedingly creditable appearance.

Plating Zinc, Iron, &c.—To coat these metals with silver, it is best to first give them a slight coating of brass or copper, in an alkaline solution, which does not occupy much time, neither is it a costly proceeding. Both these metals adhere pretty firmly to zinc, iron, and steel, while silver attaches itself freely to brass and copper. If hot solutions of copper or brass are used, the trifling deposit required to enable the subsequent coating of silver to adhere to the zinc, &c., can be obtained in a very few minutes. Each operation, however, should follow in quick and unbroken succession, for if the brass or copper-coated article be allowed to remain, even for a few seconds, in the air before being placed in the silver bath, it will rapidly oxidise, and render the deposited silver liable to strip when the article is scratch-brushed. Moreover, if the brassed or coppered articles are allowed to remain for a short time in the air while in a moist condition, voltaic action will be set up between the zinc and the metallic covering, by which the latter will become loosened, and will readily peel off under the action of the scratch-brush. Each article, after being brassed or coppered, should, after rinsing, be placed at once in the silvering-bath.

Replating Old Work.—Under this head must be considered not only the old Sheffield and Birmingham ware, the manufacture of which became superseded by the electro-plating process, but also the more modern article known as "electro-plate" (the basis of which is German silver), which has, by domestic use, become unsightly in consequence of the silver having worn off the edges and other prominent parts most subject to friction in the process of cleaning. In the business of replating, there must ever be a constant if not a growing trade, if we consider the enormous quantity of plated goods which annually flow into the market, and which must—even the best of it—require resilvering at some time or other, while the inferior classes of goods may require the services of the electro-plater at a much earlier period than the purchaser of the articles expected.

Preparation of Old "Plated" Ware for Resilvering.—These articles, whether of Sheffield or Birmingham manufacture, have a basis of copper. The better class of plated ware, which was originally sold at about half the price of standard silver, and some of which may be occasionally met with, though doubtless becoming rarer every year, is of most excellent quality, both as to design and workmanship, and when properly prepared for replating, and well silvered and finished after, is well worthy of being replaced upon the table by the side of the more modern articles of electro-plate. Such articles, however, should never be replated with an insignificant coating of silver, since the copper surface beneath would soon reappear and expose the indifferent quality of the plater's work. It may be well to state, however, that by far the greater proportion of old "plated" articles are not of the same quality as the old Sheffield plate and the equally admirable work formerly manufactured by the distinguished firm of Boulton and

Watt, of Birmingham, some specimens of which may also be occasionally met with ; but a very inferior class of goods, which may generally be recognised by their having lost nearly the whole of their silver covering—which was never very much—whereas in the better class of old plated ware the silver has worn off chiefly at the extreme edges, while the remainder of the article retains a sound coating of silver.

In preparing old plated cruet frames, &c., for replating, the wires, which are generally attached by soft solder to the stands, must be separated by first scraping the solder clean, and then applying a hot soldering-iron (using a little powdered resin), which must be done very carefully, otherwise the solder which connects the feet of the stand may become melted, causing them to drop off ; it is safer, when applying the hot iron, to have an assistant at hand, who with a brush or hare's foot should wipe away the solder from the joint when it is melted. All the joints being treated in this way, in the first instance, the ground is cleared, when by a fresh application of the soldering-iron the legs of the wire may be loosened, one at a time, until the whole series have become partially displaced, after which, by again applying the hot iron, the legs, one after another, may be forced out. If the two parts of the frame are not taken asunder in this cautious way, the workman may involve himself in much trouble from the melting of the lead mounts (called " silver " mounts), the dropping off of legs, feet, &c., all of which may be avoided in the way we have suggested. It must be understood that our suggestions are specially made for the guidance of those who, though good platers, may not be experts in the application of the soldering-iron. It is usually the practice to remove what silver there may be upon old plated articles by the process termed " stripping." This consists in immersing the article in a hot acid liquid which, while dissolving the silver from the surface, acts but little upon the underlying metal, whether it be of copper, brass, or German silver. The process of *stripping* being an important auxiliary in connection with the replating of old work, as also in cases in which an unsuccessful deposit has been obtained upon new work, we may advantageously describe the process at once ; but previous to doing so, we may state that the silver removed by stripping from the better class of old plated articles is sometimes an important gain to the electro-plater, if he be fortunate enough to receive a liberal amount of such work, while, on the other hand, the inferior qualities of plated ware will yield him no such satisfaction.

Stripping Silver from Old Plated Articles.—A *stripping-bath* is first made by pouring a sufficient quantity of strong oil of vitriol into a suitable stoneware vessel, which must be made hot, either by means of a sand bath, or in any other convenient way. To this must be added a small quantity of either nitrate of potash, or nitrate of soda, and the mixture stirred with a stout glass rod until the salts are dissolved. The article to be stripped is first slung upon a stout copper wire; it is then to be lowered in the liquid, being held by the wire, until wholly immersed. Leave the article thus for a few moments, then raise it out of the solution, and observe if the silver has been partially removed; then redip the article and leave it in the bath for a short time longer, then examine it again; if the action appears rather slow, add a little more nitre, and again immerse the article. When the silver appears to be dissolving off pretty freely, the operation must be watched with care, by dipping the article up and down in the solution, and looking at it occasionally, and the operation must be kept up until all the silver has disappeared, leaving a bare copper surface. When a large number of articles have to be stripped, a good many of these may be placed in a hot acid bath at the same time, but since they will doubtless vary greatly in the proportion of silver upon them, they should be constantly examined, and those which are first stripped, or *desilvered*, must be at once removed and plunged into cold water. When all the articles are thoroughly freed from silver, and well rinsed, they are to be prepared for plating by first *buffing* them, as described in the chapter on polishing, after which they are cleaned and quicked in the same way as new work.

A Cold Stripping Solution, which is not so quick in its action as the former, is made by putting in a stoneware vessel a quantity of strong sulphuric acid, to which is added concentrated nitric acid in the proportion of 1 part of the latter acid to 10 parts of the former (by measure). In this mixture the articles are suspended until they give signs of being nearly deprived of their silver, when they are somewhat more closely attended to until the removal of the silver is complete, when they are at once placed in cold water. The articles must be perfectly dry when placed in this stripping liquid, since the presence of even a small quantity of water will cause the acid to attack the copper, brass, or German silver, of which the articles may be made. The vessel should also be kept constantly covered, since sulphuric acid attracts moisture from the air. The silver may be recovered from old stripping solutions by either of the methods described elsewhere.

Buffing Old Work after Stripping.—The stripped articles, after being thoroughly well rinsed and dried, are sent to the polishing shop, where they are buffed and finished, and the cavities, caused by the action of vinegar or other condiments upon the base of cruet stands, as far as possible removed. Sometimes these depressions are so deep that they cannot be wholly removed without rendering the surface so thin that, in burnishing this portion of the article, it is liable to warp or become stretched, rendering the flat surface unsightly for ever after. The back of the stand, which is usually coated with tin, should be roughly "bobbed" with sand until all the tin is removed. The next items, which usually give some trouble, are the so-called "silver mounts," which are commonly of two kinds. The edge, or border of the stand, being originally a shell of silver foil, struck in design, and filled or backed up with lead or solder, is generally more or less free from silver, except in the hollows ; and since the soft metal does not receive the silver deposit so favourably as the metal of which the rest of the article is composed, these edges must receive special treatment, otherwise the silver deposited upon them will be brushed off in the after-process of scratch-brushing. There are several ways of treating "lead edges," as they are properly called. Some persons remove them altogether, and replace them by brass mounts, which are specially sold for this purpose. If this plan be not adopted, we must endeavour to induce the silver to adhere to the lead mounts by some means or other. The edge of the article, after being cleaned, may be suspended, one angle at a time, in a brassing bath, or alkaline coppering solution, until a film of either metal is deposited upon the leaden mount, when, after being rinsed, a second angle may be treated in the same way, and so on, until the entire edge is brassed or coppered. The small amount of brass or copper, as the case may be, which may have deposited upon the plain portions of the work, may be removed by means of a soft piece of wood, powdered pumice, and water. Edges treated in this way generally receive a good adherent coating of silver. Sometimes, but not always, the ordinary "quicking" will assist the adhesion of the silver to the lead mounts. Another method of depositing a firm coating of copper upon lead edges is to put a weak acid solution of sulphate of copper in a shallow vessel, and having a small piece of iron rod in one hand, to lower one portion of the edge of the cruet bottom into the solution ; then touching the article under the liquid, in a short time a bright coating of copper will be deposited upon the leaden surfaces, by means of the voltaic action thus set up, when this portion may be rinsed, and the remainder treated in the same way. Or take a small piece of copper, and connect it by a wire to the positive electrode of a battery, envelop this copper in a piece of chamois leather or rag, then put the article in connection with the negative electrode. By dipping the pad, or "doctor," in either an acid or an alkaline solution of copper, or in a warm brassing solution, and applying it to the part required to be coated, a deposit will at once take place, which may be strengthened by repeatedly dipping the pad in the solution and applying again. In this way, by moving the pad containing the small anode of copper or brass along the edge, the required deposit may be effected in a very short time with a battery of good power—a Bunsen cell, for example.

Old "plated" tea and coffee pots are invariably coated inside with tin ; and if this part of the article is required to be silvered—which is sometimes, though not always, the case—the tin should first be removed by dissolving it in some menstruum which will not dissolve the copper beneath. For this purpose either hydrochloric acid or a solution of caustic potash may be used. If the former, the inside of the vessel should first be filled with a boiling hot solution of potash, and after a time the liquid is to be poured out and thoroughly rinsed. It must then be filled with strong muriatic acid, and allowed to rest until the upper surface, upon being rubbed with a strip of wood, exposes the copper, when the acid is to be poured out, and the vessel again rinsed. The inside must now be cleaned by brushing with silver sand and water as far as the brush will reach, when the bottom and hollow parts of the body may be scoured with a mop made with rag or pieces of cloth and silver sand. If it is preferred to dissolve the tin from the inside of the vessel by means of potash, the hot liquid must be poured in as before, and the vessel placed where the heat can be kept up until the desired object—the removal of the tin—is attained, when the vessel must be cleaned as before. Dissolving the tin from the inside of such old plated articles should be the first preparatory process they are subjected to ; indeed, the interiors of all vessels to be electro-plated should be attended to first, in all the preliminary operations, but more especially in the operations of scouring, in which the *handling* of the outside, though a necessity, is liable to cause the work to strip (especially in nickel-plating), unless the hands are kept well charged with the pumice or other gritty matter used in scouring. To remove tin from copper surfaces, a hot solution of *perchloride of iron* may also be used, for although this iron salt acts freely upon copper, voltaic action is at once set up when the two metals, tin and copper, come in contact with the hot solution of the perchloride, which quickly loosens the tin so that it may be brushed away with perfect ease. From the rapidity of its action, we should prefer to adopt the latter mode of de-tinning copper articles, but either of the former would be safest in the hands of careless or inexperienced manipulators.

Old "plated "—we use the term in reference to Sheffield ware more especially—sugar-bowls, cream-ewers, mugs, goblets, &c., which have been gilt inside, should have what gold may still remain upon the article " stripped off " before other operations are proceeded with ; and since these articles were originally *mercury gilt*, in which a liberal amount of gold was often employed, it is frequently worth while to remove this by dissolving it from the insides of the vessels ; and the same practice should be adopted with all silver-gilt articles which are merely required to be whitened, to which we shall refer in another place.

Stripping Gold from the Insides of Plated Articles.—The sugar-bowl or other vessel is placed on a level table or bench, and put in connection with the positive electrode of a battery. A strip of sheet copper or platinum foil is next to be attached to the negative electrode, and placed inside the vessel, without touching at any point. By this arrangement the article becomes an *anode*. The vessel must now be filled with a moderately strong solution of cyanide of potassium, consisting of about 4 ounces of cyanide to 1 quart of water. Since the metal beneath will also dissolve in the cyanide solution, the operation must be stopped as soon so the gold has disappeared from the surface. The solution should then be poured out, and bottled for future use. When the stripping solution, from frequent use, has acquired sufficient gold to make it worth while to do so, the metal may be extracted by any of the processes given in another chapter.

Old plated table candlesticks, some of which are of admirable design and well put together, may be occasionally met with, as also a very inferior article, the parts of which are mainly held together by a lining or "filling" of pitch, or some resinous compound. In treating old plated candlesticks, the removal of the *filling* should be the first consideration, since it will give the plater a vast amount of after trouble if he attempts to plate them while the resinous or other matter remains in the interior. In the first place, the silver solution will be sure to find its way into the hollow of the article, from which it will be next to impossible to entirely extract it when the article is plated, for the liquid will continue to slowly exude for days, or even weeks, after the article is finished. Again, if the article be plated without removing the filling material, this, being freely acted upon by the cyanide solution, will surely harm it. After removing the socket, the green baize or cloth should be removed from the base of the candlestick, when it should be placed before a fire until the whole of the resinous matter or pitch has run out. To facilitate this, the article should be slightly inclined in an iron tray or other vessel, so that the resinous matter may freely ooze out and be collected. In dealing with the inferior varieties of candlesticks—which may be known by *all* or nearly all the silver having worn from their surface—the plater may find, to his chagrin, that before all the stuffing has run out the candlestick will have literally fallen to pieces. The various parts, not having been originally put together with solder, but held in position merely by the filling material, readily come asunder when the internal lining is loosened. In such a case as this he should, without losing his temper (if possible), determine to prepare and plate all the parts separately (keeping the parts of each " stick " together), and after scratch-brushing, carefully put them together again. The candlestick should now be turned upside down, and held in this position by an assistant, while a sufficient quantity of pitch (previously melted in an earthenware pipkin) is poured in. The candlestick must be left in the erect position until the filling has nearly set, when the hollow formed by the contraction of this substance must be filled up with the same material, and the article then left until quite cold, when it may be handed over to the burnisher. When burnished, the surface of the pitch should be levelled with a hot iron, and then at once brought in contact with a piece of green baize, placed upon a table, and gentle pressure applied to cause the uniform adhesion of the two surfaces. When cold, the remainder of the baize is cut away by means of a sharp pair of scissors, when, after being wiped with a clean or slightly rouged chamois leather, the article is finished.

Hand Polishing.—When the electro-plater is unprovided with a proper polishing lathe and the various appliances ordinarily used in polishing metals, he must have recourse to the best substitute he can command for polishing by hand. To aid those who may be thus circumstanced, and who may have no special knowledge of the means by which the rough surfaces of old work may be rendered sufficiently smooth for replating, we will give the following hints : Procure a few sheets of emery-cloth, from numbers 0 to 2 inclusive ; one or two lumps of pumice-stone ; a piece of Water-of-Ayr stone, about ¾ inch square and 5 inches long ; also a little good rottenstone, and a small quantity of sweet oil. Suppose it is necessary to render smooth the base, or stand, of an old cruet frame, deeply marked on its plane surface by the corrosive, or, rather, voltaic action of the vinegar dropping from the cruets upon the plated surface. The article, after being *stripped*, as before, should be laid upon a solid bench, and a lump of pumice (previously rubbed flat upon its broadest part) frequently dipped in water and well rubbed over the whole surface, that is, not merely where the cavities are most visible, but all over. After thus rubbing for some time, the stand is to be rinsed, so that the operator may see how far his labour has succeeded in reducing the depth of the "pit-marks." The stoning must then be resumed, and when the surface appears tolerably uniform, the article should be well rinsed, dried, and again examined, when if the marks are considerably obliterated, a piece of No. 2 emery-cloth may be briskly applied to the surface by being placed over a large cork or bung, after which a finer emery-cloth should be applied. The article should next be *thoroughly* rinsed, and brushed with water to remove all particles of emery ; and while still wet, the Water-of-Ayr stone must be rubbed over the surface. The stone should be held in an inclined position, frequently dipped in water, and passed from end to end of the article. The effect of this will be—and *must* be—to remove all the scratches or marks produced by the pumice and emery-cloth. Until these have disappeared, the smooth but keenly-cutting stone must be applied. After having rendered the surface perfectly smooth, the article is to be again rinsed and dried.   It must now be briskly rubbed with rotten-stone, moistened with oil, and applied with a piece of buff, or belt (such as soldiers' belts are made of), glued to a piece of wood.   When sufficiently rubbed or buffed with the rottenstone, the surface will be bright, and in order to ascertain how the work progresses, it should occasionally be wiped with a piece of rag.   In very old plated articles, the pit-holes are frequently so deep that to entirely obliterate them would render the metal so thin as to spoil the article.   It is better, therefore, not to go too far in this respect, and to trust to the cruets, when in their places, disguising whatever remains of the blemishes, after the foregoing treatment, rather than to endanger the solidity of the stand itself.   By employing pieces of pumice of various sizes (keeping the flattened piece for *plane* surfaces), strips of emery-cloth folded over pieces of soft wood, Water-of-Ayr stone, and ordinary hand " buff-sticks " of various kinds, the "wires" of old cruet and liqueur frames may be rendered smooth enough for plating.   With perseverance, and the necessary labour, many old articles may be put into a condition for plating by hand labour with very creditable results ; and it may be some consolation to those living at a distance from large towns, if we tell them that during the first ten years of the electro-plating art, the numerous host of " small men " had no other means of preparing their work for plating than those we have mentioned, many of whom have since become electro-depositors upon an extensive scale.

Resilvering Electro-plate.—This is quite a distinct class of ware from the preceding, inasmuch as the articles are manufactured from what is called *white metal*, in contradistinction to the basis of Sheffield plate, which, as we have said, is the red metal copper.   The better class of electro-plate is manufactured from a good quality of the alloy known as German silver, which, approaching nearly to its whiteness, does not become very distinctly visible when the silver has worn from its surface.   Inferior qualities of this alloy, however, are extensively used for the manufacture of cheap electro-plate, which is very little superior, as far as colour goes, to pale brass, while the latter alloy is also employed in the production of a still lower class of work.   The comparatively soft alloy, of a greyish-white hue, called *Britannia metal*, is also extensively adopted as a base for electro-plate of a very showy and cheap description, of which enormous quantities enter the market, and adorn the shop-windows of our ironmongers and other dealers in cheap electro-plated goods.   To determine whether an electro-plated article has been manufactured from a hard alloy, such as German silver, or from the soft alloy Britannia metal, it is only necessary to strike the article with any hard substance, when a ringing, vibratory sound will be produced in the former case, while a dull, unmusical sound, with but little vibration, will be observed in the latter.

Characteristics of Electro-Plate.—Electro-plated articles of the best quality are invariably *hard-soldered* in all their parts ; the wires of cruet and liquor frames are attached by German silver nuts to the screwed uprights, or feet of the wires, instead of pewter solder, as in plated ware, and the bottoms of the stands are coated with silver, instead of being tinned, as in the former case. The mounts are of the same material as the rest of the article, and the handles and feet of cream-ewers and sugar-bowls are frequently of solid cast German silver. With these advantages the electro-plater should have little difficulty, if the articles have received fair treatment in use, in re-plating them and turning them out nearly equal to new, which it should be his endeavour to do. It sometimes occurs that "ship plate"— that is, plated work which has been used on board ship—when it reaches the hands of the electro-plater exhibits signs of very rough usage ; corner dishes are battered and full of indentations, while the flat surfaces of the insides are scored with cuts and scratches, suggestive of their having been frequently used as plates, instead of mere receptacles for vegetables ; the prongs of the forks, too, are frequently notched, cut, and bent to a deplorable extent. All these blemishes, however, must be removed by proper mechanical treatment, after the remaining silver has been removed by the stripping-bath. It is not unusual for those who contract for the replating of ship work to pay *by the ounce* for the silver deposited, in which case they will not allow the electro-plater to reap the full advantage of the old silver removed by stripping, but will demand an allowance in their favour, which, if too readily agreed to by an inexperienced plater, might greatly diminish his profit if the cost of buffing the articles happened to be unusually heavy ; he must, therefore, be upon his guard when undertaking work of this description for the first time, since, otherwise, he may suffer considerable loss, for which the present rate of payment for each ounce of silver deposited will not compensate.

After stripping and rinsing, the articles require to be well polished, or buffed, and rendered as nearly equal to new work as possible ; they are then to be potashed, quicked, plated, and finished in the same way as new goods. Since there is now a vast quantity of nickel-plated work in the market, some of which is exceedingly white even for nickel, inexperienced or weak-sighted platers must be careful not to mistake such articles for silver-plated work. When in doubt, applying a single drop of nitric acid, which blackens silver while producing no immediate effect upon nickel, will soon set the mind at rest upon this point.

Electro-tinned articles, which very much resemble silvered work, may also be detected in this way. We are tempted to make one other suggestion upon this subject, which may not be deemed out of place, it is this: a considerable quantity of nickel-plated German silver spoons and forks are entering the market, which, should they eventually fall into the hands of the electro-plater to be coated with silver, may cause him some trouble if he inadvertently treats them as *German silver work*, which in his haste he might possibly do, and attempts to render them smooth for plating by the ordinary methods of hand or lathe-buffing; the extreme hardness of nickel—even as compared with German silver—will render his work not only laborious, but unnecessary, for if he were aware of the true nature of the surface he would naturally remove the nickel by means of a stripping solution, and then treat the article as ordinary German silver work. The stripping solution for this purpose will be given when treating of nickel *re*-plating. It must be understood that in making suggestions of this nature, in passing, that they are intended for the guidance of those who may not have had the advantages of much practical experience, of whom there are many in every art.

Depositing Silver by Weight.—In this country the silver deposit is frequently paid for by weight, the articles being carefully weighed both before and after being placed in the plater's hands. The price charged for depositing silver by the ounce was formerly as high as 14s. 6d.; at the present period, however, about 8s. per ounce only could be obtained, and in some cases even less has been charged. But unless dynamo-electricity be employed this would be about as profitable as giving ten shillings for half-a-sovereign. In France electro-plating is regulated by law, all manufacturers being required to weigh each article, when ready for plating, in the presence of a comptroller appointed by the Government, and to report the same article for weighing again after plating. In this way the comptroller knows to a fraction the amount of precious metal that has been added, and puts his mark upon the wares accordingly, so that every purchaser may know at a glance what he is buying. In Birmingham there is a class of electro-depositors called "electro-platers to the trade," who work exclusively for manufacturers of plated goods and others who, though platers, send a great portion of their work to the "trade" electro-platers, whose extensive and more complete arrangements enable them to deposit large quantities of the precious metals with considerable economy and dispatch.

In depositing silver at so much per ounce, the weighed articles, after being cleaned, quicked, and rinsed, are put into the bath, in which they are allowed to remain until the plater deems it advisable to re-weigh them, when they are removed from the bath, rinsed in hot water, and placed in boxwood sawdust; they are then lightly brushed over to remove any sawdust that may adhere to them, and carefully weighed.   If still insufficiently coated the articles are again scratch-brushed, quicked, rinsed, and replaced in the bath ; the re-weighing and other operations being repeated as often as is necessary until the required deposit is obtained.   This is a tedious and trouble-some method, and is sometimes substituted by the following : Suppose a certain number of spoons and forks have been weighed for the plating-bath, one of these articles is selected as a test sample, and is weighed separately ; being placed in the bath with the others, it is removed from time to time and re-weighed, to determine the amount of silver it has acquired in the bath.   Thus if 24 dwts. of silver are required upon each dozen of spoons or forks, when the test sample has received about 2 dwts. of silver it is known that the rest have a like proportion, provided, of course, that each time it has been suspended in the bath the slinging wire and that part of the conduct-ing rod from which it was suspended were perfectly clean ; it is obvious, however, that even this method is open to a certain amount of doubt and uncertainty, if the workmen are otherwise than very careful.   To render the operation of depositing by weight more certain and less troublesome, some electro-platers in France adopt what is termed a " plating balance."   The articles are suspended from a frame connected to one end of the beam, and a scale pan, with its weights from the other end ; the balance, thus arranged, is placed in communi-cation with the negative electrode of the electric generator, and the anodes with the positive electrode.   When the articles, as spoons and forks, for example, are suspended from the frame, and immersed in the bath, counter-balancing weights are placed in the scale-pan.   A weight equivalent to the amount of silver to be deposited is then put into the pan, which, of course, throws the beam out of balance ; when the equilibrium becomes restored, by the weight of deposit upon the articles in solution, it is known that the operation is complete.   The plater usually employs scales for each bath, especially when silvering spoons and forks.   If preferred, the supporting frame may be circular, so that the soluble anode may be placed in the centre of the bath, and at equal distance from the articles.   The centre anode need not prevent the employment of other anodes round the sides of the vessel, so that the articles receive the action of the current in front and behind them. A sounding bell may be so connected that it will indicate the precise moment when the equilibrium of the scale takes place.   In working the silver baths for this purpose, the anode surface immersed in solu-tion is much greater than that of the articles.   When the solution loses its activity additions of cyanide of silver are given to it, and when the cyanide is found to have become partially converted into carbonate of potassa, hydrocyanic acid is added, which combines with carbonate, and liberates carbonic acid gas. This method is preferred to that of adding fresh cyanide, since an accumulation of the carbonated alkali retards the conductivity of the solution, as also does the hydrocyanic acid when added in excess.

Roseleur's Argyrometric Scale.—This is an automatic apparatus and is designed for obtaining deposits of silver "without supervision and with constant accuracy, and which spontaneously breaks the electric current when the operation is terminated." The apparatus is made in various sizes, suitable for small or large operations; Fig. 98 represents the apparatus to be employed for the latter purposes.

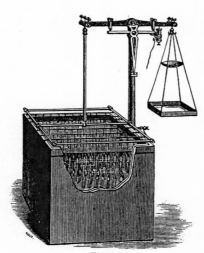

Fig. 98.

It consists of :—1. A wooden vat, the upper ledge of which carries a brass winding rod, having a binding screw at one end to receive the positive conducting wire of the battery; from this rod the anodes are suspended, which are entirely immersed in the solution, and communicate with cross brass rods by means of platinum wire hooks. These cross rods are flattened at their ends so that they may not roll, and at the same time have a better contact with the "winding rod." 2. A cast-iron column screwed at its base to one of the sides of the bath, carries near the top two projecting arms of cast iron, the extremities of which are vertical and forked, and may be opened or closed by iron clamps, these forks being intended to maintain the beam and prevent the knives from leaving their bowls when the beam oscillates too greatly. In the middle of the two arms are two bowls of polished steel, hollowed out wedge-shaped, to receive the beam knives. One arm of the pillar has at its end a horizontal iron ring, in which is fixed a heavy glass tube which supports and insulates a polished iron cup to contain mercury; beneath this cup is a small pad of india-rubber, which, by means of a screw beneath, may be raised or lowered, by which means the mercury in the cup is levelled. A second lateral binding screw connects the negative electrode of the battery. 3. A cast-iron beam, carrying in its centre two sharp polished steel knives ; at each end are two parallel steel bowls, separated by a notch, intended for the knives of the scale pan and of the frame for supporting the articles. One arm of the beam is furnished with a stout platinum wire, placed immediately above and in the centre of the mercury cup, and as the beam oscillates it dips into, or passes out of, the cup. The scale pan is furnished with two cast-steel knives fixed to the metallic bar, which is connected to chains supporting the lower wooden box for the *tare ;* the smaller pan, for the weight representing the amount of silver to be deposited, is placed between these two. 4. The frame for supporting the work is also suspended by two steel knives, the vertical of which is of stout brass tubing, and is equal in size to the opening of the bath, and supports the rods to which the articles are suspended. The slinging wires are formed into a loop at one end for supporting the spoons or forks, and the vertical portion of each wire is covered with india-rubber tubing, to prevent it from receiving the silver deposit.

In adjusting the apparatus, the pillar must be set perfectly upright by aid of a plumb line ; the clamps are then withdrawn from the *forks,* and the beam is carefully put in its place, care being taken to avoid injuring the knives that rest in the bowls in the centre of the pillar. The clamps are now replaced, and the beam should oscillate freely upon the knives without friction. The knives of the frame are next put in their places, as also those of the scale pan ; mercury is then poured into the six bowls, where the knives rest, until all the polished parts of the latter are covered. The insulated steel cup is then filled with mercury so high that the point of the platinum wire just touches it, when the beam is level ; the small elastic pocket is used for raising and lowering the mercury cup, so as to place it at the proper height for bringing the mercury in contact with the end of the platinum wire. When the articles have received the amount of silver corresponding to the weight in the pan at the opposite side of the beam, the equilibrium will be established, and the platinum wire will then leave the mercury, and thus break the circuit and stop the opera-tion. By this automatic arrangement the operation needs no atten-tion, since the moment the platinum wire loses contact with the mercury electricity ceases to pass ; if, however, the articles are allowed to remain in the bath after they have received the proper amount of silver, a portion of this metal may be dissolved by the free cyanide in the solution, in which case the end of the platinum wire would again dip into the mercury and complete the circuit, when deposition would be renewed and continue until the increased weight of silver again caused the platinum wire to lose contact with the mercury.

Solid Silver Deposits.—Although it is possible to deposit silver, from a cyanide solution rich in metal (say eight ounces of silver per gallon), upon wax or gutta-percha moulds, this method is not practically adopted. The usual method is to first obtain a copper electrotype mould or shell of the object in the ordinary way; silver is then deposited within the mould (supposing it to be a hollow object) until of the required thickness; the copper is afterwards dissolved from the silver either by boiling the article in hydrochloric acid, or, still better, a strong solution of perchloride of iron, either of which substances will dissolve the copper mould without in any way injuring the silver. The perchloride of iron for this purpose may be readily formed by dissolving peroxide of iron (commercial " crocus ") in hot hydrochloric acid. The method of dissolving the copper recommended by Napier, is as follows : " An iron solution is first made by dissolving a quantity of copperas in water ; heat this till it begins to boil; a little nitric acid is then added—nitrates of soda or potash will do ; the iron which is thus peroxidised may be precipitated either by ammonia or carbonate of soda ; the precipitate being washed, muriatic acid is added till the oxide of iron is dissolved. This forms the solution for dissolving the copper. When the solution becomes almost colourless, and has ceased to act on the copper, the article is removed, and the addition of a little ammonia will precipitate the iron along with a portion of the copper ; but after a short exposure the copper is redissolved. The remaining precipitate is washed by decantation ; a little ammonia should be put into the two first waters used for washing. When washed, and the copper dissolved out, the precipitate is redissolved in hydrochloric acid, and the silver article returned until the copper is all dissolved off. It is convenient to have two solutions of perchloride of iron, so that while the iron in the one is being precipitated, the article is put into the other. The persalt of iron will be found to dissolve the copper more rapidly than muriatic acid alone ; persulphate of iron must not be used, as it dissolves the silver along with the copper.

" The silver article is now cleaned in the usual way, and heated to redness over a clear charcoal fire, which gives it the appearance of dead silver, in which state it may be kept, or, if desired, it may be scratched and burnished." A very simple and economical method of producing perchloride of iron is to reduce the native peroxide of iron, known as " redding," to a powder, and digest it in hot hydrochloric acid, by which the salt is obtained at a cost but little exceeding that of the acid employed, the native ore being worth only about 25s. per ton.

One great objection to solid electro-deposits of silver (and gold) is that the articles have not the metallic "ring," when struck with any hard substance, as silver ware of ordinary manufacture. "This disadvantage," says Napier, " is no doubt partly due to the crystalline character of the deposit, and partly to the pure character of the silver, in which state it has not the sound like standard or alloyed silver. That this latter cause is the principal one appears from the fact that a piece of silver thus deposited is not much improved in sound by being heated and hammered, which would destroy all crystallisation." This is quite true, but when electro-deposited silver has been *melted*, and cast into an ingot, by which its crystalline character is *completely* destroyed, and which is only partially affected by simply annealing and hammering, the characteristic " ring " of the pure metal is restored. The absence of a musical ring in electro-deposited silver is not of much consequence, however, since this method of reproduction would only be applied to rare works of art, such as antique figures, and richly chased articles kept solely for ornament.

On the Thickness of Electro-Deposited Silver.—This may be considered a somewhat delicate theme to expatiate upon when we reflect that some articles of commerce, but more especially export goods and articles sold at mock auctions, frequently receive a coating of silver which not only defies measurement by the most delicate micrometer, but also renders estimation by any other means all but impossible. This class of work includes spoons and forks, cruet-frames, toast-racks, &c., manufactured from a very inferior description of German silver or brass, while Britannia metal tea services, salt-cellars, and many other articles made from the same alloy enter the market in enormous quantities, with a mere blush of silver upon them, the thickness of which might be more readily estimated by imagination than by any practical test. As to the amount of silver which should be deposited upon articles of domestic use, to enable them to withstand ordinary wear and tear for a reasonable period, from 1 to 3 ounces per dozen for spoons and forks may be deposited. Taken as a guide, with the smaller quantity of silver upon them, such articles, with careful usage, should present a very creditable appearance after five years' use ; with the larger proportion, the articles should look well, though probably somewhat bare upon those parts most subject to friction, at the end of twenty years. The same articles, if used in hotels or on board ship, would become unsightly in less than half the periods named. German silver tea and coffee services, to be *fairly* well plated or silvered, should not have less than 2 ounces of silver upon the four pieces, which may be distributed in about the following proportions : for a 5-gill coffee-pot 12 dwts. ; 5-gill teapot 12 dwts. ; sugar-basin 10 dwts. ; cream-ewer 6 dwts.

When the same articles are required to be *fully* well plated, the proportions should be about as follows : For coffee and teapot, about 1½ ounce of silver each ; sugar-basin 1 ounce, and cream-ewer about 10 to 15 dwts.

The proportion of silver which should be deposited per square foot, for plating of good quality, is from 1 to 1½ ounce. With the latter proportion the electro-silvered work would nearly approach in quality the old Sheffield plate, and would last for a great number of years without becoming bare, even at the most prominent parts, unless the article were subjected to very severe treatment in use.

Pyro-plating.—It is well known that when a silver-gilt article—as a watch-chain, for example—has been broken, and afterwards repaired by hard soldering, that the film of gold almost entirely disappears from each side of the soldered spot, under the heat of the blow-pipe flame, to the extent of 1 or 2 inches on either side of the joining. The film of gold has, in fact, sunk into the body of the silver, as though it had become alloyed with this metal. By some persons this is really believed to be the case. We are, however, disposed to think that the absorption of the gold under these circumstances is due, not to an actual alloying of the two metals in the ordinary sense, but to the expansion of the silver by the heat, by which its molecular structure becomes disturbed, and the film of gold, being thus split up into infinitely minute particles, these become absorbed by the silver as the metal contracts on cooling, and consequently disappear from the surface. We hold this view because we do not think that the heat of the blowpipe flame required to fuse the solder would be sufficient to form an alloy in the proper sense ; indeed, the heat required to " run " silver solder would not be sufficiently high even to " sweat " the silver of which the article is composed. The fact of a film of metal becoming absorbed by another metal under the influence of heat has been taken advantage of, and a process termed " pyro-plating " has been introduced, and has been worked to some extent in Birmingham. The process, which has been applied to coating articles—of steel and iron more especially—with gold, silver, platinum, aluminium, copper, &c., may be thus briefly described : The article is first steeped in a boiling solution of caustic potash ; it is then brushed over with emery-powder, and afterwards with a steel brush and a solution of common soda, in which it is allowed to remain for some time. It is next connected to the negative electrode of a strong battery, and immersed in a hot solution of caustic potash, abundance of hydrogen being evolved, and is allowed to remain until it has a " silvery " appearance. After rinsing, it is suspended in a silver bath, with a previously weighed metal plate of the same amount of surface placed as a cathode by its side ; this plate is taken out and weighed from time to time until sufficient silver has been deposited, which indicates approximately the amount of deposit upon the article itself. The article is then removed and rinsed, and afterwards heated in a furnace until the silver is "driven" into the surface of the metal. If the steel article requires to be *tempered*, it is quenched in water, and then brought to the proper temper in the usual way.

Whitening Electro-plated Articles.—It is well known that articles which have been electro-plated tarnish more rapidly than silver goods ; and while this has by many persons been attributed to the extreme purity of the electro-deposited metal, which, it was believed, was more susceptible of being attacked by sulphurous fumes and other impurities in the air, by others it is believed to be due to a small quantity of undecomposed salt remaining in the pores of the deposited metal, which undergoes decomposition, and causes the work to tarnish. In order to render electro-plate less liable to discoloration, the following method has been adopted, but, as will readily be seen, it could not be applied to all classes of work : The article is first dipped in a saturated solution of borax and then allowed to dry, when a thin layer of the salt remains upon the surface ; the article is then dipped a second or even a third time (drying after each dipping) until it is completely covered with a layer of borax. When large articles are to be treated this way, the borax may be applied with a soft brush. The article is next to be heated to a dull red heat, or until the borax *fuses*. When cold, it is to be put into a pickle of dilute sulphuric acid, which rapidly dissolves the borax ; after rinsing in hot water it is placed in hot boxwood sawdust, and then treated in the usual way.

Whitening Silver Work.—Articles of silver which in their original finished state were left either wholly or in part a dead white, and have lost this pleasing effect by wear or oxidation, may be restored to their original condition by the process termed *whitening*. The article is first brought to a dull red heat (not sufficient to melt the solder) over a charcoal fire—if it be a brooch, watch-dial, or other small silver article, by means of the blowpipe flame, the article being placed on a large and flat piece of charcoal. When the piece of work has thus been heated uniformly all over, it is allowed to become cool, after which it is placed in a glazed earthenware vessel (an ordinary white basin will do), containing a sufficient quantity of very dilute sulphuric acid. In a short time the acid will dissolve the oxide from the surface, together with a small quantity of oxide of copper derived from the copper with which the silver was alloyed, and which, with the silver, becomes oxidised by the heat and subsequent action of the atmosphere. When the article is removed from the pickle—in which it should remain for at least twenty minutes to half an hour—if not of a sufficiently pure whiteness it may be heated and pickled again. When the whitening is properly effected, the surface should present a beautiful pearl-white appearance, and be perfectly uniform in its lustrous dulness. Directly the article is removed from pickle, it should be rinsed in two separate waters, the last water (which should be distilled water, by preference) being boiling hot. The article, after being removed from the rinsing-bowl, should be allowed to dry spontaneously, which it will do if the water is boiling hot. It is not a good plan, though it is frequently done, to put work which has been whitened in boxwood sawdust, since if it has been much used it is liable to produce stains.

CHAPTER XVII.

IMITATION ANTIQUE SILVER.

Oxidised Silver.—Oxidising Silver.—Oxidising with Solution of Platinum.—
Oxidising with Sulphide of Potassium.—Oxidising with the Paste.—
Part-gilding and Oxidising.—Dr. Elsner's Process.—Satin Finish.—Sul-
phuring Silver.—Niello, or Nielled Silver.—Pink Tint upon Silver.—
Silvering Notes.

Oxidised Silver.—Soon after the art of electro-plating had become
an established industry, the great capabilities of the "electro" process, as it was called, received the serious attention of the more gifted
and artistic members of the trade, who, struck with the great beauty
of electro-deposited silver, and the facilities which the process offered
for the reproduction of antique works, induced some electro-platers of
the time to make experiments upon certain classes of work with a view
to imitate the effects seen upon old silver ; some of the results were
highly creditable, and in a short time after "oxidised" silver became
greatly in vogue, and has ever since been recognised as one of the
artistic varieties of ornamental silver or electro-plated work. We
scarcely think we shall err, however, if we venture to say that much
of the "oxidised" silver-plated work of the present time is far inferior
in beauty and finish to that with which our shops and show-rooms
were filled some thirty years ago. Indeed, when visiting the Paris
Exhibition of 1878, we were much displeased with the very slovenly
appearance of some of the plated goods which had been part-gilt and
oxidised in the exhibits of some of the larger English and French
firms. The specimens referred to had the appearance of having done
duty as specimens in all the exhibitions since 1851, and had suffered
by being repeatedly "cleaned up" for each occasion ; they were certainly far from being creditable.

"Oxidising" Silver.—This term has been incorrectly applied, but
universally adopted, to various methods of darkening the surface of
silver in parts, by way of contrast to burnished or dead-white surfaces of an article. Oxygen, however, has little to do with the
discoloration, as will be seen by the following processes, which are
employed to produce the desired effect. The materials used are various,
and they are generally applied with a soft brush, a camel-hair brush

being suitable for small surfaces. In applying either of the materials the article should be quite dry, otherwise it will spread over portions of the work required to be left white, and thus produce a patchy and inartistic effect. The blackening substances are generally applied to the hollow parts or groundwork of the object, while the parts which are in relief are left dead, or burnished according to taste.

Oxidising with Solution of Platinum.—Dissolve a sufficient quantity of platinum in aqua regia, and carefully evaporate the resulting solution (chloride of platinum) to dryness, in the same way as recommended for chloride of gold. The dried mass may then be dissolved in alcohol, ether, or water, according to the effect which it is desired to produce, a slightly different effect being produced by each of the solutions. Apply the solution of platinum with a camel-hair brush, and repeat the operation as often as may be necessary to increase the depth of tone ; a single application is frequently sufficient. The ethereal or alcoholic solution of platinum must be kept in a well-stoppered bottle, and in a cool place. The aqueous solution of platinum should be applied while hot.

Oxidising with Sulphide of Potassium.—Liver of sulphur (sulphide of potassium) is often used for producing black discoloration, erroneously termed *oxidising*. For this purpose four or five grains of the sulphide are dissolved in an ounce of hot water, and the solution applied with a brush, or the article wholly immersed if desired. The temperature of the solution should be about 150° Fahr. After a few moments the silver surface assumes a darkened appearance, which deepens in tone to a bluish-black by longer treatment. When the desired effect is produced the article is rinsed and then scratch-brushed, or burnished if required, or the blackened hollow surfaces are left dead according to taste. When it is desired to produce a *dead* surface upon an article which has been electro-silvered, the article may be placed in a sulphate of copper bath for a short time, to receive a slight coating of copper, after which it is again coated with a thin film of silver in an ordinary cyanide bath. It has then the dead-white appearance of frosted silver. Where portions of the article are afterwards *oxidised* a very fine contrast of colour is produced. In using the sulphide of potassium solution it should be applied soon after being mixed, since it loses its activeness by keeping. Fresh solutions always give the most brilliant results. Since the sulphide dissolves the silver, it is necessary that it should be applied only to surfaces which have received a tolerably stout coating of this metal, otherwise the subjacent metal (brass, copper, or German silver) will be exposed after the sulphide solution has been applied.

Oxidising with the Paste.—For this purpose a thin paste is formed by mixing finely-powdered plumbago with spirit of turpentine, to which mixture is sometimes added a small quantity of red ochre or jewellers' rouge, to imitate the warm tone sometimes observed in old silver articles. The paste is spread over the articles and allowed to dry, after which the article is brushed over with a long-haired soft brush, to remove all excess of the composition. The parts in relief are then cleaned by means of a piece of rag, or chamois leather, dipped in spirit of wine. This method of imitating old silver is specially applicable to vases, tankards, chandeliers, and statuettes. In case of failure in the manipulation, the dried paste may be readily removed by placing the article in a hot solution of caustic potash or cyanide, when, after rinsing and drying, the paste may be reapplied. To give the old silver appearance to small articles, such as buttons, for example, they are first passed through the above paste, and afterwards revolved in a barrel or " tumbler " containing dry sawdust, until the desired effect is produced.

Part-gilding and Oxidising.—To give this varied effect to work, the articles are first gilt all over in the usual way ; certain parts are then *stopped off*, as it is termed, by applying a suitable varnish. When the varnish has become dry, the article is placed in the silvering bath until a sufficient coating, which may be slight, has been obtained. After rinsing, the object is immersed in a solution of sulphide of potassium until the required tone is given to the silvered parts, when the article is at once rinsed, carefully dried, and the protecting varnish dissolved off, when it is ready to be finished.

Dr. Elsner's Process.—A brownish tone is imparted to plated goods by applying to the surface a solution of sal-ammoniac, and a still finer tone by means of a solution composed of equal parts of sulphate of copper and sal-ammoniac in vinegar. To produce a fine black colour, Dr. Elsner recommends a warm solution of sulphide of potassium or sodium.

Sulphide of Ammonium.—This liquid may also be applied to the so-called oxidation of silver, either by brushing it over the parts to be oxidised, or by immersion. It may also be applied, with plumbago, by forming a thin paste with the two substances, which is afterwards brushed, or smeared over the surface to be coloured, and when dry a soft brush is applied to remove the excess of plumbago. If preferred, a little jewellers' rouge may be added to the mixture.

Satin Finish.—This process is thus described by Wahl : The sand-blast is in use in certain establishments to produce the peculiar dead, lustrous finish, known technically as *satin finish*, on plated goods ; a templet of some tough resistant material, like vulcanised india-rubber, is made of the proper design, and when placed over the article, protects the parts which it is desired to leave bright from the depolishing action of the sand, while the only open portions of the templet are exposed to the blast. The apparatus employed for this purpose consists of a wooden hopper, with a longitudinal slit below, through which a stream of fine sand is allowed to fall, by opening a sliding cover. Closely surrounding the base of the hopper is a rectangular trunk of wood, extending some distance below the base of the hopper, and tapering towards the bottom, to concentrate the sand-jet. This trunk is closed about the sides of the hopper, and open below, and is designed to direct the stream of sand upon the surface of the article presented beneath its orifice. To increase the rapidity of the depolishing action of the sand, a current of air, under regulated pressure, is admitted into the upper part of the trunk, which, when the sand-valve is opened, propels it with more or less accelerated velocity upon the metallic surface below. For this purpose, either a " blower," or an air-compressor with accumulator, may be used : and the pressure may be regulated at will. The sand is thus driven with more or less velocity down the trunk by the air-blast admitted above, and, falling upon the surface of the article presented at the bottom, rapidly depolishes the exposed parts, while those protected by the templet are not affected. The articles are presented at the orifice of the trunk by the hands of the operator, which are suitably protected with gloves ; and as rapidly as the depolishing proceeds, he turns the article about till the work is done. The progress of the work is viewed through a glass window, set in a horizontal table, which surrounds the apparatus and which forms the top of a large box, into which the sand falls, and which is made tight to prevent the sand from flying about. A portion of this box in front, where the workman stands, is cut away, and over the opening is hung a canvas apron, which the operator pushes aside to introduce the work. The sand that accumulates in the box below is transferred again to the hopper, as required, and is used over and over again. The satin-finish produced by the sand-blast is exceedingly fine and perfectly uniform, and the work is done more rapidly than with the use of brushes in the usual way.

Sulphuring Silver.—A very fine blue colour, resembling " blued " steel, may be imparted to silver or plated surfaces, by exposing the article to the action of sulphur fumes. For this purpose, the article should be suspended in an air-tight wooden box ; a piece of slate or a flat tile is laid upon the bottom of the box, and upon this is placed an iron tray, containing a small quantity of red-hot charcoal or cinders ; about a teaspoonful of powdered sulphur is now quickly spread over the glowing embers, and the lid of the box immediately closed. After about a quarter of an hour, the lid may be raised (care being taken not to inhale the sulphur fumes) and the article promptly withdrawn ; if the article is not sufficiently and uniformly blued, it must be again suspended and a fresh supply of hot charcoal and sulphur introduced. It is necessary that the articles to be treated in this way should be absolutely clean.

Niello, or Nielled Silver.—These terms* are applied to a process which is attributed to Maso Finniguerra, a Florentine engraver of the fifteenth century, and somewhat resembles enamelling. It consists, essentially, in inlaying engraved metal surfaces with a black enamel, being a sulphide of the same metal, by which very pleasing effects are produced. The "nielling" composition may be prepared by making a triple sulphide of silver, lead, and copper, and reducing the resulting compound to a fine powder. The composition is made as follows : A certain proportion of sulphur is introduced into a stoneware retort, or deep crucible. In a second crucible, a mixture of silver, lead, and copper is melted, and when sufficiently fused the alloy thus formed is added to the fused sulphur in the first vessel, which converts the metals into sulphides ; a small quantity of sal-ammoniac is then added, and the compound afterwards removed from the retort or crucible and reduced to a fine powder. The following proportions are given by Mr. Mackenzie :—

Put into the first crucible—

Flowers of sulphur	750 parts.
Sal-ammoniac	75 ,,

Put into the second crucible—

Silver	15 parts.
Copper	40 ,,
Lead	80 ,,

When fused, the alloy is to be added to the contents of the first crucible. Roseleur recommends diminishing the proportion of lead, which impairs the blue shade of the nielling, and corrodes too deeply.

To apply the powder, obtained as above, it is mixed with a small quantity of a solution of sal-ammoniac. After the silver work is engraved, the operator covers the entire surface with the nielling composition, and it is then placed in the muffle of an enamelling furnace, where it is left until the composition melts, by which it becomes firmly attached to the metal. The nielling is then removed from the parts in relief, without touching the engraved surfaces, which then present a very pleasing contrast, in deep black, to the white silver surfaces. This process, however, is only applicable to engraved work.

Wahl describes a cheaper process of nielling, which consists "in engraving in relief a steel plate, which, applied to a sheet of silver, subjected to powerful pressure in a die, reproduces a faithful copy of the engraving. The silver sheet thus stamped is ready to receive the

---

* The art was formerly called *working in niello.*

nielling. A large number of copies may be obtained from the same matrix. Such is the method by which a quantity of nielled articles are manufactured, as so-called Russian snuff-boxes, cases for spectacles, bon-bon boxes, &c.

Roseleur suggests the following to produce effects similar to nielling : A pattern of the design, cut out of thin paper, such as lace paper, is dipped into a thin paste of nielling composition, or into a concentrated solution of some sulphide, and then applied upon the plate of silver, which is afterwards heated in the muffle. The heat destroys the organic matter of the paper, and a design remains, formed by the composition which it absorbed.

A solution of chloride of lime (bleaching powder) will blacken the surface of silver, as also will nitric acid. For all practical purposes, however, chloride of platinum and weak solutions of the sulphides before mentioned will be found to answer very well if applied with proper judgment.

Pink Tint upon Silver.—Fearn recommends the following for producing a fine pink colour upon silver : Dip the cleaned article for a few seconds in a strong hot solution of chloride of copper, then rinse and dry it, or dip it in spirit of wine and ignite the spirit.

Silvering Notes.—1. The anodes, if of rolled silver, should always be annealed before using them. This may easily be done by placing them over a clear charcoal, or even an ordinary clear fire, until they acquire a cherry-red heat : when cold, they are ready for use. If convenient to do so, it is a good plan to hard solder a short length of stout platinum wire (say about three inches long) to the centre of one edge of the anode, which may be united to the positive electrode of the battery, or other source of electricity, by a binding screw, or by pewter solder. The object of attaching the platinum wire is to enable the anode to be wholly immersed in the bath, and thus prevent it from being *cut through* at the *water line*, which is generally the case where anodes are only partially immersed.

2. *Worn Anodes.*—When the anodes have been long in use, their edges frequently become *ragged*, and if these irregularities are not removed fragments of the metal will fall into the bath, and, possibly, upon the work, causing a roughness of deposit. It is better, therefore, to trim the edges of anodes whenever they become thin and present a ragged appearance.

3. *Precautions to be Observed when Filling the Bath with Work.*—Assuming the suspending rods to have been cleaned, the battery connections adjusted, and the preparation of the articles to be plated commenced, some means must be adopted to prevent the articles *first* put into the bath from receiving too quick a deposit while others are being got ready. In the first instance, the full force of the current must be

checked, which may be done by exposing a small surface of anode in solution or suspending a plate of brass or a small silver anode as a " stop," or check, to the negative rod, until a sufficient number of articles (say spoons or forks, for example) have been suspended, when the stop may be removed and the remainder of the articles immersed until the conducting rod is full; the rest of the suspending rods should then be treated in the same way. When magneto or dynamo-electric machines are employed, the full strength of the current is checked by the employment of the *resistance coil*, a description of which will be found in Chapter III. A simple way of diminishing the amount of current, when filling the bath with work, is to interpose a thin iron wire between the positive electrode and the suspending rod, which must be removed, however, when the cathode surface (the articles to be plated) in the bath approaches that of the anode surface.

4. *Plating Different Metals at the Same Time.*—It is not good practice to place articles composed of different metals or alloys indiscriminately in the bath, since they do not all receive the deposit with equal facility. For example, if two articles, one copper or brass, and another Britannia metal or pewter, be immersed in the solution simultaneously, the former will at once receive the deposit of silver, while the latter will scarcely become coated at all, except at the extremities. Since the best conductor receives the deposit most freely, the worst conductor (Britannia metal, pewter, or lead) should first be allowed to become completely coated, after which copper or brass articles may be introduced. It is better, however, if possible, to treat the inferior conductors separately than to run the risk of a defective deposit.

5. *Excess of Cyanide.*—When there is a large excess of cyanide in the plating bath, the silver is very liable to strip, or peel off the work when either scratch-brushed or burnished; besides this, the anodes become dissolved with greater rapidity than is required to merely keep up the proper strength of the bath, consequently the solution becomes richer in metal than when first prepared. The depositor must not confound the terms " free cyanide " with " excess " of cyanide : the former refers to a small quantity of cyanide beyond that which is necessary to convert into solution the precipitate thrown down from the nitrate, which is added to the solution to act upon and dissolve the anode while deposition is going on ; the latter term may properly be applied to any quantity of cyanide which is in excess of that which is necessary for the latter purpose.

In preparing plating solutions from the double cyanide of silver and potassium, great care must be taken, when precipitating the silver from its nitrate solution, not to add the cyanide in excess, other-wise a portion of the precipitated cyanide of silver will be re-dissolved, and probably lost when decanting the supernatant liquor from the precipitate. When the precipitation is nearly complete, the last additions of cyanide solution should be made very cautiously, and only so long as a turbidity, or milkiness, is produced in the clear liquor above the precipitate. Instances have been known in which not only silver, but gold precipitates also, have been partially redissolved by excess of cyanide and the solutions thrown away by ignorant operators as waste liquors. If by accident an excess of cyanide has been used during the precipitation of gold or silver solutions, the difficulty may be overcome by gradually adding a solution of the metallic salt until, in its turn, it ceases to produce turbidity in the clear supernatant liquor. Again, in dissolving precipitates of silver or gold, care is necessary to avoid using a large excess of cyanide ; a moderate excess only is necessary.

6. *Articles Falling into the Bath.*—When an article falls into the bath, from the breaking of the slinging wire or otherwise, its recovery generally causes the sediment which accumulates at the bottom of the vat to become disturbed, and this, settling upon the work, produces roughness which is very troublesome to remove. If not immediately required, it is better to let the fallen article remain until the rest of the work is plated ; or if its recovery is of immediate importance, the rod containing the suspended articles should be raised every now and then during about half an hour, in order to wash away any sediment that may have settled on the work. By gently lifting the rod up and down, or raising each piece separately, the light particles of sediment may readily be cleared from the surface of the work. When very large articles, as salvers, for example, are immersed in the bath, they should be lowered very gently, so as not to disturb the sediment referred to ; if this precaution be not rigidly followed, especially if the vat be not a very deep one, the lower portion will assuredly become rough in the plating, which the most skilful burnishing will be incapable of removing. We have frequently known it to be necessary to strip and replate articles of this description from the cause referred to. It must also be borne in mind that when anodes become very much worn minute particles of silver fall to the bottom of the vessel, which, when disturbed in the manner indicated, rise upward, settle upon the work, and become attached, by what may be termed *electrosoldering,* to the work, causing the deposit to be rough, and when such surface is afterwards smoothed by polishing, the part exhibits numerous depressions, or is " pitted, " as the Sheffield burnishers term it.

7. *Cleaning Suspending Rods.*—It is a very common practice with careless workmen to clean the suspending rods with emery cloth while they are in their places across the sides or ends of the plating vat.

This is a practice which should be strictly disallowed, for it is evident that the particles of brass, emery, and metallic oxides which become dislodged by the rubbing process, must enter the solution, and being many of them exceedingly light, will remain suspended in the solution for a considerable time, and finally deposit upon the articles when placed in the vat, while some portions of the dislodged matter will become dissolved in the bath. All suspending rods should be cleaned at some distance from the plating vat, and wiped with a clean dry rag after being rubbed with emery cloth before being replaced across the tank.

8. *Electro-silvering Pewter Solder.*—Besides the methods recommended elsewhere, the following may be adopted : After thoroughly cleaning the article, apply to the soldered spot with a camel-hair brush a weak solution of cyanide of mercury ; or if it be a large surface the soldered part must be dipped for a short time in the mercury solution. In either case the article must be well rinsed before being immersed in the silver bath.

9. *Metal Tanks.*—When working solutions in iron tanks, the plater should be very careful not to allow the anodes, or the work to be coated, to come in contact with the metallic vessel while deposition is taking place, since this will not only cause the current to be diverted from its proper course, but will also cause the anodes, especially if there be a large excess of free cyanide in the bath, to become eaten into holes, and fragments of the metal will be dislodged and fall to the bottom of the vat, and possibly small particles of the metal will settle upon the work. We remember an instance in which several wooden nickel-plating tanks, lined with stout sheet lead, coated with pitch, yielded very poor results from some cause unknown to the plater. Having been consulted on the matter, the author soon discovered the source of mischief : the copper hooks supporting the heavy anodes had become imbedded in the pitch, and were in direct communication with the lead lining, from which a greater portion of the pitch had scaled off, leaving the bare metal exposed below the surface of the solution. Upon applying a copper wire connected to the negative electrode of the large Wollaston battery, at that time used at the establishment, to the leaden flange of each tank the author obtained brilliant sparks, to the great astonishment of the plater and his assistants, and subsequently caused strips of wood to be placed between the side anodes and the lead lining, after which nickel-plating proceeded without check.

10. *Bright Plating.*—Even in the most skilful hands the bright solution is very liable to yield ununiform results. When the solution has remained for some time without being used it is apt to give *patchy* results, the work being bright in some parts only ; if the solution is

disturbed, by taking out work or by putting in fresh work, sometimes the latter will refuse to become bright, and the remainder of the work in the bath will gradually become dull. To obviate this the bath should be well stirred over night, and all the work to be plated at one time put into the bath as speedily as possible, and all chances of disturbance avoided. When the work is known to have a sufficient coating of the bright deposit, the battery connection should be broken and the articles then at once removed from the bath. On no account must an excess of the " bright " liquid be allowed to enter a bath.

11. *Dirty Anodes.*—When the anodes, which should have a greyish appearance while deposition is taking place, have a pale greenish film upon their surface, this indicates that there is too little free cyanide in the bath, or that the current is feeble ; the battery should first be attended to, and if found in good working order, and all the connections perfect, an addition of cyanide should be made ; this, however, should only be done the last thing in the evening, the bath then well stirred and left to rest until the following morning.

12. *Dust on the Surface of the Bath.*—Sometimes in very windy weather the surface of the bath, after lying at rest all night, will be covered with a film of dust ; to remove this spread sheets of tissue paper, one at a time, over the surface of the liquid, then take the sheets up one by one and place them in an earthen vessel; the small amount of solution which they have absorbed may be squeezed from the sheets, passed through a filter, and returned to the bath, and the pellets of paper may then be thrown amongst waste, to be afterwards treated for the recovery of its metal.

13. *Old Slinging Wires.*—It is not a good plan to use a slinging wire, one end of which has received a coating of silver or other metal more than once, without first stripping off the deposited metal , in the first place the coated end of the wire becomes very brittle, and is liable to break when twisting it a second time, possibly causing the article to fall into the bath, or on a floor bespattered with globules of mercury and other objectionable matter , again, the broken fragments of silver-covered wire, if allowed to fall carelessly on the floor, get swept up with the dirt, and the silver thus wasted. The wires which have been used once should be laid aside, with the plated ends together, and at a convenient time these ends should be dipped in hot stripping solution, until all the silver is dissolved off, and after rinsing, the ends should be made red hot, to anneal them ; the wires may then be cleaned with emery cloth and put in their proper place to be used again. These minor details should always be attended to, since they do not necessarily involve much time and are assuredly advantageous from an economical view. It is too commonly the practice with careless operators to neglect such simple details, but the consequence is that their

plating operations are often rendered unnecessarily troublesome, while their workshops are as unnecessarily untidy.

14. *Battery Connections.*—Before preparing work for the bath, the binding screws, clamps, or other battery connections should be examined, and such orifices or parts as form direct metallic communication between the elements of the battery and the anodes and cathodes should be well cleaned if they have any appearance of being oxidised or in any way foul. The apertures of ordinary binding screws may be cleaned with a small rat-tail file, and the flat surfaces of clamps rubbed with emery cloth laid over a flat file. When binding screws, from long use or careless usage, become very foul, they should be dipped in dipping acid, rinsed, and dried quickly. Previous to putting work in the bath, a copper wire should be placed in contact with the suspending rod and the opposite end allowed to touch the anode, when the character of the spark will show if the current is sufficiently vigorous for the work it has to do : if the spark is feeble, the connections should be looked to, and the binding screws tightened, if necessary ; the hooks and rods supporting the anodes should also be examined, and if dirty, must be well cleaned, so as to insure perfect contact between the metal surfaces.

15. *Gutta-percha Lining for Plating Tanks.*—This material should never be used for lining the insides of tanks which are to contain cyanide solutions, since the cyanide has a solvent action upon it, which, after a time, renders the solution a very bad conductor. The author once had to precipitate the silver from an old cyanide solution which had remained for a long period in a gutta-percha lined bath, and soon after the acid (sulphuric) had been applied to throw down the silver, there appeared, floating upon the surface of the liquid, numerous clots of a brown colour, which proved to be gutta-percha, although greatly altered from its original state.

CHAPTER XVIII.

ELECTRO-DEPOSITION OF NICKEL.

Application of Nickel-plating.—The Depositing Tank.—Conducting Rods.—
Preparation of the Nickel Solution.—Nickel Anodes.—Nickel-plating by
Battery.—The Twin-Carbon Battery.—Observations on Preparing Work
for Nickel-plating.—The Potash Bath.—Dips or Steeps.—Dipping Acid.
—Pickling Bath.

Application of Nickel-plating.—When applied to purposes for
which it is specially adapted, nickel-plating may be considered one of
the most important branches of the art of electro-deposition. In the
earlier days of nickel-plating too much was promised and expected
from its application, and, as a natural consequence, frequent disap-
pointments resulted from its being applied to purposes for which it
was in no way suited. For example, it was sometimes adopted as a
substitute for silver-plating in the coating of mugs or tankards used
as drinking vessels for malt liquors, but it was soon discovered that
those beverages produced stains or discolorations upon the polished
nickel surface, which were not easily removed by ordinary means,
owing to the extreme hardness of the metal as compared with silver or
plated goods. Again, nickel-plated vegetable-dishes became stained
by the liquor associated with boiled cabbage or spinach, rendering the
articles unsightly, unless promptly washed after using—a precautionary measure but seldom adopted in the best-regulated sculleries.
It was also found that polished nickel-plated articles when exposed
to damp assumed a peculiar dulness, which after a time entirely
destroyed their brilliant lustre, whereas in a warm and dry situation
they would remain unchanged for years, a fact which the mullers of
our restaurants and taverns which were nickel-plated many years ago
bear ample testimony at the present day.

While practical experience has taught us what to avoid in connection with nickel-plating, it has also shown how vast is the field of
usefulness to which the art is applicable, and that as a protective and
ornamental coating for certain metallic surfaces, nickel has at present
no rival. Its great hardness—which closely approximates that of
steel—renders its surface, when polished, but little liable to injury
from ordinary careless usage ; while, being a non-oxidisable metal,
it retains its natural whiteness, even in a vitiated atmosphere.

The metals ordinarily coated with nickel by electro-deposition are copper, brass, steel, and iron, and since these require different *preparatory* treatment, as also different periods of immersion in the nickel bath, they will be treated separately. The softer metals, as lead, tin, and Britannia metal, are not suited for nickel-plating, and should never be allowed to enter the nickel bath.

The Depositing Vat, or Tank.—The depositing vessel may be made from slate or wood, but the following method of constructing a vat is that most generally adopted, and when properly carried out produces a vessel of great permanency. The tank is made from $2\frac{1}{2}$-inch deal, planed on both sides, the boards forming the sides, ends, and bottom being grooved and tongued, so as to make the joints, when put together, water-tight ; they are held together by long bolts, tapped at one end to receive a nut. The sides and ends, as also the bottom, are likewise secured in their position by means of screw-bolts, as seen in Fig. 99. When the tank is well screwed together, as in the

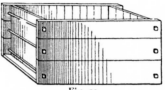

Fig. 99.

engraving, the interior is to be well lined with *pure* thin sheet lead. It is of great importance that the lead used for this purpose be as pure as possible, for if it contain zinc or tin it will be liable to be acted upon by the nickel solution which it is destined to hold, and pin-holes will be formed, through which the solution will eventually escape. The joints of the leaden lining must not be united by means of solder, but by the *autogenous process*, or "burning," as it is called, that is, its seams are *fused* together by the *hydrogen flame*—an operation with which intelligent plumbers are well acquainted. If solder were used for this purpose voltaic action would soon be set up between the lead and the tin of the solder by the action of the nickel solution, and in time a series of holes would be formed, followed by leakage of the vat. When the lead lining is complete the vessel must be lined throughout with matched boarding, kept in its position by a rim of wood fastened round the upper edge of the tank. These tanks are usually 3 feet wide, 3 feet deep, and about 6 feet long, and hold about 250 gallons.

Before using the tank it should be well rinsed with clean water. It is a good plan to quite fill the tank with water, and allow it to remain therein for several hours, by which time the pressure of the liquid will soon indicate if there be a leakage at any part ; it should then be emptied and examined, to ascertain if thoroughly water-tight.

U

We will assume that it is desired to make up 100 gallons of nickel solution—in which case the depositing tank should be capable of holding not less than 120 gallons, to allow for the displacement of liquid by the anodes and articles to be immersed, as also to allow sufficient space—say 3 inches—above the solution to prevent the liquid from reaching the hooks by which the anodes are suspended, when the bath is full of work. Although we have taken 100 gallons of solution as a *standard*, we may state that, for large operations, tanks capable of holding 250 up to 300 gallons, or even more, are commonly employed.

Conducting Rods.—These rods, which are used for supporting the nickel anodes, as also the articles to be nickeled, generally consist of 1-inch brass tubing, with a core of iron rod ; they are commonly laid across the bath, lengthwise, extending about 3 inches beyond the extreme ends of the vessel. Sometimes, however, shorter rods are employed, and these are laid across the bath from side to side. For a nickel bath of 100 gallons and upwards three such suspending rods are used, one rod being laid from end to end, close to each side of the tank, upon which the requisite number of anodes are suspended by their hooks ; a third rod is laid, also longitudinally, along the centre

of the tank, midway between the other two, for suspending the articles to be nickeled ; the anode rods are to be connected together by a stout copper wire at one end by soldering. These rods are termed respectively the *positive* and *negative* conducting rods, the former receiving the anodes, and the latter the work to be nickeled. Fig. 100 represents a cast nickel anode and its supporting hook of stout copper wire, which latter should not be less than ¼ inch in thickness. In order to insure a *perfect* connection between the copper hook and the anode, the author has found it

Fig. 100.

very advantageous to unite the two by means of pewter solder, in the following way,* and which it may be useful to quote here : The holes

---

* " Electro-Metallurgy Practically Treated." By Alexander Watt.

being cleaned with a rat-tail file, the hooks were dipped into ordinary dipping acid (sulphuric and nitric acid) for one instant, and rinsed. One end of each hook was then moistened with chloride of zinc, and immediately plunged into a ladle containing molten tin or pewter solder. The tinned hook was next inserted into the hole in the anode, and a gentle tap with a hammer fixed it in its place. The anode being laid flat on a bench, with a pad of greased rag beneath the hole, the next thing to do was to pour the molten solder steadily into the hole, and afterwards to apply a heated soldering iron. It is better, however, before pouring in the solder, to heat the end of the anode, so as to prevent it from chilling the metal, and a little chloride of zinc solution should be brushed over the inner surface of the aperture, so as to induce the solder to "run" well over it, and thus insure a perfect connection between the hook and the anode. The importance of securing an absolutely perfect connection between these two surfaces will be recognised when we state that we have known instances in which more than half the number of anodes, in a bath holding 250 gallons, were found to be quite free from direct contact with the supporting hooks, owing to the crystallisation of the nickel salt within the interior of the perforation having caused a perfect separation of the hooks from their anodes. It was to remedy this defect that the author first adopted the system of soldering the connections.

Preparation of the Nickel Solution.—The substance usually employed is the *double sulphate of nickel and ammonia* (or "nickel salts," as they are commonly called), a crystalline salt of a beautiful emerald green colour. This article should be *pure.* For 100 gallons of solution the proportions employed are :—

Double sulphate of nickel and ammonia   .   .   75 lbs.
Water   .   .   .   .   100 gallons.

Place the nickel salts in a clean wooden tub or bucket, and pour upon them a quantity of hot or boiling water ; now stir briskly with a wooden stick for a few minutes, after which the green solution may be poured into the tank, and a fresh supply of hot water added to the undissolved crystals, with stirring, as before. This operation is to be continued until all the crystals are dissolved, and the solution transferred to the tank. A sufficient quantity of cold water is now to be added to make up 100 gallons in all. Sometimes particles of wood or other floating impurities occur in the nickel salts of commerce ; it is better, therefore, to pass the hot solution through a strainer before it enters the tank. This may readily be done by tying four strips of wood together in the form of a frame, about a foot square, over which a piece of unbleached calico must be stretched, and secured either by means of tacks or by simply tying it to each corner of the frame with string.

Nickel Anodes.—It is not only necessary that the nickel salts should be perfectly pure—which can only be relied upon by purchasing them at some well-known, respectable establishment—but it is equally important that the nickel plates to be used as anodes—which may be either of cast or rolled nickel—should be of the best quality. A few years ago there was no choice in this matter, for rolled nickel was not then obtainable. Now, however, this form of nickel can be procured of almost any dimensions, of excellent quality, and any degree of thinness, whereby a great saving may be effected in the first cost of a nickel-plating outfit. Again, some years ago it was impossible to obtain cast nickel anodes of moderate thickness, consequently the outlay for this item alone was considerable. Such anodes can now be procured, however, and thus the cost of a nickel-plating plant is greatly reduced, even if cast anodes are adopted instead of rolled nickel.

Nickel-plating by Battery.—For working a 100-gallon bath, four cells of a 3-gallon Bunsen battery will be required, but only two of these should be connected to the conducting rods until the bath is about half full of work, when the other cells may be connected, which should be done by uniting them for *intensity;* that is, the wire attached to the carbon of one cell must be connected to the zinc of the next cell, and so on, the two terminal wires being connected to the positive and negative conducting rods. If preferred, however, the batteries may be united in series, as above, before filling the bath with work, in which case, to prevent the articles first placed in the solution from "burning," as it is termed—owing to the excess of electric power—it will be advisable to suspend one of the anodes temporarily upon the end of the *negative* rod farthest from the battery, until the bath is about half filled with work, when the anode may be removed, and the remainder of the articles suspended in the solution. In working larger arrangements with powerful currents—to which we shall hereafter refer—resistance coils are employed, which keep back the force of the electric current while the bath is being supplied with work, and even when such coils are used it is usual to suspend an anode or some other "stop," as it is called, from the negative rod during the time the work is being put into the solution.

Twin-Carbon Battery.—A very useful modification of the Bunsen battery, and well suited for nickel-plating upon a small scale, is the American twin-carbon battery, introduced by Condit, Hanson and Van Winkle, of New Jersey, U.S.A., which, in its dissected condition, is represented in Fig. 101. A pair of carbon plates are united by a clamp, with binding screw attached, as shown at A. A plate of stout sheet zinc, is cut out so as to leave a projecting piece, to which a binding screw is also connected, as at B, and the zinc is turned up into an oval form to admit the porous cell, D. The zinc being put into the outer cell, c (which is made of stoneware), the porous cell is

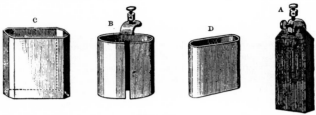

Fig. 101.

placed within the zinc cylinder, and the twin carbons then deposited in the porous cell. The exciting fluids are, for the zinc, which must of course be well amalgamated, 1 part oil of vitriol to 12 parts water. The porous cell is filled to the same height with a mixture composed of equal measures of oil of vitriol and water, to which 2 ounces of nitric acid are added. This is an exceedingly useful and compact battery, and is specially serviceable in nickel-plating upon a moderate scale. When great electro-motive force is required, strong nitric acid is used instead of the above mixture in the porous cell.

Observations on Preparing the Work for Nickel-plating.—For several reasons, it is of greater importance that the articles to be coated with nickel should be what is termed *chemically clean*, than in any other branch of electro-deposition. The excess of cyanide used in gilding, silvering, and brassing solutions is capable of dissolving from the work such slight traces of organic matter as might be accidentally communicated by the hands, and being a powerful solvent of metallic oxides, the delicate film of oxide which quickly forms upon the surface of recently scoured work becomes at once dissolved in a cyanide solution. In the case of a nickel solution, however, which is prepared from a neutral salt, no such solvent action would take place, and the slightest trace of organic matter or of oxide resulting from the action of the air upon the prepared article, would prevent the adhesion of the nickel to the underlying metal, and the work would consequently *strip*. In some establishments, to prevent the possibility of direct contact of the hands with the work while being scoured, the men are required to hold the work with a clean piece of rag, which is frequently dipped in water during the operation of scouring; a good substitute for this is to keep the hand holding the work, while

brushing it with powdered pumice or other material, well charged
with the substance by dipping the fingers occasionally in the powder.
Before explaining the operation of scouring, it will be necessary to
describe the various solutions, or "dips," as they are termed, in
which the articles are immersed before and after being scoured. The
first and most important of these is the potash bath, in which all
articles to be nickel-plated are immersed before undergoing any other
treatment.

The Potash Bath.—The vessel in which the solution of potash is
kept for use generally consists of a galvanised wrought-iron tank
capable of holding from 20 to 150 gallons, according to the requirements of the establishment. An iron pipe, or *worm*, is placed at the
bottom of the tank, one end of which communicates with a steam
boiler, a stopcock being connected at a convenient distance for turning the steam on or off; or the tank may be heated by gas jets, by

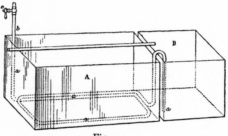

Fig. 102.

means of perforated piping fixed beneath it. An ordinary form of
potash tank is shown at A in Fig. 102, in which the worm-pipe is
indicated by the dotted lines, &c., *a a*, the vertical pipe *b*, with its
stopcock *c*, being conveniently placed at one corner of the tank, as
shown in the engraving. The waste steam from the worm-pipe
escapes into a second tank B, partly filled with water, which thus
becomes heated, and is used for rinsing. A rod of iron, or brass tube
with an iron core, rests upon the bath, longitudinally, for suspending
the articles in the caustic liquor.

The potash solution is made by dissolving half a pound of American
potash in each gallon of water required to make up the bath, and the
solution is always used *hot*. The object of immersing the work to be
nickeled in the potash bath, is to render soluble any greasy matter
which may be present, as, for example, the oil used in the various
processes of polishing. In a freshly made solution (which must

always be kept hot), the work will only require to be immersed for a few minutes, by which time the greasy matter will have become converted into soap, and being thus rendered soluble, may easily be removed by the subsequent operations of brushing with pumice, &c.; but we must bear in mind that the *causticity* of the solution (and consequently its active property) gradually becomes diminished, not only in consequence of the potash having combined with the greasy matter, but also owing to its constantly absorbing carbonic acid from the air. When the bath has been some time in use, therefore, it will be necessary to add a fresh quantity of potash, say about a quarter of a pound to each gallon. It is easy to ascertain if the potash has lost its caustic property by dipping the tip of the finger in the solution, and applying it to the tongue. As the bath becomes weakened by use, the articles will require a longer immersion, and, with few exceptions, a protracted stay in the bath will produce no injurious effect. Articles made from Britannia metal, or which have pewter solder joints, should never be suffered to remain in the potash bath longer than a few minutes, since this alkali (caustic potash) has the power of dissolving tin, which is the chief ingredient of both. Again, articles made from brass or copper should never be suspended from the same rod as steel and iron articles, in case the potash solution should have become impregnated with tin dissolved from solder, &c. ; for if this precaution be not observed it is quite likely (as we have frequently seen in an old bath) that the steel articles will become coated with tin, owing to voltaic action set up in the two opposite metals by the potash solution. Cast-iron work, in which oil has been used in the finishing, should, owing to its porous character, be immersed in the potash bath for a longer period than other metals in order to thoroughly cleanse it from greasy matter.

Dips, or Steeps.—Besides the potash solution, certain other liquids are employed in nickel-plating after the work has been " potashed " and scoured, which may be properly described in this place ; and we may here remind the reader that the employment of these *dips*, as they are called, is based upon the fact that the *neutral* solution of nickel has no power (unlike cyanide solutions) of dissolving even slight films of oxide from work which, after being scoured, has been exposed to the air and become slightly oxidised on the surface. In order, therefore, to remove the faintest trace of oxidation from the surface of the work—the presence of which would prevent the nickel from adhering—it is usual to plunge it for a moment in one or other of the following mixtures after it has been scoured, then to rinse it, and *immediately* suspend it in the nickel bath.

The Cyanide Dip.—This solution is formed by dissolving about half a pound commercial cyanide of potassium in each gallon of water ; for operations on a moderate scale, a stoneware vessel capable of holding about fifteen gallons may be supplied with about twelve gallons of the solution. Baths of the form shown in A, Fig. 103, and which are to be obtained at the Lambeth potteries, are well suited to this purpose. Another form of stoneware vessel is seen in Fig. 104, which, being

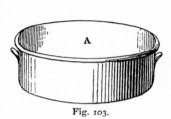

Fig. 103.   Fig. 104.

deeper, is useful for certain classes of work. In applying the *cyanide dip* to articles of great length, it is commonly the practice to employ a common earthenware jug, kept near the dipping bath ; this, being filled with the cyanide solution, is held above the highest point of the article (a brass tube, for instance) and tilted so that its contents may flow downward and pass all over the tube, which is then quickly taken to the water trough or tray and well rinsed, when it is at once placed in the nickel bath. On using the cyanide dip, it must be remembered that its only object is to dissolve from the surface of the recently scoured work an *almost imaginary film* of oxide , therefore the mere contact of the cyanide solution is amply sufficient to accomplish the object ; on no account should brass or copper articles be exposed to the action of the dip for more than *a few seconds ;* indeed, if the solution is in an active condition, the quicker the operation is conducted the better. It will readily be understood, however, that the weak cyanide bath will gradually lose its activity, when the dipping may be effected somewhat more leisurely. It is a common fault, however, to use these dips long after they have yielded up their active power, and we have frequently known them to be employed, and relied upon, when they were utterly useless.

The Acid Dip.—This solution, which is used for dipping steel and iron articles after they have been scoured, is composed of hydrochloric (muriatic) acid and water, in the proportion of half a pound of the acid to each gallon of water. The solution is generally contained in a shallow wooden tub, which may conveniently be the half of a brandy cask or rum puncheon ; but since the acid eventually finds its way to the iron hoops by which such vessels are held together, it is a good plan, in the first instance, to have a couple of wooden hoops, secured by copper rivets, placed over the vessel so as to prevent it from leaking in the event of the iron hoops giving way in consequence of the corrosive action of the acid liquor. Precautions of this nature will prevent leakage and the inconvenience which it involves.

Dipping Acid.—This name is given to a mixture which is frequently used for imparting a bright surface to brass work, and which is variously composed according to the object to be attained. When required for dipping brass work preparatory to nickel-plating, it is commonly composed of—

Sulphuric acid	4 lbs.
Nitric acid	2 ,,
Water	4 pints.

In making up the above mixture, the nitric acid is first added to the water, and the sulphuric acid (ordinary oil of vitriol) is then to be gradually poured in, and the mixture stirred with a glass rod. When cold, it is ready for use. The mixture should be made, and kept, in a *stoneware* vessel, which should be covered by a sheet of stout glass each time after using, to prevent its fumes from causing annoyance and from injuring brass work within its vicinity. The "dipping" should always be conducted either in an outer yard, or near a fire-place, so that the fumes evolved during the operation may escape, since they are exceedingly irritating when inhaled by the lungs. When it is convenient to do so, it is a good plan to have a hood of wrought iron, painted or varnished on both sides, fixed above an ordinary fireplace in the workshop, and to have a hole made in the brickwork above the mantelpiece to conduct the fumes into the chimney ; this arrangement, however, will be of little use, unless there is a good draught in the chimney. It is well to ascertain this, therefore, before the *dipping* is proceeded with, which may be readily done by holding a large piece of ignited paper above the grate, when, if the flame persistently inclines towards the chimney, the draught may be considered perfect ; if, however, it shows any inclination to come *forward*, it may be assumed that the draught is imperfect, owing to the chimney being filled with *cold* air. In this case lighted paper should be applied as before, until the flame and smoke of the ignited material have a *direct* tendency upward, or in the direction of the chimney. We are induced to give these precautionary hints more especially for the guidance of those who may be necessitated to work in apartments of limited space. In all cases, a vessel of clean water should be placed close to the dipping bath, into which the articles are plunged *the instant after* they have been removed from the dipping acid.

Pickling Bath.—Cast iron, before being nickeled, requires to be placed in a cold acid solution, or *pickle*, as it is called, to dissolve or loosen the oxide from its surface. The pickle may be prepared in a wooden tub or tank, from either of the following formulæ :—

> Sulphuric acid (oil of vitriol) . . . . ½ lb.
> Water . . . . 1 gallon.

Cast-iron work immersed in this bath for twenty minutes to half an hour will generally have its coating of oxide sufficiently loosened to be easily removed by means of a stiff brush, sand, and water.

When it is desired that the articles should come out of the bath bright, instead of the dull black colour which they present when pickled in the plain sulphuric acid bath, the following formula may be adopted :—

> Sulphuric acid . . . . 1 lb.
> Water . . . . 1 gallon.

Dissolve in the above two ounces of zinc, which may be conveniently applied in its *granulated* form. When dissolved, add half a pound of nitric acid, and mix well.

CHAPTER XIX.

ELECTRO-DEPOSITION OF NICKEL (*continued*).

Preparation of Nickeling Solutions.—Adams' Process.—Unwin's Process—Weston's Process.—Powell's Process.—Potts' Process.—Double Cyanide of Nickel and Potassium Solution.—Solution for Nickeling Tin, Britannia Metal, &c.—Simple Method of preparing Nickel Salts.—Desmur's Solution for Nickeling Small Articles.

Preparation of Nickeling Solutions.—Although many solutions have been proposed, we may say, with confidence, that for all practical purposes in the electro-deposition of nickel, a solution of the double sulphate of nickel and ammonium, with or without the addition of common salt, will be found the most easy to work and the most uniform in its results, while it is exceedingly permanent in character if worked with proper care and kept free from the introduction of foreign matter. The preparation of a nickel bath from the pure double salt is exceedingly simple, as we have shown, and only needs ordinary care to keep such a solution in good working order for a very considerable period. In order that the reader may, however, become conversant with the various solutions and modifications which ingenious persons have from time to time introduced, we will, as briefly as possible, explain such of these processes as may appear to deserve attention, if not adoption. Boettger's original process having been already referred to, we will now describe Mr. Adams' modification of it, for which he obtained patents in this country, in France, and the United States, and which, after much costly litigation, and consequent loss to those who had become possessed of them, were proved to be unnecessary to the successful deposition of nickel by electrolysis. When the ordinary simple methods of preparing the double salts of nickel and ammonium are taken into consideration, it seems marvellous that Adams' exceedingly roundabout process—which no one with practical chemical knowledge would dream of following—should have been considered worth contesting ; not to defend the process as such, which no one infringed, but to secure the sole right to deposit nickel by electro-chemical means, by any process whatever. And what was the real " bone of contention "? It was based upon the most absurd " claim " ever allowed to become attached to a patent, which runs as follows :—

"'The electro-deposition of nickel by means of a solution of the double sulphate of nickel and ammonia, or a solution of the double chloride of nickel and ammonium, prepared as [below] described, and used for the purposes [below] set forth, in such a manner as to be free from the presence of potash, soda, alumina, lime, or nitric acid, or from any acid or alkaline reaction."

According to this, if any solution of nickel, no matter how prepared, which could be proved by analysis to be free from the substances named (not one of which would be a *necessary* associate of nickel or of its double salts), such solution, if used in nickel-plating, would be an infringement of the patent! This *we know* was the impression of those who held the English patent, and we vainly endeavoured to show its fallacy. " Any solution of nickel which is free from these substances and used for plating purposes is an infringement of our patent." That was the contention, and the owners of this patent believed themselves entitled to an absolute monopoly of the right to nickel-plate within the four quarters of the United Kingdom.

Adams' Process.—In preparing the solution, the inventor prefers to use pure nickel, but commercial nickel may be used. "Commercial nickel," says the patentee, " almost always contains more or less of the reagents employed in the purification of this metal, such as sulphate of lime, sulphide of calcium, sulphide of sodium or potassium, chloride of sodium, and alumina. When any of these substances are present, it is necessary to remove them. This can be done by melting the nickel, or by boiling it in water containing at least 1 per cent. of hydrochloric acid. The boiling must be repeated with fresh acid and water until the wash-waters give no indication of the presence of lime when treated with oxalate of ammonia. When the metal is purified by melting, the foreign substances collect on the top of the metal in the form of slag, which can be removed mechanically. If the nickel contains zinc, it should be melted in order to volatilise the zinc and drive it off. The crucible in such a case must not be closed so tightly as to prevent the escape of the zinc fumes. If copper, arsenic, or antimony be present in the nickel, they can be removed, after the nickel is dissolved, by passing sulphuretted hydrogen through the solution. The acid to be used in dissolving the metal consists of 1 part strong nitric acid, 6 parts muriatic acid, and 1 part water. Nitric acid or muriatic acid may be used separately, but the above is preferred. A quantity of this acid is taken sufficient to dissolve any given amount of the metal, with as little excess of the former as possible ; a gentle heat is all that is required. The resulting solution is filtered ; and to prepare the solution of the double sulphate of nickel and ammonium, a quantity of strong sulphuric acid, sufficient to convert all the metal into sulphate, is added, and the solution is then evaporated to dryness. The mass is then again dissolved in water, and a much smaller quantity than before of sulphuric acid is added, and the whole again evaporated to dryness, the temperature being raised finally to a point not to exceed 650° Fahr. This temperature is to be sustained until no more vapours of sulphuric acid can be detected. The resulting sulphate of nickel is pulverised, and thoroughly mixed with about one-fiftieth of its weight of carbonate of ammonia, and the mass again subjected to a gradually increasing temperature, not to exceed 650° Fahr., until the carbonate of ammonia is entirely evaporated. If any iron is present, the most of it will be converted into an insoluble salt, which may be removed by filtration. The resulting dry and neutral sulphate of nickel is then dissolved in water by boiling, and if any insoluble residue remains, the solution is filtered. From the weight of nickel used before solution, the amount of sulphuric acid in the dry sulphate can be calculated. This amount of sulphuric acid is weighed out and diluted with four times its weight of water, and saturated with pure ammonia or carbonate of ammonia —the former is preferred. This solution, if it is at all alkaline, should be evaporated until it becomes neutral to test-paper. The sulphate of ammonia of commerce may likewise be used, but pure sulphate of ammonia is to be preferred. The two solutions of the sulphate of nickel and sulphate of ammonia are then united, and diluted with sufficient water to leave $1\frac{1}{2}$ to 2 ounces of nickel to each gallon of solution, and the solution is ready for use. The object of twice evaporating to dryness and raising the temperature to so high a degree is, in the first place, to drive off the excess of sulphuric acid ; and secondly, to convert the sulphate of iron, if it exists, into basic sulphate, which is quite insoluble in water.

" In order to give the best results, it is necessary that the solution should be as nearly neutral as possible, and it should in no case be acid. The inventor prefers to use the solution of a specific gravity of about 1·052 (water 1·000), though a much weaker or still stronger solution may be used. At temperatures above the ordinary the solution still gives good results, but is liable to be slowly decomposed. An excess of sulphate of ammonia may be used to dilute the solution, in cases where it is desirable to have it contain much less than 1 ounce of nickel to the gallon.

" In preparing the solution of double chloride of nickel and ammonium, the nickel is to be purified and dissolved in the same manner as is described for the previous solution ; and it is to be freed from copper and other foreign matters in the same manner. The solution is then evaporated to dryness ; it should be rendered as anhydrous as possible. The salt is then placed in a retort, and heated to a bright red heat. The salt sublimes, and is collected in a suitable receiver, the earthy matter being left behind. The salt, thus purified, is dissolved in water, and to the solution is added an equivalent quantity of pure chloride of ammonium. The solution is then ready for use ; it may have a specific gravity of 1·050 to 1·100."*

The repeated evaporations recommended by Adams are wholly unnecessary in the preparation of the double sulphate of nickel and ammonium or the double chloride, for if the nickel be pure (and there is no difficulty in obtaining it in this condition), the ordinary method of dissolving the metal or its oxide, and subsequent addition of the ammonia salt and careful crystallising the double salt, would give the same result, with far greater economy, both of time and trouble.

Unwin's Process.—This ingenious process, for which Mr. Unwin obtained a patent in 1877, and in which crystallisation of the salts is rendered unnecessary, is conducted as follows : — He first prepares the sulphate of nickel " by taking three parts of strong nitric acid (sp. gr. about 1·400), one part of strong sulphuric acid (sp. gr. about 1·840), and four parts of water, all by measure, mixing them cautiously, and about half filling an open earthenware pan with the mixture. To every gallon of this mixed acid, I then add about two pounds of ordinary grain or cube nickel, and I heat the liquid by a sand-bath or other suitable means. If during the process of solution the action becomes inconveniently violent, I moderate it by the addition of a little cold water. If the nickel entirely dissolves (except a small quantity of black matter), I add more of it, in small portions at a time, and continue the addition at intervals until it is in excess. When the production of red fumes has nearly, or entirely, ceased, or when the liquid becomes thick and pasty, from the separation of solid sulphate of nickel, I add a moderate quantity of hot water, and boil and filter the solution ; the deep green liquid so obtained is a strong solution of sulphate of nickel. If, from the circumstance of its production, I consider that it requires purification, I concentrate the solution by evaporation, until on cooling it yields a considerable percentage of crystals of sulphate of nickel ; these crystals I collect, wash with a little cold water, and redissolve in a moderate quantity of hot water, filtering again if necessary. When cold, the liquid is ready for further treatment.

" I next prepare a strong solution of sulphate of ammonia, by dissolving the salt in hot water, in the proportion of about four pounds of the salt to each gallon of water, and then filter the liquid if necessary, and allow it to become cold. I then obtain the pure double sulphate of nickel and ammonia by adding the above solution of sulphate of ammonia to that of the sulphate of nickel ; but I do not

---

* For further remarks upon Adam's process, see pp. 460, 461.

stop the addition of the solution of sulphate of ammonia, when sufficient has been added to combine with all the sulphate of nickel present, but I continue to add a large excess. I do this because I have discovered that the double sulphate of nickel and ammonia is far less soluble in the solution of sulphate of ammonia than in pure water, so that it is precipitated from its solution in water on adding sulphate of ammonia. I therefore continue adding the solution of sulphate of ammonia, continually stirring, until the liquid loses nearly all its colour, by which time the double sulphate of nickel and ammonia will have been precipitated as a light blue crystalline powder, which readily settles to the bottom of the vessel. I then pour off the liquid from the crystalline precipitate of double sulphate of nickel and ammonia, and wash the latter quickly with a strong, cold solution of sulphate of ammonia, as often as I consider necessary for its sufficient purification ; but I do not throw away this liquid after use, but employ it at my discretion for combining with fresh sulphate of nickel, instead of dissolving a further amount of sulphate of ammonia. If I desire to make a further purification of the double sulphate of nickel and ammonia, I make a strong solution of it in distilled water, and add to the liquid a strong solution of sulphate of ammonia, by which means the double sulphate is precipitated in a very pure condition, and is separated from the liquid by filtration, or by other convenient means, and then dried, or used direct as may be desired ; the liquid strained away can be employed, instead of fresh solution of sulphate of ammonia, for combining with more sulphate of nickel, or for washing the precipitate of the double sulphate.''

Weston's Process.— Mr. Edward Weston, of Newark, N.J., having observed that boric acid, when added to nickel solutions, produced favourable results in the electro-deposition of nickel, obtained a patent for '' the electro-deposition of nickel by means of a solution of the salts of nickel containing boric acid, either in its free or combined state. The nickel salts may be either single or double.'' The advantages claimed for the boric acid are that it prevents the deposit of sub-salts upon the articles in the bath, which may occur when the bath is not in good condition. Mr. Weston further claims that the addition of this acid, either in its free or combined state, to a solution of nickel salts renders it less liable to evolve hydrogen when the solution is used for electro-deposition ; that it increases the rapidity of deposition by admitting the employment of a more intense current, while it also improves the character of the deposit, which is less brittle and more adherent. Mr. Wahl, after extended practical trials of Mr. Weston's formula, states that they have '' convinced him of the substantial correctness of the claims of the inventor,'' and he adds, '' Where the double sulphate of nickel and ammonia is used, the addition of boric acid, in the proportion of from 1 ounce to 3 ounces to the gallon of solution, gives a bath less difficult to maintain in good working order, and affords a strongly adhesive deposit of nickel. The deposited metal is dense and white, approaching in brilliancy that obtained from the solution of the double cyanide." The formula for preparing the solution is—

Double sulphate of nickel and ammonia . . .	10 parts.
Boric acid, refined	2½ to 5 „
Water	150 to 200 „

Powell's Process.—This inventor claims to have discovered that benzoic acid, added to any of the nickel salts, arrests, in a marked degree, the tendency to an imperfect deposit, prevents the decomposition of the solution, and consequent formation of sub-salts. The proportion of benzoic acid to be added to the bath is said to be one-eighth of an ounce to the gallon of solution. This bath has been favourably spoken of. Powell also gives the following formulæ for nickel baths:—

1. Sulphate of nickel and ammonia	10 parts.
Sulphate of ammonium	4 „
Citric acid	1 „
Water	200 „

The solution is prepared with the aid of heat, and when cool, a small quantity of carbonate of ammonia is added until the solution is neutral to test paper.

2. Sulphate of nickel	6 parts.
Citrate of nickel	3 „
Phosphate of nickel	3 „
Benzoic acid	1½ „
Water	200 „

3. Phosphate of nickel	10 parts.
Citrate of nickel	6 „
Pyrophosphate of sodium	10½ „
Bisulphite of sodium	1½ „
Citric acid	3 „
Liquid ammonia	15 „
Water	400 „

These solutions are said to give good results, but the very complicated nature of the latter almost takes one's breath away.

Potts' Process.—In 1880, Mr. J. H. Potts, of Philadelphia, patented an improved solution for the electro-deposition of nickel, which consists in employing acetate of nickel and acetate of lime, with " the addition of sufficient free acetic acid to render the solution distinctly acid." The formula is given below :—

Acetate of nickel	2¾ parts.
Acetate of calcium	2½ „
Water	100 „

To each gallon of the above solution is added 1 fluid ounce of acetic acid of the sp. gr. 1·047. Mr. Potts first precipitates the carbonate of nickel from a boiling aqueous solution of the sulphate, by the addition of bicarbonate of soda, then filters and dissolves the well-washed precipitate in acetic acid, with the aid of heat.

" To prepare this bath, dissolve about the same quantity of the dry carbonate of nickel as that called for in the formula (or three-quarters of that quantity of the hydrated oxide) in acetic acid, adding the acid cautiously, and heating until effervescence has ceased and solution is complete. The acetate of calcium may be made by dissolving the same weight of carbonate of calcium (marble dust) as that called for in the formula (or one-half of the quantity of caustic lime), and treating it in the same manner. Add the two solutions together, dilute the volume to the required amount by the addition of water, and then to each gallon of the solution add a fluid ounce of free acetic acid as prescribed."

In reference to the above solution, Wahl says that he has worked it under a variety of circumstances, and has found it, in many respects, an excellent one. "It gives satisfactory results," he states, "without that care and nicety in respect to the condition of the solution and the regulation of the current which are necessary with the double sulphate solution. The metallic strength of the solution is fully maintained, without requiring the addition of fresh salt, the only point to be observed being the necessity of adding from time to time (say once a week) a sufficient quantity of acetic acid to maintain a distinctly acid reaction. It is rather more sensitive to the presence of a large quantity of free acid than to the opposite condition ; as in the former condition it is apt to produce a black deposit, while it may be run down nearly to neutrality without notably affecting the character of the work. The deposited metal is characteristically bright on bright surfaces, and requiring but little buffing to finish. It does not appear, however, to be so well adapted for obtaining deposits of extra thickness as the commonly used double sulphate of nickel and ammonium. On the other hand, its stability in use, the variety of conditions under which it will work satisfactorily, and the trifling care and attention it calls for, make it a useful solution for nickeling."

Double Cyanide of Nickel and Potassium Solution.—This was

X

one of the earliest solutions used for depositing nickel, and is capable of yielding an exceedingly white deposit.   Though neither so economical nor so susceptible of yielding stout deposits of nickel as the ordinary double sulphate or double chloride, it may be advantageously employed when only a thin coating of a fine white colour is desired. It is stated to be somewhat extensively used in some large nickel-plating works in the United States.   To prepare the solution, pure nickel or oxide of nickel is dissolved in either of the mineral acids ; a mixture of hydrochloric and nitric acids, in the proportion of four parts of the former to one of the latter, may be used, an excess of the metal being taken to fully neutralise the acid.   The solution is then evaporated and set aside to crystallise.   The crystals, after being well drained and quickly rinsed in cold water, are next dissolved in water by the aid of heat, and when the solution has become cold a solution of cyanide of potassium is carefully added, with stirring, until all the metal has been thrown down in the form of cyanide of nickel.   Care must be taken not to add an excess of cyanide.   The supernatant liquor is now to be poured off, and the precipitate washed repeatedly with water. A strong solution of cyanide is next added, with stirring, until all the cyanide of nickel is dissolved.   A small excess of cyanide is then to be added, when a reddish-brown solution of double cyanide of nickel and potassium will result, which, after filtering, is ready for use.   The solution should be as concentrated as possible, almost to the point of saturation.

Solution for Nickeling Tin, Britannia Metal, &c.—The following formula has been recommended for coating tin, Britannia metal, lead, and zinc, as also brass and copper :—

Sulphate of nickel and ammonium	. . .	10 parts.
Sulphate of ammonium	. . . . .	2 „
Water .	. . . .	. 300 „

The salts are to be dissolved in boiling water, and when cold the solution is ready for use.   For nickeling cast and wrought iron and steel the following bath is recommended :—

Sulphate of nickel and ammonium	. . .	10 parts.
Sulphate of ammonium	. . . .	1½ „
Water .	. . . . . .	. 250 „

Simple Method of Preparing Nickel Salts.—To make the double chloride of nickel and ammonia take, say, 2 ounces of pure cube nickel, or oxide of nickel, and dissolve in hydrochloric acid, to which a little nitric acid may be added, taking care not to have an excess ; apply gentle heat to assist the chemical action.   When the evolution of gas has ceased dilute the resulting solution with cold water to make about 1 quart of liquor ; now add liquid ammonia gradually, stirring after each addition, until the solution is neutral to test-paper ; now dissolve 1 ounce of chloride of ammonium (sal-ammoniac) in sufficient water, and add this to the former solution ; evaporate the mixture until crystals begin to form, then allow it to cool and crystallise gradually ; next pour off the clear liquor, and repeat the evaporation to obtain a second batch of crystals ; in the latter operation the solution may be evaporated to dryness. Finally, mix all the resulting products together and dissolve in about three pints of hot water, filter, and make up to about one gallon by the addition of cold water. The solution should have a specific gravity of 1·050 to 1·075.

The Double Sulphate of Nickel and Ammonium may readily be formed by dissolving oxide or carbonate of nickel in dilute sulphuric acid (1 part acid to 2 parts water). The resulting solution is then to be neutralised with ammonia and crystallised. To each pound of the dry crystals add 1 pound of pure sulphate of ammonia, dissolve the mixed salts, evaporate the solution, and re-crystallise. Cube or grain nickel may also be dissolved in a mixture composed of 1 part sulphuric acid and 2 parts water, with the addition of a small quantity of nitric acid, moderate heat being applied as before. The solution is then to be evaporated and set aside to crystallise, and to convert the sulphate of nickel into the double salt, sulphate of ammonia is to be added in the same proportion as before : the mixed salts must be dissolved, filtered, and crystallised. In making up a bath from the double sulphate prepared by either of the above methods, about 12 ounces of the dry crystals are to be taken for each gallon of bath, and the crystals should be dissolved in sufficient hot water, the solution filtered, and the requisite quantity of cold water added to make up the full quantity of the solution in the proportions given. At the temperature of 60° Fahr. the bath should have a specific gravity of about 1·052. It is necessary to state that the nickel employed should be *pure*, which can only be relied upon by obtaining it from some well-known respectable house.

Desmur's Solution for Nickeling Small Articles.—The author is indebted to M. Desmur for the following formula for coating small articles, which we recommend to the attention of those whose trade chiefly lies in nickeling struck work, such as umbrella-mounts, and the like :—

Double sulphate of nickel and ammonium . 7 kilogrammes
Bicarbonate of soda . . . . . 800 grammes.
Water . . . . 100 litres.

The bicarbonate of soda must be added when the nickel solution is warm, in small quantities at a time, otherwise the effervescence which occurs may cause the solution to overflow. The bath is to be worked up to nearly boiling point. If, after working for some time, the deposit becomes of a darkish colour, add a small lump (about the size of a nut) of sulphide of sodium, which will remedy it. "Of all the solutions of nickel which I have tried," says M. Desmur, "this has, without doubt, given me the best results, both as to quickness of working and whiteness of deposit, which is equal to that of silver. Nickel deposited from this solution can be burnished. If the nature of the articles to be nickeled will not allow them to be either polished or burnished, they may be rendered bright by first dipping them in nitric acid and afterwards passing them rapidly through a mixture of old nitric acid dip (already saturated with copper), sulphuric acid, greasy calcined soot, and common salt."

CHAPTER XX.

ELECTRO-DEPOSITION OF NICKEL (*continued*).

Preparation of the Work for Nickel-plating.—Since the various
metals ordinarily coated with nickel require different treatment, it
will be more convenient to
treat them under their respective heads, by which
the intending nickel-plater
will become more readily
conversant with the mani-
pulation requisite in each
particular case. All the
preliminary arrangements

Fig. 105.

of nickel bath, batteries, dips, &c., being complete, the work, as it
is received from the polishing shop, should be placed in regular
order upon a bench, the name of each customer being indicated by a
ticket for each group of work, so as to prevent confusion. Small
work is generally handed into the plating-room upon shallow trays,
of the form indicated in Fig. 105. These trays are usually about
2 feet long by 15 inches wide, and about 3 inches deep ; they are
made of ordinary inch deal, planed on both sides, and the corners are
bound with stout sheet iron. The trays are made of various sizes to
suit the different classes of work to be conveyed in them. The
reader is referred to another chapter for a description of the process
of polishing.

The Scouring Tray.—This apparatus, which has to be subjected
to much wear and tear, requires to be well put together, and must be

thoroughly water-tight. A sketch of the scouring tray generally adopted is shown in Fig. 106. It is usually made from two-inch deal, planed on both sides ; the joints are rendered water-tight by means of india-rubber, and the various parts are well bound together by screwed bolts and nuts. The dimensions may be 6 or 8 feet long (inside), 2 feet 6 inches wide, and about 15 or 18 inches deep. It is divided into two equal compartments by a wooden partition, and a stout shelf is fixed across one compartment, upon which is a small block of wood —about 7 or 8 inches long, and 2 inches square, secured to the shelf, by screws, from beneath, for scouring small articles. A water-tap,

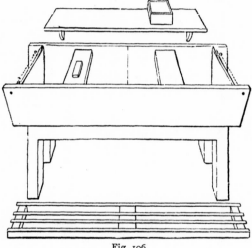

Fig. 106.

with india-rubber hose, is placed at a convenient distance above the tray, by which means either compartment may be filled at pleasure. At the corner of either compartment of the scouring tray is a flanged exit pipe, let into the bottom at the far corner, to allow the tray to be emptied when required. The second compartment is used as a rinsing trough. The exit pipes are furnished with a wooden plug, which the workman withdraws when he desires to run off the water from either compartment. A wooden shelf is generally fixed at a convenient distance above the back of the scouring tray, to hold various brushes; pumice-box, or other tools required in preparing work for the bath.

Brass and Copper Work.—The articles are first suspended, by means of short lengths of copper wire, in the hot potash bath, where they are allowed to remain until ready for scouring. The "slinging wires" for this purpose, as also for suspending the articles in the nickel bath, should be of various thicknesses, according to the weight they have to sustain, and it is a good plan to keep bundles of these wires, cut up into regular lengths, bound together by a piece of the same wire, so that they may be readily withdrawn as required. The articles being taken out of the potash bath, one by one, or a few at a time, according to their size, are at once plunged into the water in either compartment of the scouring tray. They are next subjected to the operation of *scouring*.

Scouring.—This usually consists in well brushing the work with finely powdered pumice and water, by means of hog-hair brushes. Some platers prefer a mixture of pumice and rottenstone for brass work, as being rather less *cutting*, and therefore less liable to scratch the work so severely as the pumice and water alone. The author's son, Mr. A. N. Watt, has succeeded well in employing ordinary whiting in scouring brass and copper work, which, while sufficiently cleaning the articles, enables them to come out of the nickel bath in a much brighter condition than when pumice is used, and as a natural consequence the work requires less time and trouble in finishing. We believe that recently slaked lime, either alone or mixed with whiting, would be better still, were it not for the fact that the caustic lime would be injurious to the hands of the workmen.

In scouring the work it is placed on the shelf across the scouring tray : the brush is then dipped in water and afterwards in the powdered pumice, or other material—which is kept in a wooden box upon the back shelf —and the article is well brushed all over, beginning at one end, and then turning the article round to brush the other ; a final brushing is then given all over, as quickly as possible, so as to render the surface uniform. As each article is brushed it is rinsed in clean water, the slinging wire is then attached, and the article next dipped, for an instant, in the cyanide dip, again well rinsed, and immediately after suspended in the nickel bath, where it is allowed to remain from four to eight hours, according to whether the work is to be moderately or thoroughly coated with nickel.

As we have before observed, all work which is to be bright when finished must be polished before being nickel-plated. If, however, it were to be immersed in the nickel bath without any further preparation (unless a *very* slight coating of nickel were given), even if perfectly free from greasy matter and oxide upon its surface, the nickel would surely strip or peel off. Hence the operation of scouring is adopted—not alone to render the surface of the metal absolutely clean, but to give it an almost imaginary degree of roughness. It is a fact well known to electro-depositors that when a surface of metal is perfectly bright any other metal deposited upon it will readily separate. The surface may be *all but bright*, and the two metals will adhere more or less firmly; but if it is *absolutely bright*, the metals have little or no cohesion. In scouring, therefore, great care must be taken that the application of the brush and pumice has been perfectly uniform all over the work, and that the bright lustre given to it by the polisher has been thoroughly removed. To produce this result, the work does not entail laborious *scrubbing*, but is accomplished by a brisk brushing, taking care to keep the brush well charged with the pumice. We have seen men, improperly instructed in this respect, who have first dipped the brush in water, then in the pumice powder, and finally in the water again, before applying it to the work, whereby they actually washed away the material before the brush was applied! Again, it is a common error to dip the brush in the pumice before shaking the superfluous water from it, which not only causes the powder to become deluged with water, and a considerable portion of it to be wasted, but in this extremely wet condition it has little effect upon the surface to be cleaned. The brush should only be *moist* when dipped in the powder, in which state it will take up sufficient material to spread over a considerable surface, and will then do its required work effectually, with very little waste. Some scourers are very wasteful in this respect, and as a rule their work is never properly cleaned, or *pumiced*. The brushes employed in scouring are made from hog bristles, and are supplied, for the general purposes of the plater, of various widths, and are known as one-rowed, two-rowed, three-rowed, and four-rowed brushes, each terminating in a suitable handle (see Fig. 94). The brushes, in their separate sizes, may be laid upon the shelf behind the scouring tray, so as to be ready to hand when required for use, and they should on no account be used for any other purpose than scouring the work. New brushes may be dipped *for an instant* in the potash bath, and immediately rinsed, by which any greasy matter communicated to the hairs or bristles during the manufacture will be rendered soluble, and will afterwards wear away in use. This precaution is not altogether unnecessary, since these brushes have frequently been used by workmen for brushing their clothes, and sometimes even their hair.

Immersing the Work in the Bath.—When we bear in mind that the nickel anodes have a stationary or fixed position in the bath, and that consequently a very large surface of the positive electrode is exposed, it will be at once apparent that some means must be adopted, when the first batch of articles are being placed in the bath, to prevent the deposit from taking place too rapidly (owing to the excess of anode surface), and thereby causing the work to "burn," as it is called.

When dynamo or magneto-electric machines are employed, *resistance coils* are used to regulate or control the force of the current, as we shall explain hereafter ; but although such coils are less necessary when depositing by battery power, some other equally effective means must be adopted. The most simple plan is to hook one of the anodes on the negative conducting rod, at its farthest end from the battery, and there to leave it until the rod is nearly supplied with work, when it may be removed and put in its proper place on the positive rod. By adopting this practice with each suspending rod in turn the " burning " of the work is prevented, and deposition takes place gradually.

When work of moderate dimensions—as brass taps, for example—and very small articles are in hand for nickeling, the larger work should be put into the bath first, and the smaller work then introduced between other pieces of larger work. It is also usual to commence suspending the work from the end of the rod nearest the battery (where the power is weakest) rather than from the opposite end. Small articles—such as screws, for example —should not be slung singly, but several of them suspended from the same wire, as in Fig. 107, in such a way as not to be in contact with each other.

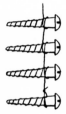

Fig. 107.

Nickeling Small Steel Articles. *—This class of work, after cleaning, immersion in the acid dip, and rinsing, should be suspended in the bath, if practicable, between other articles of larger dimensions, so that deposition may take place slowly and gradually ; otherwise the articles are very liable to strip. This precaution is specially necessary in nickeling small dentists' tools, as excavators, &c., which, when exposed to too strong a current, are apt to burn at the lower end and strip. In nickeling such work the rule is, after the article has become " struck " (that is, coated all over), to allow the deposit to take place very slowly, especially during the first half-hour's immersion. When battery power is used, from one to two hours' immersion will be sufficient for a serviceable coating upon the smaller dental tools, but a somewhat longer period—say, up to three hours—should be given to dental forceps. When a dynamo-machine is employed, about half this time will be sufficient. It is very important that steel work should be placed in the bath *immediately* after being cleaned, since even a few moments' exposure to the air or immersion in water will cause an invisible film of oxide to form on the surface, which will prevent the nickel from firmly adhering to the steel. After nickeling, the articles are rinsed in hot water and handed to the finisher, who gives them the

---

* See also pp. 461 *et seq.*

necessary high polish. Small steel or iron articles which are not required to receive a stout coating of nickel are first steeped for a short time in the potash bath, and after being rinsed are dipped for a moment in the hydrochloric acid dip, again rinsed, and put into the nickeling bath, without any preparatory scouring, and given a short immersion only—say, half-an-hour. Such work is generally finished by being *dollied* only, which brings up the surface to its proper brightness.

Nickeling Small Brass and Copper Articles.—When these have to receive a good coating and afterwards to be finished bright, they must be scoured after polishing, and treated in all respects the same as larger work. Articles which are not required to be stoutly nickeled, however, but only moderately well coated with this metal, may be polished with the rouge composition referred to in another chapter, instead of with lime in the usual way, and then placed in the bath without previous scouring. When they have received a moderate coating of nickel they are rinsed in hot water, and afterwards finished with the mop, or dolly, with the aid of the same composition. This method of treating small brass work—which we believe is of American origin—is especially suitable for umbrella mounts, reticule and purse frames, cheap fancy work, and such articles as are not liable to much friction in use. Small brass articles which are not required to be bright are first put into the potash bath for a short time, and after rinsing they are dipped in ordinary dipping acid, again well rinsed in several waters, and then put into the nickel bath, in which they receive a deposit according to the nature and quality of the work and the price to be paid for it, a short immersion, in many cases, being all that is given when the price is low.

Nickeling by Dynamo-electricity.—Although a very considerable amount of work of all kinds is coated with nickel by battery current, by far the greater portion passes through the hands of those who adopt dynamo or magneto-electric machines as the source of electricity. Indeed, if it were not for the great advantages which these machines present in the deposition of this metal, the art of nickel-plating would never have attained its present magnitude. In arranging a nickeling plant upon a large scale, the baths should be placed parallel to each other, having sufficient space between each vat for the free passage of the workmen ; and the dynamo-machine should be stationed conveniently near the vats, so as to be under the immediate control of the plater. The conducting wires should be so arranged that the current may be applied to one or more of the baths, as occasion may require, and this may be most conveniently effected by fixing two stout brass or copper rods, by means of insulating brackets, to the wall of the apartment nearest the nickel tanks ; these leading

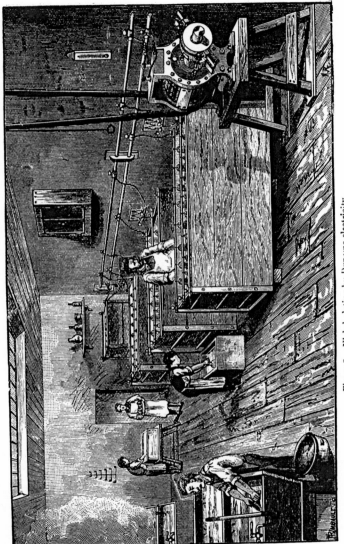

Fig 108.—Nickel-plating by Dynamo-electricity.

wires or conducting rods must have attached to them a series of binding screws, corresponding in number to the connecting screws of the suspending rods. A large form of nickel tank, capable of holding from 250 to 500 gallons of solution, is shown in Fig. 109. To connect the machine with the leading rods stout copper wire is used, the thickness of which is regulated according to the power of the machine. For a medium-sized Weston, half-inch copper wire is generally used, but for larger machines the wire employed is usually three-fourths to one inch in thickness. To convey the current from the leading rods to the baths, the wire need not be so stout as in the former case, about one-half the thickness being sufficient. To give motion to the machine a counter-shaft is usually fixed overhead, with its driving pulley immediately in a line with the pulley of the machine, the two being connected by a belt in the usual way. The counter-shaft, an im-

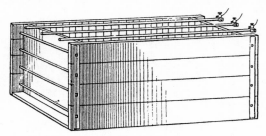

Fig. 109.

proved form of which has been introduced by Carlyle, is shown in Fig. 110, is furnished with a long iron handle within reach, by the raising or lowering of which the belt is placed on the fast or loose pulley of the shaft, according to whether the machine is required to run or stop. To regulate the amount of current entering the respective baths, a resistance coil is either attached to the end of each bath or fastened to the wall facing the end of each vat, and these coils are interposed in the circuit by means of short conducting wires. (See Fig. 108.)

In working large tanks of nickel solution for coating articles of moderate size, as taps, spurs, bits, table lamps, &c., three rows of anodes are generally used, which are thus disposed : one row of anodes is suspended from a conducting rod on each side of the tank, and the third row is placed in the centre of the bath, midway between the other two. Two rods for supporting the work to be nickeled are placed between the side and centre rows of anodes, by which arrangement the suspended articles will be exposed to the action of two anode surfaces. The three anode rods must be united at their ends by means of thick copper wire, in which case one binding screw only, attached to the end of one of the side rods, will be necessary to connect the anodes with the positive leading wire of the machine ; or a separate binding screw may be connected to the end of each rod, and the connection with the leading rods completed with short lengths of stout wire. The latter plan is the best, since one or more rows of anodes can be more readily thrown out of circuit by simply disconnecting the wire from the binding screw at the end of the rod.

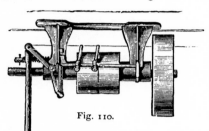

Fig. 110.

Nickeling Mullers, Sausage Warmers, &c.—Large brass and copper articles—such as beer and wine mullers, for example—owing to the extent of surface they present, and their peculiar form, require a different arrangement of the anodes to that which is adopted for ordinary work, and for this reason : it is well known that all metals receive the deposit most freely upon the surfaces facing the anode ; and although in gilding, silvering, and coppering the deposit takes place to a moderate extent upon those surfaces of an article which do not directly face the anode, in the case of nickel it is quite different, for under the same conditions little or no deposit would take place at the opposite parts of the article unless an anode were suspended on each side of it, as in the arrangement we have described. Since mullers, and articles of the class to which they belong, present an extensive convex surface, it is necessary, in order to secure a uniform coating of nickel, to *surround* such work with anodes as far as is practicable. This is ordinarily done in the following manner :— The centre row of anodes is first removed ; two short brass rods are then placed across the other positive rods, about 2 feet apart, and upon each of these is suspended one or more anodes, according to their width. The centre conducting rod, lately occupied by the anodes, is now used as the suspending rod for the muller. Where more than one nickel bath is employed, it is best to keep one of these specially for mullers and other large work, in which case two rows of anodes only and one centre negative rod, should be applied. The bath used for nickeling mullers should be kept covered with a frame, upon which oiled calico is stretched, to protect the work from dust. The drawing (Fig. 111) shows the relative position of the muller and the surrounding anodes. When the article has been in the bath some time, its position must be reversed—that is to say it must be *inverted*—so as to equalise the coating as far as possible, since the deposit always occurs most energetically at the *lower* surface of the article in the solution. In a 100-gallon bath only a single muller, or similar article, could be nickeled at one time ; in other words, it should have the whole bath to itself. When dynamo-machines are employed, however, the baths seldom consist of less than 250 gallons of solution.

In nickeling the above class of work great care and smartness of manipulation are necessary. The work requires to be well and briskly brushed with pumice after removal from the potash bath, and after being rinsed is passed through the cyanide dip for a moment, and again well rinsed, and no time should now be lost in getting it into the nickel bath, and connecting it to the conducting rod. Soon after immersion the characteristic whiteness of the nickel should be visible

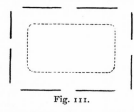 upon its surface, as evidence that current is sufficiently strong to do the work required of it. Such being the case, the article must be left for awhile, after which it may be gently moved up and down by its slinging wires (but not out of the solution) to disperse any dusty particles that may have settled upon its upper surface, since these, however slight and imperceptible, will

Fig. 111.

sometimes cause a rough and irregular deposit, which will give some trouble to the polisher when finishing the article. When plating work of this description in a bath which has been long in use, the anodes should be arranged as described some time before a muller is placed in the bath, so that the sediment (which always accumulates at the bottom of the vessel), if disturbed, may have time to subside ; and in placing the article in the bath care should be taken to lower it gently, so as not to disturb the "mud," if we may call it so, at the bottom. The opposite of this careful treatment needs only to be tried once to make the plater exceedingly particular thereafter.

However careful the operator may be, it sometimes happens that certain parts of the article will become bare, or "cut through," as it is termed, during the process of finishing, in which case it is sent back from the polishing-room to the plater, who must in some way deposit nickel upon the exposed surface. This is accomplished by applying the "doctor," by which means a coating of metal is deposited upon the naked spot in the following way : a piece of stout copper wire, about a foot in length, is bent in the form of a hook at each end ; a small piece of nickel, about an inch and a half square, is attached to one of the hooks, and this is wrapped up in several folds of rag, secured to the wire by twine. The other hook is connected by a long copper wire to the anode rod, and the article to be *doctored* connected in the same way to the negative rod. Now dip the rag-end of the wire in the nickel bath, and apply it to the bare spot (which should be previously brushed over lightly with pumice), keeping it in contact for a few seconds, then dip it in the bath again and apply as before, repeating the operation every half-minute or so, until a sufficient deposit of nickel has been given to the spot to enable the finisher to apply the "dolly," and thus render this part as bright as the rest of the article. Although this may not be considered a very conscientious method of getting over the difficulty, unless performed with patience, so as to impart something more than a mere film upon the bare place, it must be borne in mind that if the entire article were to be re-nickeled this would involve an amount of trouble and expense of labour which would never be compensated for. The "doctoring," however, should always be done *well*, and since the articles to which it is usually applied are rarely subjected to such friction as would affect so hard a metal as nickel, the defective portion of the work may cause little or no annoyance to the owner.

Nickeling Bar Fittings, Sanitary Work, &c.—Articles of this description require to be thoroughly well coated with nickel, and finished in the best possible manner. Before submitting such work to any preparatory process, the plater should carefully examine each piece to ascertain if it has been properly polished and all scratches and file marks obliterated, since if any of these be present after the article is nickeled and finished they will greatly impair the appearance of the work, while the finisher will be quite powerless to remove them without cutting through the nickel deposit. A careful examination of all work should be made by the plater before allowing it to enter the potash bath, and since in most establishments the polishing and finishing are done on the establishment, a proper understanding should exist between the finisher and plater as to the absolute necessity of having the work finished in the best possible manner. The articles, after being approved by the plater, are handed to the scourer, who connects a stout copper wire to each piece, and slings them to the suspending rod of the potash bath ; after a short time these are removed, one or more at a time, according to their size, and after rinsing are taken to the scouring tray, where they are well brushed with pumice, then well rinsed in the water-trough of the scouring tray, and dipped for a moment in the cyanide dip. After being again well rinsed they are promptly suspended in the nickel bath. The articles should be thoroughly rinsed after being in the cyanide dip to prevent the introduction of this substance into the nickel bath. About two and a half hours' immersion in the bath, when a dynamo-machine is used, will be sufficient time to obtain a good deposit. It may be well to remark here, that however desirous a nickel-plater may be to give a good thick coating to his work, there is a limit, as far as nickel is concerned, which must on no account be exceeded ; otherwise the deposited will strip or peel off the work, even without touching. Indeed, we have known the nickel, when the articles have been too long in the bath, to separate from the work and curl up in flakes, while a second deposit has taken place upon the parts thus deprived of metal.

Nickeling Long Pieces of Work.—Hand-rails, cornice poles, the framework of shop fronts, and other long pieces of work, require to be nickeled in a bath of suitable dimensions : for this purpose a tank of the form shown in Fig. 112 is generally used, which is supplied with a series of short anodes to suit the form of the vessel. Such a tank should be about 12 feet long, 20 inches deep, and about 18 inches wide. Since articles of this character do not very fre-

Fig. 112.

quently come into the hand of the plater, it is unnecessary to employ special dipping baths for potash and cyanide ; the hot potash liquor may be poured over the article from a jug, beginning at one end, and continuing the operation until the whole surface has been washed over with the hot liquor. After rinsing and scouring, the cyanide dip may be applied, quickly, in the same way. The article must now be well rinsed and got into the bath as promptly as possible.

Dead Work.—Ships' deck lamps, and many other classes of work, which are not required to be polished, but left *dead*—that is, just as they come out of the nickel bath—are potashed, as usual, and after scouring and rinsing placed in the bath and allowed to remain until sufficiently coated. Since work of this kind should look as white as deposited nickel is capable of becoming, it is necessary, more especially during the last few minutes' immersion, to employ a strong current. When the articles are sufficiently coated they must be taken out of the bath, one at a time, and at once plunged into *perfectly clean* hot water for a few moments, and then placed aside to dry spontaneously. Since dead nickel is very readily stained or soiled even when touched with clean hands, the work should be handled as little as possible before being sent home to the customer. We have known instances in which dead nickel work, which from its silvery whiteness was pleasing to behold after removal from the bath, looked dirty and patchy before delivery to the customer, merely through the careless fingering to which it had been subjected by the warehouseman and others.

Nickeling Stove Fronts, &c.—These are usually nickeled in a vat specially constructed for the purpose, the form of which is shown in Fig. 113. It consists of a wooden vessel about 5 feet deep, 4 feet long, and about 18 inches wide, held together by bolts and screws, and is sometimes lined with pitch owing to the difficulty of lining such a vessel with lead. The anodes should be at least 30 inches long, and since only the fronts of stoves require to receive the coating of nickel, a single row of anodes only is necessary. This class of work is usually sent to the plating works in a polished state—that is, such parts as are to be bright are put in this condition by the manufacturers. To prepare the front for plating, it is first put into the potash bath as usual, and after a short immersion is well rinsed and scoured with pumice ; it is next dipped in the hydrochloric acid dipping bath, again rinsed, and then put into the nickel vat with all possible despatch. After about two to three hours' immersion, the article is steeped in hot water, and when dry is handed to the finisher. Large pieces of ornamental iron work which have to be left dead may be also nickeled in the "stove bath."

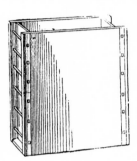

Fig. 113.

Nickeling Bicycles, &c.—When this class of work is sent to the plater in parts, these may be nickeled in the ordinary bath with the exception of the rim of the wheel, and all parts must be polished and treated in the same way as other steel work. A convenient method of suspending bicycle spokes in the bath is shown in Fig. 114. The copper slinging wire is simply coiled into a series of equidistant loops, through which the wires of the spokes pass freely, and when a sufficient number have been wired in this way the two ends of the slinging wire are pulled with both hands, by which the loops become tightened and the spokes held firmly. They are then lowered into the bath and suspended from the negative rod as shown in the engraving. After a short immersion, each spoke is shifted a little, so as to allow the wire mark to be coated, and this operation is repeated several

Y

times during their immersion in the bath, so that the coating may be as regular as possible. With a dynamo-machine a sufficient coating will be obtained in about an hour and a half. In nickeling the backbone and fork of a bicycle, and the larger parts of tricycles, these pieces should be frequently shifted in the bath to ensure uniformity of deposit, for it must be borne in mind that from the peculiar curved form of the backbone, for example, the parts farthest from

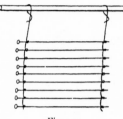

the anodes will receive the least deposit. In cases where the bath is not large enough to take in the entire rim of a large bicycle wheel, it is usual to nickel one-half at a time ; when this has to be resorted to, great care must be taken to well clean the line where the deposited metal and the bare steel meet, otherwise, when depositing upon the second or third portion of the rim, the nickel will strip at the junction of the separate deposits. In each case, a

Fig. 114.

portion of the nickeled part should be immersed in the bath with the uncoated surface. When finishing the rim the polisher should be particularly careful with these junctions of the separate deposits, otherwise he may readily cut through the nickel and expose the underlying metal. In establishments where the nickeling of bicycles forms a special branch of the business, baths of suitable dimensions are employed for depositing nickel upon the larger pieces of bicycle and tricycle work.

Nickeling Second-hand Bicycles.—Some few years ago, when nickel-plated bicycles first appeared in the market, the whole bicycle fraternity, who had been accustomed to plain steel or painted wheelers, looked with admiration, if not with envy, upon those who appeared amongst them upon their brilliant and elegant nickel-plated roadsters. At the time we speak of there was a rush of bicyclists at the various nickel-plating works, and anxious inquiries were made as to the possibility of nickeling bicycles which had become hideously rusty from neglect, or even those which had been more carefully treated. Could not a bicycle be popped in the solution, or whatever it was, and covered with the stuff, so as to come out bright like those in the shop windows ? Questions such as these were asked, even with apparent seriousness. One firm, after consulting the foreman, determined to undertake the task of nickeling one of these second-hand bicycles, and after a good deal of trouble—since it was probably the first time such a thing had been attempted—the task was accomplished with considerable success, and the owner cheerfully paid the cost of its transmutation, three pounds ten shillings—a price that in these days of brisk competition would scarcely be thought of. Since the period referred to, the nickeling of bicycles has become an ordinary matter of detail in most nickel-plating works.

In preparing a bicycle for nickeling, the principal parts must first be taken asunder. The head nut is first unscrewed to liberate the backbone ; the bolt which runs through the fork of the backbone must next be removed, by which the small wheel becomes dislodged ; the bolt is next withdrawn from the hub of the large wheel, which liberates the fork ; the spring is next disconnected by removing the screws at the head and back of the spring. All these parts, with the exception of the wheels, must pass through the hands of the polisher. It is not usual to remove the spokes, which in the case of a much-used machine would entail considerable risk, since much difficulty would occur not only in removing but in replacing them. The wheels are, therefore, nickeled entire, but before doing so they must be polished in the best way possible by hand, since it would not only be dangerous, but impracticable, to polish them at the lathe. The spokes and other parts of the wheel are first well rubbed with emery-cloth of various degrees of fineness, and then hand-buffed with chamois-leather, first with trent-sand, and afterwards with lime, as good a surface as possible being produced by these means. The wheels and other parts, when polished, are placed in the hot potash bath, where they are allowed to remain for a considerable time to remove the large amount of grease which invariably hangs about this class of work. To assist in the removal of this, the pieces are brushed over while in the potash tank : it is important that the potash liquor be in an active condition—that is, rich in the caustic alkali—or it will fail to *kill* the grease, as it is termed, or convert it into soap. After being thus cleansed in the potash bath, the work is removed piece by piece and rinsed, after which it is briskly scoured, and, after again rinsing, is passed through the acid dip for an instant, again well rinsed, and put into the nickel tank. When all parts of the machine are nickeled they are handed to the finisher, who " limes " them ; that is, the backbone, fork, and other pieces, excepting the wheels, are polished and dollied with Sheffield lime at the lathe. The wheels, as before, are finished with lime, applied, by means of chamois-leather, by hand. The various parts are then readjusted, the machine carefully wiped all over, and it is then ready for the customer. Should the india-rubber tyre come off the wheel after being in the nickel bath, it may be replaced by fusing india-rubber cement upon the periphery of the wheel by heating over a gas-burner. While the cement is hot the tyre should be replaced in its position.

Nickeling Sword Scabbards, &c.—It not unfrequently occurs that a nickel-plater receives a sword and sheath with instruction to nickel the latter only. When such is the case, the sword should be withdrawn and placed where it cannot become moistened by the steam from the potash tank or otherwise injured. To prepare the scabbard for plating, the thin laths of wood with which it is lined must first be removed, since if the sheath were placed in the nickel bath without doing so, these pieces of wood, by absorbing the nickel solution, would become so completely saturated that much difficulty would afterwards occur in drying them. We have heard of instances in which this precaution has not been observed, and as a consequence the sword, after being sheathed for some time—probably for some months—was not only thickly coated with rust, but deeply corroded, owing possibly to voltaic action set up by the nickel solution absorbed by the wooden lining ; in one such instance the sword had become so firmly fixed in the scabbard, through the oxidation of its blade, that it was unsheathed with great difficulty, and when at last withdrawn it was thickly coated with rust. The strips of wood referred to must, therefore, in all cases be removed before the sheath is immersed in either of the liquids employed. To do this, remove the screw which unites the collar to the upper part of the sheath ; remove the collar, and with the blade of a knife loosen the strips of wood and withdraw them from the sheath, taking care to remove all of them. The two parts of the sheath and the screw must then be handed to the polisher, and when returned to the plating-shop they are first to be potashed, and afterwards scoured, passed through the acid dip, and after well rinsing put into the nickel bath, in which the scabbard should be slung horizontally, so as to get as uniform a deposit as possible. The collar and screw, slung upon separate wires, should then be placed in the bath, care being taken that the latter does not receive too heavy a coating, or some difficulty may arise in replacing it. To avoid this, the head of the screw only should be put into the bath. To prevent the nickel deposit from entering the screw-hole of the scabbard, a small plug of wood may be forced into the hole before the latter is put into the bath. When the several parts are sufficiently coated, which occupies about two hours, they are removed from the bath, rinsed in hot water, and *well dried ;* they are then sent to the finisher, after which any lime that may have got into the screw-hole must be removed with a brush ; the strips of wood and collar are then readjusted, the scabbard carefully wiped with a chamois-leather, and the sword replaced.

Nickeling Harness Furniture, Bits, Spurs, &c.—This class of work, when properly nickeled, may be considered one of the most useful applications of the nickel-plating art, but unfortunately—as is also the case with many other articles—a good deal of indifferent nickeling, the consequence, in a great measure, of unwholesome competition, has appeared from time to time, which has had the effect of shaking the confidence of manufacturers who were at one time much disposed to encourage this branch of electro-deposition. That competition may be carried too far is evidenced by the extremely low prices which are asked for nickeling articles at the present time, as compared with, say, five years ago; in many instances (if the work were done conscientiously) below the fair cost of polishing. When it is borne in mind that bits, spurs, stirrups, and all kinds of harness work are necessarily subjected to severe treatment in use, and that to nickel-plate such articles badly, for a temporary advantage, has a positive tendency, if not to close this market entirely against nickel-plating, at least to confine it solely to those who have a known reputation for doing their work properly, and can therefore be relied upon. We are led to make these remarks, *en passant*, because we have an earnest desire that nickel-plating should not lose its character for absolute usefulness for the temporary advantages of competition. We say *temporary*, because we know that much mischief has accrued to the art generally in consequence of work undertaken at prices that could not yield a profit being so badly nickel-plated, that some manufacturers have ceased to avail themselves of this branch of industry except in cases of absolute necessity.

In nickeling the class of work referred to, all the parts which are to be bright when finished must, as in all other cases, be previously well polished. Sometimes the articles are sent from the manufactory in this condition, but when such is not the case the pieces must be first handed to the polisher, and when returned to the plater they are to be potashed, scoured, and passed through the acid dip, and rinsed as before, and then placed in the nickel vat, where they should remain (with an occasional shifting) for about an hour and a half, by which time, with a good dynamo, they will have acquired as thick a coating as may be given without fear of peeling. After removal from the bath and rinsing in hot water, the articles are placed in the finisher's hands, and when finished, the lime which lodges in the crevices should be brushed away and the articles then wiped with a chamois-leather and wrapped up. The brushing of work after finishing is too often neglected, and we have known of many complaints having been made by customers of the "filthy state" in which nickel-plated work has been received, owing to the lime falling out of tubes and hollows and from other parts of articles when they have been unpacked and examined on the counter. All work, after lime-finishing, should be well brushed, and wiped with a leather; it does not occupy much time, and should be considered a necessary detail of the business.

Nickeling Cast-iron Work.—Articles of this class—as kilting machines, for example—are first potashed in the usual way, and after rinsing they are immersed in a pickle composed of half-a-pound of sulphuric acid to each gallon of water used to make up the bath. In this they are allowed to remain for about half-an-hour, when they are removed, well rinsed, and scoured : for this purpose the author prefers sand to pumice powder, from the fact that when the former is used the articles have a brighter or more lustrous appearance when nickeled than if pumice be employed, besides which sand is cheaper. It frequently occurs, in cast-iron work, that numerous cavities, or " sand-holes," of greater or less magnitude, become visible after pickling and scouring the work, and since the nickel will probably refuse to enter these hollows—which is generally the case—it may be advisable in the first instance to give the article a coating of copper in an alkaline coppering bath, by which these cavities, if they are clean after sand-brushing, will become coppered with the rest of the article and the nickel will follow. Sometimes, however, the sand-holes are filled with flux or oxide of iron, in which case the former must be picked out with a hard steel point, and the hollow discoloured by oxide of iron should be scraped out with a small steel scraper. This being done, the article must be again sand-brushed and put into the coppering bath until coated all over with a slight film of copper. We have seen large iron castings in which the sand-holes have been so large and deep that the workmen at the foundry have been compelled to plug them with lead. Such defects as these should be looked for by the plater, and if any of these leaden stoppings appear it will be undoubtedly advisable to coat the article with copper before nickeling it, otherwise the nickel will not firmly adhere to the leaden stoppings. We should in all cases prefer to give a coating of copper to cast-iron work in the alkaline bath, since the cast metal is a very indifferent conductor, and requires, when not coated with copper, a very strong current ; indeed, a few tolerably large pieces of cast iron uncoppered will often monopolise the whole of the current from a dynamo-electric machine, and thereby hinder the progress of the other work.

Nickeling Chain Work.—It sometimes happens that steel, iron, and brass chains of considerable length are required to be nickeled, in which case the object must be treated according to the directions given for the respective metals. A convenient method of slinging a chain in the nickel bath is shown in Fig. 115. A number of pieces of stout copper wire, of uniform length, are cut while the chain is being

Fig. 115.

scoured, and both ends of the wires are dipped in dipping acid for a moment, and then well rinsed. The wires are then turned up into

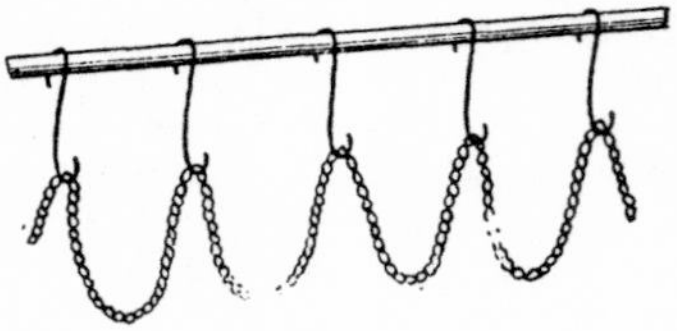

the form of a hook at one end, and when the chain is ready for slinging, the hooks are passed through the links one at a time and at equal
distances apart, each portion being lowered into the bath and suspended by bending the end of the wire over the conducting rod, as in
the figure : in this way two men can immerse a chain of considerable
length in a very few moments. After a short immersion, each hook
may be shifted one link, to allow the wire mark to be nickeled, or the
same link may be inverted, as preferred.

Re-Nickeling Old Work.—When goods which have been nickel-
plated require to be re-nickeled, it is always better to first remove the
old coating by means of a stripping solution, for the reason, as we have
before remarked, that nickel will not adhere to a coating of the same
metal. A stripping bath for nickel may be composed as follows :—

Oil of vitriol	. 16 pounds.
Nitric acid	. 4 ,,
Water	. 2 quarts.

Add the oil of vitriol to the water (not the reverse, which it is dangerous to do) gradually, and when the mixture has cooled down add the
nitric acid, and stir the mixture with a glass rod. When cold, it is
ready for use. The articles to be stripped should be attached to a
piece of stout brass or copper wire and placed in the stripping liquid,
and after a few moments they should be lifted by the wire and
examined. If the articles are of a cheap class of work, the small
amount of nickel upon them may become dissolved off in less than half
a minute : this is generally the case with American, French, and
German goods. The better qualities of English nickel-plating will
sometimes occupy many minutes before the whole of the nickel will
come off. This great difference in the thickness of the nickel-plating
necessitates much caution and judgment on the part of the workman,
for if he were to treat all classes of work alike, the metal of which the
thinly-coated articles are made would become severely acted upon if
left in the stripping bath while work of a better class was being *de-
nickeled*, as we may term it. The operation of stripping should be
conducted in the open air, or in a fire-place with good draught, so
that the acid fumes may escape through the chimney. From the
moment the articles are immersed in the stripping bath they should be
constantly watched, being raised out of the bath frequently to see
how the operation progresses, and they should not on any account be
allowed to remain in the liquid one moment after the nickel has been
dissolved from the surface, but should be immediately removed and
plunged into cold water. On the other hand care must be taken to
remove *all* the nickel, for if patches of this metal be left in parts it
will give the polisher some trouble to remove it, owing to the great

hardness of nickel as compared with the brass or copper of which the article may be composed. When the stripping of brass work has been properly conducted, the surface of the stripped article presents a smooth and bright surface, but little affected by the acid bath.

Nickel may be removed from the articles by means of the battery or dynamo-machine, by making them the anodes in a nickel bath ; but in this case a separate solution should be employed for the purpose ; or a bath may be made with dilute sulphuric or hydrochloric acid ; the stripping solution, however, when in good condition and used with care, is not only quick in its effect, but comparatively harmless to the underlying metal, if proper judgment and care have been exercised. Work which is in any way greasy should be steeped in the potash bath before stripping.

After the work has been stripped and thoroughly well rinsed, it should be dipped in boiling water, and then laid aside to dry spontaneously ; it is next sent to the polishing room, where it must be polished and finished in the same way as new work, and afterwards treated in the nickeling room with as much care and in the same way as new goods.

Nickel-facing Electrotypes.—In printing from electrotypes with coloured inks, but more especially with vermilion inks, which are prepared from a mercurial pigment, not only is the surface of the electrotype injuriously affected by the mercury forming an amalgam with the copper, but the colours are also seriously impaired by the decomposition which is involved. To avoid this it is frequently the practice to give electrotypes to be used for such purposes a coating of nickel, which effectually protects the copper from injury. In some printing establishments a nickel bath is kept specially for this purpose. The electrotypes, after being backed up and prepared for mounting in the usual way, are lightly brushed over with a ley of potash, and after well rinsing are suspended in the nickel bath for about an hour or so, by which time they generally receive a sufficient coating of nickel. Great care should be taken, however, not to employ too strong a current, lest the lower corners of the electrotype should become *burnt*, as it is called, by which a rough surface is produced, from which the ink, in subsequent printing, would fail to deliver properly ; this defect, however, is readily avoided with care, and by occasionally reversing the position of the plate while in the bath.

For nickel-facing electros of moderate dimensions, an oval stoneware pan, capable of holding about ten to fifteen gallons of solution, may be used. The nickeling bath should consist of about three-quarters of a pound of good nickel salts (double sulphate of nickel and ammonia) to each gallon of water. The salts should be dissolved in hot water and filtered into the containing vessel through a piece of unbleached calico. The anodes may consist of two plates of rolled nickel, each about 12 inches long by 6 inches wide, these being suspended in the bath by hooks from a brass rod laid across the vat. A Bunsen battery of about one gallon capacity will give a current sufficient for nickeling electros of moderate size. The positive electrode (the wire proceeding from the carbon of the battery) is to be connected to the brass rod supporting the anodes, and a similar rod, connected to the zinc of the battery, is to be laid across the vat in readiness to receive the prepared electros to be nickeled. The suspending rods and all binding screw connections must be kept perfectly clean.

When putting an electro in the bath, care must be taken to expose its face to the anodes, otherwise little or no deposit will take place upon this surface. If desired, a second row of anodes and an additional negative rod for supporting electrotypes may be employed, in which case the electros must be all suspended back to back, so as to face the anodes. An additional battery will be required. The faces of the electros may be placed within 3 or 4 inches of the anodes, and each should be supported by two wires passed through the nail holes in the backing metal which are nearest the corners.

Nickeling Wire Gauze.—Messrs. Louis Lang & Son obtained a patent in 1881 for a method of nickeling wire gauze, or wire to be woven into gauze, more especially for the purposes of paper manufacture. These wires, which are generally of copper or brass, are liable to be attacked by the small quantities of chlorine which generally remain in the paper pulp, by which the gauze wire eventually suffers injury. To nickel wire before it is woven, it is wound on a bobbin, and immersed in a nickel bath, in which it is coated with nickel in the usual way ; it is then unwound and re-wound on to another bobbin, and re-immersed in the nickel bath as before, so as to coat such surfaces as were in contact with each other and with the first bobbin. To deposit nickel on the woven tissue, it may either be coated in its entire length, as it leaves the loom, or in detached pieces. For this purpose the wire gauze is first immersed in a pickle bath, and next in the nickel solution. On leaving the latter it is rinsed, and then placed in a hot-air chamber, and when thoroughly dry may be rolled up again ready for use.

Nickeling Printing Rollers.—Mr. Appleton obtained several patents in 1883 for coating with nickel the engraved rollers used for printing and embossing cotton and other woven fabrics, to protect them from the chemical action of the various colours and chemical matters used in calico printing, &c., by which the copper rollers become deteriorated. Nickel-plated rollers, moreover, presenting a much harder surface than copper, are far more durable. The rollers are first engraved as usual, after which they are immersed in any ordinary nickel bath. The inventor finds it advantageous, in order to secure a uniform coating of nickel, to "vibrate, agitate, oscillate, or rotate the roller continuously, or intermittently," while the deposition is taking place. He has found, however, some difficulty in obtaining a firm deposit, "owing to the formation of gas bubbles upon the surface of the roller, and the difficulty in dislodging them. To obviate this he finds it advantageous to employ "a brush, which is in contact with the roller during the plating operation, the roller being rotated continuously, or intermittently, as preferred." The brush is suspended from the cathode rod, so that the bristles may touch the surface of the roller, and thus remove any adhering bubbles. He prefers a brush made with vegetable fibre or spun glass, or other substance not liable to be acted upon by the solution.

Nickeling Notes.—1. It may be taken as a rule that only a limited quantity of nickel can be deposited upon either brass, copper, steel, or iron ; if this limited amount of metal be exceeded the deposited metal will assuredly separate from the underlying metal. It has also been found in practice that a greater thickness of nickel can be deposited upon brass and copper without spontaneously peeling off than upon steel or iron. Since nickel, however, is an exceedingly hard metal, and will bear a considerable amount of friction, a very thin coating indeed is all that is necessary for most of the articles to which nickel-plating is applied. We may, however, state that too much advantage has been taken of this fact, for many articles of American and continental manufacture enter the market upon which a mere film of nickel has been deposited, and consequently they soon become unsightly from the rapidity with which the flimsy coating vanishes with even moderate wear. As a rule, the nickel-platers of this country deposit a very fair, and in many instances a very generous, coating of nickel upon their work, which has caused the home nickel-plating industry to hold a high position both as regards the *finish* of the work and its durability. It will be a thing to be regretted if price-competition should cause this useful branch of electro-deposition to become degraded by coating well-manufactured articles with a mere skin of nickel !

2. *Nickeling Steel Articles.*—When small steel work, such as purse mounts, book-clasps, &c., have to be nickeled, it is better first to suspend larger articles of brass or copper upon one end of the conducting rod, and to reserve the other end of the rod for the steel articles, or to sling them between the larger pieces of work ; when it is not convenient to do this, one of the anodes should be slung from the end of the rod farthest from the battery, as a cathode, so as to take up a portion of the current. When steel articles are placed in the bath they should become "struck," as it is termed—that is, receive a slight coating of

nickel—almost immediately after immersion, but from that moment the deposition must be allowed to progress slowly, otherwise the work will surely *strip*, and this it will sometimes do even while in the bath: we have known steel work peel, when removed from the bath, by simply striking it gently against a hard substance. It is also of much importance that steel work should be placed in the bath *directly after* it has been passed through the hydrochloric acid pickle and rinsed, since even a few moments' exposure to the air—especially if there be any acid fumes given off by the batteries—will cause a film of oxide to form on the surface and render the deposit liable to strip.

3. *Rinsing the Articles.*—It will be readily understood that if articles are imperfectly rinsed after dipping, the acid or cyanide, as the case may be, which may still hang about them must be a source of injury to the nickel bath. It is therefore advisable not to depend upon one rinsing water only, but to give the work a second rinsing in perfectly clean water. It is very commonly the practice to give the final rinsing in one division of the scouring tray, the water of which can be readily changed by simply removing the plug and turning on the tap when it is replaced.

4. *Lime used in Finishing Nickel-plated Work.*—The lime used for finishing work which has been nickel-plated is generally obtained from Sheffield, and since this substance becomes absolutely useless after it has been exposed to the air—by which it attracts carbonic acid and falls to an impalpable powder possessing little or no polishing effect upon nickel—it must be preserved in air-tight vessels. For this purpose olive jars, or large tin canisters such as are used by grocers, answer well. Small quantities may be preserved in stone jars, covered with a well-fitting bung. The general practice is to take a lump of lime from the jar, cover the vessel immediately, and after breaking off a sufficient supply from the selected lump, to return it to the jar, which is again securely covered. The fragments of lime are then powdered in a mortar, and after sifting through a fine sieve or muslin bag, the powder is handed to the finisher, who informs the assistant (generally a boy) a short time before he requires a fresh supply of the powdered lime. By this arrangement the lime then always gets into the hand of the finisher in good condition for his purpose.

5. *Nickeling Dental Work.*—One of the most successful purposes to which nickeling has been applied for many years, is in coating dentist's tools, including forceps, excavators, and other implements used in dental practice. These articles, which are made from fine steel, are usually sent by the makers to the nickel-plater in a highly-finished condition, and therefore require but a moderate amount of labour in the plating and polishing shops to turn them out of hand.

To prepare this class of work for the bath, the pieces are first wired, after which they are suspended in the potash bath for a short time, or until required to be scoured. They are now removed, a few at a time, and rinsed, after which they are taken to the scouring bench, where they are brushed over with pumice and water ; each piece, after rinsing, is dipped for a moment in the hydrochloric acid dip, again rinsed, and immediately suspended in the vat. To prevent these small pieces from receiving the deposit too quickly and thus becoming " burnt " they are usually suspended between articles of a larger size which are already in the bath. When battery power is used for coating articles of this class (with larger work), from two to three hours' immersion in the bath will be required to obtain a fair coating ; with a dynamo about half that period will be sufficient. Dental forceps require a somewhat longer immersion than the smaller tools. When the work is sufficiently nickeled, it is removed from the bath, rinsed in hot water, and sent into the polishing room to be lime-finished, after which it should be thoroughly well brushed to remove the lime, especially from the interstices. Some packers, or warehousemen, are apt to be rather careless in this respect, and are satisfied with giving nickel-plated and finished work a slight rub up with a leather, so that when the articles are received by the customer, the first thing that attracts his attention, when unpacking the work, is the appearance of a quantity of dirty lime which has fallen from the goods after they were wrapped in paper. This negligence has often been the cause of complaint, and since it can be so readily avoided by a little extra care, this should always be impressed upon the packer of finished work.

6. *Recovery of Nickel from Old Solutions.*—This is most readily effected by following Mr. Unwin's ingenious method of preparing the double salts of nickel and ammonia, namely, by taking advantage of the insolubility of the double sulphate of nickel and ammonia in concentrated solutions of the sulphate of ammonia. To throw down the double salts from an old solution, or from one which fails to yield a good deposit, prepare a *saturated* solution of sulphate of ammonia, and add this, with constant stirring, to the nickel solution, when, after a little while, a granular deposit of a green colour will form, which will increase in bulk upon fresh additions of the sulphate being given. The effect is not immediate, on adding the sulphate of ammonia solution, but after a time the green deposit will begin to show itself, and when a sufficient quantity of the ammonia salt has been added, the supernatant liquor will become colourless, when the operation is complete. The additions of sulphate of ammonia should be gradually made, and the mixture allowed to rest occasionally, after well stirring, to ascertain if the green colour of the nickel solution has disappeared. The clear liquor is to be poured off the granular deposit—which is

pure double sulphate of nickel and ammonia—and this should be allowed to drain thoroughly. It may afterwards be dissolved in water and used as a nickel-plating bath. The solution of sulphate of ammonia may be evaporated, and the salt allowed to crystallise ; and if the crystals are afterwards re-dissolved and again crystallised, the resulting product will be sufficiently pure for future use.

7. "*Doctoring.*"—This term is applied to a system of patching up an article which has been "cut through," or rendered bare, in the process of lime-finishing, and it is adopted to avoid the necessity of re-nickeling the whole article, which would often entail considerable loss to the plater. When the faulty article is sent back from the polishing room the first thing to do is to arrange the "doctor," which is performed as follows :—A piece of stout copper wire is bent in the form of a hook at each end ; a piece of plate nickel, about one and a half inch square (or a fragment of nickel anode) is now bound firmly to one of the hooks with a piece of twine ; the lump of nickel is then wrapped in several folds of calico, or a single fold of chamois-leather. The second hook is now to be connected by a wire to the anode rod of the bath, and the article put in contact with the negative electrode. The rag end is now to be dipped in the nickel bath, applied to the defective spot (which should be first lightly scoured with pumice and water) and allowed to rest upon it for a few moments, then dipped again and reapplied. By repeatedly dipping the rag in the nickel bath and applying it in this way a sufficient coating of nickel may be given in a few minutes to enable the finisher to apply the "dolly" to the re-nickeled spot, and thus render it as bright as the rest of the article. When the operation is skilfully performed, both by the plater and finisher, no trace of the patch will be observable.

8. *Common Salt in Nickel Solutions.*—Owing to the inferior conductivity of nickel baths, various attempts had been made to improve the conducting power of these solutions by the addition of other substances, but the most successful of these, of French origin, was the introduction of chloride of sodium (common salt), which is a very good conductor of electricity. The addition of this substance was subsequently adopted by a well-known London firm, the character of whose nickel-plated work was much admired for its whiteness as compared with some other specimens, of a more or less yellow tone, which appeared in the market at that time. The advantages to be derived from the addition of common salt to nickel solutions have been very clearly demonstrated by M. Desmur, who, in a communication to the author, in June, 1880, made the following interesting statement,[*] which he

* Electro-Metallurgy, Practically Treated. By Alexander Watt. Eighth edition, p. 229.

deems it advisable to reproduce in this place from its importance to those who follow the nickel-plating industry :—

9. *Augmentation of the Conductivity of Nickel Baths*—M. Desmur says : " The resistance of nickel baths as they are usually prepared, *i.e.* by dissolving double sulphate of nickel and ammonia in water, is very great. I would advise persons engaged in the trade to introduce into their baths ten per cent. of chloride of sodium (common salt). I have observed, by means of a rheostat, that the addition of this salt augments the conductivity by thirty per cent., and that the deposit is much whiter and obtained under better conditions. The diminution of resistance is in proportion to the quantity of chloride of sodium added, for the conducting power of a solution of this salt increases with its degree of concentration up to the point of saturation. I mention this fact because it is not the case with all saline solutions. For example, saturated solutions of nitrate of copper, or sulphate of zinc, have the same conductive power as more diluted solutions, because the conductibility of these solutions increases as the degree of concentration reaches its maximum, and diminishes as the concentration increases."

In our own experience we have observed that not only is the nickel deposit rendered much whiter by the addition of chloride of sodium, but it is also tougher and more reguline ; indeed, we have known a stout deposit of nickel upon sheet brass or copper to allow the metal to be bent from its corners and flattened without the least evidence of separation or even cracking—a condition of deposit not often obtained in plain double sulphate solutions.

10. *Nickeling Small Articles by Dynamo-electricity.*—Small steel pieces, such as railway keys, for example, should not be kept in the bath longer than an hour, or an hour and a half at the most. About twice this period will be necessary when battery power is employed. Brass and copper work, as a rule, may remain in the bath about double the length of time required for steel work.

11. *Nickeling Small Screws*—When a large number of small screws have to be nickeled, they may be placed in a brass wire-gauze basket, made by turning up a square piece of wire-gauze in the form of a tray, and overlapping the corners, which must then be hammered flat and made secure by soldering. A piece of stout copper wire, bent in the form of a bow and flattened at each end, is then to be soldered to the centre of each side of the tray, forming a handle, by which it may be suspended in the bath by the negative wire of the battery or other source of electric power. The screws, having been properly cleaned, are placed in the basket, which is then immersed in the bath, and while deposition is taking place the basket must be gently shaken occasionally to allow the parts in contact to become coated : this is espe-

cially necessary during the first few moments after immersion. When nickeling such articles in the wire basket, they should be placed in a single layer, and not piled up one above another, since nickel has a strong objection to deposit *round the corner*. It is better, however, to sling screws by thin copper wire than to use a basket; and though the operation is a rather tedious one, a smart lad can generally "wire" screws, after a little practice, with sufficient speed for ordinary demands. The simple method of wiring screws before described will be found very useful, and if the necessary twist is firmly given there need be no fear of the screws shifting : the wire used for this purpose should never be used a second time without stripping the nickel from its surface and passing it through a clear fire to anneal it. When nickeling screws, it is best to sling them between other work of a larger size, otherwise they are liable to become *burnt*, which will necessitate stripping off the deposited nickel or facing them upon an emery wheel.

12. *Dead Nickel-plating.*—Certain classes of work, as ship deck lamps, kilting machines, and various cast-iron articles, are generally required to be left *dead*—that is, just as they come out of the nickel bath. All such work, when removed from the bath, should be at once rinsed in very hot and perfectly clean water. Care should be taken not to allow the work to be touched by the fingers at any part that catches the eye, since this handling invariably leaves an unsightly stain. Cast-iron work, when properly nickel-plated, presents a very pleasing appearance, which should not be marred by finger-marks before it reaches the hands of the customer.

13. "*Dry*" *Nickel-plating.*—This method, which is of American origin, has sometimes been adopted in this country for umbrella mounts and other small work, but it is only applicable to very cheap work, upon which the *quantity* of nickel is of secondary importance. Work of this character is generally dollied with a " composition " consisting of crocus (oxide of iron) mixed up into the form of a hard solid mass with tallow. The workman takes a lump of the composition, which he presses against the revolving dolly until it has acquired a small amount of the composition upon its folds (as in lime-finishing). He now holds the piece of work to the dolly, which quickly becomes brightened. When a sufficient number of pieces have been prepared in this way, they are suspended by any suitable means and at once placed in the bath, and so soon as they have become sufficiently coated for this class of work, that is in about half-an-hour or so, the articles are removed, rinsed, and dried, and after a slight dollying are ready for market. A convenient arrangement for suspending umbrella mounts, and articles of a like description, is shown in Fig. 116.

14. *Removing Nickel from Suspending Appliances.*—When wire·

gauze trays, wire suspenders, and other contrivances by which articles have been supported in the bath have been used many times, they

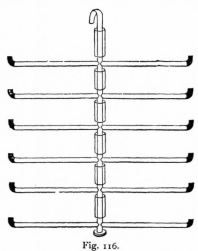

Fig. 116.

naturally become thickly coated with nickel, and since this metal when deposited upon itself has no adhesion, the various layers of

Fig. 117.

nickel which the tray, &c., have received from time to time generally curl up and break off with the slightest touch, and the fragments are liable not only to fall into the bath, but upon any work which may be in the solution at the time. It is better, therefore, to remove the nickel from these appliances, either by means of a *stripping solution* or by connecting them to the positive electrode of a battery and dissolving the metal off by electrolysis, for which purpose a small bath may be specially kept.

15. *Recovery of Dropped Articles from the Bath.*— When an article is accidentally dropped into the nickel vat, the workman should have at hand a ready means of recovering it without resorting to the unhealthy practice of plunging his bare arm into the solution. Many contrivances have been adopted for this purpose, amongst which may be mentioned an instrument of which a sketch is shown in Fig. 117. This simply consists of a per-

forated iron plate, fitted with a suitable handle, which may be conveniently attached by means of a socket brazed on to the perforated plate. If this tool, or *lift*, be gently lowered into the bath, in the direction in which the article is supposed to lie, and carefully moved about, so as not to disturb the sediment more than can be avoided, the lost article will probably soon come in contact with the lift, which should then be guided so as to draw the article to the side or end of the bath, when it may be shovelled on to the perforated plate and gradually lifted to the surface of the bath and taken off the plate, and the instrument hung up in its proper place ready for use another time. When small steel or iron articles fall into the bath, they may be recovered by means of a horse-shoe magnet. For this purpose a tolerably large magnet, having a cord attached to its centre and allowing the poles to hang downward, may be employed, and if allowed to drag along the bottom of the vat slowly, so as to avoid disturbing the sediment as much as possible, the lost article may generally be recovered and brought to the surface, even when the bath is full of work, without stirring up the sediment to any serious extent. When the recovery of the dropped pieces is not of any immediate consequence, this is better left till the evening, after the last batch of work has been removed.

16. *Rolled Nickel Anodes.*—The cost of a nickel-plating outfit, when *cast* anodes are employed, is in this item alone excessively heavy, since in many cases such anodes, for large operations, frequently weigh more than a quarter of a hundredweight each ; and when it is borne in mind that for a 250-gallon bath from sixteen to twenty-four anodes would be required—except when a dynamo or magneto-electric machine is employed, when about half that number would be sufficient—it will be at once seen that the aggregate weight of metal would be considerable. Since *rolled* nickel anodes can now be obtained of almost any required thinness, from one-fourth to one-eighth of the quantity of metal only would be required to that of the cast metal. It is a common fault with cast nickel anodes that after they have been in use a short time they become soft and flabby while in the depositing vat, and will even fall to pieces with the slightest handling and become deposited—not in the electrolytic sense—at the bottom of the vat. It is not an uncommon circumstance, moreover, to find a considerable percentage of loose carbon—graphite—interspersed with the badly-cast nickel, and which, of course, if paid for *as nickel*, entails a loss upon the consumer. We have seen samples of such anodes containing nearly thirty per cent. of graphite, which could easily be scooped out with a teaspoon ! Some very good specimens of cast nickel, however, enter the market in which neither of the above faults are to be found ; indeed, we have examined samples containing 99 per cent. of nickel,

Z

which for all practical purposes may be said to be pure. We should, in any case, give our preference to *rolled* nickel anodes ; and for the following reasons :—They are less costly ; they become more uniformly dissolved in the bath ; they are generally more pure ; they do not soften in the solution, and are less cumbersome to handle than cast anodes, which is an advantage when these require to be shifted, as in plating mullers and other large pieces.

17. *Nickeling Cast Brass Work.*—It sometimes occurs that work of this description is full of sand-holes ; when such is the case, the polisher should receive instructions to obliterate these as far as possible, for nothing looks more unsightly in nickel-plated and finished articles than these objectionable cavities. It not unfrequently happens, however, that some sand-holes are too deep to be erased by the polishing process, with any amount of labour, while sometimes, in his endeavour to obliterate these defects the polisher finds that they extend in magnitude, and are found to enter deep into the body of the work. In such cases all attempts to eradicate them will be futile, and must therefore be abandoned. Polishers and finishers accustomed to prepare work for nickel-plating are fully aware of the importance of a fair face on the work, and they generally do their best to meet the requirements of the nickeling process, and many of them are exceedingly careful to prepare the work so that, when nickeled and finished, it shall look creditable

CHAPTER XXI.

DEPOSITION AND ELECTRO-DEPOSITION OF TIN.

THERE are three different methods of coating brass and other metals with tin in what is termed *the wet way*, in contradistinction to the ordinary method of tinning by immersion in a bath of molten metal. By two of these methods a beautifully white film of tin is deposited, but not of sufficient thickness to be of a durable character. By the third method, a deposit of any required thickness may be obtained, although not with the same degree of facility as is the case with gold, silver, and copper.

Deposition by Simple Immersion, or "Dipping."—For this purpose, a saturated solution of cream of tartar is made with boiling water : in this solution small brass or copper articles, such as brass pins, for example, are placed between sheets of grain tin, and the liquid is boiled until the desired result is obtained—a beautifully white coating of tin upon the brass or copper surfaces. Ordinary brass pins are coated in this way. Some persons add a little chloride of tin to the bath to facilitate the *whitening*, as it is termed. The articles are afterwards washed in clean water, and brightened by being shaken in a leathern bag with bran, or revolved in a barrel.

Tinning Iron Articles by Simple Immersion.—A solution is first made by dissolving, with the aid of heat, in an enamelled pan—

Protochloride of tin (fused)	. . .	2½ grammes.
Ammonia alum	. . . . .	75 ,,
Water	. . . . . .	5 litres.*

---

* Tables of French weights and measures are given at the end of the volume.

The chloride of tin (which may be obtained at the drysalters) is readily made by dissolving grain tin in hydrochloric acid, with the aid of heat, care being taken to have an excess of the metal in the dissolving flask. When the bubbles of hydrogen gas which are evolved cease to be given off the action is complete. If the solution be evaporated at a gentle heat until a pellicle forms on the surface, and the vessel then set aside to cool, needle-like crystals are obtained, which may be separated from the "mother liquor" by tilting the evaporating dish over a second vessel of the same kind. When all the liquor has thoroughly drained, it should in its turn be again evaporated, when a fresh crop of crystals will be obtained. The crystals should, before weighing, be gently dried over a sand bath.

The ammonia alum is an article of commerce, and is composed of ammonia, 3·75 ; alumina, 11·34 ; sulphuric acid, 35·29 ; and water, 49·62, in 100 parts. It may be prepared by adding crude sulphate of ammonia to a solution of sulphate of alumina.

When the solution of tin and alum has been brought to a boil, the iron articles, after being well cleaned and rinsed in water, are to be immersed in the liquid, when they quickly become coated with a delicately white film of a dead or matted appearance, which may be rendered bright by means of bran in a revolving cask, or in a leathern bag shaken by two persons, each holding one end of the bag. The scratch-brush is also much used for this purpose. To keep up the strength of the tinning or *whitening* bath small quantities of the fused chloride of tin are added from time to time. Articles which are to receive a more substantial coating of tin by the separate battery may have a preliminary coating of tin in this way.

Tinning Zinc by Simple Immersion.—To make a bath for tinning zinc by the dipping method, the ordinary alums of commerce (potash and soda alums) may be used. In other respects, the solution is prepared and used in the same way as the above ; and it may be stated that the proportions of the tin salt and ammonia in water need not of necessity be very exact, since the solution, after once being used, becomes constantly weakened in its proportion of metal, still giving very good results, though somewhat slower than at first.

For coating articles made of brass, copper, or bronze, a boiling solution of peroxide of tin in caustic potash makes a very good bath, yielding a coating of extreme whiteness. A still more simple solution may be made by boiling grain tin, which should first be *granulated*, in a moderately strong solution of caustic potash, which in time will dissolve sufficient tin to form a very good whitening solution.

Tinning by Contact with Zinc.—Deposits of tin upon brass, copper, iron, or steel may easily be obtained from either of the following solutions by placing the articles, while in the hot tinning bath, in contact with fragments of clean zinc, or with *granulated* zinc. To granulate zinc, tin, or other metals, have at hand a deep jar, or wooden bucket, nearly filled with cold water, upon the surface of which spread a few pieces of chopped straw or twigs of birch. When the metal is melted, let an assistant stir the water briskly *in one direction only*, then, holding the ladle or crucible containing the molten metal high up above the moving water, pour out *gradually*, shifting the position of the ladle somewhat, so that the metal may not all flow down upon the same part of the vessel's bottom. When all the metal is poured out, the water is to be run off, and the granulated metal collected and dried. It should then be put into a wide-mouthed bottle or covered jar until required for use.

Roseleur's Tinning Solutions.—Roseleur recommends either of the two following solutions for tinning by contact with zinc: 1. Equal weights of distilled water, chloride of tin, and cream of tartar are taken. The tin salt is dissolved in one-third of the cold water; the remaining quantity of water is then to be heated, and the cream of tartar dissolved in it; the two solutions are now to be mixed and well stirred. The mixture is clear, and has an acid reaction. 2. Six parts of crystals of chloride of tin, or 4 parts of the fused salt, and 60 parts of pyrophosphate of potassium or sodium are dissolved in 3,000 parts of distilled water, the mixture being well stirred; this also forms a clear solution. Both the above solutions are to be used hot, and kept constantly in motion. The articles to be tinned are immersed in contact with fragments of zinc, the entire surface of which should be equal to about one-thirtieth of that of the articles treated. In from one to three hours the required deposit is obtained. To keep up the strength of the bath equal weights of fused chloride of tin and pyrophosphate are added from time to time. Roseleur gives the preference to this latter solution if the pyrophosphate is of good quality. He also prefers to use coils of zinc instead of fragments of the metal, as being less liable to cause markings on the articles than the latter, which expose a greater number of points. It is evident from this that granulated zinc should not be used with these solutions, since metal in this form would exhibit an infinite number of points for contact. For tinning small articles, such as nails, pins, &c., these are placed in layers upon perforated zinc plates or trays, which allow of the circulation of the liquid; the edges of the plates are turned up to keep the articles from falling off the zinc surfaces. These plates are placed upon numbered supports, in order that they may be removed from the bath in the inverse order in which they were immersed. The plates are scraped clean each time before being used, in order that a perfect metallic contact may be insured between the plates and the articles to be tinned. During the tinning the small articles are occasionally stirred with a three-pronged iron fork, to change the points of contact. After the articles have been in the bath from one to three hours an addition of equal parts of pyrophosphate and fused chloride is made, and the articles are then subjected to a second immersion for at least two hours, by which they receive a good deposit. Large articles (as culinary utensils, &c.) coated in the above solution are scratch-brushed after the first and second immersions. The final operations consist in rinsing the articles, and then drying them in warm sawdust

In reference to the working of the above solutions, Roseleur says :— " If we find that the tin deposit is grey and dull, although abundant, we prepare [? strengthen] once or twice with the acid crystallised protochloride of tin. With a very white deposit, but blistered, and without adherence or thickness, we replace the acid salt, by the fused one. In this latter case we may also diminish the proportion of tin salt, and increase that of the pyrophosphate." For tinning zinc in a pyrophosphate bath, the following proportions are recommended :—

Protochloride of tin (fused)	.	.	1 kilogramme.
Pyrophosphate of soda	.	.	5 kilogrammes.
Distilled water	. .	.	300 litres.

Deposition of Tin by Single Cell Process.—Weil makes a tinning solution by dissolving a salt of tin in a strong solution of caustic potash or soda ; a porous cell nearly filled with the caustic alkali (without the tin salt), and in this metallic zinc, with a conducting wire attached, is placed, the end of the wire being put in contact with the articles to be tinned. The solution of zinc formed in the porous cell during the action is revived by precipitating the zinc with sulphide of sodium.

Dr. Hillier's Method of Tinning Metals.—A solution is prepared with 1 part chloride of tin dissolved in 20 parts of water ; to this is added a solution composed of 2 parts caustic soda and 20 parts of water ; the mixture being afterwards heated. The articles to be tinned are placed upon a perforated plate of block tin and kept in a state of agitation, with a rod of zinc, until they are sufficiently coated.

Heeren's Method of Tinning Iron Wire.—This consists in first cleaning the wire in a hydrochloric acid bath in which a piece of zinc is suspended. The wire thus cleaned is then put in contact with a plate of zinc in a bath composed as follows :—

Tartaric acid	.	.	.	.	.	2 parts.
Water	.	.	.	.	.	100 „

To this is added, 3 parts of each, chloride of tin and soda. After remaining in the above bath about two hours, the wire is brightened by drawing it through a hole in a steel plate.

Electro-deposition of Tin.—Although the deposition of tin by simple dipping, or by contact with zinc, is exceedingly useful for small articles, and may be pursued by persons totally ignorant of electro-deposition, the deposition of this metal by the direct current is far more reliable when deposits of considerable thickness are desired, besides being applicable to articles of large dimensions. There have been many different processes recommended—some of which have been patented—for the electro-deposition of this metal, and several of these have been worked upon a tolerably extensive scale. For many purposes, this exceedingly pretty metal, when properly deposited by electrolysis, is very useful, but more especially for coating the insides of cast-iron culinary vessels, copper preserving pans, and articles of a similar description. There is one drawback connected with the electro-deposition of this metal, however, which stands much in the way of its practical usefulness, and renders its deposition by separate current more costly than would otherwise be the case, namely, that the anodes do not become dissolved in the bath in the same ratio as the deposit upon the cathode, consequently the strength of the bath requires to be kept up by constant additions of some salt of the metal to the solution while deposition is taking place. If this were not done, the bath would soon become exhausted, and cease to work altogether. To overcome this difficulty, and to keep up a uniform condition of the bath, the author proposed in his former work * the following method :—Arrange above the depositing tank a stone vessel, capable of receiving a tap (Fig. 118) ; to this connect a vulcanised india-rubber tube, reaching nearly to the surface of the solution. Let this jar be nearly filled with concentrated solution of the tin salt employed, made by dissolving the salt in a portion of the main solution.

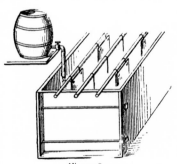

Fig. 118.

When the bath is being worked, let the tap be turned slightly, so that the concentrated solution may drip or flow into the depositing bath. When the stone vessel has become empty, or nearly so, a fresh concentrated solution should be made, using the liquor from the bath to a certain extent in lieu of water, so as not to increase the bulk of

---

* " Electro-Metallurgy." Eighth edition, p. 241.

the bath more than is absolutely necessary. By this method several advantages are gained—(1) By using the weakened bath each time to make the concentrated solution, there will be but trifling addition to the bulk of the solution ; (2) By allowing the concentrated solution to *continually* enter the bath while deposition is taking place, there will be no necessity to disturb the bath by stirring in a larger quantity of the solution all at one time. In cases in which it is necessary to make additions of two separate substances, these may be introduced by employing two tapped vessels instead of one.

Roseleur's Solution.—The bath which this author recommends as possessing all the conditions desired by the operator, is composed of :—

Protochloride of tin (in crystals)	.	. 600 grammes.
Pyrophosphate of soda or potassa	.	. 5 kilogrammes.
Distilled or rain water	. .	. 500 litres.

Instead of employing crystals of the tin salt, the fused substance is to be preferred, 500 grammes of which take the place of the former. In making up the bath, the water is put into a tank lined with anodes of sheet tin, united together, and put in connection with the positive electrode of the battery or other source of electricity. The pyrophosphate salt is then put into the tank, and the liquid stirred until this is dissolved. The protochloride is placed in a copper sieve, and this half immersed in the solution. A milky-white precipitate is at once formed, which becomes dissolved by agitation. When the liquid has become clear and colourless, or slightly yellow, the bath is ready for use. The cleaned articles are now to be suspended from the negative conducting rods as usual.

" The anodes," says Roseleur, " are not sufficient to keep the bath saturated ; and when the deposit takes place slowly, we add small portions of equal weights of tin salt and pyrophosphate. The solution of these salts should always be made with the aid of the sieve, for if fragments of the protochloride of tin were to fall on the bottom of the bath they would become covered with a slowly soluble crust, preventing their solution." It is stated that any metal may be coated in this solution with equal facility, and that a good protective coating may be obtained with it, while the metal has a dead white lustre resembling that of silver, which may be rendered bright either by scratch-brushing or by burnishing. An intense current is necessary in working this solution.

Fearn's Process.—This process, for which a patent was granted in 1873, includes four different solutions, which may be thus briefly described : No. 1. A solution of chloride of tin (containing but little free acid) is first prepared, containing 3 ounces of metallic tin per gallon : 30 pounds of caustic potash are dissolved in 20 gallons of water ; 30 pounds of cyanide of potassium in 20 gallons of water ; and 30 pounds of pyrophosphate of soda in 60 gallons of water. 200 ounces (by measure) of the tin solution are poured slowly, stirring with a glass rod, into the 20 gallons of potash solution, when a precipitate is formed, which quickly redissolves ; into this solution is poured first all the cyanide solution, then all the pyrophosphate, and the mixture well stirred. No. 2. 56 pounds of sal-ammoniac are dissolved in 60 gallons of water ; 20 pounds of pyrophosphate of soda in 40 gallons of water ; into the latter is poured 100 ounces by measure of the chloride of tin solution, and the mixture well stirred, when the precipitate formed redissolves as before. Lastly, the sal-ammoniac solution is added, and the whole well stirred together. No. 3. 150 pounds of sal-ammoniac are dissolved in 100 gallons of water ; into this 200 ounces by measure of the tin solution are poured, and well stirred in. No. 4. To make this solution, 400 ounces of tartrate of potash are dissolved in 50 gallons of water ; 1,200 ounces of solid caustic potash in 50 gallons of water ; 600 ounces by measure of the tin solution are then added slowly, with stirring, to the tartrate solution ; the caustic potash solution is next added, the stirring being kept up until the precipitate which forms has become entirely redissolved.

In using the above solutions, No. 1 is to be worked at a temperature of 70° Fahr., with a current from two Bunsen batteries; No. 2 is used at from 100° to 110° Fahr., with a weaker current ; No. 3 is to be worked at 70° Fahr. ; and No. 4 may be used cold. It is stated that solutions 1 and 4 yield thick deposits without requiring alternate deposition and scratch-brushing. Since during the working less tin is dissolved from the anodes than is deposited, the oxide or other salt of the metal must be added from time to time, except in the case of No. 3, which acts upon the anode more freely than the others. In tinning cast iron in these solutions, they require first to have a deposit of copper put upon them. For tinning zinc articles, No. 1 solution is employed.

Steele's Process.—This process is applied to coating articles of copper, brass, steel, iron, and zinc with tin. The solution is prepared thus : Dissolve 60 pounds of common soda, 15 pounds of pearlash, 5 pounds of caustic potash, and 2 ounces of cyanide of potassium in 75 gallons of water, then filter the solution ; next add 2 ounces of acetate of zinc, 16 pounds of peroxide of tin, and stir the mixture until all is dissolved, when the solution is ready for use. The solution is to be worked at about 75° Fahr.

In preparing articles for electro-tinning, they must be rendered perfectly clean, either by scouring or dipping. Articles of cast iron may advantageously be first coppered in an alkaline coppering bath. Sometimes a deposit of tin is given in a boiling-hot solution by the zinc-contact method, and a stouter deposit afterwards obtained by the separate current in either of the foregoing solutions. The process of electro-tinning has been much adopted in France, and during the past few years there has been considerable attention paid to it in this country. It has yet to be developed into a really extensive industry.

Electro-Tinning Sheet-Iron.—Spence's Process.—This inventor says :—" When it is desired to make tin plates as cheaply as possible, I first place the plates in a solution of zinc, and deposit that metal on the surface ; and then put them in a solution of tin, and deposit a coating of that metal. In manufacturing these plates, I coat the sheet iron with zinc, as before, and then deposit a coating of lead by electricity." By this method he reduces the quantity of tin usually required ; and in regard to terne plates, he dispenses with the use of tin altogether. When removed from the bath, the electro-tinned plates are brightened by being placed in a stove heated to a temperature slightly above that at which tin melts (442° Fahr.). As the plates are taken out of the tinning bath, they are placed in a rack capable of containing 24 pieces. These racks, as they are filled, are placed in the stove, where they are allowed to remain until the tin melts on the surface. The plates are afterwards passed through rollers, with that edge first which was at the bottom of the rack. To avoid the employment of heat, one or more pairs of polished steel rollers may be used in succession, and so adjusted as to bear on the plate with some pressure. On removing the plates from the bath, they are passed through the rollers, which remove inequalities of the tin surface. To give the necessary polish, the plates are then placed on a table, on which is a pair of rolls rotating at high speed, and coated with cloth or other suitable material. These rolls are so arranged " as to rotate in the reverse direction to the transverse of the plate, and hence the plate has to be pushed through them."

Recovery of Tin from Tin Scrap by Electrolysis.—Dr. J. H. Smith, in a paper read before the Society of Chemical Industry, described a method for working up tin scrap which he found to be successful. The scrap to be dealt with had, on an average, about 5 per cent. of tin and there was a supply of some 6 tons a week, for which quantity the plant was arranged. It was designed to convert the tin into chloride of tin for dyers' use, the iron scrap being utilised as copperas. On the recommendation of Messrs. Siemens and Halske, of Berlin, one of their dynamos (C. 18), was used. The machine in question was stated to give a current of 240 ampères, with an electro-motive force of 15 volts, and an expenditure of 7-horse power. Eight baths were used, made of wood lined with rubber. They were

1½ mètres long, 70 centimètres wide, and 1 mètre deep. The anodes were, of course, formed of the tin scrap, which was packed in baskets made of wood, and of a size to hold 60 kilos to 70 kilos of the scrap. There was an arrangement for constantly agitating these baskets by raising and lowering them, thus promoting circulation of the solution and regularity of action. The cathodes were copper plates 1½ millimètres thick and 120 centimètres long by 95 centimètres broad. There were sixteen of these, placed two in each tank, one on each side of the basket. The electrolyte used was sulphuric acid, diluted with 9 volumes of water. The tin precipitated was rather over 2 kilos per hour ; it was very pure, easily melted when required, and in a form very suitable for solution in acid for preparation of tin salts. Dr. Smith having worked his process in a district in Germany,*where probably tin scrap was obtainable at a low price, was enabled to show that a profit could be obtained upon the working. The same results might possibly be obtained in Birmingham, London, and other districts where large quantities of sheet tin are used. There have been many patents taken out for the electrolytic treatment of tin scrap, but the expense of collecting the scrap has always been the chief difficulty in rendering such processes commercially available.

---

* For further remarks on this process (which was not conducted in Germany, as here stated, but in Milan), see p. 578.

CHAPTER XXII.

ELECTRO-DEPOSITION OF IRON AND ZINC.

Electro-deposition of Iron.—*Facing Engraved Copper-plates.*—The extreme hardness of electro-deposited iron as compared with copper and type metal has caused the electro-deposition of iron to be applied to the *facing* of printers' type and engraved copper-plates, by which their durability is greatly augmented. The importance of protecting the surface of engraved copper-plates from the necessary wear and tear of the printing operations can scarcely be over-estimated, and a deposit of iron answers this purpose admirably. Another great advantage of the iron or "steel facing," or, as it is termed in France, *acierage*, is that when the deposited metal begins to wear off, the old coating is readily removed from the surface by means of dilute sulphuric acid, and another deposit given in its place in a very short time. In this way copper-plates may be preserved almost for an indefinite period, while each impression from the plate is as sharp and distinct as another even after a vast number of copies have been printed from the same plate. This system of facing printers' type and engraved copper-plates—which was originally suggested by Boettger—and the plates used for printing bank notes, has been much adopted by several large firms, including the eminent firm by whom this work was printed.

Iron may readily be deposited from a solution of its most common salt, the protosulphate, or green copperas, but for this purpose the salt should be as pure as possible. A depositing solution may also be prepared by passing a strong current through a large iron anode suspended in a tolerably strong solution of sal ammoniac. After the electrolytic action has been kept up for an hour or so, a cathode of clean sheet brass or copper should be substituted, which, if the solution has become sufficiently impregnated with metal, will at once receive a coating of iron, of a good white colour, though not perhaps quite so bright as the deposit obtained from a solution of the proto-sulphate of iron. An iron-depositing solution may be made in the same way by employing a moderately strong solution of either acetate of ammonium or acetate of potassium. A mixture of two parts sulphate of iron and one of sal ammoniac dissolved in water may also be employed, but the solution should not be too strong, otherwise the deposit is apt to be irregular, and of an indifferent colour. In making up iron baths for the electro-deposition of this metal, it has commonly been the practice to employ somewhat concentrated solutions, but the author, in the course of a long series of experiments, found that in most cases such a condition was far from being necessary and that weaker baths frequently yielded better results.

The author has obtained deposits of iron from most of its salts, including those prepared with the vegetable acids, as the acetate, citrate, tartrate of iron, &c., from some of which exceedingly interest-ing results were obtained, but possessing, however, no practical significance. The results of some of the more useful experiments are given in the Appendix.

Klein's Process for depositing Iron upon Copper.—This process, which from its successful results obtained the recognition and support of the Russian Government, is specially applicable to the production of electrotypes, as a substitute for those produced from copper, and is stated to be eminently successful in bank-note printing. The solution is prepared in a very simple way, as follows :—A solution of sulphate of iron is first made, and to this is added a solution of carbonate of ammonia until all the iron is thrown down. The precipitate is then to be washed several times, and afterwards dissolved by sulphuric acid, care being taken not to use an excess of acid. The solution is to be used in as concentrated a state as possible. To prevent the iron bath from becoming acid by working, a very large iron anode is employed—about eight times larger in surface than that of the copper cathode to be coated. After working this bath for some time, M. Klein found that the deposition became defective, and this he dis-covered was due to the presence of acid in the bath, owing to the anode not having supplied the solution with its proper equivalent of iron to replace that which had been deposited. To overcome this, he attached a copper or platinum plate to the anode, by which the two plates formed a separate voltaic pair in the liquid, causing the iron (the positive metal) to become dissolved, while the battery current was not passing through the bath. It is stated that the iron deposited by this process is as hard as tempered steel. but very brittle ; it may, however, be rendered malleable by annealing, when it may be engraved upon as easily as soft steel. The following process is given for copying engraved metal plates in electrotype, and then giving them a surface of iron.

To Copy Engraved Metal Plates and Face them with Iron.—" If the plate be of steel, boil it one hour in caustic potash solution. Brush and wash it well. Wipe it dry with a rag, and then with one moistened with benzine. Melt six pounds of the best gutta-percha very slowly indeed, the gum being previously cut up into very small pieces. Add to it three pounds of refined lard, and thoroughly incorporate the mixture. Pour the melted substance upon the centre of the plate. Allow it to stand twelve hours, and then take the copy off.

" *Phosphoric Solution.*—Dissolve a fragment of phosphorus half-an-inch in diameter in one teaspoonful of bisulphide of carbon, add a similar measure of pure benzine, three drops of sulphuric ether, and half-a-pint of spirit of wine. Wash the mould twice with this solution, allowing it to dry each time.

" *Silver Solution.*—Dissolve one-sixth of an ounce of nitrate of silver, in a mixture of half-a-pint of strong alcohol and half a teaspoonful of acetic acid ; wash the mould once with this liquid, and allow it to dry.

" *Copper Solution.*—Dissolve fifty-six pounds of sulphate of copper in nineteen gallons of water, and add one gallon of oil of vitriol. Deposit a plate of copper upon the mould in this solution.

" *Iron Solution.*—To coat the copper plate with a surface of iron, dissolve fifty-six pounds of carbonate of ammonium in thirty-five gallons of water. Dissolve iron into the liquid, by means of a clean anode of charcoal iron and a current from a battery. Clean the anode frequently, and add one-pound of carbonate of ammonium once a week. The copper plate, before receiving the deposit, should be cleansed with pure benzine, then with caustic potash, and thoroughly with water. Immerse the cathode in the iron solution for four minutes, take it out, wash, scrub, replace in the vat, remove and brush it every five minutes, until there is a sufficient deposit ; then wash it thoroughly, well dry, oil, and rub it, and clean with benzine. If it is not to be used at once, coat it with a film of wax."

Jacobi and Klein's Process.—For depositing iron upon moulds for reproducing engraved surfaces and for other useful purposes, the following process is given. A bath is prepared with a solution of sulphate of iron, with the addition of either sulphate of ammonia, potash, or soda, which form double salts with the salt of iron. The bath must be kept as neutral as possible, though a small quantity of a weak organic acid may be added to prevent the precipitation of salts of peroxide of iron. A small quantity of gelatine improves the texture of the deposit. To accelerate the rapidity of the deposit, and favour its uniform deposition, the solution should be warm. The anodes employed are large iron plates, or bundles of iron wire, and since it is found that the anodes do not dissolve with sufficient rapidity to keep up the normal metallic strength of the bath, the inventors have found it useful to employ anodes of gas carbon, copper, or platinum—or any metal which is electro-negative to iron—as well as the iron anodes ; or these auxiliary anodes may be placed in separate porous cells, excited by dilute sulphuric or nitric acid, or the nitrates or sulphates of potash or soda. The current employed is either from one or two Daniell cells only, or from a single Smee, the size of which is proportionate to the surface of the cathode. The Daniell cells should have a large surface, and the zinc be excited by a solution of sulphate of magnesia instead of dilute sulphuric acid. It is said to be " indispensable that the current should be regulated and kept always uniform with the assistance of a galvanometer having but few coils, and therefore offering only a small resistance. The intensity of the current ought to be such as to admit only of a slight evolution of gas bubbles at the cathode ; but it would be prejudicial to the beauty of the deposit if gas bubbles were allowed to adhere to its surface." In working this process, the same moulds used for electrotyping may be employed ; but it is advisable in using lead or gutta-percha moulds to first coat them with a film of copper in the usual way, and after rinsing to place them at once in the iron solution. The film of copper may be afterwards removed, either by mechanical means or by dipping in strong nitric acid.

The following formula is given for the composition of the iron bath :—

Sulphate of iron . . . . . 139 parts.
      ,,   magnesia . . . . 123 ,,

These substances are to be dissolved together in hot water, with the addition of a little oxalic acid and some iron shavings. This solution should be kept, in its concentrated condition, in well-stoppered glass bottles or carboys, and when required for use must be diluted until it has a specific gravity of 1·155 (water being 1·000). When working this solution, the oxide of iron which appears at the surface of the liquor must be skimmed off, with some of the solution, and shaken up in a bottle with a little carbonate of magnesia, and after settling, the clear liquor may be returned to the bath. To prevent air-bubbles from adhering to the mould, while in the bath, the mould may be first washed with alcohol, and afterwards with water ; it is then to be placed in the bath before it has time to become dry. It is said that the iron deposited by this process is very hard and brittle, therefore much care must be taken to avoid breaking the electro-deposit when separating it from the mould. When annealed, however, the iron acquires the malleability and softness of tempered steel, and has a remarkably fine appearance when brushed with carbonate of magnesia.

Amongst the numerous solutions recommended for the electro-deposition of iron, we select the following :—

Ammonio-sulphate of Iron Solution.—This double salt, which was first proposed by Boettger for depositing this metal, may be readily prepared by evaporating and crystallising mixed solutions of equal parts of sulphate of iron and sulphate of ammonia ; a solution of the double salt yields a fine white deposit of iron with a moderate current, and has been very extensively employed in "facing" engraved copper-plates. When carefully worked, this is one of the best solutions for the deposition of iron upon copper surfaces.

Boettger's Ferrocyanide Solution.—This solution, which is considered even better than the former for coating engraved copper-plates with iron, is formed by dissolving 10 grammes of ferrocyanide of potassium (yellow prussiate of potash) and 20 grammes of Rochelle salt in 200 cubic centimètres of distilled water. To this solution is added a solution consisting of 3 grammes of persulphate of iron in 50 cubic centimètres of water. A solution of caustic soda is next added, drop by drop, with constant stirring, to the whole solution, until a perfectly clear light yellowish liquid is obtained, which is then ready for immediate use.

Mr. Walenn obtained good results from a slightly acid solution of sulphate of iron (1 part to 5 of water). Sulphate of ammonia, however, was found to increase the conductivity of the solution.

Ammonio-chloride of Iron Solution, made by adding sal-ammoniac to a solution of protochloride of iron, may also be used for depositing iron, a moderately strong current being employed. When carefully prepared and worked, this solution is capable of yielding very good results, but it has these disadvantages : the solution becomes turbid, and a shiny deposit is apt to form upon the electrodes. It is a common defect in iron solutions that they are liable to undergo change by absorbing oxygen from the air. To overcome this, Klein adopted the ingenious expedient of adding glycerine to the solution, by which he was enabled to keep his solution bath tolerably clear, except on the surface, upon which a shiny foam accumulated, which became deposited upon the articles in solution. To prevent the air from injuriously affecting the baths, it is advisable that the depositing vessel should be kept covered as far as practicable.

Sulphate of Iron and Chloride of Ammonium Solution.—The addition of chloride of ammonium (sal-ammoniac) to sulphate of iron solution improves the character of the deposit while improving the conductivity of the solution. Meidinger found that engraved copper-plates coated with iron in a bath thus composed were capable of yielding from 5,000 to 15,000 impressions. *

Electro-deposition of Zinc.—*Watt's Solution.*—The deposition of this metal has never attained the dignity of a really practical art. In the earlier periods of electro-deposition many iron articles, including the sheet metal, were coated with zinc by this means, to protect them from rust, or oxidation, but it was soon found that the porous and granular nature of the coating, instead of acting as a preservative from rust, greatly *accelerated* the action of moisture upon the underlying metal (iron) by promoting electro-chemical action.

The process of *galvanising* iron, as it is fancifully termed—by which articles of this metal are dipped into a bath of molten zinc—soon proved, although not wholly faultless, so superior to that of electro-zincing, that it became generally adopted to the entire exclusion of the latter. There are many purposes—gauze wire, for example—to which the process of "galvanising" is inapplicable, and for which a good electro-deposit of zinc would be specially serviceable. To obtain a solution which would give a good reguline deposit of zinc suitable for such purposes, the author, after a long series of experiments, succeeded in forming a solution, for which he obtained a patent in 1855, from which he obtained some exceedingly beautiful deposits, possessing the fullest degree of toughness which this metal exhibits when in a perfectly pure state. The most satisfactory result was obtained by dissolving the best milled zinc in a strong solution of cyanide of potassium, with the addition of liquid ammonia, by means of a strong voltaic current. The process is briefly as follows : 200 ounces of cyanide of potassium are dissolved in 20 gallons of water ; to this solution is added 80 ounces, by measure, of strong liquid ammonia. The whole are then well stirred together. Several large porous cells are then filled with the solution, and these are placed upright in the vessel containing the bulk of the solution, the liquids in each vessel being at an equal height. Strips of copper are then connected by wires to the negative pole of a compound Bunsen battery of two or more cells, and these strips are immersed in the porous cells. A large anode of good milled zinc, previously well cleaned, is now connected to the positive pole of the battery, and the plate suspended in the larger vessel. The voltaic action is to be kept up until the zinc has become dissolved to the extent of about 60 ounces, or 3 ounces to each gallon of solution. To this solution is added 80 ounces of carbonate of potash, by dissolving it in portions of

---

* For further remarks on the electro-deposition of iron, see pp. 442-446.

A A

the solution at a time, and returning the dissolved salt to the bath. The porous cells being removed, the solution is allowed to rest for about twelve hours, when the clear liquor is to be transferred to another vessel, the last portions, containing sedimentary matter, being filtered into the bath.

Preparing Cast and Wrought Iron Work for Zincing.—The articles require to be first dipped for a short time in a hot potash bath, after which they are to be well rinsed. They are next steeped in a "pickle" composed of oil of vitriol half-a-pound, water 1 gallon. As soon as the black coating of oxide yields to the touch the articles are removed and plunged into clean cold water ; they are then taken out one by one and well brushed over (using a hard brush) with sand and water ; if any oxide still remains upon the surface, the articles must be immersed in the pickle again, and allowed to remain therein until, when the brushing is again applied, they readily become cleaned. They are now to be well rinsed, and at once suspended in the zincing bath, in which they should remain for a few minutes, then taken out and examined ; and if any parts refuse to receive the deposit, these must be again well sand-brushed or scoured, the article being finally brushed all over, again rinsed, and placed in the depositing bath, where they are allowed to remain until sufficiently coated. An energetic current from at least two 3-gallon Bunsen cells, where a dynamo-machine is not used, is necessary to obtain a good deposit. The articles may be rendered bright by means of the scratch-brush, but large articles may be sufficiently brightened by means of sand and water, with the assistance of soap. When finished they should be dipped into hot water, and may then be further dried by means of hot sawdust. The anodes should be of the best milled zinc, and well cleaned before using.

Zincing Solutions.—For the electro-deposition of zinc, solutions of the sulphate, ammonio-sulphate, chloride, and ammonio-chloride may be employed, as also alkaline solutions prepared by dissolving zinc oxide or carbonate in a solution of cyanide of potassium or caustic potassa ; the deposit from either of these alkaline solutions is generally of very good quality, and if too strong a current be not employed, the deposited metal is usually very tough.

Person and Sire's Solution.—This consists of a mixture of 1 part of oxide of zinc dissolved in 100 parts of water, in which 10 parts of alum have been previously dissolved at the ordinary temperature. The current from a single battery cell is employed, and the anode surface should be about equal to that of the articles to be coated, when, it is stated, the deposition proceeds as easily as that of copper, and takes place with equal readiness upon any metal.

Deposition of Zinc by Simple Immersion. — According to

Roque, cast and wrought-iron articles may be coated with zinc in the following way :—A mixture is first made consisting of (by measure) hydrochloric acid 550 parts, sulphuric acid 50 parts, water 1,000 parts, and glycerine 20 parts. The iron articles are first pickled in this mixture and then placed in a solution composed of carbonate of potassa 1 part and water 10 parts. The articles are next to be immersed from three to twelve hours in a mixture composed as follows :—water 1,000, chloride of aluminium 10, bitartrate of potassium 8, chloride of tin 5, chloride of zinc 4, and acid sulphate of aluminium 4 parts. The thickness of the deposit is regulated by the length of the immersion.

Hermann's Zinc Process.—By this process, which was patented in Germany in 1883, zinc is deposited by electrolysis from dilute solutions of sulphate of zinc with the aid of sulphates of the alkalies, or alkaline earths—potassium, sodium, ammonium, strontium magnesium, or aluminium—either added singly, or mixed together. The addition of these salts is only advantageous when dilute solutions of sulphate of zinc are to be treated. According to Kiliani, during the electrolysis of a solution of sulphate of zinc of $1\cdot33$ specific gravity (the anodes and cathodes consisting of zinc plates), the evolution of gas is greatest with a weak current, diminishing with an increasing current, and ceasing when on one square centimètre electrode surface, three milligrammes of zinc are precipitated per minute. The deposit obtained with a strong current was very firm. From a 10 per cent. solution the deposit was best with a current yielding from $0\cdot4$ to $0\cdot2$ milligramme of zinc. From very dilute solutions the zinc was always obtained in a spongy condition, accompanied by copious evolutions of hydrogen. With a weak current and from a 1 per cent. solution, oxide of zinc was also precipitated, even with an electro-motive force of 17 volts, when only $0\cdot0755$ milligramme of zinc per minute was deposited on one square centimètre of cathode surface. The size of the electrode surfaces must therefore be adjusted according to the strength of the current and the degree of concentration of the electrolyte.*

---

* For further details concerning the electro-deposition of zinc, see p. 631.

CHAPTER XXIII.

ELECTRO-DEPOSITION OF VARIOUS METALS.

Electro-deposition of Platinum. — Electro-deposition of Cobalt. — Electro-deposition of Palladium. — Deposition of Bismuth. — Deposition of Antimony.—Deposition of Lead.—Metallo-Chromes. — Deposition of Aluminium.—Deposition of Cadmium.—Deposition of Chromium.— Deposition of Manganium.—Deposition of Magnesium.—Deposition of Silicon.

THERE are many metals which have been deposited by electrolysis more as a matter of fact than as presenting any practical advantage in a commercial sense ; others, again, possessing special advantages which would render their successful deposition a matter of some importance, have been the subject of much experiment, in the hope that the difficulties which stood in the way of their being practically deposited for useful purposes could be overcome. Of these latter, the intractable but most valuable metal, platinum, may be considered the most important.

Electro-deposition of Platinum.—The peculiar attributes of this interesting metal—its resistance to the action of corrosive acids, and of most other substances, render it invaluable in the construction of chemical apparatus, while its high cost, its infusibility, and the great difficulty experienced in giving this metal any required form, greatly limit the area of its usefulness. If, however, articles of copper, brass, or German silver—metals which may be so readily put into shape by casting, stamping, or by any ordinary mechanical means—could be successfully and economically coated with platinum, this branch of the art of electro-deposition would soon meet with considerable support from the manufacturers of chemical apparatus, as also from opticians, who would gladly adopt electro-platinised * articles for many purposes of their art.

---

* To contradistinguish the art of depositing bright reguline platinum upon metals from the process of *platinising*, devised by Smee for imparting a black powdery film upon silver for the negative plates of voltaic batteries, the term *platinating* has been proposed, but we would suggest that a simpler term would be *platining*. Electro-platinising would be a more correct term than platinating.

One great difficulty that stands in the way of depositing platinum economically and of any required thickness is that the anodes do not dissolve in the solutions which have as yet been adopted for its deposition ; consequently, unless repeated additions of a platinum salt are made as the solution becomes exhausted, it is impossible to obtain a coating of sufficient thickness for any practical purpose. In order to meet this difficulty in some degree, the author suggested in his former work that the strength of the solution may be kept up in the same way as he recommended for electro-tinning; that is to say, a reservoir, containing concentrated platinum solution, is placed upon a shelf a little above the electro-platinising bath (Fig. 118), and the strong liquid is allowed to drip or flow out through a tap in the reservoir, and trickles at any required speed, into the solution bath, *while deposition is going on*, and in this way the strength of the bath may be kept up to any desired density. For small quantities of solution, the funnel arrangement shown in Fig. 119 may be adopted. The concentrated platinum solution being made in part from a portion of the larger solution, instead of with water, the original quantity of the liquid may be very fairly balanced. For example, if we take, say, one quart of the platinising solution and add to this a considerable proportion of platinum salt and the solvent employed in its preparation, so as to make as strong a solution as possible, when this is added and returned to the bath in the way above indicated, it will not add much to its original bulk. By weighing the articles before and after immersion, the weight of metal deposited may soon be ascertained (the time occupied being noted), and if the exact percentage of metal in the concentrated liquor is previously determined, there will be no difficulty in determining at what speed the strong solution should be allowed to flow into the bath to keep it up to the proper strength. Another suggestion we have to make is this: since the platinum anode does not become dissolved during electrolysis, a carbon anode may be substituted, which, in large operations, would add much to the economy of the process.

Fig. 119.

Preparation of Chloride of Platinum.—As in the case of gold, this metal must first be dissolved in *aqua regia*, to form chloride of platinum, previous to making up either of the baths about to be described. For this purpose, fragments of platinum, which may be pieces of foil or wire, are put into a glass flask, and upon them is to be poured two to three parts of hydrochloric acid and one part nitric acid ; the flask is then placed on a sand bath, and gently heated until the red fumes at first given off cease to appear in the bulb of the flask. A solution of a deep red colour is formed, which must now be carefully poured into a porcelain evaporating dish, placed on the sand bath, and heated until nearly dry, moving the vessel about, as recommended in treating chloride of gold, until the thick blood-red liquor ceases to flow, at which period the vessel may be set aside to cool. Any undissolved platinum remaining in the flask may be treated with nitro-hydrochloric acid as before until it is all dissolved. The dry mass is to be dissolved in distilled water, and the subsequent solution, after evaporation, added to it. If the original weight of the platinum is known, it is a good plan to dissolve the dried chloride in a definite quantity of distilled water, so that in using any *measured* portion of the solution, the percentage of actual metal used may be fairly determined when making up a solution.

Cyanide of Platinum Solution.—Take a measured quantity of the chloride of platinum solution representing about five pennyweights of metal, and add sufficient distilled water to make up one pint. Now add of strong solution of cyanide sufficient to precipitate and redissolve the platinum ; add a little in excess, filter the solution, and make up to one quart with distilled water. The solution must be heated to about 130° Fahr. when using it. A rather weak current from a Wollaston or Daniell battery should be used ; if too strong a current be applied, the deposit will probably assume the form of a black powder.

Deposition by Simple Immersion.—Platinum readily yields itself up when brass, copper, German silver, &c., are immersed in its solutions, but the deposit is of little or no practical use. It may also be deposited from its solution by contact with zinc as follows :—Powdered carbonate of soda is added to a strong solution of chloride of platinum until no further effervescence occurs ; a little glucose (grape sugar) is then added, and lastly, as much common salt as will produce a whitish precipitate. The articles of brass or copper are put into a zinc colander and immersed in the solution, heated to about 140° Fahr., for a few seconds, then rinsed and dried in hot sawdust.

Deposition by Battery Current.—Roseleur describes a solution from which he obtained platinum deposits of considerable thickness. The solution is prepared as follows :—

> Platinum, converted into chloride .   .   10 parts.
> Distilled water .   .   .   .   .   500 „

Dissolve the chloride in the water, and if any cloudiness appears in the solution, owing to the chloride having been over-heated during the last stage of the evaporation, it must be passed through a filter.

Phosphate of ammonia (crystallised) . 100 parts.
Distilled water . . . . . . 500 „

Dissolve the phosphate in the above quantity of water, and add the liquid to the platinum solution, with brisk stirring, when a copious precipitate will be formed. To this is next added a solution of—

Phosphate of soda . . . . . 500 parts.
Water (distilled) . . . . . 1,000 „

The above mixture is to be boiled until the smell of ammonia ceases to be apparent, and the solution, at first alkaline, reddens blue litmus paper. The yellow solution now becomes colourless, and is ready for use. This solution, which is to be used hot, with a strong battery current, is recommended for depositing platinum upon copper, brass, and German silver, but is unsuited for coating zinc, lead, or tin, since these metals decompose the solution and become coated in it by simple immersion. Since the platinum anode is not dissolved in this solution, fresh additions of the chloride must be made when the solution has been worked.

Boettger's Solution for Depositing Platinum consists of a boiling-hot mixture of chloride of platinum solution and chloride of ammonia (sal-ammoniac), to which a few drops of liquid ammonia are added. The solution, which is weak in metal, requires to be revived from time to time by additions of the platinum salt. *

Electro-deposition of Cobalt.—Until somewhat recently the electro-deposition of cobalt had chiefly been of an experimental character, based upon the belief, however, that this metal, if deposited under favourable conditions, was susceptible of some useful applications in the arts. The difficulty of obtaining pure cobalt anodes—as was the case with nickel until a comparatively few years ago—as a commercial article, stood in the way of those practical experimentalists who would be most likely to turn the electro-deposition of this metal to account. Moreover, the extremely high price of the metal, even if rudely cast in the form of an ingot, rendered its practical application all but impossible. That unfavourable epoch is now passed, and we have cobalt anodes and " salts " in the market, as easily procurable, though not, of course, at so low a price, as the corresponding nickel products. The author is indebted to the courtesy of the enterprising firm of cobalt and nickel refiners, Messrs. Henry Wiggin & Co., of Birmingham, for some excellent examples of their single and double cobalt salts and rolled cobalt anodes, and is thus enabled to state that those who may desire to embark in the electro-deposition of this metal can readily obtain the chief requisites, the salts and anodes, in

---

* For further details concerning the electro-deposition of platinum, see pp. 441, 442.

any desired quantity from this firm at the following rates:—Rolled and cast cobalt anodes, 16s. per lb.; single cobalt salts, 5s. 6d. per lb.; double cobalt salts, 4s. 6d. per lb.

Characteristics of Cobalt.—Believing that this metal is destined to take an important position in the art of electro-deposition at no distant period, a few remarks upon its history, and the advantages which it presents as a coating for other metals, may not be unwelcome. Cobalt, like its mineral associate, nickel,* was regarded by the old German copper miners with a feeling somewhat akin to horror, since its ore, not being understood, frequently led them astray when searching for copper. Brande says, " The word *cobalt* seems to be derived from *Cobalus*, which was the name of a spirit that, according to the superstitious notions of the times, haunted mines, destroyed the labours of the miners, and often gave them a great deal of unnecessary trouble. The miners probably gave this name to the mineral out of joke, because it thwarted them as much as the supposed spirit, by exciting false hopes, and rendering their labour often fruitless ; for as it was not known at first to what use the mineral could be applied, it was thrown aside as useless. It was once customary in Germany to introduce into the church service a prayer that God would preserve miners and their works from *kobalts* and *spirits*. Mathesius, in his tenth sermon, where he speaks of *cadmia fossilis* (probably cobalt ore) says, ' Ye miners call it *cobalt :* the Germans call it the black devil, and the old devil's hags, old and black *kobel*, which by their witchcraft do injury to people and to their cattle.' "

In chemical works cobalt is generally described as a reddish-grey metal, and this fairly represents the tone of its colour, though a *warm steel grey* would perhaps be a more appropriate term. When deposited by electrolysis under favourable conditions, however, cobalt is somewhat whiter than nickel, but it acquires a warmer tone after being exposed to the air for some time. Becquerel states that cobalt, deposited from a solution of its chloride, " has a brilliant white colour, rather like that of iron ; " while Gaiffe says that, when deposited from a solution of the double sulphate of cobalt and ammonium, it is " superior to nickel, both in hardness, tenacity, and beauty of colour." Wahl remarks, " The electro-deposits of this metal which we have seen equal, if indeed they do not surpass, those of nickel in whiteness and brilliancy of lustre." Much of the beauty of electro-deposited cobalt depends, not only upon the electrolyte employed, but also upon the quality of the current, as is also the case with nickel, and indeed most other metals and their alloys.

---

* Nickel was called, by the old German miners, *kupfernickel*, or "false copper."

According to Deville, cobalt is one of the most ductile and tenacious of metals, its tenacity being almost double that of iron. It is fused with great difficulty, but more readily when combined with a little carbon, in which respect, as in many other characteristics, it bears a close resemblance to its mineralogical associate, nickel. It is soluble in sulphuric and hydrochloric acids, but more freely in nitric acid.

Cobalt Solutions.—The salts most suitable for making up cobalt baths are :—1. *Chloride of cobalt*, rendered neutral by ammonia or potash ; 2. the *double chloride of cobalt and ammonium* ; and 3, the *double sulphate of cobalt and ammonium*.

Chloride of Cobalt.—The single salt (chloride) may be prepared by dissolving metallic cobalt or its oxides (the latter being the most readily soluble) in hydrochloric acid, and evaporating the solution to dryness. The residuum is then heated to redness in a covered crucible, when a substance of a bright blue colour is obtained, which is pure chloride of cobalt. When this *anhydrous* (that is without water) chloride of cobalt is dissolved in water it forms a pink solution, which, by careful evaporation, will yield crystals of a beautiful red colour. This is *hydrated* chloride of cobalt, from which various cobalt baths may be prepared according to the directions given below.

Becquerel's Solution.—This is formed by neutralising a concentrated solution of the chloride of cobalt by the addition of ammonia or caustic potash, and adding water in the proportion of 1 gallon to 5 ounces of the salt. The bath is worked with a very weak current, and the deposit is in coherent nodules, or in uniform layers, according to the strength of the current. The deposited metal is brilliantly white, hard, and brittle, and may be obtained in cylinders, bars, and medals, by using proper moulds to receive it. The deposited rods are magnetic,[*] and possess polarity. If an anode of cobalt be used, the solution is of a permanent character. A portion of the chlorine is disengaged during the electro-deposition, and if iron be present in the solution, the greater portion of it is not deposited with the cobalt.

Beardslee's Solution.—The following has been recommended by Mr. G. W. Beardslee, of Brooklyn, New York, and is stated to yield a good deposit of cobalt, which is "very white, exceedingly hard, and tenaciously adherent." Dissolve pure cobalt in boiling muriatic acid, and evaporate the solution thus obtained to dryness. Next dissolve from 4 to 6 ounces of the resulting salt in 1 gallon of distilled water, to which add liquid ammonia until it turns red litmus paper blue. The solution, being thus rendered slightly alkaline, is ready for use. Battery power of from two to five Smee cells will be sufficient to do good work. Care must be taken not to allow the solution

---

[*] Faraday says that perfectly pure cobalt is not magnetic.

to lose its slightly alkaline condition, upon which the whiteness, uniformity of deposit, and its adhesion to the work greatly depend.

Boettger's Solution.—Boettger states that from the following solution a brilliant deposit of metallic cobalt was obtained by means of a current from two Bunsen cells.

Chloride of cobalt	40 parts.	
Sal-ammoniac	20	„
Liquid ammonia	20	„
Water	100	„

By another formula it is recommended to dissolve five ounces of dry chloride of cobalt in one gallon of distilled water, and make the solution slightly alkaline by means of liquid ammonia. A current from three to five Smee cells is employed, with an anode of cobalt. The solution must be kept slightly alkaline by the addition of liquid ammonia whenever it exhibits an acid reaction upon litmus paper. Since these solutions are liable to become acid in working, it is a good plan to keep a strip of litmus paper floating in the bath, so that any change of colour from blue to red may be noticed before the altered condition of the bath has time to impair the colour and character of the deposited metal: if some such precaution be not adopted, the deposit may assume a black colour and rescouring be necessary.

Double Sulphate of Cobalt and Ammonia.—Cobalt is freely deposited from a solution of the double salt, of a fine white colour, provided that an excess of ammonia be present in the bath. From four to six ounces of the double salt may be used for each gallon of water in making up a bath, according to the strength of current employed. The solution of this salt and that of the double chloride more readily yield up their metal than the corresponding salts of nickel, therefore a proportionately smaller quantity of the metallic salts are required to make up a cobalting bath.*

Electro-deposition of Palladium.—This metal may be deposited more freely from its solution than platinum; it is dissolved in *aqua regia* and treated in the same way as the latter metal, and the dry salt dissolved in distilled water. This palladium is then precipitated by means of a solution of cyanide of potassium, and the precipitate redissolved by an excess of the same solution. Since a palladium anode becomes dissolved in the cyanide bath, deposits of any required thickness may be obtained. This metal may also be deposited from a solution of the ammonio-chloride, using a palladium anode, and a current from two or three Smee cells. M. Bertrand advises a neutral solution of the double chloride of palladium and ammonium for the electro-deposition of this metal either with or without the use of a

---

* For further remarks on the electro-deposition of cobalt, see p. 464 *et seq.*

voltaic battery. The deposition of palladium is, however, more interesting as a fact than of any practical use.*

Deposition of Bismuth.—This metal may be dissolved in dilute nitric acid (2 parts acid to 1 part water) with moderate heat, and the solution evaporated and allowed to crystallise. The resulting salt is known as *acid nitrate of bismuth,* which may be dissolved in a very small quantity of distilled water; but if the solution, even when acid, be poured into a large quantity of water it becomes decomposed, and forms a white, somewhat crystalline precipitate, commonly called *subnitrate of bismuth, basic nitrate,* or *pearl white.* If strong nitric acid be poured upon powdered bismuth the chemical action is intensely violent, and ignition sometimes results. *Chloride of bismuth* is formed by dissolving the metal in 4 parts of hydrochloric acid and 1 part nitric, by measure, the excess acid being expelled by evaporation.

To deposit bismuth upon articles of tin by simple immersion, Commaille employs a solution form by dissolving 10 grains of nitrate of bismuth in a wineglassful of distilled water, to which two drops of nitric acid have been added. After the article is immersed the bismuth will be deposited in very small shiny plates. The metal may also be deposited by means of the separate battery. The deposited metal is said to be explosive when struck by a hard substance.

Bismuth may be deposited from a cyanide solution, but since the anode is not freely acted upon by the cyanide the solution soon becomes exhausted. M. A. Bertrand states that bismuth may be deposited upon copper or brass from a solution consisting of 30 grammes of the double chloride of bismuth and ammonium, dissolved in a litre of water, and slightly acidulated with hydrochloric acid. A current from a single Bunsen cell should be used.

Deposition of Antimony.—Chloride of antimony, terchloride of antimony, or, as the ancients termed it, *butter of antimony,* is thus prepared, according to the pharmacopœias: 1 lb. of prepared sulphuret of antimony is dissolved in commercial muriatic acid, 4 pints, by the aid of gentle heat, gradually increased to ebullition. The liquid is filtered until quite clear, then boiled down in another vessel to 2 pints; it is then cooled, and preserved in a well stoppered bottle. The solution has a specific gravity of 1·490. It is highly caustic.

A solution of chloride of antimony may also be prepared by the electrolytic method, that is by passing a moderately strong current through an anode of antimony, immersed in hydrochloric acid, employing a plate of carbon as the cathode, the action being kept up until a strip of clean brass, being substituted for the carbon plate, promptly receives a coating of antimony.

---

* For further details concerning the electro-deposition of palladium, see p. 472.

Another solution may be prepared by digesting recently precipitated teroxide of antimony in concentrated hydrochloric acid, and then adding an excess of free acid to the solution thus obtained. This solution yields a very quick and brilliant deposit of antimony upon brass or copper surfaces with a current from three small Daniell's cells, arranged in series, with electrodes of about equal surface. The fresh teroxide may be readily formed thus : Dissolve 4 ounces of finely powdered tersulphuret of antimony in 1 pint of muriatic acid, by the aid of gentle heat, by which a solution of terchloride of antimony is obtained ; filter the liquid, and then pour it into 5 pints of distilled water. By this dilution a greater part of the terchloride is decomposed, the chlorine unites with the hydrogen of the water, forming hydrochloric acid, and the oxygen of the water, being set free, unites with the antimony, forming a teroxide, that is, an oxide containing three equivalents of oxygen. The teroxide thus obtained is separated by filtration, and washed, to free it from acid. It is then washed with a weak solution of carbonate of soda, which decomposes any terchloride present, leaving the teroxide free. It is then dried over a water bath, and preserved in a well-stoppered bottle.

Antimony may also be deposited from a solution prepared by dissolving oxychloride of antimony in strong hydrochloric acid, the latter being in excess. The oxychloride may be obtained by largely diluting with water a solution of the chloride of antimony, when a white precipitate falls, which is insoluble in water. The liquor is now. to be poured off, and hydrochloric acid added until the precipitate is entirely dissolved. The resulting solution, which must not be diluted with water—which decomposes it—should be used with a moderate current and rather small anode surface, and the articles to be coated in it must be perfectly dry, and when the required deposit is obtained the article should be dipped in a strong solution of hydrochloric acid before being rinsed in water, otherwise a white insoluble film of oxychloride will form on the surface.

A depositing bath may also be formed by mixing equal parts, by measure, of a solution of commercial chloride of antimony and salammoniac. The solution thus formed is a very good conductor, deposits freely a good reguline metal, and is not so liable to yield deposits upon the baser metals by simple immersion as the former solution.

A very good antimony bath may be made by dissolving tartar emetic (potassio-tartrate of antimony) in 2 parts hydrochloric acid and 1 part water, by measure ; or, say, tartar emetic 8 lbs., hydrochloric acid 4 lbs , and water 2 lbs., a larger proportion of water being added if desired. The resulting solution forms a very good bath for the deposition of antimony, and yields up its metal very freely. With the current from two to three Daniells the metal is deposited very quickly, and in a good reguline condition. To insure the adherence of the deposit, however, the anode surface should at first be small, until a film of moderate thickness has been obtained, after which it may be gradually increased until both electrodes are of about equal surface. The above solution is not affected by atmospheric influence nor by continual working, and would be very useful for small operations for producing thick deposits of antimony; but the cost of the mixture would preclude its adoption except for experiment.

Deposition by Simple Immersion.—The acid solution of chloride of antimony readily yields up its metal to brass by simple immersion, and by this means brass articles are coloured of a lilac tint. A solution is made for this purpose by adding a large quantity of water to a small quantity of chloride of antimony, when a dense white precipitate of oxychloride of antimony is formed. The mixture is boiled until this is nearly redissolved, when more water is added, and the boiling resumed. The liquor is then filtered, and the clear liquor heated to boiling ; into this the cleaned brass articles are placed, when they at once receive a coating of antimony of a lilac colour, being kept in the boiling solution until the desired shade of colour is obtained. After rinsing in clean water, the articles are dried in hot sawdust, then brushed clean and lacquered.

Commercial chloride of antimony (butter of antimony) is also used for bronzing or *browning* gun-barrels, and when used for this purpose it is known as *bronzing salt*. To apply it for bronzing gun-barrels the chloride is mixed with olive-oil, and rubbed upon the barrel, slightly heated ; this is afterwards exposed to the air until the requisite tone is obtained ; a little aquafortis is rubbed on after the antimony to hasten the operation. The browned barrel is then carefully cleaned, washed with water, dried, and finally burnished or lacquered.

When a piece of clean zinc is immersed in a solution of chloride of antimony the metal becomes reduced to a fine grey powder, which is employed to give the appearance of grey cast-iron to plaster of Paris casts.

Deposition of Lead.[*]—Lead is readily produced from a solution of its nitrate or acetate—as exemplified in the production of the well-known *lead-tree;* it may also be deposited upon zinc or tin from a solution formed by dissolving litharge (oxide of lead) in a solution of caustic potash. Iron articles will become coated, by simple immersion, in a solution of sugar of lead (acetate of lead). Becquerel[†] deposited lead upon a bright, cleaned surface of copper, in contact with a piece of zinc, in a solution of chloride of lead and sodium. This metal may

---

[*] See also pp. 446 *et seq.*
[†] "The Chemist," vol. v. p. 408.

also be deposited, by means of the battery, from dilute solutions of acetate or nitrate of lead, or from a solution formed by saturating a boiling solution of caustic potash with litharge, employing a lead anode. The deposition of this metal is not, however, of any commercial importance. The electrolysis of salts of lead under certain conditions are, however, exceedingly interesting in what is termed *metallochromy*, as will be seen below.

Metallo-Chromes.—A remarkably beautiful effect of electro-chemical decomposition is produced under the following conditions : A concentrated solution of acetate of lead (sugar of lead) is first made, and after being filtered is poured into a shallow porcelain dish. A plate of polished steel is now immersed in the solution, and allowed to rest on the bottom of the dish (see Fig. 120). A small disc of sheet copper

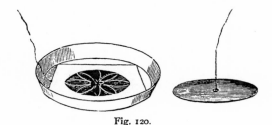

Fig. 120.

is then to be connected to the wire proceeding from the zinc element of a constant battery of two or three cells, and the wire connected to the copper element is to be placed in contact with the steel plate. If now the copper disc be brought as close to the steel plate as possible, without touching it, in a few moments a series of beautiful prismatic colorations will appear upon the steel surface, when the plate should be removed, and rinsed in clean water. These colorations are films of lead in the state of peroxide, and the varied hues are due to the difference in thickness of the precipitated peroxide of lead, the light being reflected through them from the polished metallic surface beneath. By reflected light, every prismatic colour is visible, and by transmitted light a series of prismatic colours complementary to the first series will appear, occupying the place of the former series. The colours are seen to the greatest perfection by placing the plate before a window with its back to the light, and holding a piece of white paper at such an angle as to be reflected upon its surface. The colorations are not of a fugitive character, but will bear a considerable amount of friction without being removed. In proof of the lead oxide being deposited in films or layers, if the deposit be allowed to proceed a few seconds beyond the time when its greatest beauties are exhibited, the coloration will be less marked, and become more or less red, green, or brown. If well rubbed when dry with the finger or fleshy part of the hand a rich blue-coloured film will be laid bare, by the removal of the delicate film above it.

The discovery of this interesting electrolytic phenomenon is due to Nobili, who in the year 1826 discovered that when a solution of acetate of lead was electrolysed by means of a current from four to six Grove cells, a large platinum anode and a platinum wire cathode being employed, prismatic colours were produced upon the anode surface ; and when the platinum anode was placed horizontally in the acetate solution and the negative wire held vertically above it, a series of rings in chromatic order were produced. These effects subsequently took the name of "Nobili's rings," and the interesting discovery induced Becquerel, Gassiot, and others to experiment in the same direction by varying the strength of the current and employing other solutions than the acetate of lead.

Becquerel's Solution.—The following formula was suggested by Becquerel : * Dissolve 200 grammes of caustic potash in 2 quarts of distilled water, add 150 grammes of litharge, boil the mixture for half an hour, and allow to settle. Then pour off the clear liquor, and dilute it with its own bulk of water.

The plan recommended by Mr. Gassiot to obtain the *metallo-chromes* is to place over the steel plate a piece of card, cut into some regular device, as shown in the illustration, and over this a rim of wood, the copper disc being placed above this. We have found that very beautiful effects are obtained when a piece of fine copper wire is turned up in the form of a ring, star, cross, or other pattern, and connected to the positive electrode as before ; indeed, this is one of the simplest and readiest methods of obtaining the colorations upon the polished metal. A few examples of metallo-chromes obtained in this way are shown in the frontispiece of this work. Metallo-chromy, as it is termed, is extensively employed in Nuremberg to ornament metallic toys, the solution used being that suggested by Becquerel, namely, a solution of the oxide of lead in caustic soda or potash. Metallo-chromy has also been adopted in France for colouring bells, and in Switzerland for colouring the hands and dials of watches. In using the lead solutions to produce metallo-chromes it must be remembered that metallic lead becomes deposited upon the cathode, consequently the solutions in time become exhausted, and must therefore be renewed by the addition of the lead salt.

---

* " The Chemist," vol. iv. p. 457.

Metallo-chromes on Nickel-plated Surfaces.—It will be obvious that if metallo-chromy were only applicable to platinum or steel surfaces—which has generally been the case heretofore—that the usefulness of the process as a means of ornamentation for industrial purposes would be greatly restricted. While the production of these colorations upon platinum foil would only be effected for experimental purposes, the application of the process to steel surfaces would necessarily be of a limited character, owing to the unsuitableness of this metal as compared with brass, German silver, and copper, for the manufacture of many articles of utility or ornament. With a view to extend the usefulness of these very beautiful colorations, and thus, to a certain extent, open up a new field for their application, the author some time since turned his attention to polished nickel-plated surfaces, as being, of all others, the most suitable, from their extreme brilliancy, to exhibit the rainbow tints of metallo-chromy. His first experiments were upon highly-polished surfaces of nickel-plated brass, and the results obtained were exceedingly satisfactory. The experiments were subsequently pursued under varied conditions of working, until the most satisfactory method of procedure was arrived at.

The reader is referred to pp. 448, *et seq.*, for a detailed description of the method adopted to produce these colorations upon nickel-plated surfaces, and for some particulars as to salts of lead applicable to the purpose.

Deposition of Aluminum or Aluminium.—This remarkable metal, which in an oxidised state (alumina) occurs most abundantly in nature as a constituent of all clays in combination with silica, was first obtained in the metallic state by Wöhler in the following way: Chloride of aluminium and pure potassium are heated in a small platinum or porcelain crucible, the heat of a spirit-lamp being sufficient, for when the substances begin to react upon each other the temperature suddenly rises to redness. When the crucible is cold, its contents are well washed with cold water, by which a finely divided grey substance with a metallic lustre is obtained, which is pure aluminium. About the year 1854, Sainte-Claire Deville, of Paris, devoted his attention to this subject, substituting chloride of sodium for potassium, and heating the chloride of aluminium with this salt in a porcelain crucible to bright redness,* by which the excess of chloride of aluminium was disengaged, and in the middle of the resulting saline mass larger or smaller globules of perfectly pure aluminium were found.

In reference to the characteristics of this metal, Deville says:

---

* "The Chemist." Edited by John and Charles Watt. Vol. i., new series, 1854.

—" It is completely unalterable, either in dry or humid air; it does not tarnish; and remains brilliant where freshly-cut zinc or tin lose their polish. Sulphuretted hydrogen has no action upon it; neither cold nor boiling water will tarnish it; nitric acid, whether weak or concentrated, or sulphuric acid employed cold, will take no effect upon it. Its real solvent is hydrochloric acid. . . . It will be easily understood that a metal as white [?] and as unalterable as silver, which does not tarnish in the air, which is fusible, malleable, ductile, and yet tough, and which has the singular property of being lighter than glass, would be most useful if it could be obtained. If we consider, besides, that this metal exists in considerable proportions in nature, that its ore is argil [clay], we may well desire that it should become of general use. I have much hope that it may be so, for chloride of aluminium is decomposed with remarkable facility by common metals at a high temperature, and a reaction of this nature, which I am now endeavouring to realise on a larger scale than a mere laboratory experiment, will decide this question in a practical point of view."

Not long after the above announcement was made, Sainte-Claire Deville, supported in the practical development of his ingenious process by the late Emperor of the French, succeeded in producing aluminium in abundance, and bars of this useful metal entered the market as a commercial product to the great surprise and delight, not only of scientists, but of those workers in metals who know how to appreciate the importance of a metal possessing such remarkable characteristics as aluminium. We all know now what an important position it has taken in the arts; but its usefulness may yet receive further development, it is hoped, by some successful process of electro-deposition. That point, however, has not yet been fully reached, although the metal has been deposited with sufficient success to warrant the belief that still more satisfactory results will be obtained by a further investigation of the subject.

Speaking upon the separation of aluminium by electrolysis, Deville observes:*—" It appeared to me impossible to obtain aluminium by the battery in aqueous liquids. I should believe this to be an impossibility if the brilliant experiments of M. Bunsen on the production of barium did not shake my conviction. Still, I must say that all processes of this description which have recently been published for the preparation of aluminium have failed to give me good results. It is of the double chloride of aluminium and sodium, of which I have already spoken, that this decomposition is effected. The bath is composed of 2 parts by weight of chloride of aluminium, with the addition of 1 part of dry and pulverised common salt. The whole is

---

* " The Chemist," new series, vol. ii. p. 12, 1855.

mixed in a porcelain crucible, heated to about 392° Fahr. The combination is effected with disengagement of heat, and a liquid is obtained which is very fluid at 392° Fahr., and fixes at that temperature. It is introduced into a vessel of glazed porcelain, which is to be kept at a temperature of about 392° Fahr. The cathode is a plate of platinum, on which the aluminium (mixed with common salt) is deposited in the form of a greyish crust. The anode is formed of a cylinder of charcoal, placed in a perfectly dry porous vessel, containing melted chloride of aluminium and sodium. The densest charcoal rapidly disintegrates in the bath and becomes pulverulent ; hence the necessity of a porous vessel. The chlorine is thus removed, with a little chloride of aluminium, proceeding from the decomposition of the double salt. This chloride would volatilise and be entirely lost, if some common salt were not in the porous vessel. The double chloride becomes fixed, and the vapours cease. A small number of voltaic elements (two are all that are absolutely necessary) will suffice for the decomposition of the double chloride, which presents but little resistance to the electricity. The platinum plate is removed when it is sufficiently charged with the metallic deposit. It is suffered to cool, the saline mass is rapidly broken off, and the plate replaced."

Bunsen electrolysed the fused chloride of aluminium and sodium in a deep covered porcelain crucible, divided by a partition of porous porcelain, which extended half-way down the vessel. Carbon electrodes were used, and these were introduced through openings in the cover. He used a current from ten cells of his zinc and carbon battery. The salt fused at 662° Fahr. (the boiling point of mercury), and readily yielded the metal. The temperature of the liquid should then be raised to nearly the melting point of silver, when the particles of the liberated aluminium fuse, uniting together into globules, which, being heavier than the fused salt, fall to the bottom of the crucible.

Corbelli has deposited aluminium by electrolysing a mixed solution of rock alum (sulphate of alumina) and chloride of sodium or calcium with an anode of iron wire, coated with an insulating material, and dipping into mercury deposited at the bottom of the solution ; a zinc cathode is immersed in the solution. Aluminium deposits upon the zinc, and the chlorine set free at the anode unites with the mercury, forming chloride of mercury (calomel).

Thomas and Tilley's Process, for which a patent was obtained in 1854, consists in forming a solution composed of freshly precipitated alumina dissolved in a boiling solution of cyanide of potassium. By another process, patented in 1855, calcined alum is dissolved in a solution of cyanide of potassium. Several other solutions are included in the same specification, and the invention includes the deposition of alloys of aluminium with silver, silver and copper with tin, silver and tin, etc.

Jeancon's Process (American) consists in depositing aluminium from a solution of a double salt of aluminium and potassium, of the specific gravity 1·161, employing a current from three Bunsen cells.

M. Bertrand states that he has deposited aluminium upon a plate of copper in a solution of the double chloride of aluminium and ammonium by using a strong current.

Goze's Process.—Mr. Goze obtained a deposit of aluminium by the single cell method from a dilute solution of the chloride. The liquid was placed in a jar, in which was immersed a porous cell containing dilute sulphuric acid; an amalgamated zinc plate was immersed in the acid solution, and a plate of copper in the chloride solution, the two metals being connected by a copper conducting wire. At the end of some hours the copper plate became coated with a lead-coloured deposit of aluminium, which, when burnished, presented the same degree of whiteness as platinum, and did not appear to tarnish readily when immersed in cold water, or in the atmosphere, but was acted upon by dilute sulphuric and nitric acids.*

Deposition of Cadmium.—This metal is readily soluble in dilute nitric, sulphuric, and hydrochloric acids, with disengagement of hydrogen, and the respective salts may be obtained in the crystalline form by concentrating the acid solutions by evaporation. The *hydrated oxide*, in the form of a gelatinous precipitate, is produced when a solution of the alkalies, soda, potassa, &c., is added to a solution of a salt of cadmium. The hydrate is white, but becomes brown from loss of water when dried by heat. Respecting the electro-deposition of cadmium, Smee states that it is difficult to obtain firm, coherent deposits from solutions of the chloride or sulphate, but that it may be easily deposited in a reguline and flexible condition from a solution of the ammonio-sulphate, prepared by adding sufficient liquid ammonia to sulphate of cadmium to redissolve the precipitate at first formed. Napier recommends the following : " A solution of cadmium is easily prepared by dissolving the metal in weak nitric acid, and precipitating it with carbonate of soda, washing the precipitate, and then dissolving it in cyanide of potassium. A battery power of three or four pairs is required, and the solution should be heated to at least 100° Fahr. The metal is white, and resembles tin ; it is very soft, and does not present many advantages to the electro-metallurgist."

Russell and Woolrich's Process.—This process, for which a patent was obtained in 1849, is thus briefly described: " Take cadmium, and dissolve it in nitric acid diluted with five or six times its bulk of water,

---

* For further remarks on aluminium deposits, see pp. 476 *et seq.*

at a temperature of about 80° or 100° Fahr., adding the dilute acid by degrees until the metal is all dissolved ; to this solution of cadmium one of carbonate of sodium (made by dissolving 1 lb. of crystals of washing soda in 1 gallon of water) is added until the cadmium is all precipitated; the precipitate thus obtained is washed four or five times with tepid water. Next add as much of a solution of cyanide of potassium as will dissolve the precipitate, after which one-tenth more of the solution of the potassium salt is added to form free cyanide. The strength of this mixture may vary ; but the patentees prefer a solution containing six troy ounces of metal to the gallon. The liquid is worked at about 100° Fahr., with a plate of cadmium as an anode.''

For depositing cadmium M. A. Bertrand recommends a solution of the bromide of cadmium, containing a little sulphuric acid, or a solution of sulphate of cadmium. He states that the deposit obtained is white, adheres firmly, is very coherent, and takes a fine polish.*

Deposition of Chromium.—In his investigation concerning the electrolysis of metallic salts Bunsen determined the causes which most influence the separation of the metal ; these causes are two in number, the principal of which is owing to the *density* of the current, and the other to the greater or less concentration of the electrolyte. By density he means the concentration to a single point of "the electrical undulations, in a manner analogous to the concentration of luminous or calorific rays in the focus of a concave mirror. Let us take, for example, a charcoal crucible in communication with the positive pole of the battery, and place in it a small capsule of glazed porcelain, containing the liquid to be decomposed ; the space between the crucible and the capsule is filled with hydrochloric acid, and the liquid of the small capsule is put in communication with the battery by means of a thin sheet or wire of platinum.* The current is then established between a large surface, the charcoal crucible, and a fine platinum wire, in which it is concentrated ; the effects are added in this direction, and the fluid becomes capable of overcoming affinities which have hitherto resisted powerful batteries.'' The apparatus is placed is a porcelain crucible, which is kept warm in a sand bath.

By the above arrangement Bunsen succeeded in separating chromium with perfect facility from a concentrated solution of its chloride ; the deposited metal, which was chemically pure, presented the appearance of iron, but was less alterable in moist air. It resisted

---

* For further remarks on the electro-deposition of cadmium, see "Arcas" silver plating, p. 473 of this volume.

† For this purpose the platinum wire must be exactly in the centre of the crucible ; if not, by virtue of its tendency to take the shortest road, the current is established in preference between the nearest points.

the action of even boiling nitric acid, but was acted upon by hydrochloric acid and dilute sulphuric acid. Bunsen found that when the current was diminished, the metal ceased to be deposited in the metallic state, but appeared as a black powder consisting of protoxide and sesquioxide of chromium.*

Deposition of Manganium, or Manganese.—The same eminent chemist succeeded in obtaining metallic manganese by the method above described from a concentrated aqueous solution of chloride of manganese. The metal was separated with the greatest facility with a powerful current, but when the current was weakened black oxide of manganese was obtained.

Deposition of Magnesium.—Bunsen electrolysed fused chloride of magnesium at a red heat by the same method as that adopted for the separation of aluminium. Magnesium, being a very light metal, is liable to rise to the surface of the fused mixture and ignite in the air; to prevent this, as far as possible, the carbon cathode was notched, so that the metal could collect in the notches. M. Bertrand says that from an aqueous solution of the double chloride of magnesium and ammonium, a strong current will deposit magnesium upon a sheet of copper in a few minutes, the deposit being homogeneous, strongly adherent, and easily polished.

Deposition of Silicon.—Mr. Goze reduced the metal *silicium* from a solution of monosilicate of potash, prepared by fusing one part of silica with $2\frac{1}{4}$ parts of carbonate of potash, the same voltaic arrangement being adopted, except that a small pair of Smee batteries were interposed in the circuit. With a very slow and feeble action of the current, the colour of the deposit was much whiter than aluminium, closely approximating that of silver.

We have given the foregoing details concerning the deposition of some of the less tractable metals, more with a view to show what ingenious methods have been devised for their extraction, or separation, than as presenting any absolute practical advantage. As interesting electrolytic facts they are valuable to the student, while to the more practical operator who may devote a portion of his spare time to electrolytic experiments, Chevalier Bunsen's methods of conducting the electrolysis of salts which do not readily yield up their metal from aqueous solutions will prove not only interesting but highly instructive. It will not be in accordance with the object of this work, however, to enter further into the deposition of metals which have no practical significance in the arts.

---

* " The Chemist," new series, vol. i. p. 685.

CHAPTER XXIV.

ELECTRO-DEPOSITION OF ALLOYS.*

Electro-deposition of Brass and Bronze.—The deposition of two metals in combination by electro-chemical means, although perfectly practical, is far more difficult to accomplish satisfactorily than to deposit a single metal. For example, the two metals zinc and copper are so widely different in all their characteristics—in their melting point, ductility, electric relàtion, and conductivity—that when in a state of solution great care is necessary to enable us to bring them together in the uniform condition of what is termed an *alloy*. Even when these two metals are alloyed in the ordinary way, by fusion, great care must be exercised, or the zinc, being a *volatilisable* metal, will pass away into the air instead of uniting with the copper to form *brass*. The copper, melting at a far higher temperature than zinc, is fused or melted first, and the zinc gradually added, until the desired object is obtained—a bright yellow alloy, the tone or colour of which may be varied according to the proportion of either metal.

In depositing brass from its solution, the *nature* and strength of the electric current are of the greatest importance, for if the electro-motive force be too weak copper only will be deposited, and if too strong zinc alone will be precipitated upon the receiving metal. Again, if too great a surface of anode be exposed in the bath in proportion to the size of the article to be coated zinc alone will deposit, the reverse being the case, that is copper alone, if the surface of anode is too small. A medium between these conditions is absolutely necessary (all other things being equal) to ensure a coating of brass of good colour upon any given article. To make this more clear to the less experienced, we may state, for instance, that a battery composed of the two elements, zinc and copper, as the Wollaston and Daniell batteries,

---

* See also pp. 451-454 and pp. 473 *et seq.*

are far less *intense*, that is to say, they possess feebler electromotive force, than Bunsen's battery, with carbon and zinc elements. The latter battery, therefore, is more suited to the electro-deposition of brass, and is indeed preferable to any other. The quality of the deposit is also much influenced by the temperature of the solution and the materials with which it is prepared, some formulæ yielding solutions which are better conductors than others, and consequently offer less resistance to the current.

In making up solutions for the deposition of alloys, as brass, bronze, and German silver, for example, the author prefers to prepare them in what may be termed *the direct way* ; that is to say, instead of forming the depositing solution from a mixture of the metallic salts and their solvents, according to the usual method of preparing such solutions, he first dissolves the metallic alloy in its acid solvent—nitric or nitro-hydrochloric acid (*aqua regia*)—and from the acid solution thus obtained he forms the depositing bath by either of the methods given below. It may be well to remark, however, that in making up a brass bath upon this system, metal of the very best quality should be employed, and the solution should be formed from the identical sample of brass which is to be used as an anode in the depositing tank. The proportions given are for one gallon of solution, but it will be readily understood that, adopting the same proportions of the materials, a bath of any desired quantity can be prepared.

Brassing Solutions.—*No. I.*—Take of

Good sheet brass	1 ounce.
Nitric acid (by measure) about . . .	4 ounces.
Water	2 „

Cut up the sheet brass into strips, and put them carefully into a glass flask, then pour in the water and acid. To accelerate the chemical action the flask should be gently heated over a sand bath, and the fumes must be allowed to escape through the flue of the chimney. When the red fumes, liberated during the decomposition, cease to be visible in the bulb of the flask the chemical action is at an end, provided a portion of undissolved brass remains in the flask. If such be not the case a few fragments of the metal should be put into the flask and the heat continued, when, if red fumes are again given off, the heat should be kept up until the fumes disappear while a portion of undissolved metal still remains in the flask. The reasons for giving these precautionary details are—1, that it is important there should be as little excess of acid as possible in the solution ; and 2, that the strength of commercial nitric acid is very variable, and therefore chemically minute proportions cannot advantageously be given. We may say, moreover, that the *exact* quantity of brass per gallon of solution is of no consequence ; if the proportion nearly approaches that given in the formula, it will be quite near enough for all practical purposes. While touching upon this subject we may also state that the active quality of commercial cyanide of potassium also varies greatly ; consequently it may be necessary to apply either more or less than the quantity specified below, according to the quality of the article that may fall into the hands of the user. Upon this subject we shall, however, say more hereafter.

The acid solution of brass must next be poured into a vessel of sufficient capacity, and diluted with about three or four times its bulk of water. Then add liquid ammonia (specific gravity ·880°), gradually to the green solution of the metals, stirring with a glass rod, when a pale green precipitate will be formed, which will afterwards become dissolved by adding an excess of ammonia, forming a beautiful deep blue solution. This solution should become perfectly clear when the necessary quantity of ammonia has been added, but if such be not the case, a little more must be added, with brisk stirring, until the precipitate is quite dissolved and a clear solution obtained. The exact quantity of ammonia required will depend upon the amount of free acid remaining in the metallic solution first prepared. A moderately strong solution of cyanide must now be added to the blue solution, with constant stirring, until the blue colour entirely disappears. When sufficient cyanide has been added to destroy the blue colour, the solution will acquire a pinkish tinge, and on the application of a little more cyanide solution this will in its turn disappear, and the liquid will assume a yellowish tint, when a moderate excess of the cyanide must be given as "free cyanide," and the solution then made up to the full quantity (one gallon) by the addition of water. The solution should then be set aside to rest for a few hours, when the clear liquor may be poured into the depositing vessel. The last portion of the liquor should be passed through a filter, to separate any impurities (chiefly derived from the cyanide) which may be present. If convenient, the entire bulk of the solution may be filtered, which is in all cases preferable. A brassing bath always works most satisfactorily if not used for at least twenty-four hours after being prepared, although it may, if required, be used directly after being filtered. If to be used hot, the solution may be further diluted.

Solution II.—One ounce of brass being dissolved as before, the solution is to be diluted with about three pints of cold water ; a solution of carbonate of potash (about half-a-pound to a quart of water) is to be gradually added, with frequent stirring, until no further precipitation takes place. The precipitate formed should next be put into a filter of unbleached calico stretched over a wooden frame, and when the liquor has ceased to drain from it, hot water should be poured on to the mass, which must be stirred with a wooden spoon, or flat strip of wood, so as to assist the *washing* of the precipitate with the water. When the precipitate is thoroughly drained, it is to be transferred to a convenient vessel, and redissolved by liquid ammonia, which is to be added gradually, and constantly stirred in until the whole is dissolved, and a dark blue solution formed. After reposing for a few minutes, the clear liquor may be poured off, and should any undissolved green precipitate remain at the bottom of the vessel, ammonia must be added to this until dissolved, when the resulting blue liquor is to be added to the bulk. A strong solution of cyanide is now to be added to the blue liquor, until its characteristic colour has entirely disappeared, after which a moderate excess of the cyanide solution is to be added, and the solution then made up to 1 gallon (according to the proportion of metal dissolved) with cold water. The repose or filtration as before should be again resorted to.

Solution III.—

Acetate of copper	. . . . . .	5 ounces.
Sulphate of zinc	. . . . . .	10 "
Caustic potash	. . . . . .	4½ lbs.
Liquid ammonia	. . . . . .	1 quart.
Cyanide of potassium	. . . . .	8 ounces.

The acetate of copper should be first powdered, and then dissolved in about 2 quarts of water. To this add one-half of the ammonia (1 pint). Now dissolve the sulphate of zinc in 1 gallon of water at a temperature of 180° Fahr.; to this add the remaining pint of the ammonia, constantly stirring while the liquid is being added. The potash is next to be dissolved in 1 gallon of water, and the cyanide in 1 gallon of hot water, after which the several solutions are to be mixed as follows : The solutions of copper and zinc are to be first mixed, the solution of potash then added, and lastly the cyanide. The whole must now be well stirred, and then allowed to repose for a short time, when the agitation may be resumed and repeated at intervals during a couple of hours or so. Water, to make up 8 gallons in all, is now to be added, and the solution then allowed to rest for a few hours, when the clear liquor is to be decanted into the bath. This solution should be worked with a strong current, with additions of liquid ammonia and cyanide from time to time when the anode becomes foul. It is important in working this, as in all other brassing solutions, that the anode should be kept clean, a condition which is not possible with these solutions unless there be an excess of the solvents, cyanide and ammonia.

Brunel, Bisson, & Co.'s Processes.—1. The brassing solution is formed from the following ingredients, which should each be dissolved in separate vessels :—

Carbonate of potassa (salt of tartar) . .	10 pounds.
Cyanide of potassium	1½ pound.
Sulphate of zinc	1¼ „
Chloride of copper	10 ounces.
Water	12½ gallons.

A sufficient quantity of the potash solution is to be added to the sulphate of zinc and chloride of copper solutions to precipitate all the metal in the form of *carbonates*. Liquid ammonia (specific gravity ·880°) is now to be poured into each vessel, being well stirred in to dissolve the respective precipitates, when the solutions are to be added to the cyanide solution ; the remainder of the potash solution is next to be added, and the whole well stirred ; water is then to be added to make up a bath of 12½ gallons. The solution is to be worked with two or more Bunsen batteries, with a large brass anode. As before recommended, the solution should not be worked until some hours after being made, and the clear liquid must be decanted, so as to separate it from any sedimentary matter that may be present from impurities in the cyanide or otherwise. After using the bath for some time, it will require moderate additions of cyanide and liquid ammonia, to keep the anode free from the white salt of zinc which forms upon its surface when the excess of these substances has become exhausted. In adding fresh cyanide, a portion of the solution may be taken out of the bath with a jug, and a few lumps of cyanide (say half a pound) added, and as this becomes partially dissolved, the liquid is to be added to the bath, and the jug again filled with the solution as before ; in this way the bath may be strengthened with cyanide without employing water to dissolve it. In warm weather, however, when the bath loses water by evaporation, the cyanide may be dissolved in water before adding it to the bath. When either liquid ammonia or cyanide are to be added to the solution, this may be conveniently done overnight, and the bath well stirred, when by the following morning the disturbed sediment, which always accumulates at the bottom of depositing vessels, will have had time to settle.

Solution 2.—This is prepared from the following ingredients :—

Sulphate of zinc	2 pounds.
Chloride of copper	1 pound.
Carbonate of potassa	25 pounds.
Nitrate of ammonia	12½ „

The chloride of copper is to be dissolved in half a gallon of water, the carbonate of potash in 6 gallons of water, and the sulphate of zinc in half a gallon of hot water. These three solutions are now to be mixed, and the nitrate of ammonia added, when the whole are to be well united by stirring. Sufficient water is next to be added to make up about 20 gallons of solution, which must be allowed to rest for some hours before using it. After working this solution for some time it will be necessary to add moderate quantities of liquid ammonia and cyanide of potassium, otherwise the anode will become foul and thus incapable of becoming dissolved in the solution.

De Salzede's Processes.—1. This is prepared from the following formula :—

Cyanide of potassium	12 parts.
Carbonate of potassa	610 ,,
Sulphate of zinc	48 ,,
Chloride of copper . . - .	25 ,,
Nitrate of ammonia	305 ,,
Water	5000 ,,

The cyanide is to be dissolved in 120 parts of the water, and the carbonate of potash, sulphate of zinc, and chloride of copper are next to be dissolved in the remainder of the water, the temperature of which is to be raised to about 150° Fahr. When the salts are well dissolved, the nitrate of ammonia is to be added, and the mixture well stirred until the latter is all dissolved. The solution should be allowed to stand for several days before using, and the clear liquor separated from any sediment that may have deposited at the bottom of the vessel.

Solution 2.—

Cyanide of potassium	50 parts.
Carbonate of potassa	500 ,,
Sulphate of zinc	35 ,,
Chloride of copper	15 ,,
Water	5000 ,,

This solution is to be made up in the same way as No. 1.

Solution 3. Bronzing Solution.—This solution is the same as No. 1, except that 25 parts of *chloride of tin* are substituted for the sulphate of zinc.

Solution 4. Bronzing Solution.—This solution is the same as No. 2, with the exception that 12 parts of *chloride of tin* are substituted for the sulphate of zinc. This solution is worked warm, that is, at about 97° Fahr.

Newton's Processes consist in forming solutions for depositing brass or bronze. He mixes chloride of zinc with the chloride of ammonium (sal-ammoniac), chloride of sodium (common salt), or chloride of potassium, dissolved in water. Or he makes a mixture of acetate of zinc dissolved in water and acetate of ammonia, soda, or potassa. In making up a *brassing* solution, Newton adds to either of the above solutions a proportion of the corresponding salt of copper : for example, with the acetate of zinc he would unite acetate of copper, and so on. In making a *bronzing* solution, he dissolves the double tartrate of copper and potassa, and double tartrate of the protoxide of tin and potassa. He deposits an alloy of zinc, tin, and copper by employing a solution composed of the following : double cyanide of copper and potassium, "zincate" of potassa, and stannate of potassa. The *zincate of potassa* he forms by fusing oxide of zinc with caustic potassa, and the *stannate* of potassa, either by fusing oxide of tin with caustic potassa, or by dissolving it in a solution of potassa. To form a brassing bath, he also employs a solution consisting of a given quantity of oxide of copper dissolved in an excess of cyanide of potassium ; oxide of zinc and a little liquid ammonia are then added, and the solution heated from 120° to 140° Fahr. Water is then added to allow the solution to contain 3 ounces of the metallic oxides to each gallon of the solution, that is, 2 ounces of zinc oxide to 1 ounce of copper oxide, being the proportions to form brass.

Russell and Woolrich's Process.—A solution is made of the following :—

Acetate of copper	10 pounds.
„ zinc	1 pound.
„ potassium	10 pounds.
Water	5 gallons.

The salts are to be dissolved in the water, and as much of a solution of cyanide added as will first precipitate the metals, and afterwards redissolve the precipitate. An excess of cyanide is then to be added, and the solution set aside to settle as before. A brass anode, or one of zinc and another of copper, may be used.

Wood's Process consists in making a solution as follows :—

Cyanide of potassium (troy weight) . .	1 pound.
„ copper	2 ounces.
„ zinc	1 ounce.
Distilled water	1 gallon.

When the ingredients are dissolved, add 2 ounces of sal-ammoniac. For coating smooth articles, it is recommended to raise the temperature of the solution to 160° Fahr., using a strong current.

Morris and Johnson's Process.—A solution is made by dissolving in 1 gallon of water—

Cyanide of potassium	1 pound.
Carbonate of ammonia	1 „
Cyanide of copper	2 ounces.
„ zinc	1 ounce.

The solution is to be worked at a temperature of 150° Fahr., with a large brass anode, and a strong current.

Dr. Heeren's Process.—According to this authority,* a brassing solution may be prepared by employing a large excess of zinc to a very small proportion of copper, as follows:—Take

Sulphate of copper	1 part.
,, zinc	8 parts.
Cyanide of potassium	18 ,,

The ingredients are to be dissolved in separate portions of warm water. The copper and zinc solutions are now to be mixed, and the cyanide solution then added, when 250 parts of distilled water are to be added, and the mixture well stirred. The bath is to be used at the boiling temperature, with two Bunsen cells. By this process it is said that very rapid deposits of brass have been obtained upon articles of copper, zinc, Britannia metal, &c.

Roseleur's Processes.—1. Dissolve in 1,000 parts of water, 25 parts of sulphate of copper and from 25 to 30 parts of sulphate of zinc ; or, $12\frac{1}{2}$ parts of acetate of copper and $12\frac{1}{2}$ to 15 parts of fused chloride of zinc. The mixture is to be precipitated by means of 100 parts of carbonate of soda previously dissolved in plenty of water, with constant stirring. The precipitate is to be washed several times, by first allowing it to subside and then pouring off the supernatant liquor (which may be thrown away), when fresh water is to be poured on the precipitate, and after again stirring it is allowed to subside, the washing to be repeated two or three times. After pouring off all the water the last time, a solution composed of 50 parts of bisulphite of sodium and 100 parts of carbonate of soda dissolved in 1,000 parts of water, is to be added, stirring well with a wooden rod. A strong solution of commercial cyanide of potassium is now to be added until the precipitate becomes just dissolved. From $2\frac{1}{2}$ to 3 parts of cyanide in excess are now to be added with stirring, when the solution is complete.

Solution 2.—To form a cold bath for brassing all metals, dissolve 15 parts of sulphate of copper and 15 parts of sulphate of zinc in 200 parts of water ; now add a solution made by dissolving 40 parts of carbonate of soda in 100 parts of water, and stir the mixture well. The precipitate is allowed to subside, as before, when the clear liquor is to be run off, and fresh water added, to wash the precipitate, the washing to be repeated several times. To the drained precipitate add 20 parts of bisulphite of sodium dissolved in 900 parts of water. Now dissolve 20 parts of cyanide of potassium and two-tenths of a part of arsenious acid (white arsenic) in 100 parts of water, and add this to the former liquor. This decolours the mixture and completes the

---

* " The Chemist," 1855, p. 345.

brassing solution. The effect of the arsenious acid is to render the deposit bright. We were long accustomed to employ small quantities of white arsenic with our brassing solutions, and when used with moderation considered the addition highly favourable to a good deposit of brass. Roseleur recommends, in working this bath, to add a little cyanide when the deposit looks earthy, or ochreous, and arsenic when it yields a dull deposit ; if too red, a little zinc and cyanide are to be added ; if too white, a little copper and cyanide ; if the solution works tardily, add both zinc and copper salts, and more cyanide ; and since the anode does not dissolve freely enough to keep up the strength of the solution, these additions of the metallic salts and cyanide must be made from time to time whenever the bath works tardily. The same remedy should be applied to all brassing solutions when they work sluggishly. When the above solution, by the additions of the metallic salts, reaches a higher specific gravity than 1·091, water must be added, but the specific gravity must not be lower than 1·036.

Solution 3.—The following solution is recommended for coating steel, cast iron, wrought iron and tin :—Dissolve 2 parts of bisulphite of soda, 5 parts of cyanide of potassium (of 75 per cent.), and 10 parts of carbonate of soda in 80 parts of distilled water, and add to the mixture 1 part of fused chloride of zinc and 1¼ parts of acetate of copper, dissolved in 20 parts of water.

Solution 4.—For coating zinc articles, the following solution is recommended :—20 parts of bisulphite of soda and 100 of cyanide of potassium (of 75 per cent.) are dissolved in 2,000 parts of water. Then dissolve 35 parts of chloride of zinc, 35 parts of acetate of copper, and 40 parts of liquid ammonia in 500 parts of water. The solutions are now to be mixed and the compound solution passed through a filter.

In working these solutions, if too strong a current be employed, or too large a surface of anode exposed in the solution, zinc only will be deposited ; if the current be feeble, or if the articles are kept in motion while deposition is taking place, the deposit will be chiefly or wholly copper. If a white deposit of oxide of zinc appears upon the anode, a small quantity of liquid ammonia should be added to the bath.

Walenn's Processes.—A solution for depositing brass is made as follows :—Crystallised sulphate of zinc 1 part, and crystallised nitrate of copper 2 parts, are dissolved to saturation. Strong liquid ammonia is then added in sufficient quantity to precipitate the oxides and redissolve them. Cyanide of potassium is then added until the purple liquid is completely decoloured. The resulting solution should be left to repose for a day or two, and may be worked with from 1 to 3 Smee cells, using heat if a brass anode be employed. It is preferred, however, to work the solution by a "porous cell arrangement, in which the surface of the solution next the zinc or other dissolving plate is at a greater elevation than that of the external or depositing solution." In working the solution, the hydrated oxides of copper and zinc are added from time to time, and, if necessary, ammoniuret of copper also.

For a hot brassing solution, Walenn gives the following formula :— A " solvent solution " is first made, consisting of—

Cyanide of potassium (standard solution) .	.	6 parts.
Nitrate of ammonium ,, ,,	.	1 part.
Sulphate of ,, ,, ,,	.	2 parts.

The standard solution of each salt consists of the solid salt dissolved in five times its weight of water. The ingredients being mixed, the whole is divided into three parts :

Free solvent solution	1 part.
Solution to dissolve cupric cyanide . . .	5½ parts.
,, ,, zinc ,, . . .	2¾ ,,

When the respective cyanides have been dissolved to saturation in the above proportions, the free solution is added, and the whole well mixed ; ammoniuret of copper is then added, and the solution set aside for a day or two. Walenn prevents the evolution of hydrogen (or nearly so) during deposition by adding the hydrated oxides of copper, or ammoniuret of copper and zinc, in sufficient quantity for the purpose.

By another process he employs solutions of cyanide of potassium and tartrate of ammonium, in equal proportions. In this menstruum he dissolves cyanides, tartrates, carbonates, &c. of copper and zinc, and the solutions thus formed may be worked either hot or cold. The proportions of the various salts must be varied according to the strength of the current employed.

Walenn makes the following observations on the electro-deposition of copper and brass :—" A solution containing one pound of cupric sulphate, and one of sulphuric acid to a gallon of water, deposits the metal in a solid and compact mass, with a somewhat botryoidal* surface. The addition of one ounce of zinc sulphate (as recommended by Napier) prevents this botryoidal form, and renders the deposit tough, compact, and even. From a solution containing a greater proportion of zinc sulphate, copper is deposited in tufts or needles,

---

* *Botryoidal*, resembling a bunch of grapes ; referring to the *nodular* or knotty form which copper assumes at the back of electrotypes.

standing at right angles to the surface of the metal. Ordinary
electro-brassing liquids [deposits from] show the same peculiarity in
even a more marked degree, and this makes it impossible to produce a
good deposit of more than ·01 to ·03 inch in thickness. This form of
deposit is owing chiefly to a copious evolution of hydrogen taking
place during its formation." While not disagreeing with Mr.
Walenn's views, the author may state that he has found that a small
quantity of arsenious acid (previously mixed with a strong solution of
cyanide) added to brassing baths had generally rendered the deposit
smooth and compact; the quantity, however, must be small, otherwise the deposit is liable to be of a brittle character. About one
drachm of arsenious acid to each gallon of bath will be sufficient. He
has usually noticed that brassing solutions evolve hydrogen most
freely when poor in metal, and when containing a large excess of
cyanide. A solution richer in metal, and containing but a moderate
excess of cyanide, generally yields better results, both as to colour
and general character of the deposit. A great deal depends, however,
upon the amount of current and its tension, and also upon the temperature of the bath. A solution rich in copper and zinc is best
worked at about 130° Fahr., or even higher. When the solution
becomes partly exhausted of its metals, owing to the brass anodes not
becoming freely dissolved in the solution, it is always advisable to add
fresh *concentrated* cyanide solutions of the zinc and copper salts from
time to time, taking care, however, only to add them in sufficient
quantity to obtain the desired effect—a coating of good colour with
but trifling evolution of hydrogen at the negative electrode.

Bacco's Solution.—The following solution is said to yield a brass
deposit upon zinc work that will stand burnishing, and the deposit
may be obtained either by simple immersion or by the battery. A
solution is first prepared by dissolving equal parts of sulphate of zinc
and sulphate of copper in water. A strong solution of cyanide of
potassium is then added in sufficient quantity to redissolve the precipitate formed ; to the resulting solution one-tenth to one-fifth of
liquid ammonia is added, and the solution is then diluted with water
until it stands at about 8° Baumé. For a light-coloured deposit of
brass 2 parts sulphate of zinc to 1 part sulphate of copper are used.
In adding cyanide to the solution of the sulphates, great care must be
taken to avoid inhaling the cyanogen fumes that are liberated, which
are highly poisonous. A solution of this character should only be
prepared by a person well accustomed to chemical manipulations.

Winckler's Solution.—Saturated solutions of chloride of zinc and
sulphate of copper are first prepared, in separate vessels. A solution
of cyanide of potassium, consisting of cyanide 100 parts in water
1,000 parts, is next prepared, and this is added to the solution of

sulphate of copper until the precipitate at first formed is redissolved, when a grass-green liquid results; into this the solution of zinc is gradually introduced, with constant stirring, until the solution exhibits a white turbidity. The solution is then diluted with 2,000 parts of water, and heated to the boiling point in an enamelled vessel, and then allowed to cool. It is next filtered, when it is ready for use. The bath is worked at the ordinary temperature, with a brass anode.

Brass Solution for Rough Cast Iron.—The following formula has been given for brassing cast-iron work, and is said to yield a good colour:—

Soft water	14 pints.
Bisulphite of soda	7 ounces
Cyanide of potassium	17 „
Carbonate of soda	34 „

To which is added—

Acetate of copper	4½ ounces.
Neutral chloride of zinc	3½ „
Water	3½ pints.

American Formulæ for Brassing Solutions. — The *Scientific American* publishes the following formulæ for brass solutions:—1. When the ordinary commercial cyanide is employed, the following is said to answer very well:—

Sulphate of copper	4 ounces.
Sulphate of zinc	4 to 5 „
Water	1 gallon.

Dissolve and precipitate with 30 ounces of carbonate of soda; allow to settle, pour off the clear liquid, and wash the precipitate several times in fresh water. Add to the washed precipitate—

Carbonate of soda	15 ounces.
Bisulphite of soda	7½ „
Water	1 gallon.

Dissolve the above salts in the water, assisting the solution by constant stirring; then stir in ordinary cyanide of potassium until the liquid becomes clear and colourless. Filter the solution, and to improve its conductivity, an additional half-ounce of cyanide may be given.

Cold Bath for all Metals.—

Carbonate of copper (recently prepared) . .	2 ounces.
„ zinc	2 „
„ soda . . .	4 „

Bisulphite of soda	4 ounces.
Cyanide of potassium (pure)	4 „
Arsenious acid	$\frac{1}{20}$ „
Water	1 gallon.

Dissolve, precipitate, and redissolve as before, and filter if necessary. The arsenious acid is added to brighten the deposit; an excess is apt to give the deposited metal a greyish-white colour.

Thick Brass Deposits.—MM. Person and Sire patented a process for obtaining stout coatings of brass upon steel or iron by depositing alternate layers of zinc and copper upon the objects, and then submitting them to heat until the metals become alloyed with each other.

Brass Solution Prepared by Battery Process.—A brassing solution may be prepared by the battery method by suspending a large brass anode in a strong and warm solution of cyanide of potassium, to which liquid ammonia is added; about 1½ pound of good cyanide and 10 ounces of strong liquid ammonia to the gallon of water will be about the best proportions. A strong current must be employed. Some persons recommend the addition of hydrocyanic acid; this will not be necessary if good cyanide be used. In preparing solutions by the battery process, or indeed by the ordinary chemical methods, it is far better to employ really good cyanide of a guaranteed strength than to call in the assistance of hydrocyanic acid, which, even in the most careful hands, is a hazardous substance to deal with in what may be termed *practical* quantities. All the best results in electro-deposition have been obtained without the direct aid of this volatile and highly-poisonous acid, and its employment should never be attempted by inexperienced persons under any circumstances whatsoever.

CHAPTER XXV.

ELECTRO-DEPOSITION OF ALLOYS (continued).

Electro-brassing Cast-iron Work.—Scouring.—Electro-brassing Wrought-iron Work.—Electro-brassing Zinc Work.—Electro-brassing Lead, Pewter, and Tin Work.— Observations on Electro-brassing.—Bronzing Electro-brassed Work.—French Method of Bronzing Electro-brassed Zinc Work. Green or Antique Bronze.—Bronze Powders.—Dipping Electro-brassed Work. - Lacquering Electro-brassed Work. — Electro-deposition of Bronze.—Electro-deposition of German Silver.—Morris and Johnson's Process.—Deposition of an Alloy of Tin and Silver.—Deposition of Alloys of Gold, Silver, &c.—Deposition of Chromium Alloys.—Slater's Process.—Deposition of Magnesium and its Alloys.—Alloy of Platinum and Silver.—New White Alloys.—Notes on Electro-brassing.

Electro-brassing Cast-iron Work.—Owing to the porous nature of this class of work, and its liability to present certain unavoidable defects of casting known as sand-holes, the articles to be coated with brass require to be prepared with some care before being immersed in the depositing bath. Moreover, it is necessary to remove the coating of oxide from the surface of the work previous to submitting the articles to the processes of scouring or cleaning. Cast-iron work should first be placed in a "pickle" composed of the following mixture, in sufficient quantity for the work in hand :—

Sulphuric acid	$\frac{1}{2}$ pound.
Water	1 gallon.

The articles being placed in the above pickle are allowed to remain therein for about twenty minutes to half an hour, when they are taken out, one at a time, and examined; if the oxide has become sufficiently loosened to readily rub off with the fingers, the articles are to be at once placed in clean cold water to rinse them ; they are then to be scoured with a hard brush, coarse sand, and water. If after rinsing any black oxide obstinately refuses to be brushed away, the work must be returned to the pickle for a short time longer, or until the objectionable matter readily yields to the brush, leaving a clean surface beneath. Some articles require but a short immersion in the acid pickle, while others need a much longer steeping. When the articles are coated with rust (oxide of iron) this may be removed by brushing them over with strong hydrochloric acid, after which they should be immersed in the sulphuric acid pickle until it is found that sand and water, applied with a very hard brush, will clean them. A solution for brassing cast-iron work should be very rich in metal.

Scouring.—When the articles are sufficiently pickled, they are to be removed from the bath and *well* rinsed in clean water; they are then taken to the "scouring tray," and being placed on the horizontal board, are to be well rubbed with the hard brush and coarse sand moistened with water, until they are perfectly bright and clean and free from all traces of oxide on their surfaces. They are now to be thoroughly rinsed in clean water, and are then ready for the brassing-bath. Some operators prefer to give them a momentary dip in a weak and cold potash bath, and then rinse them before placing the articles in the depositing bath. The work should be suspended in the bath by stout copper wires, and in the case of large pieces of work several such slinging wires should be employed, not only to give support to the articles, but to equalise, as far as possible, the action of the current; since it must be remembered that cast iron is but an indifferent conductor as compared with other metals.

Electro-brassing Wrought-iron Work.—This class of work is more readily coated with brass (and copper) than the former, the metal being less porous and the articles generally in a smoother condition. The work is first to be pickled as before, and afterwards well scoured with sand and water, and then rinsed. The solution in which wrought-iron goods are brassed may have rather less metal (that is, zinc and copper) than is necessary for cast iron. When the articles have been in the bath a few moments, they should be tinted with the characteristic colour of brass. If, however, the colour is of too *red* a tone (showing an excess of copper in the deposit) the brass anode should be lowered a little further into the solution until the deposit is of the proper colour. If, on the contrary, the deposit is *pale*, or whitish (indicating an excess of zinc), the anode must be raised out of the solution to a slight extent. By regulating the *anode surface in solution* the colour of the deposit may be greatly varied; the current of electricity must also be regulated according to the surface of work in the bath and the character of the metal to be coated—cast iron, for example, requiring a current of greater electromotive force than wrought iron.

Electro-brassing Zinc Work.—This metal receives the brass deposit very freely, and the articles made from it are generally prepared for the bath with very little trouble as compared with iron work. Zinc goods should first be steeped in a pickle composed of dilute sulphuric acid, or the following mixture :—

Sulphuric acid	.	.	.	.	.	1 ounce.
Hydrochloric acid	.	.	.	.	.	2 ounces.
Water	.	.	.	.	.	.	1 gallon.

The work should be immersed in the above bath from ten to twenty minutes, then well rinsed, and next scoured with a hard brush, silver sand, and water ; after being again rinsed, the article is to be immersed in the brass bath, and it will generally become coated all over in a few moments, providing the bath be in good condition, the current of sufficient power, and the proper surface of anode exposed in the solution. If the deposit does not take place within a few seconds after immersion, the anode should at once be lowered in the bath until the yellow colour has *struck* well over the article, after which it may be again raised, and the operation then allowed to proceed, undisturbed, until a coating of sufficient thickness is obtained. Since this class of work is not generally required to be subjected to friction (being chiefly castings of an artistic or ornamental character) a stout coating of brass is unnecessary ; moreover the zinc work, when electro-brassed, is usually required either to be *bronzed*, electro-gilt, lacquered, or finished in some other way. Zinc and iron articles should not be suspended in the bath at the same time if it be possible to avoid it ; if, however, this rule cannot be conveniently followed, the iron articles should enter the bath first, and when these have become coated with brass the zinc work may then be introduced. When zinc goods have received the required deposit, they should be well rinsed in hot but not boiling water, and then placed in hot mahogany or boxwood sawdust. After being well brushed with a soft, long-haired brush to remove the sawdust, the piece of work should be rubbed with a clean diaper or chamois-leather, and may then be lacquered or *bronzed*, as desired. If the article is required to be gilt, it is to be simply well rinsed after removing from the bath, and at once placed in the gilding bath.

Electro-brassing Lead, Pewter, and Tin Work.—Brass does not deposit upon lead so readily as upon zinc ; but pewter, however, receives the deposit pretty freely. Lead and pewter articles should be pickled for about half an hour in a dilute solution of nitric acid, consisting of about eight ounces of the acid to each gallon of water, then scoured with silver sand and water, a *soft* brush being employed. They are then to be rinsed and placed in the brass bath, a rather large surface of anode being immersed in the solution when they are first suspended. The current should be strong, otherwise lead is very apt to receive the coating only in parts. It is also best to employ a warm solution for brassing lead and pewter, but more especially the former. Articles of tin, or tinned iron articles, should also be brassed in a warm solution, and treated in other respects in the same way as the former metals, except that the preliminary pickling (which is not absolutely necessary in either case) may be dispensed with.

Stereotype-plates may advantageously receive a deposit of brass, and, indeed, this method of producing a hard surface upon these plates has been to some extent adopted, a warm brassing solution being employed, with a strong current. For plates which have to be used for printing with *vermilion* ink neither a brass nor a copper facing should be adopted, since this colour, being a *mercury* preparation, would be liable to attack both the copper and its alloy (brass), and thus not only injure the metal facing, but also affect the colour of the ink itself.

Observations on Electro-brassing.—After an electro-brassing bath has been used for some time, its whole character becomes greatly changed, unless, indeed, it has been worked under exceptionally favourable conditions, and even with such advantages it will invariably yield results far different from those at first obtained. There are several causes for this, and when these are fully understood it will be more readily seen how tolerable uniformity of action may be secured, absolute uniformity being, so far as we are aware, impossible in the deposition of this or any other alloy. The principal causes of change in the condition of brass deposits are—(1) The anode, being composed of two metals of unequal solubility in cyanide of potassium, does not become freely and uniformly dissolved in the solution, consequently the latter becomes partially, and indeed greatly, deprived of its metallic constituents ; (2) If an excess of ammonia be employed as one of the solvents, this volatile substance by constantly evaporating alters the condition of the bath in proportion to its volatilisation ; (3) The oxide of zinc formed at the brass anode being less soluble than the oxide of copper evolved at the same electrode, the free cyanide in the solution is largely taken up by the latter to the exclusion of the less soluble zinc oxide, and as a consequence this latter substance hangs upon the surface of the anode, or falls to the bottom of the bath as an undissolved mass. In this way the solution becomes more freely supplied with copper than with zinc, and therefore becomes altered from its original condition ; (4) When a current of low electro-motive force is used in depositing brass, the copper of the alloy is more readily deposited than the zinc, which requires a current of higher electro-motive force than copper for its deposition, and as a consequence the solution soon becomes altered in its constitution. For example, when a Bunsen battery is employed in the deposition of this alloy, a very good quality of brass is obtained from a well-made brassing solution ; if for this battery a single Wollaston battery were substituted, all other conditions being the same, copper alone would be deposited on the cathode ; (5) The amount of anode surface immersed in solution in proportion to that of the cathode affects the deposition of brass in a sensible manner. To illustrate this, a very important lesson may be learnt in a very simple way : Take a small steel article, say a pocket latch-key, for example, and connecting it with the negative electrode of a Bunsen battery, suspend it in the brass bath, having previously immersed a large surface of the brass anode. We shall at once observe that a deposit of zinc only has taken place upon the key. Let the key be scratch-brushed, and the anode raised out of the bath so that only a very small portion of one of its corners remains in the solution ; if now the key be suspended in the liquid we shall soon observe that it becomes coated with copper only. If the anode be now cautiously lowered in the bath, we find that the coating gradually assumes the characteristic colour of brass ; and if we take care to estimate the approximate amount of anode surface which yields the yellow alloy, we may in this way form a tolerable notion of the surface of this electrode which it is necessary to expose to a given surface of cathode with the electric power employed. As a further illustration of the caprice which attends the deposition of brass from its solutions, we have known instances in which a steel rod suspended horizontally in the bath has exhibited the following varieties of deposit : at one end zinc alone was deposited ; at the opposite end, copper ; and midway between the two, brass of various tints, from pale straw or lemon colour to a rich golden yellow. *Motion* also affects the character of the deposit, copper being deposited instead of brass when a brisk motion is given to the article while in the bath.

When a much-used bath has a tendency to deposit copper alone, from the cause above stated, its condition may be improved in several ways—(1) By adding liquid ammonia to dissolve the deposited oxide of zinc ; (2) By adding a strong cyanide solution of zinc until the required yellow deposit is obtained ; and (3) By syphoning off the clear solution, and adding ammonia to the deposit at the bottom of the depositing vat, and then returning the clear liquor, when after a few hours' repose the bath will generally work well. A moderate addition of cyanide may also be necessary. When it is found that the bath exhibits general signs of weakness owing to the anode failing to keep up its *metallic* strength, an addition of a concentrated solution of the metals, copper and zinc, must be made, in which there should be an excess of the metal most needed to bring the bath up to its proper condition. In working brassing solutions, it is always advisable to keep in hand a quantity of very concentrated brass solution, so that this may be added to the bath from time to time as required, and thus prevent the annoyance which attends the working of a sluggish and defective solution. Another way of strengthening an exhausted bath is the following : Take a large porous cell, and about three parts fill it with a strong solution of cyanide of potassium ; now connect a long and broad strip of copper to the negative pole of the battery ; immerse the porous cell in the brassing bath, either by standing it upright or by suspending it, according to the depth of the vessel. Now connect a large brass anode to the positive pole of the battery, and allow the current to pass through the solution for a few hours, by which time it will have taken up a considerable amount of brass if the current was sufficiently strong. Two or more 3-gallon Bunsen cells should be employed for a 100-gallon bath of brass solution. If a white deposit appears upon the brass anode, liquid ammonia should be added and well stirred into the solution.

Bronzing Electro-brassed Work.—By the term *bronzing* is meant the application of one or other of the numerous methods of *staining* brass, or imparting to the metal an antique or artistic appearance. When the electro-deposited metal is required to assume the appearance of solid brass, the process of *lacquering* is applied ; but for cast zinc or iron work electro-brassed, another system of ornamentation is adopted, which is known by the name of *bronzing*. For example, if a dilute solution of chloride of platinum be brushed over a brass surface, a *black stain* is produced, the depth or tone of which may be heightened by a second application of the solution. In this case the metal platinum is reduced by electro-chemical action, and becomes deposited upon the more positive metal. The stain, however, produces an effect of contrast which, when artistically applied, is exceedingly pleasing to the eye ; and by this means ornamental brass and also electro-brass work is greatly enriched by art-metal workers and others to meet the requirements of the public. Since it is important that electro-depositors of brass should be well acquainted with the methods of producing upon their work the varied effects which are applied to the solid alloy, we will now explain some of the many processes adopted.

Black Bronze.—This is produced, as before observed, by means of a dilute solution of chloride of platinum. For this purpose, the platinum salt may be dissolved in spirits of wine, methylated spirit, or distilled water, and a few drops of the concentrated solution added to a small quantity of water, and applied with a camel-hair brush or by dipping. For large pieces of work, a sufficient quantity of the dilute solution should be prepared to finish the piece in hand, so as to ensure uniformity of tone throughout the entire piece. When bronzing very large articles, as stove fronts, fenders, &c., it is sometimes the practice to mix a little of the dilute platinum solution with plumbago, made into a thin paste with water, and to brush this over the entire ornamental surface of the article, and when nearly dry, the article is well brushed with a rather soft, long-haired brush until quite bright; the *high lights*, or prominent points of the article, are then gently rubbed with a piece of chamois leather moistened with spirit of wine or rubbed on a lump of chalk, the object being to remove the black stain from these points, so as to show the yellow metal with which the work is coated. Instead of employing the platinum solution for this purpose, a small quantity of sulphide of ammonium may be mixed with the plumbago paste ; this latter is most frequently adopted for the sake of economy. The platinum salt, however, produces the most brilliant and lasting effect.

Warm Bronze Colour.—When it is desired to give a warm chocolate tone to an electro-brassed article, a mixture of jewellers' rouge and black-lead, in varying proportions according to the tone required, is first made with water ; to this a few drops of chloride of platinum solution or sulphide of ammonium are added and intimately mixed. This bronzing paste is spread over the article with a soft brush, and allowed to become nearly dry, as before, when the surplus powder is brushed away by polishing with a long-haired brush. The high lights are then touched up as before to expose the metal. The article should now be made moderately warm, and then brushed over quickly with a very thin, hard, and quick-drying varnish ; when this is done the work is complete. A bronzing composition for imparting a warm chocolate tone to electro-brassed work may be made by mixing into a paste with water the following ingredients : black-lead, 1 ounce ; Sienna powder, 2 ounces ; rouge, $\frac{1}{2}$ ounce. To this may be added a few drops of sulphide of ammonium. Or a mixture of black-lead and rouge or crocus may be employed. In each of these cases the formulæ may be varied at the will of the operator ; indeed, the tone or bronze effect is so greatly a matter of taste that the proportions of the various materials may properly be left to the discretion of the electro-bronzer.

Green Bronze.—Mix into a paste with water the following substances, varying the proportions, as before suggested, according to taste :—

Chromate of lead (chrome yellow) . .	2 ounces
Prussian blue	2 „
Plumbago	$\frac{1}{2}$ pound
Sienna powder	$\frac{1}{4}$ „
Lac carmine	$\frac{1}{4}$ „

When applying the above composition, a small quantity of sulphide of ammonium or chloride of platinum solution may be added. It should be mentioned that the particles set free by brushing off the superfluous portions of the above mixture would be unwholesome to breathe, on account of the chromate of lead present in the composition, the other substances are virtually innocuous, though the inhalation of small particles of mineral, or indeed any other substance whatever, should be avoided as much as possible. In polishing work which has been coated with bronzing powders, therefore, it would be well if the workpeople could be induced to protect the nose and mouth by a thin piece of muslin, more especially in the earlier stages of the polishing operation, when the great bulk of the superfluous material has to be brushed away.

French Method of Bronzing Electro-brassed Zinc Work.—If a warm tone is desired, the electro-brassed article is first dipped in a weak solution of sulphate of copper and then dried. It is next moistened with sulphide of ammonia or a solution of liver of sulphur ; after again drying the surface is brushed over with a mixture of hematite or jewellers' rouge and black-lead, the mixture being made according to the tone required. The brush should be slightly moistened with turpentine to assist the adhesion of the powder. The parts in relief are then to be " set off," that is, well rubbed, to disclose the metal, and give it the appearance of having been subjected to wear. The object is then to be coated with a thin colourless varnish.

Green or Antique Bronze.—Dissolve in 100 parts of acetic acid of moderate strength, or in 200 parts of good vinegar, 30 parts of carbonate of ammonia or sal-ammoniac, and 10 parts each of common salt, cream of tartar, and acetate of copper, and add a little water ; mix well, and smear the object with it, and then allow it to dry, at the ordinary temperature, from twenty-four to forty-eight hours. At the end of that time the article will be found to be entirely covered with verdigris, which presents various tints. It is then to be brushed, but more especially the prominent parts, with a waxed brush, that is, a brush passed over a lump of yellow beeswax. The relief parts may then be " set off " with hematite, chrome yellow, or other suitable colours. Light touches with ammonia impart a blue shade to the green parts, and carbonate of ammonia deepens the colour of the parts to which it is applied.

Steel Bronze.—This is obtained by moistening the articles with a dilute solution of chloride of platinum, and slightly heating them. Since this bronze is liable to scale off with friction, it should not be applied in successive doses, but the solution used should be of such a strength that the desired effect may be obtained, if possible, by a single application. Copper bronze, that is, electro-brass with an excess of copper, may be darkened by dipping it into a weak and warm solution of chloride of antimony (butter of antimony) in hydrochloric acid. Sometimes, however, the coloration will be violet instead of black.

Bronze Powders, as the Bessemer bronzes, for instance, are largely used for imparting a metallic appearance to plaster casts and ceramic wares, and also for ornamenting cast-iron work, to give it the appearance of bronze. The mode of application is as follows : The article, after being cleaned, is coated with a fatty drying varnish, which is allowed to become nearly dry. The bronze powder is then applied with a badger-hair brush, when it firmly adheres to the sticky varnish. After drying, the article is coated with a hard, colourless varnish, which fills up the details. This process is chiefly applied to metals for such work as cheap iron fenders, cast-iron dogs for fireplaces, umbrella-stands, and other coarse work, and is in no degree suitable for articles which have been electro-brassed, in which a more artistic finish is required.

Dipping Electro-brassed Work.—When steel or iron articles have received a good coating of brass, but of an indifferent colour, they may be greatly improved in colour by being *dipped* in the ordinary dipping liquids used for brass. The dippings, however, must be done with great promptitude, otherwise the coating will either be dissolved off, or at least much reduced in thickness. If the operation is conducted with smartness, and the articles at once plunged into cold water, the desired result may be obtained without risk—provided, of course, that a tolerably stout coating has been deposited upon the work. This method of improving the colour of the deposit we have successfully adopted, and, indeed, frequently made it a practice to give to the work an extra strong coating to allow for the reduction of its thickness in the acid dip. By this means we were enabled to produce results in electro-brassing which were acknowledged to be fully equal in colour to the finest specimens of solid brass, and not unfrequently superior.

Lacquering Electro-brassed Work.—After being worked for some time, a brassing bath is liable to give deposits which are either too red, or *coppery*, or they may assume a sickly pale colour, which, after scratch-brushing, is of too light a colour to fairly represent brass. Articles in this condition, although they may be greatly improved by a coating of good yellow lacquer, still fail to resemble ordinary brass. When zinc or steel articles are required to be lacquered after electro-brassing, it is a good plan to be provided with an extra brassing solution, capable of yielding a really good colour, in which the articles, after being coated in the ordinary bath and scratch-brushed, may be immersed, to give them a final coating of good yellow brass. Generally speaking, an immersion in the second bath of only a few minutes is sufficient to produce the desired effect, and the solution we should recommend for this purpose is one prepared from the sulphates of copper and zinc, precipitated by carbonate of potash, and redissolved with cyanide and liquid ammonia. A solution carefully prepared from these ingredients is capable of yielding a brass deposit of a very fine colour. The solution should be used only for giving a final coating to electro-brassed work which is of a bad colour, and when it ceases to yield a good coloured deposit, it may be added to the ordinary brassing bath and another solution prepared in its place. If the plan we have suggested be adopted, the work may then be lacquered in the same way as ordinary brass, after being thoroughly rinsed and dried. It may be mentioned that acid dipping, to improve the colour of the brass deposit, cannot so safely be applied to zinc work which has been electro-brassed.

Electro-deposition of Bronze.—The electro-deposition of the alloy of copper and tin known as bronze is less frequently practised than that of the more common alloy, brass ; indeed, the latter with an excess of copper in the solution generally answers the purpose equally well for most of the uses to which the deposited bronze alloy would be applicable as an imitation of *real* bronze. In making up a bronzing solution, it is only necessary to substitute a salt of tin (chloride of tin by preference) for the zinc salt in the preceding formulæ, but in rather less proportion than the latter, say about one-third less. The simpler method, however, is to make up a brassing bath with a slight excess of copper, and to depend upon the artificial methods of *bronzing* previously given, or such modifications of them as may suggest themselves, for producing imitations of solid bronze. A very little practice in this direction will enable the operator to meet almost every requirement as to tone or colour. In this, as in the case of electro-brassed work, the most prominent parts of the work should be rendered bright, so as to expose the deposited metal at such points by gently rubbing away the materials used in producing the artificial bronze colour. By so doing a very pleasing artistic effect may be produced, provided the removal of the bronzing material is not carried too far, but merely confined to such points as may be assumed to have been subjected to friction in use.

Electro-deposition of German Silver.—We have succeeded in depositing an alloy of copper, nickel, and zinc, forming German silver of good quality, by making a solution of the alloy in the *direct* way, as recommended for preparing brassing solutions, thus : Cut up into small pieces sheet German silver, about one ounce ; place the strips in a glass flask, and add nitric acid, diluted with an equal bulk of water. Assist the solution of the metal by gentle heat ; when the red fumes cease to appear in the bulb of the flask, decant the liquor, and apply fresh acid, diluted as before, to the undissolved metal, taking care to avoid excess ; it is best to leave a small quantity of undissolved metal in the flask, by which an excess of acid is readily avoided. The several portions of the metallic solutions are to be mixed, and diluted with about three pints of cold water in a gallon vessel. Next dissolve about four ounces of carbonate of potash in a pint of water, and add this gradually to the former, with gentle stirring, until no further precipitation takes place. The precipitate must be washed several times with hot water, and then redissolved by adding a strong solution of cyanide with stirring, and about one ounce of liquid ammonia. To avoid adding too great an excess of cyanide, it is a good plan, when the precipitate is nearly all dissolved, to let it rest for half an hour or so, then decant the clear liquor, and dissolve the remainder of the precipitate separately. A small excess of cyanide solution may then be added, as " free cyanide," and the whole mixed together and made up to one gallon with cold water. The solution should then be filtered, or allowed to repose for about twelve hours and the clear liquor then carefully decanted from any sediment which may be present from cyanide impurities. The bath must be worked with a German silver anode, which should be of the same quality as that from which the solution is prepared ; a Bunsen battery should be employed as the source of electricity, or a dynamo-machine.

Morris and Johnson's Process.—By this process a German silver bath is prepared by the battery method. One pound each of cyanide of potassium and carbonate of ammonia are dissolved in a gallon of water, and the solution heated to 150° Fahr. A large German silver anode, connected with the positive electrode of a powerful battery, is immersed in the solution ; a small cathode of any suitable metal is connected to the negative pole of the battery and also immersed in the solution. The electrolytic action is to be kept up until a considerable amount of metal is dissolved, and a bright cathode receives a deposit of good colour, when the solution is ready for use. If the deposit is too red, carbonate of ammonia is to be added ; if too white, cyanide of potassium.

The electro-deposition of German silver may with advantage be substituted for nickel-plating for many articles of ornament and usefulness ; a coating of this alloy looks exceedingly well upon bright steel surfaces, and is, to our mind, specially suitable for revolvers, dental instruments, and scabbards, having, when deposited of a good colour, a more pleasing tone than that of nickel.

Deposition of an Alloy of Tin and Silver.—Messrs. Round and Son obtained a patent in 1879 for a process for depositing an alloy of tin and silver, which is said to be applicable to coating brass, German silver, and copper, and, if slightly covered with a film of copper, iron and steel also. The inventors state that from their solution a white reguline metal is obtained which is easily polished, and greatly resembles fine silver. The solution is prepared as follows : Dissolve 80 ounces of commercial cyanide of potassium in 20 gallons of water in a suitable vessel ; then pour in 100 ounces by measure of strong liquid ammonia, of the specific gravity of 0·880, stirring well together; next add 10 ounces of nitrate of silver ; any soluble *tin salt* may then be added at discretion ; now add 3 pounds of carbonate of potassa, and allow the compound solution to rest until all sediment has subsided, then carefully decant the clear liquor, and the bath is ready for use. It is worked with a large anode of tin and a smaller one of silver. The articles to be plated by this process are cleaned in caustic ley, and all oxide carefully removed ; they are then immersed in the bath, in connection with the negative pole of a strong voltaic battery, the two anodes being connected to the positive as usual. The articles are allowed to remain in the bath until the required thickness of deposit is obtained, when they are removed, rinsed, and dried, and may then be polished or burnished to a high degree, closely resembling unalloyed silver, but produced at far less cost.

Deposition of Alloys of Gold, Silver, &c.—These are noticed in Chapter XV. on Electro-gilding. We may, however, state that by mixing gold and silver cyanide solutions in varying proportions, gilding of various shades of colour may be obtained. The same results may be effected by blending cyanide solutions of gold and copper. The colour of the deposit is greatly influenced by the strength of the current employed, the amount of anode surface, and the temperature of the bath. With these hints to guide him, the experimentalist may obtain very interesting results by modifying the condition of the bath, strength of current, &c., at will.

Deposition of Chromium Alloys.—Slater's Process.—This process, for which a patent was obtained in March, 1884, may be thus briefly described : Anodes of chromium alloy are prepared by heating chromium compounds with charcoal in a closed crucible, and pouring upon the reduced mass $2\frac{1}{2}$ parts of fused copper, and, subsequently, from 1 to $1\frac{1}{2}$ parts of molten tin, and then granulating, re-fusing, and casting in moulds of the desired form. The plates thus formed are used as anodes in a solution made by dissolving 1 pound of cyanide of potassium and one pound of carbonate of ammonia in a gallon of water, heated to 150° Fahr., until a good deposit of alloy is formed upon the cathode ; the bath is then ready for use. To finish the articles coated in this solution, they are " coloured " in a bath composed of chloride of tin 6 to 8 parts; chloride of copper, 20 to 25 parts ; bichromate of ammonia, 10 to 15 parts : chloride of platinum, 6 to 12 parts, and water 101 to 110 parts. A very moderate current only is required.

Deposition of Magnesium and its Alloys.—Gerhard and Smith obtained a patent in December, 1884, for the following process : Ammonio-sulphate of magnesia is prepared by dissolving and crystallising together 228 parts of sulphate of magnesia (Epsom salt), and

132 parts of sulphate of ammonia. The crystals are dissolved in 35,000 parts of water, and the solution thus formed is best used at a temperature of 150° to 212° Fahr. For white metal, a nickel anode is used ; for magnesium bronze, a copper anode must be employed. In the latter case, the bath is formed of ammonio-sulphate of magnesia, 360 parts ; cyanide of potassium, 550 parts, and carbonate of ammonia, 550 parts, dissolved in 35,000 parts of water.

Alloy of Platinum and Silver.--Mr. Milton H. Campbell, of America, has taken out a patent for depositing this alloy, which is said to resist the action of nitric acid and sulphides. A bath is made by dissolving 30 parts of platinum and 70 parts of silver in aqua regia, and the metals are precipitated as a grey powder by means of chloride of ammonium. The compound chloride thus obtained is dissolved in a solution of cyanide of potassium, which constitutes the electrolytic bath. The anode is an alloy of 3 parts of platinum and 70 parts of silver, a feeble current being employed for the deposition of the alloy.

New White Alloys.—Many attempts have been made by refiners and metal workers to produce a metallic alloy to resemble silver in whiteness and texture, and sufficiently low in price for general manufacturing purposes ; but although many excellent results have been obtained, there is no doubt whatever that the new alloy introduced by Messrs. Henry Wiggin and Co., under the title of "Silveroid," is the nearest approach to silver yet produced. This pretty alloy is not only beautifully white and of close and fine grain, but has a silvery lustre which renders its commercial name exceedingly appropriate. The new alloy, moreover, files and turns well, and is susceptible of a high polish. Being readily fusible, it is admirably adapted for ornamental castings, and produces very fine work. It is, we understand, being adopted for carriage, railway, and steamship fittings, machinery-bearings, taps, &c., and is intended as a substitute for brass, bronze, and gun-metal in all cases where a brilliantly white metal would be preferred as a substitute for the commoner alloys. A specimen of rolled "silveroid" sent to us by the above firm many months ago has undergone no change in appearance, being as white and silvery as when first received. Silveroid is an alloy of copper and nickel, to which zinc, tin, or lead in varying proportions are added, according to the purpose for which it is to be used. Another alloy has been introduced by Messrs. Wiggin & Co., under the title of "cobalt bronze," which is more steel-like in colour than the former, and is also much harder. This alloy, which takes a bright polish, is suitable for all kinds of work in which a hard, white, non-tarnishable metal is required, and would, we should say, be invaluable for steamship and railway carriage fittings, and work of that class. The cost of this alloy is rather higher than "silveroid," owing to the metal cobalt being one of its necessary constituents. Since it does not much exceed the price of ordinary German silver, however, while being much whiter, we have no doubt that it will be accepted as a valuable substitute for the former for unplated spoon and fork work. A long exposure of a sample of this alloy to the atmosphere, and also to the mingled fumes of our laboratory, by which it was unaffected, establish the fact that "cobalt bronze" will resist all ordinary atmospheric influences.

Notes on Electro-brassing.—When the brass anodes become foul, owing to undissolved sub-salts of zinc or copper, or both, forming on the surface, it indicates that the bath requires an addition of cyanide and liquid ammonia. After making these additions, it is well to remove the anodes, rinse them, and scour them perfectly clean.

Colour of Bronzes.—The tone or colour of the bronzing paste applied to electro-brassed cast-iron work (as fenders, for example) should be regulated according to the colour of the deposit. For instance, for a *yellow* brass, the black or green tones will be most appropriate, while for deposits of a more coppery hue, the warmer bronzes should be used, as those containing rouge, crocus, &c.

Bronze Tone.—To deposit metal approaching the tint of real bronze, a slight excess of copper should be added to the ordinary brass solution.

Green Bronze Colour.—To impart an artificial green bronze appearance to electro-brass, the article may be placed in a closed wooden box, having a saucer containing a little chloride of lime (bleaching powder) placed at the bottom. A small quantity of hydrochloric acid is then to be poured on the powder, the lid of the box immediately closed, and the article allowed to be subjected to the chlorine fumes which are given off, for a short time, after which the article is to be exposed to the air. The process may be repeated until the desired effect is produced. The article should be well coated with brass or bronze before being submitted to the action of the chlorine, otherwise this gas will attack the underlying metal.

Fender and Stove Work, which are generally required to be electro-brassed at a very low price, should first be pickled in dilute sulphuric acid for a short time, then rinsed and briskly scoured with coarse sand and water, again rinsed, and placed in the bath. A very strong current should be employed, so that the article may receive a sufficient coating in a few minutes. After rinsing and drying quickly the bronze paste is applied, and this is to be dried on the article as quickly as possible; when nearly dry the article is polished, its prominent parts then rubbed up with a piece of chamois leather and dry whiting, and a thin coating of hard spirit varnish laid on while the article is warm; the object is now finished.

Evolution of Hydrogen during Deposition.—A great deal has been written and said concerning the vigorous evolution of hydrogen which commonly occurs with electro-brassing baths when under the influence of the current; and, while we readily agree with much that has been said upon this subject, we must frankly confess that we have generally obtained the best results when the escape of hydrogen has been most brisk. We should certainly not consider a brassing bath, in which deposition takes place directly the articles are immersed in it (the deposit being of a good colour), a defective solution, though evolving hydrogen, since some of our best results have been obtained under such conditions. In depositing very stout coatings of this alloy, however, it is certainly desirable that the evolution of hydrogen should, as far as possible, be prevented, a result which may most readily be obtained with solutions containing a considerable quantity of the metallic constituents. Such baths, however, require a frequent addition of concentrated solution to keep up their metallic strength, which the brass anodes, under the most favourable conditions, fail to do.

Keeping up the Strength of the Bath.—To keep up the strength of brassing baths, the plan suggested by the author in respect of electrotinning and platinising solutions may be adopted (see page 343). By this method a highly concentrated solution of brass, delivered from a tank above, may be allowed to trickle into the bath while deposition is going on, and thus its metallic strength fairly well kept up. Such an arrangement can be effected with very little trouble, a small barrel, furnished with a tap with a long piece of rubber tubing, being all that is necessary.

Solution for Cast-iron Work, &c.—The brassing bath for this class of work should be rich in metal, otherwise, even with a strong current, the deposit will take place chiefly, or only, at the corners or prominent portions of the articles. It is better to employ a solution containing a good percentage of metal and small quantity of free cyanide, than a great excess of the latter and a small proportion of copper and zinc. The current for depositing brass upon cast iron, especially in cold solutions, must be strong, and a large anode surface exposed in the bath. The same observations apply, to a certain extent, to lead, which requires a bath rich in copper and zinc to obtain successful results, especially when battery power is employed.

Brassing Different Metals.—It must be borne in mind, as we have hinted, that all metals do not receive the brass deposit with equal facility ; indeed, if two articles—one composed of zinc and the other of cast iron—were placed in the bath simultaneously the former would at once become coated with brass, while the cast-iron article would either remain uncoated with the alloy, or at most a slight deposit would be visible at the points nearest the anode, or at the lower parts of the article, according to its form. This being the case, the cast-iron article should first be put into the bath, and when this has become *perfectly* coated all over the zinc articles may then be suspended in the solution. It is better, however, to deposit these metals separately. Even wrought and cast iron will not receive the brass deposit with equal readiness ; the latter, therefore, should be put into the bath first, and the former only when the cast-iron piece is well coated all over.

Brassing in Hot Solutions.—When an article is first put into the bath (being connected to the negative electrode) it should be gently moved about for a few moments, to cause the deposit to take place as uniformly as possible all over the surface of the article, and when the characteristic yellow tint of the alloy appears uniformly all over the object, it may be allowed to rest in the bath for a short time, when the slinging wires should be shifted to allow the parts they have covered to become coated with the alloy. After a while the article should be inverted in the bath to equalise the deposit as far as possible. With these exceptions, it is not judicious to disturb work in brassing solutions while in circuit, as the colour of the deposit is often affected even by slight motion.

CHAPTER XXVI.

RECOVERY OF GOLD AND SILVER FROM WASTE, ETC.

Recovery of Gold from Old Cyanide Solutions.—Recovery of Silver from Old Cyanide Solutions.—Recovery of Gold and Silver from Scratch-brush Waste.—Recovery of Gold and Silver from Old Stripping Solutions.— Stripping Metals from each other.—Stripping Solutions for Silver.— Stripping Silver from Iron, Steel, Zinc, etc.—Stripping Silver by Battery. —Stripping Gold from Silver Work.—Stripping Nickel-plated Articles.— Stopping-off.—Soldering.—Removing Soft Solder from Gold and Silver Work.

Recovery of Gold from Old Cyanide Solutions.—Since the precipitation of gold (and silver) from cyanide solutions, which is effected by means of mineral acids, involves the liberation of hydrocyanic acid (prussic acid) the operations must always be conducted with the utmost caution, and should always be carried on in the open air. It is well to remark here that when acid is added to a cyanide solution, not only hydrocyanic acid but also carbonic acid is liberated ; and since this heavy gas cannot escape through the flue of an ordinary chimney, owing to its gravity, but flows over the vessel in a dense white vapour, the operator should be careful not to disturb these fumes, so as to cause them to rise upward, but to allow them to flow over the sides of the vessel and escape into the open air, where they will be dispersed by the wind. The reduction of the gold by the *dry way* is, however, less hazardous, not so offensive, and fully as economical.

To precipitate gold from cyanide solutions, hydrochloric acid is to be gradually added, until no further precipitation takes place ; the solution should then be boiled and the vessel set aside to cool. The precipitate (cyanide of gold), which is of a yellowish colour, must then be separated from the supernatant liquor by decantation, and then filtered. Since a small portion of gold, however, still remains in the liquor, it must not be thrown away, but should be heated, and zinc filings added, which will throw down the remainder of the gold ; the clear liquor is now to be poured off, and the residue boiled with dilute hydrochloric acid, to remove excess of zinc, and after washing, the deposit is to be added to the other portions. Ignite and fuse the mixture in a platinum or ordinary crucible, with an equal weight of sulphate of potassium. Dissolve the saline residue in boiling sulphuric acid, then wash it with water, when perfectly pure gold will remain.—(R. Huber.)*

Böttger recommends the following method for recovering gold from old solutions :—The solution should be evaporated to dryness, the residue then finely powdered, and intimately mixed with an equal weight of litharge (oxide of lead) and fused at a strong heat ; the lead is extracted from the resulting button of gold and lead alloy by warm nitric acid, when the gold will remain as a loose brown spongy mass. The same author says, " If we pour hydrochloric acid into a pure solution of gold in cyanide of potassium, there is slowly formed at ordinary temperatures, and immediately on the application of heat, a yellow precipitate, which is cyanide of gold ; the filtered liquid which has given this precipitate still contains a little gold in solution. On evaporating the liquid to dryness, fusing, dissolving, and filtering afresh, there remains upon the filter the remainder of the gold. Crystallised double cyanide of gold and potassium fuses and effervesces by heat, and is resolved into cyanogen gas, ammonia, and cyanide of potassium, if air be present; its complete decomposition requires a strong heat. When it is strongly ignited, mixed with an equal weight of carbonate of potash, a button of metallic gold is obtained.''

Cyanide gilding solutions, when mixed with sulphuric, nitric, or hydrochloric acid, slowly deposit cyanide of gold ; and when boiled with hydrochloric acid, it is completely resolved into cyanide of gold and chloride of potassium. The same result is obtained with nitric and sulphuric acid, and even with oxalic, tartaric, and acetic acids. When heated with sulphuric acid, it gives off hydrocyanic acid gas, and after ignition, leaves a mixture of gold and sulphate of potassium. Iodine sets free cyanogen gas, forms iodide of potassium, and throws down the cyanide of gold.—(Böttger.)

Gold may be precipitated from washing waters containing traces of the precious metal by adding a solution of protosulphate of iron, when a brown deposit of pure gold is obtained ; the subsidence of the metal may be hastened by heating the liquid.

Recovery of Silver from Old Cyanide Solutions.—Elsner made a series of important researches upon the extraction of gold and silver from cyanide solutions, the results of which he communicated in a valuable paper, from which the following extracts are taken. He prefaced his description of the practical methods recommended, by mentioning the results of some of his experiments upon which they were based.

1. " If we add hydrochloric acid to a solution of silver in cyanide of

---

potassium until the liquid exhibits an acid reaction, we obtain a white precipitate of chloride of silver, which, when submitted to heat, melts into a yellow mass. If this was cyanide of silver, the application of a red heat would have left a regulus of silver. The addition of the hydrochloric acid precipitates all the silver present in the liquid in the form of chloride of silver.

" If we evaporate a solution of silver in cyanide of potassium to dryness, and heat the residue to redness, until the mass is in a state of quiet fusion, and has assumed a brown colour, there remains, when we wash the mass with water, metallic and porous silver. The wash waters, when filtered, still contain a little silver in solution, because, if hydrochloric acid is added to them, it produces a precipitate of chloride of silver. In evaporating and calcining a solution of gold in cyanide of potassium, the result is similar, i.e. we obtain metallic gold. The wash waters, acidulated with hydrochloric acid, give, when treated with sulphuretted hydrogen, a brown precipitate of sulphide of gold ; and with the salt of tin a violet precipitate (purple of Cassius), a proof that these liquids still contain a little gold in solution.

" *Extraction of Silver by the Wet Method*—Add hydrochloric acid until the liquid exhibits a strongly acid reaction. The precipitate of chloride of silver which is thus obtained, will be, as we have already said, of a reddish-white colour, because of the cyanide of copper which is precipitated with it when the solution has been used a long time for silvering objects containing copper. In this precipitation by hydrochloric acid, there is hydrocyanic acid gas set free, therefore the operation should only be performed in the open air, or in a place where there is good ventilation ; if the precipitate is very red, it must be treated with hot hydrochloric acid, which will dissolve the cyanide of copper. The chloride of silver, having been washed with water, must be dried and then fused with pearlash in a Hessian crucible coated with borax, in the ordinary manner for obtaining metallic silver.

" This method is very simple in its application, and very economical, considering that by the aid of the hydrochloric acid all the silver contained in the solution of cyanide of potassium is precipitated, and there remains no trace of it in the liquid. But the quantity of hydrocyanic gas which is disengaged is a circumstance which must be taken into serious consideration when operating on large quantities of silver solution, the vapour of which is most deleterious, and nothing but the most perfect ventilation, combined with arrangements for the escape of the poisonous gases, will admit of the process being carried on without danger to the workmen ; when, however, we have taken the precautions dictated by prudence, the method in question may be con-

sidered as perfectly practical. The liquid should be poured into very
capacious vessels, because the addition of the acid produces a large
amount of froth.

"*Extraction of Silver by the Dry Method.*—The solution of
cyanide of silver and potassium is evaporated to dryness, the residue
fused at a red heat, and the resulting mass, when cold, is washed
with water. The remainder is the silver in a porous metallic condition. There still remains in the wash waters a little silver, which
may be precipitated by the addition of hydrochloric acid."

Recovery of Gold and Silver from Scratch-brush Waste.—The
sludge or sediment which accumulates in the scratch-brush box frequently contains a considerable quantity of gold and silver, removed
by the brushes from articles which have stripped in parts owing to defective cleaning of the work. This waste, with other valuable refuse
of a similar description, should be collected every few months, and
after being dried, should be mixed with a little dried carbonate of
potash and fused. The resulting button, being an alloy of gold,
silver, copper, &c., may be treated as follows to separate the gold and
silver :—Remelt the alloy, and granulate it by pouring the molten
metal into cold water. Place the grains in a flask and pour on them
a mixture of 2 parts nitric acid to 1 part water, and apply moderate
heat, when all but the gold will be dissolved, the latter remaining as a
brown powder at the bottom of the flask. The liquid must then be carefully poured into another vessel, and strips of clean copper immersed in
it, which will cause any silver present to be thrown down in the metallic
state. The gold and silver deposits must afterwards be well washed
with warm water, and after drying, be mixed with dried carbonate of
potash or soda, and fused as before. In fusing these fine deposits,
after they have been intimately mixed with the dried alkali, which is
to act as a flux, the mixture should be compressed as much as possible
when placed in the crucible, or melting pot, by means of an iron pestle
or other suitable tool, and the heat allowed to progress slowly at first,
and after a short time this may be increased until the contents of the
crucible assume a semi-fluid condition ; when in this state, the heat
should be moderated, to allow the metal to "gather," as it is termed,
by which the molten globules will gradually subside and unite in the
form of a liquid mass at the bottom of the pot. It is very important
at this stage to keep the fused mass in as liquid a state as possible,
taking care also not to apply too great heat, or the contents may rise
up and overflow. Should this be likely to occur, a pinch of *dried* common salt may be thrown into the pot, which will cause the fused mass
to subside. When the operation is complete, the pot is to be withdrawn from the fire and placed aside to cool ; the pot is afterwards
broken at its lower part, by a blow from a hammer, and the button

extracted. This may then be plunged into a dilute sulphuric acid pickle to remove any flux that may attach to it.

Recovery of Gold and Silver from Old Stripping Solutions. —The gold may be recovered from exhausted stripping baths by evaporating them to dryness and fusing the residue with a little carbonate of potash or soda. There are several methods of treating old silver stripping solutions, of which the following are the simplest: 1. Dilute the liquid with three or four times its bulk of water; now place in the liquid several stout plates of clean zinc, which will rapidly become covered with a spongy layer of reduced silver; the plates should be occasionally shaken in the liquid to remove the deposited metal, which will fall to the bottom of the vessel. When the zinc plates, after having been immersed for a few hours, cease to become coated with silver, the liquid may be decanted into another vessel, and a few drops of hydrochloric acid added, when, if a white cloudiness is produced, more acid should be added (or a solution of common salt) until it produces no further effect. The white precipitate, which is chloride of silver, may afterwards be collected and treated separately. The reduced silver in the first vessel should be well washed, to free it from sulphate of zinc, and afterwards dried and fused as before. 2. The silver stripping solution may be treated with a strong solution of common salt, which will throw down the metal in the form of chloride, and this, after being *well washed*, may be employed for making up a silver bath, or the chloride may be decomposed and the silver reduced to the metallic state, with or without the aid of heat, by immersing in the deposit several stout pieces of clean zinc, which after awhile will convert the deposit into metallic silver in the form of a grey powder. To facilitate the action, a few drops of sulphuric acid should be added. After well washing with hot water, this powder may be dissolved in nitric acid, to form nitrate of silver, which can then be used for making up silver baths. Or the grey powder may be dried and mixed with *dried* carbonate of potash and fused as before directed. After putting the mixture of reduced silver and carbonate of potash into the crucible, it should be compressed as much as possible, by pressing it with an iron pestle, which will greatly facilitate the " gathering " of the globules of fused silver; indeed it is a good plan, when the crucible has become fully heated, to gently press the *crust* of unfused matter with an iron rod, so as to force it to the lower part of the vessel where the heat is greatest. When the gathering of the metal is complete, a small quantity of nitre may be occasionally dropped into the crucible, which will remove any traces of iron or copper that may be present, and thus render the silver button more pure.

"Stripping" Metals from each other.—Old articles which have been gilt, silvered, or nickel-plated, or new work which has been unsuccessfully coated with these metals by electro-deposition, generally require to be deprived of the exterior coating before a proper deposit can be obtained by electrolysis. The operation of removing the exterior layer of metal is technically termed *stripping*, and the various solutions applied for the purpose are termed *stripping solutions*. It may be well to remark here that metals of a like character do not adhere firmly to each other; thus electro-deposited gold will not adhere to a gilt surface, silver to a silver-plated surface; nor will nickel attach itself to a nickel-plated article; this fact is most conspicuously observable when an attempt is made to re-nickel a nickel-plated article without previously removing the old layer of this metal, when the second coating will generally rise up from the underlying coat, even without subjecting it to any provocation by any mechanical means, such as buffing. Indeed, so persistent is this metal in refusing to accept a second coating that we have known a brass rod (placed in the bath as a " stop," to check the force of the current when first filling the bath) which had remained *in the bath* for many weeks, to become coated with countless layers of nickel, which had partially separated from each other, giving the lower end of the rod the appearance of a metallic mop. The author's impression was that every time the circuit was broken, by the stoppage of the dynamo machine, that the layer next deposited, when the machine was again in motion, *did not adhere* to its predecessor, but became a distinct and separate layer. Although this refusal to attach itself to a metal of its own kind is not so marked in the case of silver and gold as with nickel (and we might say copper also), it is unquestionably the case that the latter metals will more firmly adhere, when electro-deposited, to copper, brass, or German silver, than they will to articles composed of or coated with the same metals. The solutions employed for removing or stripping the precious metals and nickel from articles which have been coated with them will be given under separate headings, since the materials employed differ in each case.

Stripping Solution for Silver.—A quantity of strong oil of vitriol is put into a stone jar or enamelled saucepan, heated on a sand-bath or in any convenient way, and to this is added a small quantity of saltpetre (nitrate of potash). Sometimes nitrate of soda, called *Chili saltpetre*, is used in place of the other salt. When the saltpetre has become dissolved, which may be accelerated by stirring the mixture with a stout glass rod, the articles to be stripped, attached to a copper wire, are dipped into the hot liquid, and allowed to remain, with occasional motion up and down, until the whole of the silver has become dissolved off.   If the operation be carefully watched, it will be observed that the silver quickly disappears from the parts where it was thinnest, and gradually appears to fade away until not a trace is left upon the article.   The chemical action of the solution upon the German silver, brass or copper, of which the article may be composed, is very slight if the articles are withdrawn directly the silver has been removed.   It is very important that no water should be allowed to enter the stripping bath ; therefore the articles should be perfectly dry before being immersed.   In stripping spoons and forks it will generally be noticed that the last portions of silver to leave the articles are at the points of the prongs and upper part of the handle of forks, and the lower portion of the bowls and extremity of the handle of spoons, which establishes the well-known fact that these parts receive the greater thickness of deposit than other portions of the article.   The same observation applies to all projecting parts, and in order to remove the last traces of silver from such portions, when the silver has been dissolved from the main body of the work, the article should be raised out of the bath, and the projections or points dipped in separately, which will save the bulk of the article from being severely acted upon by the acid mixture.   When the solution begins to work tardily, after a certain number of articles have been dipped in it, more saltpetre must be added from time to time, and the liquid kept well heated.   Since oil of vitriol attracts moisture from the air, every time the bath is done with the vessel should be covered with a stout plate of glass.

When a stripping bath has been much used, it works slowly, and the addition of saltpetre fails to invigorate it.   When in this condition a mass of crystals will deposit at the bottom of the vessel as the liquid cools.   The bath must now be put aside and replaced by a fresh mixture.   The process for recovering the silver from old stripping solutions is described at page 407.

Cold Stripping Solution for Silver.—A large quantity of strong sulphuric acid is poured into a sound and deep stoneware vessel ; to every two parts of the acid by measure one part of strong nitric acid (also by measure) is added, and the mixture is employed in its cold state.   The process of stripping in this solution is much slower than in the former bath, and therefore requires less attention ; since, however, the thickness of silver upon plated work varies considerably, from a mere film to a good stout coating, the progress of the work must be carefully watched from time to time, and the operator's judgment will soon guide him as to the quality of the plated work under treatment.   The articles to be stripped are suspended from stout copper wires, or preferably by means of glass hooks, which may readily be formed from stout glass rods by simply bending them to the required form over an ordinary gas jet or Bunsen burner. It is very important that no water should be allowed to enter the stripping bath, otherwise the metal of which the articles may be composed, as brass, copper, German silver, etc., will be acted upon by the acid mixture. When the liquid begins to act tardily, after being worked for some time, a small quantity of strong nitric acid must be added, and this addition must be made whenever the liquid shows signs of weakness.

When stripping silver from articles which have been plated, it is necessary to remove *all* the silver, otherwise, when the work is replated, the second coating may strip or peel off such parts as may have small portions of the old coating adhering to them. After the articles have been stripped, rinsed, and dried, they should be polished, or buffed, to render them uniformly smooth for replating, after which they are treated in the same way as new work preparatory to being placed in the depositing vat.

Stripping Silver from Iron, Steel, Zinc, &c.—Articles made from these metals, as also lead, Britannia, and pewter, must not be stripped in the acid stripping solutions, but the silver upon their surface may be removed by making them the *anode* in a cyanide of silver bath, and, as we have before suggested, it is better to keep a small bath for this special purpose than to risk injuring the usual plating bath by the introduction of other metals, which will surely occur when the silver is partially removed from the plated article by the solvent action of the cyanide.

Stripping Silver by Battery.—Make a strong solution of cyanide of potassium, say about one pound to the gallon of water. Attach the article to be deprived of its silver to the positive electrode of the battery or dynamo-electric machine, and suspend a strip of platinum foil to the negative electrode. When the bath has acquired a certain amount of silver (dissolved from the plated articles) the platinum will become coated, and if the current be powerful, the silver may become deposited in a granular state, and be liable to fall from the cathode in minute grains. To prevent these from falling to the bottom of the bath, the platinum cathode may be enclosed in a muslin bag, which by retaining the particles will enable them to be readily collected. A plate of gas carbon, German silver, or brass may be employed as a cathode instead of platinum, if desired.

Stripping Gold from Silver Work.—If done with great care, the gold may be readily dissolved from the surface of solid silver articles (not electro-plated) by means of warm aqua regia, composed of 4 parts of hydrochloric acid and 1 part nitric acid. The article may be either dipped in the aqua regia, or the acid may be applied to the article by pouring it over a part at a time, from a small porcelain ladle, and allowing the liquid to flow into the vessel containing the bulk of the acid. When this method is adopted, a vessel of clean water should stand by the side of the acid bath, in which the articles should be rinsed occasionally, and then allowed to drain before again applying the acid. The operation should be conducted over a sand bath, above which is a hood to conduct the fumes given off to the flue of the chimney. As we have hinted, the operation requires care, but if properly conducted it is expeditious. It may be as well to state that silver articles which have been mercury gilt—probably more than once—cannot be wholly deprived of their gold without injury to the article, for the reason that a considerable portion of the precious metal, in the primary stages of the amalgam process, becomes *alloyed* with the silver base. Electro-gilt silver articles, on the other hand, may readily be de-gilded, or stripped, by the above plan, or by making the articles an anode in a strong cyanide bath such as we have recommended for stripping silver, and employing an active current. To remove gold from silver articles by another method, they are first brought to a cherry-red heat, and then thrown into a weak solution of sulphuric acid, by which the gold scales off in spangles, and falls to the bottom of the vessel. The process of heating and plunging into the acid pickle is repeated until all the gold is removed ; after removing from the pickle each time, the article should be rubbed with a hard brush to remove any loosened particles of gold, and rinsed before being again heated.

Stripping Nickel-plated Articles.—Bearing in mind what we have urged, that nickel will not adhere to a nickel-plated surface, it is necessary to remove the old nickel coating from all articles which have to be re-covered with this metal. In the case of German, French, and American nickel-plated articles, which are largely imported into this country, the removal of the nickel coating is by no means a troublesome task ;* trifling though the film may be, however, as, indeed, it frequently is, the film must be removed before any attempt is made to replate the article, otherwise the new coating will assuredly strip from the old one during the process of finishing, if not while it is in the bath. The stripping acid, which may be used either cold or tepid, is composed of : strong sulphuric acid, 4 lbs. ; nitric acid, 1 lb. ;

---

* We allude only to *imported* articles; doubtless those retained in these countries for home use are, like our own, better treated.

water, about 1 pint. The water should first be put into a stoneware jar, and the sulphuric acid added cautiously and a little at a time, since considerable heat is generated when this acid is mixed with water. When the entire quantity of sulphuric acid is added, the nitric acid is then to be poured in, when the bath is ready for use. In making up the stripping bath, the proportion of the acids may be varied, but the foregoing will be found to answer every purpose.

When stripping nickel-plated articles in the above bath, it is necessary to watch the operation attentively, since, as we have observed, some articles are very lightly coated, and a momentary dip is frequently sufficient to deprive them of their nickel. Other articles which having been thoroughly well nickeled require, from some accidental cause, to be stripped and re-nickeled, will need immersion for several minutes—indeed, we have known well-nickeled articles to occupy nearly half an hour in stripping before the underlying brass surface has been entirely free from nickel. The operation of stripping should be conducted in the open air, or in a fire-place, so that the acid fumes, which are very pernicious, should escape freely. The articles should be attached to a stout copper wire, aud after a few moments' immersion should be removed from the bath occasionally, to ascertain how the stripping progresses, and the moment it is found that the nickel has *quite* disappeared from every part, the article must be plunged into clean cold water. It is absolutely necessary that the work should not remain in the stripping solution one instant after the nickel is removed. When the stripping has been properly effected, the underlying metal exhibits a bright, smooth surface, giving little evidence of the mixture having acted upon it.

Nickel may be stripped from brass and copper articles, by electrolysis, in a dilute solution of sulphuric acid, making the article an anode, as in other arrangements of a similar kind ; or a small nickel bath may be kept specially for this purpose.

Stopping-off.—This term is applied to various methods of protecting certain parts of an ornamental article which are required to be part gilt and part silvered, or otherwise varied, according to taste. For this purpose, certain varnishes, called "stopping-off" varnishes, or "stopping," are employed. The materials vary in their composition according to whether they have to be used with hot or cold solutions, more especially when cyanide of potassium is the active ingredient in the depositing bath. A formula which has, with modifications, been much employed for protecting plated work, to be gilt in the hot cyanide bath, from receiving the gold deposit upon certain ornamental parts of the work, is composed of—

Clear resin	.	.	.	.	.	.	10 parts
Yellow beeswax	.	.	.	.	.		6 „
Best red sealing-wax	.	.	.	.			4 „
Jewellers' rouge	.	.	.	.	.		3 „

The three first-named substances are to be thoroughly melted, with gentle stirring, and the rouge, which is the peroxide of iron, gradually added, and incorporated by stirring.

A solution of red sealing-wax, of the finest quality, in alcohol, forms a very useful varnish for warm gilding solutions, if allowed to become thoroughly hard by drying before the article to which it is applied enters the gilding bath. Good, quick-drying copal varnish, mixed with a small quantity of jewellers' rouge or ultramarine, is also employed for hot cyanide solutions; the same varnish, mixed with chromate of lead (chrome yellow), may be used with cold solutions. Almost any quick drying and tough varnish may be used with cold solutions, and for the sake of recognising more freely the parts to which the varnish has been applied, the addition of a little mineral colouring matter, as red lead, chrome yellow, or ultramarine, should be added to the varnish. The article to which the stopping-off varnish has been applied, should never be placed either in a hot or cold bath, until it has become thoroughly dry and hard. The stopping-off varnishes will generally become sufficiently hard in from three to four hours in warm weather, or even in less time if the articles, after stopping, are placed in a lacquering stove moderately heated.

Applying Stopping-off Varnishes.—The article to be "stopped-off" must first be carefully well scratch-brushed, rinsed in hot water, and well dried by wiping with soft diaper. The parts which are to retain the silver colour (for example) are to be very carefully and neatly brushed over with the varnish, special care being taken not to spread it beyond its proper boundary, otherwise, when the article is gilt, the outlines of the various parts will exhibit a ragged and unsightly appearance; the work should be done by the steady hand of a skilful workman. In gilding the articles which have been stopped-off the temperature of the gold solution should be as low as possible, even when the most resisting varnishes are used. It is not advisable to employ too strong a current, otherwise the bubbles of gas evolved are liable to dislodge the thinner layers at the extreme edge of the varnish, whereby such parts, being denuded of the material, become coated, giving a ragged appearance to these portions of the object.

After the articles have received the required deposit, they are well rinsed and dried, and the varnish is dissolved off (if an oil varnish, like copal, for example) with warm spirit of turpentine or benzole;

sealing-wax varnish may readily be removed by methylated spirits, with the addition of heat, supplied by a hot-water bath. Another way to remove the varnish is to destroy it by plunging the article for a short time in cold concentrated oil of vitriol. In ornamenting articles, it is sometimes necessary to produce various coloured effects upon the same object, as orange yellow gold, pink and green gold, bright and dead silver, oxidised silver, etc., in which case the stopping-off needs the utmost artistic skill and delicacy of manipulation.

Soldering.—*Hard Soldering.*—It not unfrequently happens, while scratch-brushing an article of jewellery or other small article, that some portion of the work will accidentally break away ; under such circumstances it will be well if the gilder can himself repair the article instead of being compelled to return it to his customer or send it out to be repaired. With a view to furnish the operator with the means of doing repairs of this nature, the author introduces the following extract from his former work ; * and if the instructions herein given are carefully followed, the operator will have little difficulty in repairing accidental breakages. He should, however, first make himself master of the use of the blowpipe, and practise upon pieces of thin brass or copper wire before venturing to solder delicate articles of jewellery :—

" Hard soldering " consists in uniting any two metals, or parts of the same metal, by means of an alloy composed of two parts of silver to one part of brass. The silver and brass should be melted together as follows :—Having obtained a broad piece of good charcoal, scoop out a slight hollow on the flattest surface to receive the alloy. Now place the metals in the hollow, and fuse them by means of a blowpipe, using either a jet of gas or an oil lamp with a good broad wick. As soon as the metals become hot, touch them with a crystal of borax (borate of soda), which will immediately fuse and act as a flux. The jet of flame must now be vigorously employed until the metals are completely fused. The fusion may be continued for a few moments in order to insure perfect amalgamation. When the " button " of solder is well melted, the flat surface of a hammer should be placed quickly upon it, by which means it will become flattened ; in this form it may be readily beaten out (unless a pair of steel rollers are at hand) until sufficiently thin to cut with a pair of jewellers' shears. The solder can be hammered out upon an anvil or any solid iron surface ; but as each time the blow is given the alloy becomes harder, it will be necessary from time to time to *anneal* it, *i.e.* place it again upon the charcoal and apply the blowpipe flame until the alloy is of a " cherry-red " heat ; it is then to be plunged into cold water, and is ready for

---

* " Electro-metallurgy," by Alexander Watt, tenth edition, p. 153.

beating out or rolling as the case may be, the object being to make the solder as thin as an ordinary card, or even thinner. When the operator is without a pair of rollers he must use the next best substitutes—a hammer and patience. The solder before being used must be scraped with a keen steel edge, and then partly cut into thin strips, and these again cross-cut into small pieces or pellets about one-sixteenth of an inch square. These pellets may be cut when required for use, or kept in a clean box used for the purpose.

The operator should next provide himself with a clean piece of slate, say about three inches square, and a small phial filled with water, and having a cork with a small groove cut in it from end to end. The bottle is used to apply moisture a drop at a time, whilst a large crystal of borax is rubbed upon the slate. By this means a thick creamy paste of borax is obtained upon the slate, which will be used as presently directed. The parts to be united or soldered must now be scraped clean *wherever the solder is expected to adhere*, and with a camel-hair brush or feather of a quill dipped in the borax paste brush over the parts to be soldered. A few pellets of the solder may be placed on the dry corner of the slate, and with the extreme point of the brush moistened by the paste one pellet at a time may be readily taken up and placed upon the prepared surface of the article. The article should be placed upon a flat piece of charcoal (made flat by rubbing on a flagstone), and, if necessary, tied to it by thin "binding wire." A gentle blast of the blowpipe will at first dry the borax, and the flame must then be increased (holding the blowpipe some distance from the flame in order to give a broad jet), and in a few moments, if the jet is favourable, the solder will "run," as it is termed, into every crevice, when the blowpipe must be *instantly* withdrawn. A very little practice will make the operator expert in this interesting art, and it will be advisable for him to practise upon articles of little value until he has not only acquired the use of the blowpipe, but also the proper kind of flame to make the solder run freely. After an article has been hard soldered it is allowed to cool, or may be at once placed in a weak solution of sulphuric acid (a few drops of acid to an ounce of water), which, after a few moments, will dissolve the borax flux which remains after the soldering is complete. The article should now be rinsed in water and dried.

Soft Soldering.—This consists in uniting articles made of sheet tin (tinned iron), lead, zinc, and sometimes iron, with an alloy of tin and lead. It is usually performed with a tool called a *soldering-iron*, which consists of an ingot or bar of copper, riveted to a cleft iron stem terminating in a wooden handle ; the operation may, however, in some cases be accomplished by means of the blowpipe flame. In soldering, the first thing to do is to well clean the parts to be united, which is most

conveniently done by *scraping*, a three-edged tool termed a *scraper*, or
the edge of a penknife, being used for this purpose.   In applying the
solder to the two first-named metals, a little powdered resin is first
sprinkled over the cleaned surfaces to be united ; the soldering-iron
must be well *tinned* by first moderately heating it in a clear fire, then
filing the bevelled surfaces of its point until bright and clean ; it must
next be at once made to touch a lump of black resin, and then brought
in contact with a strip of solder ; care should be taken that all surrounding parts of the point of the tool are well coated with solder.
When about to apply the soldering-iron to the prepared surfaces, it
must be first made moderately hot, not on any account *red* hot ; its
point should then be wiped on a piece of cloth having a small piece of
resin upon it, and then touched with the strip of solder, when a small
globule of the metal will attach itself, and the tool may now be applied
to the object to be soldered ; at the same time the strip of solder,
being held in the other hand, should be brought in contact with the
soldering-iron and a sufficient quantity of the solder allowed to melt
while the tool is being applied.   As the soldering-iron cools, it must
be re-heated and cleaned as before.   Sometimes powdered sal-ammoniac
is employed in soft soldering, and it is a good plan to press the point of
the hot tool upon a lump of this salt occasionally, by which the oxidation of its surface becomes removed.   In passing the soldering-iron
along the parts to be joined, the solder should *run*, as it is termed,
freely and form a bright and even layer.

In soldering iron with soft solder, the surfaces, after being well
cleaned, must be brushed over with a solution of *chloride of zinc*, or
"tinning salt ; " this is made by pouring a little muriatic acid upon a
strip of clean zinc ; vigorous effervescence at once takes place, and
when this has nearly subsided the solution is ready for use.   The solution may be applied to the cleaned iron surface by means of a camel-
hair brush or the feather of a quill, when the soldering-iron is to be
employed as before ; it should, however, be rather hotter for this purpose than for soldering the more fusible metals.   In soldering zinc the
tinning salt is also used, but a little muriatic acid spread over the
surface is better, since it cleans the surface of the zinc, forming, of
course, chloride in doing so.   When it is desired to solder a wire
upon a stout zinc plate for battery purposes, it is a good plan to
moderately heat the end of the zinc to which the wire is to be attached,
then to apply a few drops of the acid, and immediately apply the
solder as before ; the end of the copper wire, being previously cleaned
and tinned, is then to be put in its place, and the hot soldering-iron
and sufficient solder applied until the end of the wire is imbedded in
the material ; a cold hammer may then be pressed on the wire, which,
by chilling the solder, will complete the operation.

Sheet lead, such as is used for lining nickel and other tanks, should not be united by soldering, since the two metals, tin and lead, when in contact with the solution (especially a nickel salt) would slowly undergo chemical action, probably resulting in perforation of the lining. It is usual, therefore, to unite the sheet lead by the *autogenous*·process, or "burning" as it is called, which consists in first scraping the surfaces clean, when a jet of hydrogen gas, or this gas mixed with common air, is applied, by which the two surfaces become fused together. This method of securing the joints of lead-lined tanks is now universally applied, and is unquestionably the best system that can be adopted.

Soldering Liquid.—As a substitute for the solution of chloride of zinc ordinarily used as a tinning salt, the following has been recommended :—Make a neutral chloride of zinc by adding strips of the metal to muriatic acid, taking care to employ an excess of the former. Then add, while the liquid is still hot from the chemical action, as much powdered sal-ammoniac as the fluid will dissolve. Instead of using water to dilute the solution, use alcohol, keeping the liquid in a well-stoppered bottle until required for use. If crystals appear when the solution is placed in an open vessel for use, add a little alcohol, which will liquefy them again.

To Remove Soft Solder from Gold and Silver Work.—This may readily be effected by placing the soldered article in a hot solution of perchloride of iron, made by dissolving crocus or jewellers' rouge in muriatic acid and diluting the solution with four times its bulk of water, and there leaving it until the solder is removed. A formula recommended by Gee* for this purpose is composed of protosulphate of iron (green copperas), 2 ozs. ; nitrate of potassa (saltpetre), 1 oz. ; water, 10 ozs. Reduce the protosulphate of iron and nitrate of potassa to a fine powder, then add these ingredients to the water, and boil in a cast-iron saucepan some time ; allow the liquid to cool, when crystals will be formed ; if any of the liquid should remain uncrystallised, pour it from the crystals, and again evaporate and crystallise. The crystallised salt should be dissolved in muriatic acid in the proportions of one ounce of the salt to eight of acid. Now take one ounce of this solution and add to it four ounces of boiling water in a pipkin, keeping up the heat as before. In a short time the most obstinate cases of soft solder will be cleanly and entirely overcome and the solder removed without the work changing colour.

According to the same authority, another solvent for the purpose may be prepared as follows : To eight ounces of muriatic acid add one ounce of crocus, and well shake it, in order that it may become

---

* "The Goldsmith's Handbook," by George E. Gee, p. 144.

E E

perfectly mixed ; of this mixture take one ounce, and add to it four ounces of hot water, place it in a pipkin, and keep up the heat by means of a gas-jet; put the articles containing the soft solder into it, and soon the desired result will be achieved.   But the former plan (adds Mr. Gee) is most to be recommended.

CHAPTER XXVII.

MECHANICAL OPERATIONS CONNECTED WITH ELECTRO-DEPOSITION.

Metal Polishing.—Brass Polishing.—The Polishing Lathe.—Brass Finishing.—Lime Finishing.—Nickel Polishing and Finishing.—Steel Polishing.—Polishing Silver or Plated Work.—Burnishing.—Burnishing Silver or Plated Work.—Electro-gilt Work.

Metal Polishing.—All articles which are required to be bright when finished are submitted to the process of polishing before they undergo the preliminary operations of cleaning, dipping, quicking, &c., to prepare them for the depositing vat. If the articles were not to be rendered perfectly smooth before being coated with other metal, it would be exceedingly difficult, if not impossible, either to polish or burnish them after being plated to such a degree of perfection as is necessary for bright work. This preliminary polishing is more especially necessary in the case of nickel-plated work, for unless the work is rendered bright and absolutely free from scratches or markings of any kind, these defects will inevitably show when the articles are finished. The extreme hardness of nickel renders the operation of polishing and finishing nickel-plated work at all times laborious, but more especially so if the work has been badly polished before it enters the nickel bath. It is also the fact that every scratch, however minute, which a careless polisher leaves upon brass, copper, or steel work, becomes plainly visible after it has been finished by the nickel-polisher. In large work, such as mullers, sausage-warmers, &c., the preparatory polishing should be of the most faultless character, since any attempt to remedy defects *after* nickel-plating would be fruitless, and probably result in *cutting through* the nickel, necessitating the replating of the article, which should under all circumstances be rigidly avoided. It should be the nickel-plater's *first* duty to examine every piece of work, to see if it has been properly polished, before placing it in the potash bath ; if badly polished, it must be sent back to the polisher again.

Brass Polishing.—These operations are performed at a lathe set in motion by steam-power. It is usually the practice for metal polishers to fix their lathes in workshops supplied with steam-power from adjacent premises, the cost of power per lathe being generally moderate.

The Polishing Lathe.—This machine ordinarily consists of a stout wooden bench set firmly in the floor. In the centre of the bench is a solid cast-iron standard—secured in its position by screwed bolts—in which runs a long double spindle, working on brass or gun-metal bearings. In the centre of the spindle are two pulleys, one fast and the other loose, by means of which it may be set in motion or stopped at will. The spindle revolves at a very high speed. A leather belt, connected to a revolving shaft, by preference below the lathe, passes over these pulleys, and either workman, by means of a stick kept for the purpose, can, by pushing the belt to left or right, set the spindle in motion or stop it as occasion may require. This arrangement is not only convenient, but absolutely necessary, since the spindle is generally worked by two men—one at each end ; and when either of them requires to change one circular buff, or " bob," for another, which very frequently happens, he takes up the short stick and pushes the belt from the *fast* pulley, which is attached to the spindle, to the *loose* pulley, which runs over it. An improved polishing lathe, with shaft-carrier and standard combined, has been produced in Birmingham, and the design is shown in Fig. 121. This arrangement obviates the necessity of fixing a wooden bench.

The polishing tools, or " bobs," as they are usually called, consist of discs of various kinds of hard leather, the stoutest of which are about three-fourths of an inch thick, and are made from walrus or hippopotamus hide ; other bobs are made from bull-neck leather, felt, &c. The materials used for brass polishing are glass-cutters' sand and Trent sand ; the former, having a sharper cut than the latter, is generally used for very rough work, such as comes direct from the founders, with the file marks extensively visible upon its surfaces. Before commencing his work, the polisher, after removing his coat and hat, envelops himself in a long, loose garment, made of brown holland, which buttons at the neck from behind, and its sleeves are secured at the wrists in the same way.

Previous to setting the lathe in motion, each polisher spreads a square piece of calico upon the bench, immediately under the point of each spindle, upon which each workman places a quantity of the sand he intends to use. The first operation, called roughing, or *rough sanding*, is generally performed by the workman standing at the right-hand end of the spindle, and the work is then passed to his mate on the left, who treats it with a finer quality of sand, or " old sand," that which has been repeatedly used, by which a much smoother surface is produced. In the process of *sanding*, as it is called, the workman, holding a piece of work in his right hand, takes up a handful of sand with his left, and holding the work up to the lower part of the revolving bob, presses it against it, while he dexterously allows the sand in his left hand to continually escape, by which it passes on to the bob while the work is being pressed against it ; the moment the handful of sand is paid out, he takes up another handful,

Fig. 121.

almost involuntarily, and keeps up this movement, at intervals of a few seconds only, with mechanical regularity. The operation of rough sanding is sometimes very laborious, as the workmen have to press with all their force upon the work, in order to obliterate deep file marks and other irregularities.

Brass Finishing.—After the work has been rough and fine sanded, it is transferred to the *finisher*, in whose hands it receives the highest degree of polish of which the metal is susceptible. As in all other branches of trade, there is much difference in the skill and judgment of those who follow the art of brass finishing. While some workmen take great delight in turning their work out of hand in the most creditable condition, others are exceedingly careless and indifferent as to whether the work be good or bad, having, perhaps, a preference for the latter. The material generally used for finishing brass work is quicklime reduced to a fine powder, and sifted through a muslin sieve. The lime preferred for this purpose is obtained from the neighbourhood of Sheffield, and is well known in the polishing trade as "Sheffield lime." This material is selected at the lime-kilns by persons who well know what the trade require, and is packed in casks and sent to the polishers in London or elsewhere, who preserve it in olive jars, or large tin chests, carefully covered with cloths to exclude the air ; if the lime be allowed to extract carbonic acid from the atmosphere it soon becomes converted into *carbonate of lime*, which is useless for polishing purposes. When the lime is required for use, a boy takes a lump or two from the jar, and removes all dirt and impurities by first scraping the lime all over, after which he breaks the lime up into small fragments, a few of which he puts into an iron mortar, and with a pestle of the same metal reduces it to a powder. He next passes the powdered lime through the sieve and hands the fine powder to the first workman who requires it. Only a small quantity of lime is thus prepared at one time, since it loses its *cutting* property if exposed to the air even for a short time, especially when in the state of powder.

Lime Finishing is generally entrusted to workmen of superior ability, since much of the beauty of the work depends upon the care and skill bestowed upon this stage of the polishing process. The lime is applied to the bobs in the same way as the sand, but a little oil is also used ; by being used over and over again, the lime becomes impregnated with particles of metal, which increases its polishing power. Indeed, we may say that the bright *polish* which metals acquire when rubbed with an impalpable powder, such as jewellers' rouge, lime, or other material, is only due to the polishing medium indirectly ; it is the metal which becomes removed from the surface of the work which produces the brilliant effect termed *finish*, or *high polish*. When the workman has carefully gone over every part of the work, changing the bobs from time to time to suit the various surfaces—plane or hollow, as the case may be—he removes the lime-bob from the spindle, and fixes a "dolly" in its place. The dolly for this purpose commonly consists of a large disc, composed of many layers of unbleached calico—the whole being about half to three-quarters of an inch in thickness. The folds of calico are first cut into a circular form by means of a chisel and mallet, and these are braced together by two discs of leather or metal, secured by copper rivets. A hole is formed in the centre to admit the screwed point of the spindle. The dolly is worked with dry lime, which is applied by frequently holding a lump of fresh lime against the revolving calico disc, by which it becomes sufficiently charged for the time. The high speed at which the dolly revolves causes the frayed edges of the cloth, when charged with dry lime, to produce an exceedingly brilliant surface in a very short time, but much judgment on the part of the finisher is needed to produce the highest finish attainable, a point which good workmen never fail to reach.

Nickel Polishing, or Finishing, is performed by aid of Sheffield lime, a little oil being applied to the bobs occasionally. Rouge and crocus compositions are used by preference by some polishers. If the work has been properly polished before plating, the nickel-finisher's task, although requiring much skill and care, is tolerably straightforward. The dull nickel deposit readily yields to the pressure upon the lime-bob, presenting to the eye of the workman that degree of brightness which he knows full well will come up to the highest possible brilliancy under the operation of the dolly. He takes good care, however, not to trust too much to the latter tool, but gives the work a brilliant surface before it is submitted to the dolly. It is his special care, moreover, by using small and thin bobs, specially reserved for such purposes, to well polish every interstice or hollow that can be reached by the smallest of his bobs, some of which are about the size of a crown piece.

Steel Polishing.—The articles are first ground upon a grindstone or emery wheel, and are afterwards *glazed*, as it is termed, which consists in submitting the steel articles to the action of round discs of wood covered with leather or metal—a mixture of lead and tin— applied with emery powder of various degrees of fineness, moistened with a little oil. By this means the work is rendered as smooth as possible, and afterwards receives a bright finish with leather-faced buffs charged with finely powdered crocus (peroxide of iron), which imparts to the surface the brilliant lustre for which good steel, as a metal, is so justly famed. Cow-hair or bristle brushes charged with crocus and oil are also used for steel polishing.

Polishing Silver or Plated Work.—The preliminary stages of the process are performed at a lathe set in motion by steam-power, or by a suitable foot-lathe ; the ordinary form of the latter machine is shown in Fig. 122. The tools used are a series of circular buffs or bobs consisting of discs of wood, faced with hard and soft leathers to suit the several stages of the process, the softer buffs being applied after the articles have received a preliminary treatment with the harder and more active tools. Circular brushes, formed of bristles set in discs of wood, are also employed, and for some purposes bobs made from bull-neck leather, &c., of various sizes and degrees of thickness, are ·used. The polisher is generally provided with the various kinds of leather required in his work, from which he cuts out his bobs to suit the particular work he may have in hand. The polishing is

Fig. 122.

effected with the material known as rotten-stone, or tripoli, moistened with oil ; the former is usually kept in a shallow tray, and the latter in a conical tin can with small tubular opening at the top ; by gently pressing upon the bottom of the can with the thumb, the oil escapes slowly, so that a single drop may be applied if necessary. Having set the lathe in motion, the workman applies a little rotten-stone and oil, in the form of a paste, to the revolving bob, and then presses the article with moderate force against it, shifting the article continu-ally, until the entire surface has been buffed. The work is carefully examined from time to time, and when sufficiently smooth for finishing, it is sent to the finishing room, where it is first cleaned by well washing with warm soap and water, with the addition of a little soda. *Finishing, or Colouring*, is performed either by lathe or by hand, according to the nature of the work. When the lathe is employed, a a mop or "dolly," made from the fabric called *swansdown*, is used, in which several layers of this material, cut into a circular form, are united by discs of wood or metal held together by screws or rivets. In the centre of these discs a hole is punched, to receive the screwed point of the spindle. The material used for finishing is the finest quality of jewellers' rouge (peroxide of iron), which is made into a paste with water, and applied, in small quantities at a time, to the face of the dolly-mop, which, by revolving at a very high speed, quickly produces a remarkably brilliant lustre. The rouge "compo" before referred to is also used for silver polishing.

Hand-Finishing, or Colouring.—This operation is best conducted by men or women whose hands are of a soft texture, or have a "velvet-hand," as it is sometimes termed in the trade. The colourer is provided with a shallow porcelain vessel, in which he puts a quantity of rouge, and pours upon it sufficient water to form a pasty mass; after having thoroughly cleaned the articles, and wiped them dry with a soft piece of diaper, he dips the tip of his finger in the rouge-paste, and smears it over a part of the work, and rubs the article briskly with the side of the hand below the little finger, or the large muscle below the thumb, according to the surface he has to treat, using moderate pressure at first, and this he diminishes as the work approaches the finish. As soon as the silver surface has acquired the black lustre for which this metal, when highly polished, is so remarkable, it is examined to ascertain if there are any scratches or other imperfections visible; if an appearance of *greyness* is noticeable upon any part of the work, such portions are again gone over until the uniform black lustre has been produced. It is of the highest importance that neither dust nor grit should get into the rouge or upon the work while being coloured, otherwise scratches difficult to obliterate will be produced. When the work is finished, the rouge is washed out of the crevices, or ornamental parts of the work by means of soap and water and a very soft long-haired brush kept specially for this purpose. The articles are then wiped dry with a soft piece of old diaper or linen, and afterwards with a soft chamois leather.

Articles which have been electro-plated should only be submitted to the process of polishing when they have received a stout coating of silver, for if but a moderate deposit has been put upon the work, the severe operation of buffing with rotten-stone will most probably cut through the silver and expose the metal beneath, more especially at the edges.

Burnishing.—Although many stoutly plated articles of electroplate—especially spoons and forks—are rendered bright by *polishing*, by far the greater proportion of this class of work is *burnished*. Burnishers are an important class of female operatives, who have regularly served an apprenticeship to the trade from an early age, and many of whom perform their allotted tasks with exquisite care and finish. The tools employed in burnishing silver and plated work are very numerous, and are made of steel for the preliminary operations of *grounding* as it is termed, and blood-stone, a hard compact variety of hæmatite (native oxide of iron) for *finishing*. A few examples of burnishing tools are shown in the accompanying engravings (Fig. 123), the cuts being about half the actual size of the tools they represent. The steel tools, which are made in various forms to suit the different surfaces to which they have to be applied, are of various degrees of thickness, the thinner or keener implements being first used to *ground the work*, as it is termed, before the stouter tools are applied. The blades of the steel burnishers are fixed into wooden hafts or handles provided with a brass or iron ferrule, to prevent the wood from splitting. The blood-stones are fitted into iron tubes, and secured by means of pewter solder, a wooden handle being inserted in the other end of the tube. Blood-stone burnishers are of several qualities, the finest material being used for *finishing*. These implements are also of different sizes, so as to be suitable for large or small surfaces.

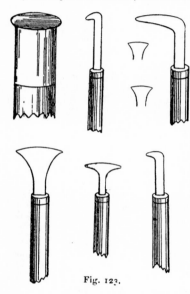

Fig. 123.

Preparing the Tools for Burnishing.—To impart a perfectly smooth and bright surface to the steel and blood-stone burnishers, each workwoman provides herself with two flat buffs, one for the steel tools and the other for the blood-stones. The steel buff consists of a piece of *buff* (Fig. 124) or belt leather, such as soldiers' belts are made from, about 10 inches long and $2\frac{1}{2}$ inches wide. The leather is first boiled for some time in water, after which it is dried as speedily as possible, by which means it becomes excessively hard ; the leather is next glued to a flat piece of wood, about three-fourths of an inch larger than itself each way, and about three-quarters of an inch in thickness. To secure the leather in its position until the glue has become hardened, a heavy weight or clamps are used. When ready for use, the burnisher forms a groove from end to end, by rubbing one of the stouter tools upon the surface of the leather, leaning with all her weight upon it, so as to form as deep a hollow or groove as possible. Having done this she places a little jewellers' rouge in the groove, and passes the tool up and down the hollow with all her force, until its face has acquired a bright black lustre. It is usually the practice to form about three such grooves in the leather with tools of various thicknesses, these being applied to stout or thin burnishers, as the case may be. The buffs for polishing the blood-stone tools are prepared in the same way as the former, a single groove or channel only being formed as evenly as possible in the centre of the leather. The material employed in polishing the face of the blood-stone is putty-powder (oxide of tin), which is used in the same way as the rouge for the steel tools.

Fig. 124.

Preparation of the Work for Burnishing.—The plated articles, after being scratch-brushed, rinsed, and dried, are transferred to the burnisher, who first scours them all over with fine silver-sand and warm soap and water, applied with a piece of soft flannel ; the work is then thoroughly well rinsed in warm water, then wiped dry with soft diaper or old linen rag. When thus prepared, the article is ready for burnishing.

Burnishing Silver or Plated Work.—After scouring the work the burnisher makes a small quantity of warm soap-suds, in a gallipot or other small vessel, by putting a few thin slices of yellow soap in the vessel, and pouring hot water over them, stirring for a few moments with one of the steel tools, until the " suds " are in a condition for use. She next selects the tool she intends to commence with, and rubs it upon the buff until the requisite surface is obtained. Having wiped the tool, she dips it in the suds, and holding it in her right hand, with the handle resting on her little finger, near the knuckle, the other three fingers being above the handle, while the thumb presses upon the top of the handle, by which means the tool is held firmly, and can be applied with the necessary pressure. The first tool employed for a large flat surface is one of the larger and thinner burnishers, whose face is of an elliptic form ; this is held in a slanting direction, and passed to and fro over the work with sufficient pressure to produce a certain degree of brightness, and every now and then the tool is wiped dry, re-buffed, and again applied, until the whole surface of the article has been gone over. A stouter steel tool is next applied, which has the effect of considerably erasing the marks left by the first implement. After going over the surface several times with steel tools of increasing thickness, the first bloodstone is next applied, and which, having a broad and highly polished surface, nearly removes all traces of the marks produced by the steel tools. The work is afterwards gone over with a *finishing stone*, which is of the finest quality of blood-stone that can be procured.

Electro-gilt Work is burnished in the same way as the preceding, but some burnishers prefer using vinegar instead of soap-suds for moistening their tools. We suggested the employment of weak ale for this purpose, and the workwomen having tried it, used it constantly with much satisfaction at our own works for many years.

CHAPTER XXVIII.

MATERIALS USED IN ELECTRO-DEPOSITION.

Acetate of Copper.—Acetate of Lead.—Acetic Acid.—Aqua Fortis.—Aqua Regia.—Bisulphide of Carbon.—Carbonate of Potash.—Caustic Potash. —Chloride of Gold.—Chloride of Platinum.—Chloride of Zinc.—Cyanide of Potassium.—Dipping Acid.—Ferrocyanide of Potassium.—Hydrochloric Acid.—Liquid Ammonia.—Mercury, or Quicksilver.—Muriatic Acid. —Nickel Anodes.—Nickel Salts.—Nitric Acid.—Phosphorus.—Pickles. — Plumbago. — Pyrophosphate of Soda. — Sal-ammoniac. — Sheffield Lime.—Solution of Phosphorus.—Sulphate of Copper.—Sulphate of Iron. —Sulphuric Acid —Trent Sand.

SINCE many persons enter into the art of electro-deposition, in one or other of its numerous branches, who have not the advantage of chemical knowledge, or even an intimate acquaintance with the substances employed in the various processes, a brief description of the chief characteristics of some of the more important materials may prove serviceable. It is frequently the case, too, that lads who have been for some time occupied as subordinate assistants in the plating room, ultimately succeed to more responsible positions, and in their turn become practical platers ; to these also some information as to the nature of the substances used in the art may prove useful, and tend to guard them against error. For the sake of easy reference, the various materials will be noticed alphabetically.

Acetate of Copper, or *Crystallised Verdigris.*—This beautiful salt of copper is in dark green crystals, which are soluble in water. The common verdigris of the shops is in the form of a powder, or soft lumps of a bluish-green colour, and is insoluble in water ; it is, however, soluble in dilute acetic acid, when it forms the same solution as that produced by the dissolved crystalline salt ; being richer in copper than the latter, it may be used with greater economy in making up copper solutions, or for other purposes in which the acetate of copper is employed.

Acetate of Lead, or *Sugar of Lead.*—This is a crystalline salt having somewhat the resemblance of crushed loaf sugar. The pure salt is wholly soluble in distilled water, but the ordinary commercial article frequently produces a slightly *milky* solution, which may be rendered clear by the addition of a small quantity of acetic or pyroligneous acid. This salt is highly poisonous.

Acetic Acid. *The Acid of Vinegar.*—A colourless liquid, having a pungent but agreeable odour. The carbonates and oxides of most bases, as those of the metals and alkalies, are soluble in the dilute acid, forming *acetates*, as acetate of copper, acetate of soda, &c. Its usual adulterant is *water*.

Aqua Fortis. *See* **Nitric Acid.**

Aqua Regia, Nitro-hydrochloric Acid.—This acid mixture, which is employed for dissolving gold and platinum, is made by mixing from two to three parts of hydrochloric acid, to one part nitric acid, by *measure*. Since aqua regia decomposes spontaneously, it should only be prepared when it is required for use.

Bisulphide of Carbon.—This highly volatile and inflammable substance must be kept in a well-corked or stoppered bottle, in a cool place, and its vapour, when the stopper is removed from the bottle, must not be allowed to approach the flame of a candle or lamp, otherwise it may take fire and ignite the contents of the bottle, even at a considerable distance, if the apartment be very warm.

Carbonate of Potash, *Pearlash, or Salt of Tartar.*—A white granular salt employed in the preparation of cyanide of potassium, in making brassing and coppering solutions, &c. It is very *deliquescent*, that is, absorbs moisture from the air, and should therefore be preserved in closely stoppered jars or bottles.

Caustic Potash.—The ordinary commercial article, used for cleaning metal work to be coated with other metals by electro-deposition, also for making up tinning solutions and for various other purposes connected with the art, is the substance known as *American potash*. This article is in the form of hard, brownish lumps, and since it readily attracts carbonic acid and moisture from the air, it must be preserved in stone jars, closed by a well-fitting bung. Caustic potash has a powerful action upon the skin, and must not therefore be handled carelessly ; when it is necessary to remove one or more lumps from the jar in which it is kept with the fingers, this should be done quickly, and the hands immediately plunged into cold water, then for a moment into a weak acid pickle, and again rinsed. A small pair of spring iron tongs should be used for taking up lumps of this caustic alkali.

It may be prepared as follows : Reduce to a powder 56 parts, by weight, of fresh lime, by slaking with water. Make a cream of the powder by adding sufficient water and stirring well. Now dissolve 138 parts of pearlash in hot water, and add the cream of lime to the solution. Boil the mixture for about half an hour, or longer, and then allow it to repose. The lime will deposit in the form of *carbonate of lime*, leaving a strong solution of caustic potash, which may be preserved in a carboy until required for use.

Chloride of Gold.—The preparation of this substance is described in the Chapter on Gilding.

Chloride of Platinum.—Small fragments of platinum are placed in a glass flask, and a mixture of three parts hydrochloric acid and one part nitric acid (by measure) added ; the flask is then to be placed on a sand bath, moderately heated, until red fumes cease to appear in the upper part of the flask. The solution is next poured into a porcelain capsule, or evaporating dish, and evaporated to near dryness, the vessel being moved about until the red mass formed sets into a solid condition. It may then be dissolved in distilled water, and bottled for future use.

Chloride of Zinc.—Granulated zinc is dissolved in hydrochloric acid ; the liquid is then evaporated, when a semi-solid hydrated mass results, which, by continuing the heat, becomes anhydrous, and may be poured on a slab to solidify. A solution of zinc, commonly called " tinning salt," which is much used in *soft soldering*, is prepared by pouring muriatic acid upon small fragments of zinc, when, after the effervescence has ceased, the solution is ready for use.

Cyanide of Potassium.—In many respects this may be considered the most important substance used in electro-deposition, it being a powerful solvent of metallic oxides and salts. The ordinary commercial article is of exceedingly variable quality, and frequently contains but a small percentage of pure cyanide. Its chief adulterant is *carbonate of potash*, one of its essential constituents, and therefore readily introduced in excess. In order that the user of cyanide of potassium may be fully acquainted with its composition, we will briefly explain the methods of preparing it ; and to enable him to determine the true value of the commercial article which may fall into his hands in the course of business, we will describe simple methods by which he may estimate the proportion of true cyanide in any given sample. Knowledge of this kind is of the utmost importance to those whose necessities or duty may require them to make up solutions of the various metals in which cyanide of potassium forms a necessary constituent.

Preparation of Cyanide of Potassium.—There are several methods of preparing this useful salt, but the process recommended by Baron Liebig is usually adopted, and is conducted as follows : 8 parts of ferrocyanide of potassium (yellow prussiate of potash) are first reduced to a powder, and then placed in a shallow iron pan, and dried at a heat not exceeding 260° Fahr., with stirring, until perfectly dry. The dry powder is next mixed with 3 parts of dried carbonate of potash, and the mixture then thrown into a red-hot crucible, and the heat kept up, with occasional stirring with an iron rod, until the whole is fused ; the fusion must be continued until the product appears perfectly white at the end of the rod, after cooling. The crucible is then removed from the fire, the contents again stirred, and after a few moments' repose the liquid salt is poured into a clean, cold, and dry iron tray, in which it quickly sets in the form of a hard cake, which should be broken in lumps while still warm. While pouring out the clear fluid salt, care must be taken to keep back the sediment, which is chiefly iron in a finely divided state (derived from the ferro-cyanide). The sedimentary matter should be knocked out of the crucible, while still hot, upon a separate slab, and the residue of cyanide which is attached to it may be separated by dissolving it out with water. Cyanide, thus prepared, contains a portion of *cyanate* of potassium, but this is not injurious to the solutions of silver or other metals in which cyanide is employed. The article prepared in this way represents ordinary commercial cyanide of good quality, and it will be readily seen (since carbonate of potash is the cheapest ingredient) that a large excess of this salt may be employed without in any way affecting the general appearance of the product ; its activeness as a solvent of metallic oxides and salts, however, will be diminished in proportion to its excess in an uncombined state. Cyanide carefully prepared by the foregoing method should contain from 70 to 75 per cent. of pure cyanide.

A pure cyanide is obtained by the following process. The requisite quantity of yellow prussiate of potash, of good quality, is powdered and dried as before ; an iron crucible, having a lip, is then made red hot ; a small quantity of the powder is now introduced into the crucible, and when this is fused, more of the powder is added, and so on, until the vessel is about three parts filled ; the iron lid of the crucible must be put on after each addition of the powdered ferrocyanide. During the fusion of the salt there is a free evolution of gas, and the fusion must be maintained for about fifteen minutes, or until a sample on the end of an iron rod dipped into it is perfectly white on cooling. The vessel should now remain undisturbed for a few minutes, to allow the iron and other impurities to subside. The clear and colourless fluid, which is nearly pure cyanide of potassium, is now to be poured upon a cold iron slab, or into an iron pan, and the black sediment, which still retains a considerable proportion of cyanide, must be scooped out of the crucible, while still in a soft and pasty condition, and carefully preserved ; the cyanide may be dissolved from this residue whenever the salt is required for future use.

It sometimes happens that the cyanide, from imperfect settling while in a fused state, assumes a grey shade instead of being purely white ; this is of no consequence, however, since the small proportion

of insoluble impurities which cause this greyness will readily subside
when the cyanide is dissolved in water for use.   To prevent the formation of *cyanate* of potassium in the above process, some persons put
a few small pieces of charcoal, and also a little powdered charcoal,
into the crucible before the ferrocyanide is thoroughly fused.   The
cyanide obtained by this method usually contains about 96 per cent.
of the pure salt.   This process, however, is not so economical as the
one previously given, inasmuch as a considerable proportion of the
cyanogen escapes in the gaseous state, whereas, when the ferrocyanide
is fused with carbonate of potash, the cyanogen unites with the
potassium (the base of this salt), and carbonic acid is liberated instead.

Grey Cyanide, as it is sometimes called, is commercial cyanide from
which the reduced iron has not been perfectly separated ; this article
is frequently preferred by some persons, since it is supposed to contain
a smaller excess of carbonate of potash ; it has generally a crystalline
fracture, when broken, while cyanide containing a large prepon-
derance of the carbonate is of a more homogeneous structure.

To Determine the Active Strength of Commercial Cyanide.—The following method was suggested by the late Thornton Herapath, in *The
Chemist*, vol. iii. p. 385 : "The first thing to be done in testing
cyanide of potassium is to prepare a standard solution of ammonio-
sulphate of copper or ammonio-nitrate of copper.*   A certain known
quantity of pure crystallised sulphate of copper, made by crushing the
pure crystals of the shops in a mortar, and pressing the powder so
obtained between folds of bibulous paper, is taken and dissolved in
water.   The solution so prepared is then to be diluted with water so
as to measure 2,000, 3,000, or more water grain measures at 60°
Fahr.   Supposing 390·62 grains of the pure sulphate to have been
taken and diluted to 2,000 grain measures, every 100 grains of such
solution will, of course, represent 5 grains of metallic copper, or 6·25
grains of the protoxide of copper ; 100 grains of each of the samples
of cyanide of potassium to be tested are then dissolved in a sufficient
quantity of water, and introduced into the colorimeters ; an excess of
ammonia is added, and the standard solution of copper is added (out
of a graduated burette), to the contents of each colorimeter in turn,
until a faint blue coloration makes its appearance in each of the
solutions.   The quantities of copper or of the solutions taken then
indicate the relative strength and money value of the samples of cya-

---

* *Ammonio-sulphate of copper* is formed by adding liquid ammonia to a
saturated solution of the sulphate until the precipitate at first produced is
redissolved, when a rich dark blue solution is obtained.   *Ammonio-nitrate of
copper* is produced by adding ammonia to a concentrated solution of nitrate
of copper.

F F

nide examined. Suppose, for instance, one specimen took 100 measures, and a second 150 measures, of the copper solution, the relative strengths and values of such specimens are, therefore, as 100 to 150, or 2 to 3." To render the above process available in the determination of the actual strength of, or proportion by weight of, pure cyanide of potassium existing in the commercial cyanides, it is only necessary to procure a small sample of *pure* cyanide, and to ascertain how much of this is required to decolorise 1 grain of copper in the form of ammonio-nitrate.

Another method of determining the proportion of pure cyanide in a given sample of cyanide of potassium is that proposed by Glassford and Napier, which is as follows : Prepare two solutions, one of cyanide, and another of nitrate of silver, each containing known weights of the respective salts, say 1 ounce of cyanide dissolved in distilled water, in a graduated glass, so as to make exactly six ounces of solution by measure ; then dissolve 175 grains of crystallised nitrate of silver in about three ounces of distilled water ; add the cyanide solution gradually and carefully to the nitrate solution, stirring continually, until the precipitate at first formed is all dissolved without any excess of the cyanide solution. The amount of the cyanide solution required to effect this, with the above quantity of nitrate of silver, will have contained 130 grains of pure cyanide, and from the quantity used may be calculated the amount of pure cyanide in the entire ounce. The authors state that " when nitrate of silver is added to a solution of cyanide of potassium, so long as the precipitate formed is all redissolved, we obtain *the whole* of the cyanide of potassium in combination with the silver ; none of the other salts in solution take any part in the action, even though they be present in a large proportion. This enables us to test the exact quantity of cyanide of potassium in any sample."

A very simple way of testing commercial cyanides where extreme accuracy is not necessary is as follows : Reduce to fine powder in a mortar about half an ounce of pure sulphate of copper ; weigh out 100 grains of the powder, and dissolve (in a half-pint vessel) in about two ounces of water ; now add liquid ammonia of sp. gr. ·880° until the precipitate first formed is all dissolved. Next dissolve 1 ounce, troy (480 grains), in about 8 ounces of water ; pour the latter solution into a tall and narrow hydrometer glass, previously *graduated* by pasting a strip of paper on its exterior surface, divided into ten equal divisions, and these again accurately subdivided into tenths ; the solution must be diluted, if necessary, until it exactly reaches the top line or *zero* of the scale. Suppose we wish to determine, roughly, the comparative value of two samples of cyanide of potassium, we prepare two copper solutions as above, each containing 100 grains of sulphate of copper, and dissolve 1 ounce of each of the cyanide samples to be tested, and, taking the first sample, dissolved and poured into the graduated glass as above, we pour it gradually into one of the copper solutions, stirring after each addition, until the blue colour of the copper solution has disappeared, and has been succeeded by a pinkish tinge; the cyanide must now be added, drop by drop, until the pink tint disappears, when the operation is complete. Now read off the balance of cyanide solution left, and make a note of it; and after emptying the vessel and rinsing it, introduce the solution of the second sample and proceed as before, and when the decolorisation of the second copper solution is effected, note the proportion of cyanide solution which has been exhausted, and compare the two results.

Dipping Acid.—This name is given to a mixture of nitric and sulphuric acids and water, with sometimes the addition of a little hydrochloric acid, nitrate of potash, &c. The composition of various dipping acids is given in another part of the work. Fuming nitric alone is frequently used for dipping articles of copper and brass, by which they assume a bright lustre suitable for certain classes of work.

Ferrocyanide of Potassium, or *Yellow Prussiate of Potash.*—This useful salt, which is chiefly used in the preparation of cyanide of potassium, occurs in large transparent crystals of a yellow colour. The crystals must be powdered and dried at a low heat before being mixed with carbonate of potash, for the preparation of cyanide.

Hydrochloric Acid, or *Muriatic Acid.*—For most purposes the commercial acid is employed, but in making aqua regia for dissolving gold and platinum, the pure acid should by preference be used.

Hydrocyanic Acid, or Prussic Acid.—This volatile and highly poisonous substance, as obtained in commerce, is in reality a solution of the acid in water (hydrated hydrocyanic acid). The strongest form of the commercial acid, known as *Scheele's Acid*, contains 5 per cent. of real acid; the dilute acid of the *London Pharmacopœia* is intended to contain only 2 per cent. of real acid. Even in its diluted forms, it is an exceedingly dangerous substance to inhale, and must therefore be used with the utmost caution. Its powerful odour, resembling the flavour of the bitter almond or young laurel leaf when chewed, always indicates its presence, and the bottle in which it is kept should be very distinctly labelled, and on no account allowed to approach the nostrils when the stopper is withdrawn. The acid is affected by light, and should therefore be kept in a stone bottle, or a glass bottle covered with yellow or brown paper, and in a cool place.

Liquid Ammonia, commonly called **Ammonia.**—This highly volatile liquid, which consists of water saturated with gaseous ammonia, should have a specific gravity of ·880. It is usually contained in Winchester bottles, which must be handled with great care, since the accidental breaking of a bottle of this capacity (about ½ a gallon) might involve most serious consequences. Ammonia should *always* be kept in a cool place ; and when pouring it from the bottle, the user should take care to stand in such a position that the full stream of its vapours may not approach his nostrils.

Mercury, or Quicksilver.—This fluid metal, when pure, is brilliantly white, and from this circumstance was called by the ancients *argentum vivum* and *quicksilver*. It emits vapour at all temperatures above 40° Fahr., and should therefore be kept in a closed bottle. It is entirely volatilised by heat, and should therefore leave no residue when evaporated from an iron spoon. If adulterated with lead, and exposed to the air, it becomes covered with a dull film of oxide, whereas if it remains bright after such exposure, the metal may be adjudged pure, since mercury in its pure state is not affected by exposure to the atmosphere.

Muriatic Acid.—Spirit of Salt.—See **Hydrochloric Acid.**

Nickel Anodes.—When we state that cast-nickel anodes, containing about 5 or 6 per cent. of added iron and a very large percentage of carbon, cost about fifteen years ago the enormous sum of 16s. 6d. per pound, and that almost pure anodes of nickel, either cast or rolled, may now be obtained for 3s. per pound, the reader will see what a great change must have taken place to bring about so marvellous a difference in the price of an article so indispensable to the nickel-plater. At the time we refer to, nickel was a comparatively scarce commodity, and was chiefly obtained from Bohemia and Germany. Since that period, however, nickel ores exceedingly rich in this metal were discovered in the French Colony of New Caledonia, and important improvements have taken place both in its extraction from the ore and in its purification from the crude metal. It is, indeed, a remarkable fact in the history of the electro-deposition of this metal, that just at the time when nickel-plating was developing into an important industry, both in the United States and in this country, New Caledonia should have given up her long-hidden treasures, and supplied our markets with an abundance of this useful metal.

While, a few years ago, even *cast* pure nickel anodes were difficult to procure, we are now able to obtain *rolled* nickel of any convenient thickness—a most important advantage to those who desire to embark in nickel-plating upon a small scale. To Dr. Fleitmann is due the credit of having been one of the most successful in this direction, and specimens of his rolled nickel which we have had in our possession were remarkable for their purity and perfect homogeneity. Being desirous of acquainting our readers with some data respecting English rolled nickel, we communicated with Messrs. H. Wiggin & Co., who kindly furnished us with the following particulars concerning this form of nickel anode, which will be useful to those who may wish to embark in the art of nickel-plating. The advantages claimed for rolled nickel anodes over the cast metal are :—"The constant and steady way in which they give off the metal ; they never become soft or fall to pieces while in the bath, as cast anodes do ; they may be light and thin to begin with (of course being far less costly in consequence) ; and they last a very long time."

Nickel Salts.—This term is applied to the double sulphate of nickel and ammonium, and the double chloride of nickel and ammonium, from which nickel-plating baths are usually prepared. The former "salts," however, are generally preferred, and may be considered the best for all practical purposes of nickel-plating. Nickel salts are, like everything else in commerce nowadays, of very variable quality and price. The finest product we have yet seen was imported from the United States about 1877—78, the price of which, however, was very high. Since that period, however, the manufacture of these salts has greatly progressed in this country, and the marvellous reduction in the cost of nickel has brought the selling price of the double sulphate of nickel and ammonia down to an exceedingly low figure. In reply to our inquiry upon this point, Messrs. H. Wiggin & Co., the eminent cobalt and nickel refiners of Birmingham, favoured us with the following quotations, which will serve as a guide to purchasers in large quantities. Single nickel salts, per pound, 1s., and double nickel salts, 9d. When we state that seven or eight years ago the price of these salts was 6s. or 7s. per pound, or even more, it will be readily seen what a remarkable change has taken place in so short a period of time in this most important item of a nickel-plating outfit.

In purchasing nickel salts, great care is necessary to avoid procuring an impure article, since this would involve the user in a great deal of trouble, either by yielding a deposit of a bad colour, or one that will not firmly adhere to the work coated with it. The double salts should be in large prismatic crystals of a fine dark green colour, and perfectly dry. A solution of the salt should not be acid to litmus paper. The double salt consists of 1 atom of sulphate of nickel, 1 atom of sulphate of ammonia, and 8 of water. When the commercial article is of doubtful quality, it may be improved by dissolving it in hot water, evaporating the liquor, and recrystallising.

Nitric Acid.—Ordinary commercial nitric acid has usually a slightly yellow tinge, but the pure acid is colourless. It is a highly corrosive acid, and freely acts upon the skin, producing yellow stains wherever it touches. This acid should never be kept in a *corked* bottle, since it readily acts upon this substance, but in a stoppered bottle, and in a cool and dark place.

Phosphorus.—This substance must be preserved *under water*, to prevent it from coming in contact with the air, in which, even by a few moments' exposure, it is liable to inflame. The sticks of phosphorus should be kept in a wide-mouthed, stoppered bottle, filled with water, and placed in a dark and cool cupboard. It should not be handled in the fingers nor cut in the open air, but when small pieces are required for use, it is a good plan to thrust the point of a penknife into one of the sticks, carefully withdraw it from the bottle, and lay it in a small dish containing sufficient cold water to cover the lump. The required piece or pieces may then be cut from the stick, and the remainder returned to the bottle by thrusting the point of the knife into it as before.

Pickles.—This term is applied to dilute solutions of the mineral acids, by means of which oxidation is removed or loosened from the surfaces of iron, silver, and other metals. The preparation of the various pickles, and the mode of using them, are described in the treatment of various articles for plating, nickeling, &c.

Plumbago, or *Graphite*, commonly called *Black Lead.*—This material, when required for electrotyping purposes, should be of the very best quality procurable. Its powder is of a dead black colour until rubbed, when it acquires a bright metallic lustre. It is commonly the practice to improve the conductibility of this substance, for electrotyping purposes, by intermixing metallic bronze powders, as copper and tin bronzes, for example, or by gilding or silvering the plumbago powder. The former is accomplished by dissolving 1 part of chloride of gold in 100 parts of sulphuric ether ; this is then to be intimately mixed with 50 parts of plumbago, and the mixture exposed to sunlight, being frequently stirred, until quite dry. It is then applied as ordinary plumbago, but is a very superior conductor.

Pyrophosphate of Soda.—This salt, which has been much used, especially in France, in the preparation of gilding baths, may be readily prepared by heating common phosphate of soda to redness in a crucible, when it parts with its waters of crystallisation, and becomes *anhydrous pyrophosphate*. Dissolved in hot water this *anhydrous* (that is, free from water) salt yields permanent crystals on cooling, which contain 10 atoms of water. These crystals are not so soluble as the common phosphate, and their solution precipitates nitrate of silver *white* (pyrophosphate of silver), and has an alkaline reaction. All the insoluble pyrophosphates, including that of silver, are soluble to a certain extent in the solution of pyrophosphate of soda, hence the usefulness of this salt in preparing gilding solutions.

Sal-Ammoniac, or *Chloride of Ammonium.*—This salt occurs in the form of crystalline lumps, which being exceedingly tough, are more readily broken by forcing a sharp steel point through the mass by means of a hammer than by the ordinary means of crushing or pulverising. When broken into small fragments in the way indicated, the salt may more easily be reduced to a more or less powdery condition.

Sheffield Lime.—This material, which is preferred by brass and nickel-plate finishers to any other kind of lime, is obtained from the neighbourhood of Sheffield, from whence the lumps are carefully selected and transported in wooden casks to London and other parts of the kingdom. The lime must be kept rigidly excluded from the air, otherwise it attracts carbonic acid, forming *carbonate of lime,* by which it loses its *cutting* property and becomes useless. Small quantities may be preserved for any length of time in stone jars, closed by a tight-fitting bung, but for larger quantities olive jars and grocers' large tin canisters have been used with advantage.

Solution of Phosphorus, or *Greek Fire.*—This preparation, which consists of phosphorus dissolved in bisulphide of carbon, is not only highly inflammable, but if any of it be accidentally dropped upon the clothes or floor, it is very liable to take fire spontaneously. It should only be prepared in small quantities, and the bottle in which it is kept should be partly immersed in sand in an earthenware vessel, covered with a metal lid, and placed in a cool situation.

Sulphate of Copper, or Bluestone.—It is of the greatest importance that this substance, whether employed in electrotyping or for any other purpose connected with electro-deposition, should be perfectly pure. The pure sulphate occurs in large crystals of a rich deep blue colour. If there be any green salt in the interstices of the crystals, this is due to the presence of sulphate of iron, or copperas, and the article should therefore be rejected. The "bluestone" of the shops is frequently contaminated with copperas. To determine the presence of iron in a sample of sulphate of copper, dissolve a small quantity of the salt in distilled water, then add liquid ammonia, stirring with a glass rod until the precipitate formed becomes entirely dissolved. Allow the blue liquid thus obtained to rest for a short time, then pour off the clear liquor and add distilled water to the sediment; after a while pour off the water and add a little hydrochloric acid to the residuum; allow the acid to react upon the deposit for a few minutes, then pour into the acid liquor a few drops of a solution of ferrocyanide of potassium, when, if a blue colour is produced (prussian blue), this proves the existence of iron in the original sample of sulphate of copper.

Sulphate of Iron, *Copperas, or Green Vitriol.*—A bright sea-green crystalline salt, readily affected by exposure to the air; it should therefore be kept in a well-corked bottle or jar. The crystals should be bright, perfectly dry, and free from red or brown powder (peroxide of iron). The presence of this powder, however, is not of much consequence, since, being insoluble in water, it will readily deposit to the bottom of the vessel when the crystals have been dissolved.

Sulphuric Acid, or *Oil of Vitriol.*—The ordinary commercial acid has usually a somewhat brownish tint, owing to small quantities of straw or other organic matters accidentally falling into the carboys in which it is conveyed. The pure acid is, however, colourless. This acid should be kept in a perfectly dry situation, since it attracts moisture from the air ; and when making a dilute solution of the acid, it should be added gradually to the water, and *not the water to the acid*, since this might cause the mixture to explode with very disastrous results.

Trent Sand.—Glass Cutters' Sand.—These materials are used by brass polishers in the earlier stages of the polishing process, to remove file-marks and other irregularities from the metal work, the latter substance being used for articles in which a very keen-cutting material is necessary.

APPENDIX (PART I.) ON ELECTRO-PLATING.

Electro-deposition of Platinum.—Platinising Silver Plates for Smee Cells.—
Electro-deposition of Iron.—Steel-facing Copper Plates.—Remarks on
the Electro-deposition of Lead.—Coloration and Staining Metals.—
Oxidising Copper Surfaces.—Electro-deposition of Alloys.—Test for
Free Cyanide.—Antidotes and Remedies in Cases of Poisoning.

Electro-deposition of Platinum* (*Thoms' Process*).—This process,
which has for some time past been carried out by the Platinum-
plating Company, has for its object the " deposition of platinum with
greater economy, solidity, and greater perfection than in the methods
hitherto adopted," and the inventor claims, as an advantage over all
other processes, that he obtains " a deposit which is so bright that it
requires no subsequent polishing. This feature constitutes in itself an
important advantage, as the cost of polishing is entirely abolished."
In the preparation of his bath the inventor prepares a solution of
chloride of platinum in the usual way, which is subsequently rendered
neutral by any ordinary method. To this solution of platinum he
adds a weak solution of common phosphate of soda, to which is next
added a weak solution of phosphate of ammonia, and the whole are
then boiled for several hours, during which operation a solution of
chloride of sodium is added. The proportions of the above ingredients
which it is preferable to use are as follows :—Chloride of platinum,
1 ounce ; phosphate of soda, 20 ounces ; phosphate of ammonia,
4 ounces ; chloride of sodium, 1 ounce. The whole of the ingredients may be varied, but the proportions given have been found to
produce good results. The solution may be worked at any suitable
temperature, but it is preferable to work it above the ordinary temperature of the atmosphere. A strong current of electricity must
be employed, and the cathode should be kept in motion while the
deposition of the platinum is going on. The same bath may be
used for a considerable time by adding fresh platinum solution and
chloride of sodium solution as required, to keep up the original strength
of the bath. When it is desired to coat surfaces to which platinum
will not adhere directly, they must be first coated in any suitable
manner with a substance to which platinum will adhere.

---

* See also p. 356, *et seq.*

Platinising.—This name is applied to coating thin silver foil with platinum, in the form of a black powder, whereby a vast number of fine points are produced, which facilitate the escape of hydrogen in the Smee battery. This ingenious method of favouring the escape of hydrogen, instead of allowing it to accumulate on the surface of the battery plates, was suggested by Smee, and was the means of rendering his admirable battery one of the most useful and popular voltaic batteries known. To platinise silver plates for the Smee battery, a solution is made by first adding to the necessary quantity of cold water one-tenth part, by measure, of sulphuric acid; after mixing these by stirring with a glass rod, add crystals of chloride of platinum with stirring, until the liquid assumes a pale yellow colour; it is then ready for use. Several Smee cells are now to be connected in series, and a couple of platinum or carbon anodes, attached to the positive pole of the compound battery, suspended in the electrolysing cell. The sheet of silver foil to be platinised should have a copper wire soldered to its upper end, and be enclosed in a frame of wood; it is then to be connected to the negative pole, and suspended between the pair of platinum or carbon plates; all being now ready, the platinum solution is to be poured into the vessel until it reaches the upper surface of the silver foil. In about fifteen or twenty minutes, the silver will become coated with a deep black film of platinum in a finely divided state, when it may be withdrawn and rinsed, and is then ready for employment as the negative element of a Smee cell. To prevent the exciting fluid of the battery from attacking the solder which connects the wire to the silver, this, and a few inches of the connecting wire, should be coated with sealing-wax varnish.

The Electro-deposition of Iron.—But little fresh information on this subject has been published which forms any very novel advance on the details given on pp. 348–353 of this volume. The experiments which have been published in the Appendix of the 1889 Edition of this work by Watt are recorded by him as follows:—

Since the publication of the former editions of this work the author made a great number of experiments in the electrolysis of iron salts, the results of which were recorded in the *Electrician.** The object of these experiments was to ascertain the behaviour of solutions of the various iron salts under electrolytic action, with a view to assist those who are engaged in depositing iron for practical purposes, and at the same time to furnish the student with information which might prove useful to him when studying this branch of electro-deposition. After having tried a considerable number of baths prepared from various

* *Electrician,* Nov. 11th and 25th, Dec. 16th and 30th, 1887, and Jan. 15th, 1888.

salts of iron, and worked at different degrees of strength, it was found that weaker solutions almost invariably yielded more uniform results than stronger ones, and it also appeared that when the iron baths were so adjusted as to the proportion of metal per gallon of solution that the anode and cathode surfaces required to be about equal, that under such conditions the most favourable results were obtained. Of the many solutions tried, those given below were found to yield the most uniform and satisfactory results.

Protosulphate of Iron Bath.—The solution was prepared from re-crystallised sulphate of iron, in the proportion of 4 ounces of the salt per gallon of water. The bath was first tried in its neutral condition, with electrodes of about equal surface, and the current from a single small Daniell cell, but as the deposit upon the copper receiving plate was of a rather dark colour, a few drops of sulphuric acid were added, and an extra cell connected, when the character of the deposit was greatly improved. A third cell was now added—the whole being connected in series—when the subsequent deposit became exceedingly white and bright, which character was maintained. At the end of about a couple of hours the copper plate was withdrawn from the bath, when its brilliant appearance resembled that of copper which had been dipped into mercury. The iron anode, which had been freely dissolved, presented a perfectly clean surface.

Ammonio-Sulphate of Iron.—For an iron-plating bath, the follow-ing proportions were found to give favourable results: ammonio-sulphate of iron, 10 ounces; water, 1 gallon. To this solution was gradually added a small quantity of acetic acid, which greatly improved the character of the deposit. It was found that with the current from 2 small Daniell cells in series a solution prepared as above gave very uniform deposits of good white colour, though not quite so bright as those obtained from the simple protosalt.

Citrate of Iron.—A solution of this salt was prepared by digesting recently precipitated carbonate of iron in a moderately strong and hot solution of citric acid, the carbonate being added a little at a time when the neutral point was nearly reached, which became known by the liquid remaining turbid after the last addition of the carbonate had been given. A moderate excess of citric acid solution was then added, when the liquid at once became clear. The solution was then set aside to cool. With the current from 3 small Daniells in series, a very white and perfectly bright film was instantly obtained upon a copper plate, the deposit retaining this character during a prolonged immer-sion. The anode was found to have been freely and uniformly dissolved. The iron deposited from this bath was exceedingly white and brilliant.

Double Sulphate of Iron and Potassa.—A bath was prepared from a mixed solution of the respective salts, the iron salt being somewhat in excess, say about 4 ounces of sulphate of iron and 3 ounces of sulphate of potash to the gallon of water. With the current from 3 Daniells, a fine white and very bright deposit of iron was obtained upon a copper plate immediately after immersion, and the deposition proceeded very uniformly for about half an hour, when the film become somewhat streaky; this was remedied, however, by adding more sulphate of potash. After this latter addition, deposition took place very freely, and the anode was considerably dissolved in the course of an hour. There is no doubt that this would be a very good working bath for the deposition of iron, and if rendered faintly acid, it appears to be less susceptible of oxidation in the atmosphere than the ammonio-sulphate of iron solution.

Persalts of Iron.—It had always been held by experimentalists that metallic deposits could not be obtained from solutions of the persalts of iron; for instance, Gore states on page 245 of his book that "solutions of persalts of iron yield no metallic deposit, but are reduced to protosalts by the passage of an electric current." The incorrectness of this statement the author fully established by a series of experiments with the persalts of iron, from each of which he succeeded in obtaining deposits of metallic iron—in some cases in a perfectly reguline condition. Thus, from a weak solution of pernitrate of iron the metal deposited in a highly comminuted condition upon the cathode, being in the form of a black gelatinous mass, which, when washed and rapidly dried, was found to be highly magnetic—iron in a state of very fine division in fact. From a solution of peroxalate of iron a reguline deposit of iron of tolerably good colour was obtained. From a weak solution of perchloride of iron (1 drachm of concentrated solution of the perchloride to 3 ounces of water) a fairly bright deposit of metallic iron was obtained, upon a copper plate, with the current from 5 small Daniells in series. A solution of persulphate of iron in about the same proportions was next tried, when a bright deposit of iron at once took place at the corners and lower edges of the cathode, which gradually extended upward in the form of a horseshoe, which character it maintained until the plate was withdrawn. Deposits of iron were also obtained from solutions of the following persalts of iron, with and without the addition of other substances: sesquicitrate of iron; sesquicitrate of iron and acetate of soda; persulphate of iron and sulphate of potassa; pertartrate of iron; persulphate of iron and sulphate of ammonium; persulphate of iron and sulphate of soda; perphosphate of iron; perlactate of iron, etc. All these solutions were used in an extremely dilute state, and it was not until this condition of the electrolyte was reached that the metal would deposit in a reguline form—if at all. It thus became evident, from the results referred to, that under suitable conditions of the electrolyte persalts of iron will yield up their metal, and that if such results had hitherto been unattainable, it is clear that such necessary conditions had not been fulfilled by those who had previously experimented with these salts.

Steel-facing Copper Plates.—A very good solution for coating copper plates with iron—commonly termed *steel-facing*—may be prepared by the battery process as follows :—The depositing vessel to be used for coating the plates is to be nearly filled with water, in which is to be dissolved sal-ammoniac, in the proportion of 1 part by weight of the latter to 10 parts by weight of the former, that is 1 lb. of the salt to each gallon of water used to make up the bath. A stout plate of sheet iron, previously pickled, scoured, and rinsed, is to be connected to the positive electrode of a battery, and immersed in the solution : a second plate of iron, about half the size of the former, and also rendered perfectly clean and bright, is to be connected to the negative pole of the battery, and suspended at a short distance from the anode, or larger plate. The battery is allowed to continue in action for two or three days, at the end of which time the iron cathode may be removed, and a strip of clean brass or copper suspended in its place, when, after a short immersion, this should become coated with a bright deposit of iron, provided that the solution has acquired a sufficient quantity of this metal during the electrolytic action. If such is not the case, the iron cathode must be again immersed, and the action kept up until a brass or copper plate promptly receives a coating of iron. When the bath is found to deposit iron freely, the copper plate to be faced with iron is to be connected to the negative pole and immersed in the bath, where it is allowed to remain until sufficiently coated. A bright deposit of iron should appear upon the plate immediately after immersion, and the plate should become quickly coated all over if the bath is in proper condition and a suitable current employed. If, after the copper plate has been in the bath a short time, the edges assume a blackish appearance, it must be at once withdrawn, and well rinsed with clean water, and this is most effectually done by holding the plate under a running stream delivered from a flexible tube connected to a water tap. The plate must then be quickly dried, and afterwards washed over with spirit of turpentine ; it is then ready to be printed from.

The following information upon the electro-deposition of iron has been more recently published :—F. Haber (*Zeitschrift für Elektrochemie*, 4 pp., 410-413, 1898) gives a description of the iron stereotype plates from which the notes of the Austro-Hungarian Bank are printed. The metal is deposited by a very feeble current (*sic*) from a solution of ferrous and magnesium sulphates. It contains about twelve times its volume of hydrogen, and this peculiarity is discussed at length in the original paper. This abstract is from *Science Abstracts*, vol. 1, p. 350.

A paper appearing in *L'Industrie Electro-chimique*, vol. 7, pp. 20—21, 1898, and abstracted in *Science Abstracts*, vol. 1, p. 436, states the electrolysis of iron salts has hitherto been made use of for either the preparation of iron in the form of powder, or for the strengthening of stereotype plates or for the steel facing (*acierage*) of small articles. According to the conditions the metal may easily be obtained in a pulverulent form or as dense and hard as tempered steel. The dense product contains a large amount of occluded hydrogen, sometimes as much as 200 times its volume. For the production of the powder, the use of citrate or oxalate of iron is to be preferred ; but the regulation of the current density is the most essential point to be observed. For the production of dense coherent deposits, sulphate of iron is almost always used. The metal deposited from solutions of the chloride has a tendency to oxidise rapidly, and the presence of chlorides should therefore be avoided. The solution of the sulphate should be neutral, and according to the majority of the various modes of procedure is mixed with a certain proportion of sulphate of ammonium, sulphate of sodium, or sulphate of magnesium. Sometimes the iron deposited is afterwards oxidised, and a fine bluish-black layer of magnetic oxide obtained, which is nearly as hard as iron. In all cases anodes of iron are employed of the same dimensions as the cathodes ; the current density varies from 1·4 to 4·2 ampères per square foot of cathode surface, and the pressure required at this current density is from 1 to 1·3 volts.

Sir William C. Roberts-Austen recently showed, at the Royal Mint, some very beautiful electrotypes in steel (so called) of designs for medals, which had a diameter of about eight inches and were in high relief. The surface of this electrotype iron was of a bright steel grey colour, and was stated to show an exceptionally satisfactory resistance to rusting action. The thickness of the iron deposit was about a tenth of an inch, and it was backed up with about a sixth of an inch of electro-deposited copper. The thickness of the iron deposit could, however, be increased by any desired amount. The importance of this application of electro-deposited iron lies in the fact that the surface is very hard and, therefore, withstands in a particularly satisfactory manner the movement of the style of the reducing machine over its surface, without wearing, when small sized copies are being produced from it.

Remarks on the Electro-deposition of Lead.—Having made a long series of experiments in the deposition of lead from the solutions of its various salts,* the author is enabled to give the details of a few

---

* *Vide Electrical Review*, April 6th, 13th and 27th, 1888.

results which may possibly be useful to the practical operator. Of the numerous lead salts from which solutions were prepared for the purposes of his experiments, the following were found to yield the most satisfactory results : the tartrate of lead ; the same prepared by electrolysis, and phosphate of lead dissolved in caustic potassa. The tartrate of lead bath was prepared by digesting recently-precipitated, and moist, hydrated protoxide of lead in a strong and hot solution of tartaric acid, in which the oxide dissolved very freely. The solution was then allowed to cool and the clear liquor afterwards poured off and moderately diluted with water. With the current from two small Daniell cells, in series, and a sheet lead anode, a brass plate received a prompt and bright deposit of lead, of good colour, and it was noticed that no hydrogen was given off at the cathode. The film of deposited metal continued bright for a considerable time, and maintained a perfectly uniform and reguline character, while being also firmly adherent. It was also observed that the lead anode, which kept clean and bright, had been freely acted upon during the electrolytic action. In this solution, brass, copper, and steel were rapidly coated with lead, which firmly adhered to the respective metals. In order to determine whether the solvent action of tartaric acid, under the influence of the current, was sufficient to maintain the tartrate bath in a uniform condition, the following method was tried. A strong solution of tartaric acid was first prepared, in which was placed a sheet lead anode, and a brass cathode, the current from two Daniells, as before, being used. Shortly after the respective electrodes had been immersed, hydrogen bubbles appeared all over the surface of the negative plate, from which they had little disposition to remove until the plate was briskly shaken in the bath. In about half an hour a bright film of lead formed upon the negative plate, at which period it was observed that the hydrogen bubbles ceased to be evolved at its surface. Since only one-half of the brass plate had been immersed in the acid liquor up to this period, the remaining portion of the plate was now lowered into the bath to ascertain whether the hydrogen would again show upon the brass surface only, and not upon the lead-coated half of the plate, and such actually proved to be the case, for while the brass surface instantly became covered with gas bubbles, the lead coating was entirely free from them. It thus became evident that when forming a lead bath by electrolysis, from a solution of tartaric acid, that both electrodes should be of lead, or that the negative plate should first be coated with a film of that metal. Having first determined the solubility of lead in the tartaric acid, under the influence of the current, it became still more apparent that the tartrate of lead could be employed with advantage in the preparation of a lead-depositing bath, and there is little doubt that a solution of this salt, of moderate strength, if worked with a feeble current, is very suitable for the electro-deposition of this metal. It was noticed, however, that the deposited metal had no disposition to assume the crystalline or spongy form in the foregoing solutions, which is too commonly the case with most solutions of lead salts.

Phosphate of Lead in Caustic Potassa.—A bath was prepared from this combination as follows :—A quantity of chloride of lead was first obtained by adding a strong solution of common salt to one of nitrate of lead, the resulting precipitate being afterwards well washed with cold water. The precipitate was next dissolved in boiling water, and to this solution was added cautiously, so as to avoid excess—as recommended by Mitscherlich, to prevent the formation of a subphosphate— a hot solution of phosphate of soda. The white precipitate of phosphate of lead thus formed was next washed repeatedly with cold water, and a strong solution of caustic potash then gradually introduced until the phosphate was entirely dissolved. The concentrated solution was next diluted with about eight volumes of water, and electrolysed with the current from a single small Daniell cell, when on immersing a clean brass plate as a cathode, this at once became coated with a bright film of lead, but after a few seconds' immersion the deposit assumed a dull appearance, while there was a copious evolution of hydrogen at the negative plate. To check this, the liquid was still further diluted by the addition of nearly an equal bulk of water, after which the subsequent deposits retained their brightness for a long period, while the anode exhibited a perfectly clean surface. The metal deposited from this bath adhered very firmly to the cathode, and there was very little disposition to sponginess—a very common characteristic of electro-deposited lead.

Coloration and Staining of Metals.—In the finishing of ornamental brass work, metal buttons, and various kinds of fancy metal goods, it is often necessary to produce certain artificial effects of colour or tone, to render the work attractive to the public eye. These effects are produced by several well-known means, but modifications are constantly sought for, with a view to giving variety to manufactured articles. From some experiments which the author tried in this direction some time ago he is now enabled to suggest one or two processes which may be found useful to persons engaged in the various branches of trade in which metal-colouring and staining are necessary operations in the workshop. The processes to which attention is now called may be best treated under separate headings, since they are each applicable to different branches of trade.

Colouring Nickel-plated Work.—Some experiments in the production of prismatic colours, called *metallo-chromes*, upon steel and platinum surfaces, have been mentioned at p. 368.

In those experiments it was the writer's desire to obtain similar effects upon polished nickel-plated surfaces, believing that the colorations would show more effectively upon the white surface of nickel-plated brass than upon steel, while at the same time the process could be applied to fancy articles which could not be conveniently made from the latter metal. In carrying out the experiments a series of polished brass plates, well nickel-plated, were employed. A nearly saturated solution of acetate of lead was next made and carefully filtered. Three small Daniell cells, in series, were used for the current, to the positive pole of which one of the nickel plates was attached, and then laid upon the bottom of a flat dish. The filtered acetate solution was next poured into the dish until the plate became immersed to the depth of about half an inch. The end of the negative wire of the battery was now brought as close to the plate as possible without touching, and there held steadily, when in a few seconds the coloration commenced, in form of the familiar ring, and quickly extended in a series of brilliant iridescent circles. Pieces of thin copper wire, formed into various designs, as a cross, an anchor, a star, etc., were subsequently used, by which very varied and pleasing effects were obtained, which looked exceedingly effective in the nickel-plated surface more especially when a sheet of white paper was reflected upon it. To render metallo-chromy practically applicable to nickel-plated articles of ornament, there are several points that must be considered: the coating of nickel must be something more than a mere film, otherwise, when placed in the acetate of lead bath, voltaic action will set up between the nickel film and the underlying brass surface, and the nickel will peel off; a fair coating of nickel therefore is absolutely essential. The process, moreover, is only effectual on plane surfaces. In producing the colorations care must be taken not to suffer the particles of spongy lead, which deposit upon the end of the negative wire, to fall upon the plate, otherwise they will cause a series of white spots, by preventing the formation of peroxide upon the surface beneath them. It is also important to prevent dust or particles of any kind from depositing upon the plate during the operation. After the bath has been worked for some time it naturally becomes acid, therefore when in this condition the articles should be quickly removed when the desired effect is obtained, and at once plunged into clean water, for if allowed to remain in the acid bath the peroxide film will rapidly dissolve. When the bath has become slightly acid it is better to pour it into a bottle containing a little carbonate or oxide of lead, and shake it frequently, when, after an hour or so, it may be filtered and again used. The end of the negative wire should be dipped into water and the deposit of lead which attaches brushed off, after each design has been produced, by doing

G G

which the chances of particles falling upon the plate are greatly diminished.

When electrolysing a series of solutions of lead salts recently, the author noticed some very interesting varieties of peroxide of lead, which formed upon the various anodes employed, and which it will not be out of place to mention here. When electrolysing a solution of the basic nitrate of lead with a platinum anode, the current from three small Daniells, in series, being used, a yellow deposit of peroxide of lead was at first formed upon its surface, which was quickly followed by a superb iridescent film on each side of the plate. After about a minute, two rainbow bands appeared upon the upper part of the anode, while the remaining immersed surface assumed, first, a deep crimson colour, which gradually changed to purple. The film was found to be very adherent. In a solution of sulphate of lead in caustic potash, platinum and steel anodes received a golden yellow film, which, even after a prolonged immersion, did not undergo any change of colour. In a solution of lactate of lead, a platinum anode at first became coated with a golden yellow film, which was succeeded by a steel grey at the lower part of the plate; in a few seconds a deep orange colour began to appear at the lower corners, followed by the primary colours in succession, which rapidly blended as in a prism, producing a remarkably fine effect, which could be modified at will by increasing or diminishing the cathode surface. In a solution of sulphocyanide of lead a platinum anode received a rich golden yellow film, which was sustained without further modification in colour during long immersion. In a moderately strong solution of nitrate of lead a very brilliant iridescent film was produced upon a platinum anode instantaneously; when the same solution was considerably diluted with water, and a small cathode surface only immersed, these effects were still more striking through developing more gradually. In a solution of pyrophosphate of lead a golden yellow film only was obtained upon a platinum anode. A solution of salicylate of lead yielded a precisely similar result. In a solution of the benzoate of lead a pale yellow colour was first produced upon a platinum plate, but this was soon succeeded by orange, then deep orange, followed by a slight tendency to iridescence. When the plate, while still wet, was rubbed with the finger, the prismatic colours at once appeared.

Colouring or Staining of Brass Surfaces.—To give a black stain to brass, a solution of chloride of platinum is usually employed, but since the platinum salt is an expensive one, its employment would necessarily be restricted to work of a special character, or to articles of a superior quality. With a view to obtain a black colour, which would be applicable to large pieces of brass work, and at the same time be less costly than the platinum salt, the author made a series of experiments, the results of which may, it is hoped, be found useful in a practical sense. Being aware that brass was far more difficult to *stain*, as it is termed, than copper, the following method was adopted to give the effect of a black stain upon brass work, as also varying shades of brown from a warm chocolate colour to a deep sepia tone. Having tried many different substances, and solutions of various degrees of concentration, the process given below was finally hit upon, and the results obtained were exceedingly satisfactory, both from an artistic and economical point of view. Knowing that the black deposit, or *stain*, as it is improperly termed, which is produced upon brass—more especially by optical instrument makers—is very easily rubbed off, it was determined, if possible, to obtain a firm adherent film which would resist ordinary rough usage, so that the staining process would be applicable to ornamental brass of all kinds, but specially so for chandelier work. Since these results could not be obtained upon brass direct, it was resolved to first give the brass work operated upon a slight coating of *copper*, in the sulphate of copper bath, and to take advantage of the ready susceptibility of this metal to become *stained* by means of solutions of alkaline sulphides. To carry out this plan, the following routine was adopted :—The brass article, after being dipped in the ordinary dipping-acid and well rinsed, was immersed in a sulphate of copper bath composed of 1 pound of sulphate of copper dissolved in 1 gallon of water, to which was added 1 pound of sulphuric acid. In from five to ten minutes, with the current from a Daniell cell, the deposit of copper was sufficiently thick for the purpose intended, when the article was removed from the bath and plunged into boiling water, the object being to allow it to dry spontaneously. A solution of sulphide of barium (5 grains of the sulphide to each ounce of water) was next poured into a deep vessel, and the coppered brass article, being held by its suspending wire, was immersed in the liquor, when it at once assumed a light brown tone, which gradually deepened, until, after a few moments' immersion, the surface acquired an intense but brilliant black. The article was then withdrawn and plunged into hot water, then into boiling water, and afterwards set aside to dry spontaneously, which occupied but a few moments. The black film thus obtained was firmly adherent, and when the smooth surfaces were rubbed with a clean chamois leather they soon acquired a brilliant polish. Other articles of various forms were subsequently treated in the same way, and when it was desired to give a chocolate brown tone to the piece of work, it was simply immersed in the sulphide bath for about half a minute or so, when a very pleasing brown tone was produced, possessing quite a metallic lustre. Solutions of the sulphides of ammonium, potassium and sodium respectively were also tried, but in some respects the barium sulphide is to be preferred. It was next resolved to treat a specimen of ornamental brasswork in the following way. The bright surfaces, or parts which were to be burnished and remain yellow, were coated with a film of paraffin, to prevent the copper from being deposited upon such parts. The piece of work was then placed in the coppering bath until a sufficient coating of the red metal was obtained; it was then plunged into hot water as before, being afterwards dipped in boiling water, which effectually removed the paraffin. When the work became sufficiently cool to handle, certain ornamental parts were painted over with a mixture formed by mixing blacklead and oxide of iron into a paste with a rather strong solution of the sulphide of barium. Soon after the last coating of the above paste had been applied the piece of work was placed under a tap of running water, and a soft long-haired brush briskly passed over the surface until the whole of the sulphide paste was removed. As it was found that even with such brisk washing the diluted sulphide had slightly discoloured the copper surfaces which had not been treated with the sulphide, a moderately strong solution of cyanide of potassium was quickly brushed over the whole piece, a portion at a time, each part thus treated being at once placed under the tap, which was kept running for the purpose. In this way the bright red of the copper became speedily restored, while the blackened surfaces were unaffected. The article was next rinsed in hot water, and finally in boiling water, as before. The plain surfaces of the article were next burnished, and a thin, colourless lacquer finally applied, when the operation was complete. It will be readily seen that ornamental brasswork—chandeliers, finger-plates and fenders, for example—may readily be treated in the manner described, and exceedingly beautiful and artistic effects produced, without involving much more cost than that of skilled workmanship.

Oxidising Copper Surfaces.—To produce a *deep black* coloration upon a clean copper surface, dissolve from 100 to 150 parts of hydrous carbonate of copper in a sufficient quantity of liquid ammonia. The cleaned articles are promptly immersed in this solution, when they immediately become coated with a fine black deposit, which, when burnished, has the appearance of a black varnish.

Sulphide of Barium.—We have found that a rich coloration may be given to clean copper articles by immersing them for a longer or shorter period, according to the effect desired, in a dilute solution of sulphide of barium. About 4 or 5 grains of the salt to each ounce of water produces an instantaneous coloration, of a warm bronze tint, which increases in vigour by longer immersion. The action is of so permanent a character that the articles will bear much friction before the coating will yield, while the warm chocolate tone which the metal assumes has a bright metallic lustre which, for some surfaces, is exceedingly pleasing.

Electro-deposition of Alloys.—In the summer of 1887 the author tried a great number of experiments in the electrolysis of mixed solutions of neutral or faintly acid salts of various metals, with a view—*first*, to determine whether alloys could be deposited from such solutions at all as a matter of fact ; and *second*, whether, if alloys could be deposited from such solutions, baths could be prepared from neutral or slightly acid solutions which might be employed, as a substitute for cyanide solutions, for the practical purpose of coating metals with alloys. The results of these experiments were published from time to time in the *Electrical Review*,* and since several of the compound solutions which were tried yielded results of a somewhat promising character, so far as relates to their application for practical purposes, it may be well to direct attention to them in the present work. It was generally understood, prior to the publication of the results of the experiments referred to, that the deposition of alloys from acid solutions was not only practically impossible but contrary to the generally accepted theory, which holds that from a solution of mixed metals the least electro-positive metal is deposited first ; that is to say, the simultaneous deposition of electro-negative and electro-positive metals had always, up to the time referred to, been considered theoretically impossible, although in the case of cyanide solutions (as those of brass and German silver) the accepted law of electrolysis, propounded by Berzelius, had in the case of these *alkaline* solutions been actually upset in practice. When experimenting with mixed solutions of neutral or acid metallic salts, the first aim of the writer was to determine whether a film of brass— an alloy of zinc and copper—could be deposited from a solution of the mixed metals under any, and if so under what, conditions. For this purpose a mixed solution of the acetates of copper and zinc was first prepared, the latter salt being in excess. The bath was tried at various degrees of strength, with the current from five small Daniell cells in series, until the desired object was obtained, the deposit of a film of unmistakable alloy in the form of brass of very good colour. In the course of this first experiment it was found that when the solution was too strong a powdery deposit of copper alone appeared upon the cathode ; the solution was then weakened by gradual additions of water, and it was noticed that a marked improvement in the character and colour of the film occurred after each successive addition of that fluid, until finally the characteristic yellow deposit of brass was obtained upon a steel surface. It was found necessary, however, in order to obtain a film of brass of good colour with the solution thus prepared to immerse an exceedingly small surface of anode, that is to say, less than one-fourth of an inch of anode surface only was required to secure

---

* *Electrical Review*, Aug. 26th, Sept. 2nd and 9th, 1887.

a coating of the alloy upon a cathode surface (a steel plate) of at least five square inches. Indeed, when the anode was allowed to merely touch the surface of the solution a steel plate of five square inches of surface instantly became coated on both sides with brass of a good yellow colour, except at the extreme end farthest from the anode, which was tipped with zinc only. In order to modify this, and place an anode of small surface opposite the entire length of the steel plate, a piece of thin brass wire was next employed as an anode, and answered the purpose admirably, for a fresh steel plate became uniformly coated with brass of very good colour, on both sides, almost immediately after immersion in the bath. Under these conditions the deposition of a true alloy was far more prompt than with any ordinary cyanide solution. A steel plate in contact with a strip of zinc immediately became coated with brass in this bath on the contact side, but copper alone was deposited on the opposite side. By simple immersion, a clean strip of zinc became immediately coated with a film of brass in the same solution. It thus became apparent that when once the proper condition of the electrolyte was reached, the deposition of an alloy from an acid solution was an exceedingly simple operation, and that an electro-negative and an electro-positive metal could be deposited conjointly from an acid bath as a veritable alloy.

If the above results had been obtainable only with mixed solutions of zinc and copper salts, while solutions of other mixed metals failed to yield similar results, an inexplicable problem would have presented itself. But that this was not the case was amply established by further experiments, in which, amongst other alloys, the following were deposited, in some cases with remarkable facility ; copper and tin, copper and lead, copper and antimony, copper and silver, copper and nickel, and even copper and platinum. In preparing a series of solutions for depositing brass, the author tried the following with success amongst others which were less successful : the sulphates of zinc and copper, citrates of zinc and copper, chlorides of zinc and copper, tartrates of zinc and copper, from each of which combinations he obtained satisfactory films of brass of a good yellow colour, but for all practical purposes he would give the preference to a solution composed of the acetates of copper and zinc. An endeavour was also made to produce a working bath from the double acetates of copper and zinc by the battery process, as it is termed, as follows :—A mixture of about equal parts of commercial acetic acid and water was first prepared, and in this was immersed a clean plate of good sheet brass, connected to the positive pole of a 5-cell Daniell battery ; a rod of carbon, connected to the negative pole of the battery, was then immersed in the liquid, which soon assumed a greenish tint near the anode, which electrode kept very bright and clean, indicating that it was being freely dissolved into the solution. The liquid was kept frequently stirred while the electrolysis was proceeding, and at the end of about twenty minutes the carbon, which exhibited traces of deposited metal upon its surface, was withdrawn, and a steel plate substituted, which became partially coated with a yellow film of brass in less than half a minute. The steel plate was then withdrawn, and the carbon again immersed, in order to allow the solution to acquire sufficient metal to form a practical working bath. In about three-quarters of an hour from the commencement of the operation the bath was found to have become sufficiently charged with the respective metals—zinc and copper—to yield a ready deposit of brass of very good colour. The plate was allowed to remain in solution for a few minutes, when it was taken out and well scoured with silver sand, moistened with soap and water, which operation it withstood without stripping in the least degree, showing that the deposit alloy was thoroughly adherent. The film was next scoured with sand and water only, with as much force as possible, to test the tenacity of the film in the severest manner, and it was found that even this severe treatment failed to remove the film, except by wearing it away by a long continued friction. The plate was next replaced in the bath and allowed to receive a fairly good coating; it was then well scoured with sand, soap and water, and after being well rinsed was burnished with a steel burnisher, but without blistering or stripping in the least degree. It was noticed, when applying the burnishing tool, that the deposited metal felt exceedingly soft under the pressure of the tool; the deposit was therefore reguline in the highest degree, while its colour was equal to that of the finest quality of brass. Since the foregoing results were published in the *Electrical Review* the author was shown by a correspondent some specimens of steel coated with brass, in a solution of the acetates of copper and zinc, as recommended by him, and these certainly gave promise that, with a little further development, the electro-deposition of brass from acid solutions would eventually become practical, in the hands of skilful manipulators, for the purposes of art. Although the author's present engagements deter him from pursuing the subject of alloy deposition for awhile, he hopes to resume his experiments at an early date, and to make known some further results which may prove interesting, if not indeed practically useful, to the electro-metallurgist.

Test for Free Cyanide.—It is sometimes useful to have at command some ready way of determining the actual quantity of free cyanide in a solution. For this purpose Mr. Sprague devised the following system, " based upon the ordinary decimal measures obtainable anywhere, and upon the basis of one ounce of cyanide per gallon of solution, from one to two ounces being the proper working strength." The method is thus described :—" One ounce per gallon is equal to

456 APPENDIX ON ELECTRO-PLATING.

62·5 grains in 10,coo; the equivalent of cyanide of potassium is 65, and it takes two of these to precipitate and redissolve cyanide of silver from nitrate of silver, the equivalent of which is 170. The test solution, therefore, is prepared from pure nitrate of silver, 81·72 grains, dissolved in a 10,000 grain flask of distilled water; 8·172 grammes in a litre make the same solution, which is equivalent, bulk for bulk, to a solution of one ounce of cyanide in a gallon, and may be used in any measure whatever, properly divided. I prefer to take 1,000 grains of it, and make it up to 10,000 again; to take 100 grains of the solution to be tested, by means of a graduated pipette, and then add this weaker solution to it from an ordinary alkalimeter. As soon as the precipitate ceases to redissolve on shaking, the test is complete. A slight cloudiness in the liquid marks this point.

"To test a sample of cyanide, dissolve 62½ grains in the 10,coo grain flask, and treat this in the same way. Thus, if a sample is so treated, 100 grains placed in a small flask or bottle, 1,000 grains of the test put into an alkalimeter and dropped into the flask as long as the precipitate disappears, and if upon adding 520 grains in this way, a permanent faint cloudiness is produced, the sample contains 52 per cent. of real cyanide. If the original test solution is preferred, 1,000 grains of that to be tested must be used, and the result is the same."

Antidotes and Remedies in Cases of Poisoning.—Since some of the substances employed in electro-deposition are of a highly poisonous nature, and the mineral acids with which we have to deal (especially nitric and sulphuric acids) are exceedingly corrosive in their action upon the skin, as also are the caustic alkalies, soda and potash, a few hints as to the best antidotes or remedies to be applied in cases of emergency will, it is hoped, prove acceptable; indeed, when we take into consideration the fatal promptitude with which hydrocyanic acid, cyanogen vapours, and cyanide of potassium will destroy life, it becomes the duty, not only of employers, but their foremen, to make themselves acquainted with such antidotes as can be applied at a moment's notice in cases of accidental poisoning or injury from corrosive substances. In the course of a long experience, the author has more than once narrowly escaped serious consequences, not only from accidental causes, but (in his youthful days) from careless disregard of the dangerous nature of cyanogen vapours. In the latter case his system was at one time so seriously affected by inhaling the vapours of cyanide baths as to partially reduce the power of the lower extremities. His esteemed friend, Mr. Lewis Thompson, a gifted scientific chemist and surgeon, having casually paid him a visit, upon observing the author's condition, and well knowing its cause, promptly prescribed *six glasses of hot brandy and water*. The advice was quickly followed, to the extent of half the prescribed dose, and by the following day the symptoms had almost entirely disappeared. Upon another occasion, the author inadvertently swallowed a quantity of old gold solution, which he had placed in a small teacup for an experimental purpose, in mistake for some coffee without milk, which it was his custom to drink when at work in his laboratory. Having discovered his mistake he at once desired his assistant to get some lukewarm water without delay; in his absence, however, he thrust his finger to the back of his throat, and by tickling the *uvula*, quickly induced vomiting; copious doses of warm water, assisted by again irritating the *uvula*, soon emptied the stomach, after which warm brandy and water completed the remedy, no ill effects from the accident being afterwards observable. Upon one other occasion, the author was ascending a spiral staircase leading to a plating room, when he suddenly felt a sensation of extreme giddiness; promptly guessing the cause, he retreated as speedily as his trembling limbs would permit and sought the open air, and as quickly as possible obtained a glass of brandy from the nearest tavern, which had the effect of checking the tremulous motion of the limbs and feeling of intense nervousness. The cause of the sensation above referred to was this: some old silver solutions had been treated with sulphuric acid, to precipitate the silver, a short time before, in an upper apartment, and the carbonic acid and cyanogen vapours liberated were descending to the base of the building at the time he ascended the staircase; it was the remembrance of this fact that prompted him to retrace his steps as quickly as the shock to his system would allow.

In mentioning the above incidents our chief object is to illustrate, from what has actually occurred, not only the sources of danger which callousness and inadvertence may invite, and to point out how such accidents may be avoided, but also to indicate how by promptitude more serious consequences may be averted.

General Treatment in Cases of Poisoning.—The first important step in all cases of poisoning, is to *empty the stomach* with all possible dispatch. This may generally be done (and should always be tried first) by thrusting the finger towards the throat, moving it about so as to tickle the parts until vomiting supervenes. While this remedy is being tried, a second person, if at hand, should hasten to procure some lukewarm water, which the sufferer should be made to swallow, whether vomiting has or has not occurred; warm mustard and water may also be tried. If these remedies fail, the stomach-pump should be applied. The vomiting should be kept up and the stomach well washed out with some bland *albuminous* or *mucilaginous* liquid, as warm milk and water, eggs beaten up in milk or water, barley-water, flour and water, or any similar substances ready at hand. After the vomiting, a brisk purgative or *enema* may be administered, and nervous irritability or exhaustion alleviated by means of opium, ether, wine, or warm spirit and water, as the case may require ; only the " domestic remedies," however, should be applied, except u der proper medical advice.

Poisoning by Hydrocyanic Acid. Cyanogen, or Cyanides.—When hydrocyanic acid has been inhaled, the vapour of ammonia or chloride of lime should be at once applied, cautiously and moderately, to the nostrils ; indeed, this highly poisonous acid should never be used, especially by inexperienced persons, without the presence of a second person, holding an uncorked bottle of liquid ammonia or chloride of lime in moderate proximity to his nostrils. In case of poisoning by this acid, *cold* water should be *at once* poured upon the head, and allowed to run down the spine of the sufferer. In the case of hydrocyanic acid or cyanides having been swallowed, four or five drops of liquid ammonia, in a large wine-glass full of water, may be administered. Mialhe recommends spreading dry chloride of lime upon a towel, folding it up in the form of a cravat, and moistening it with vinegar ; this is then placed over the mouth and nostrils of the patient, so that he may inhale the chlorine which is gradually liberated. In cases of poisoning by swallowing cyanides—as gold and silver solutions, for example—emptying the stomach by every means would undoubtedly be the most important step. The application of *very* cold water to the head and spine should not, however, be neglected in severe cases. As antidotes for cyanide poisoning, iron salts are recommended, which convert the deleterious acid into the comparatively innoxious prussian blue.

Poisoning by Corrosive Acids.—In case of either of the mineral acids —nitric, sulphuric, or hydrochloric—having been swallowed, copious doses of lukewarm water, mixed with magnesia, chalk, carbonate of soda, or potassa, should be administered at once. Milk, broth, salad oil, or oil of sweet almonds may also be given.

Poisoning by Alkalies.—Vegetable acids—as vinegar and water, dilute acetic acid, lemon-juice—should be given by preference, but if these are not at hand, very dilute hydrochloric, nitric, or sulphuric acid (about ten drops in half a pint of water) may be substituted. When the painful symptoms have subsided, a few spoonfuls of salad oil should be administered.

Poisoning by Metallic Salts.—The sufferer should be caused to drink tepid water copiously and repeatedly, vomiting being also urged by tickling the throat with the finger or a feather ; copious draughts of milk and the white of eggs (albumen) may also be given ; but flowers of sulphur or sulphuretted waters are recommended in preference, since these transform most of the metallic salts into insoluble sulphides, which are comparatively inert.

Cyanide Sores.—These painful affections may arise from two principal causes : first, from dipping the hands or arms into cyanide baths to recover articles which have dropped into them—a very common practice, and much to be condemned ; and second, from the accidental contact of the fingers or other parts of the hand, on which a recent cut or scratch has been inflicted, with cyanide solutions. In the former case, independent of the constitutional mischief which may arise from the absorption by the skin of the cyanide salts, the caustic liquid acts very freely upon the delicate tissue of the skin, but more especially upon the parts under the finger nails. We have known instances in which purulent matter has formed under the nails of both hands from this cause, necessitating the use of the lancet and poulticing. Again, when cyanide solutions come in contact with recent wounds—even very slight cuts or abrasions of the skin—a troublesome and exceedingly painful sore is sure to result, unless the part be at once soaked in warm water ; indeed, it is a very good plan, after rinsing the part in cold water, to give it a momentary dip in a weak acid pickle, then soak it for a few moments in warm water, and after wiping the part dry with a clean rag or towel, apply a drop of olive oil and cover up with a strip or thin sheet of gutta-percha.

Poisoning by Acid Fumes.—When the lungs have been affected by inhaling the fumes arising from dipping baths, stripping solutions, etc., or chlorine gas, the sufferer should at once seek the open air ; he may also obtain relief by inhaling, in moderation, the vapour of ammonia from *the stopper*, not from the bottle itself ; or a little water may be put into a glass measure and a few drops of ammonia mixed with it, which may be inhaled more freely. When an apartment has become oppressive from the fumes of acid, it is a good plan to pour *small* quantities of liquid ammonia upon the floor in several places, but the acid fumes should be expelled as quickly as possible by the opening of all windows and doors.

Caution.—Never add an acid to any liquid containing cyanide, or ferrocyanide, in a closed apartment, but always in the open air, taking care to keep to windward of the liberated gases, which are *poisonous in the highest degree.*

APPENDIX (PART II.) ON ELECTRO-PLATING.

Effect of Nitrates upon Nickel Deposits.—Dary's Barrel Method of Nickel-plating.—Employment of the Barrel Method of Plating for Metals other than Nickel.—The Electrolytic Manufacture of Metal Papers.—Electro-Deposition of Cobalt.—Professor S. P. Thompson's Process of Cobalt Deposition.—The Electrolytic Formation of Parabolic Mirrors for Searchlights.—" Arcas " Silver-plating.—An Hotel Silver-plating Plant. —Aluminium Plating by Electrolysis and Otherwise.— Plating Aluminium with other Metals.

The Effect of Nitrates upon Nickel Deposits.—In Prof. C. F. Chandler's Presidential Address to the Society of Chemical Industry, in July, 1900, he makes the following remarks concerning nickel-plating (*Journ. Soc. Chem. Ind.*, vol. 19, 1900, p. 611) :—"Nickel-plating is a most useful application of electrolysis. It was invented by Isaac Adams, of Boston. Adams wondered why all attempts to plate nickel upon other metals had proved unsuccessful, and he began a careful investigation to ascertain the cause of the difficulties experienced. He soon found that the real difficulty was due to the presence of nitrates in the solutions employed. Nickel always appearing in commerce in the metallic form it was natural when anyone desired a solution, to dissolve it in nitric acid, to precipitate the nitrate with carbonate of soda and dissolve the carbonate of nickel in the proper acid for the solution desired. No one ever washed the carbonate of nickel with sufficient care to remove the last portions of the nitrate of soda. Consequently all the nickel solutions previously experimented upon had contained nitrates, the presence of which Adams found to be fatal to successful nickel-plating. This fact having been ascertained, nickel-plating was the immediate result. Adams obtained a patent for this process, which was afterwards the subject of prolonged litigation. The novel proposition was presented to the Court of a patent for not doing something, namely, for not permitting nitrates to find their way into the nickel solutions employed in nickel-plating, and the Court held that the exclusion of nitrates was an essential condition of successful nickel-plating, and that a process involving this condition was just as patentable as a process involving any other special condition necessary for successful execution, and the patent was sustained."

It must be remarked here that Adams' English patents 3,125, of 1869 and a disclaimer 3,125* of 1869, do point out the necessity of the nickel solutions employed being pure, but do not place, as far as I can observe, any such very marked emphasis upon the necessity of the absence of nitrates from the solutions as would be supposed from Prof. Chandler's remarks quoted above.   In my opinion Adams' patents place more stress upon the necessity of the absence of impurities other than nitrates, than upon the necessity of removing the nitrates themselves.   A very fair series of extracts from Adams' patents is given by Watt on pages 300, 301 of this volume.   The only two portions of the patents mentioning nitric acid or nitrates which are omitted from Watt's extracts are in the final clauses, and are as follows :—

1. " It is important that great precaution should be used to prevent the introduction into the solution of even minute quantities of potash, soda, or nitric acid.   When an article to be coated is cleaned in acid or alkaline water, or is introduced into it for any purpose, the greatest care must be taken to remove all traces of these substances before the article is introduced to (sic) the nickel solution, as the introduction of the most minute quantities of acids or alkalies will surely be injurious. It is important that the solution be kept free from all foreign substances, but its purity from those above-named is especially important."

2. " What I therefore claim is, first, the electro-deposition of nickel by means of a solution of the double sulphate of nickel and ammonia, or a solution of the double chloride of nickel and ammonium, prepared as above described, and used for the purposes above set forth, in such a manner as to be free from the presence of potash, soda, alumina, lime, or nitric acid, or from any acid or alkaline reaction."

Dary's Barrel Method of Nickel-Plating.—G. Dary (*Electrician*, vol. 20, pp. 49, 50, July 28th, 1900) describes a method for economically and rapidly plating the innumerable small articles which are now required to be nickel-plated.   This method is patented, and the English patent is owned by the Electrolytic Plating Apparatus Company.   The process is described as follows, in an abstract of the original paper (*Science Abstracts*, vol. 3, p. 896, 1900, or *Journ. Soc. Chem. Ind.*, vol. 19, p. 1024, 1900):—The preliminary treatment of the small articles which are to be nickel-plated involves the usual boiling with strong potash solution, rinsing in water, and immersion in a nitric acid bath ; but these and the subsequent electrolytic deposition of the nickel are all carried on in the same piece of apparatus.   This consists of a hexagonal perforated wooden barrel mounted so that it may be rotated at any desired speed in the tanks containing the various cleaning and depositing solutions.   The barrel and its mountings are suspended by an axial metal rod which rests on the tanks, and the whole can be raised or lowered by means of pulleys.   The method ot working is as follows : The barrel is filled one-half or two-thirds with the articles and is then lowered into the cleaning or depositing solutions. The speed of rotation varies according to the form of the articles to be plated—a speed of fifty-five revolutions per minute being found the best for certain forms.   The electrical contact between the articles in the barrel and the negative leads is effected by means of the metal rod which carries the barrel, and by the trunnions on which it revolves. From the latter metallic strips run along the interior of the barrel, and thus make contact with all the articles which it contains.

A current of 10 ampères at 6 volts pressure will plate 20 kilograms of metallic objects in four hours, the barrel used for such a charge being 0·75 metre in length. Sawdust is used to dry the articles after they have been thoroughly washed with water. The cost of nickel-plating by this process lies between 0·80 and 2·50 francs per kilogram, according to the form of the articles requiring to be plated.

Employment of the Barrel Method of Plating for Metals other than Nickel.—The barrel method of electro-plating is not only employed for nickel-plating, and the following details of a patent trial on this method will be of interest in showing the success with which the process has been employed. The details are taken from a notice in the *Electrician*, vol. 48, p. 110 :—

On Monday, November 4th, 1901, the Electrolytic Plating Apparatus Company brought an action against Messrs. H. Holland and Company to restrain them from infringing Patent No. 5,274 of 1896, the property of the former company, which patent is the one described above, and is entitled, " An improved machine for the electro-deposition of metals in a perforated drum designed to rotate in the electrolyte, and having an axis which forms the negative pole." The defence was a denial of infringement, and a plea that the plaintiffs' patent had been anticipated by the patents of Carnforth, W. R. Lake, Zingsem, Richard Heathfield, and W. S. Rawson. Mr. Swinburne, giving evidence, said that he had read the plaintiffs' patent, which related to an improved apparatus for the electro-deposition of metals. According to their answer to interrogatories defendants had used, but had not manufactured, apparatus for the electro-deposition of metals having a perforated drum revolving in the electrolyte, such drum having an axis forming the negative pole, with which the articles to be plated came in contact, directly or indirectly, through contact with each other. If that were so an infringement of plaintiffs' patent had taken place. He had considered the anticipations alleged by defendants, but did not find anything in any of them in any way anticipating the plaintiffs' patent. He stated that Heathfield and Rawson put their anodes inside the barrel instead of outside, as the plaintiffs did, and the plaintiffs' method was more convenient, as the anodes were always wearing away, and were more easily replaced. Moreover, with Heathfield and Rawson's arrangement, mud and dirt from the anode would get into the work, whilst finally the connections were simpler with the plaintiffs' arrangements, as there was only one side of the connections to be brought to the moving parts of the apparatus ; he, however, considered that Heathfield and Rawson's apparatus would perform the plating satisfactorily. Messrs. Elkingtons' electro-plating manager, however, stated that Heathfield and Rawson's apparatus was employed by his firm and then discarded, as it was not successful in depositing various metals, the chief disadvantages being the size, the difficulty of getting more than a small quantity of work into the barrel, and the difficulty of renewing the anodes.

On the other hand the plaintiffs' apparatus worked successfully and economically, and he knew of no machine better than the plaintiffs' for doing this work. They had two of the plaintiffs' machines at work. The central cathode gave the great advantage that the current went through the work before it reached the cathode. The manager of Messrs. Brompton Bros., of Birmingham, stated that his firm had two of the plaintiffs' machines, which gave entire satisfaction, the saving effected being about 75 per cent. of the cost of production over and above the process which was in use prior to the introduction of the plaintiffs' machine. Mr. Justice Ridley delivered judgment for the plaintiffs. He held that neither Zingsem's nor Heathfield and Rawson's patents were anticipations. Plaintiffs had just hit the right thing which others had failed to do by a slight margin. The invention had proved an unqualified success.

A modification of the ordinary barrel process of plating small articles has been recently introduced by MM. Delval and Pascales, and has been described in *L'Electricien*, August 17th, 1901, vol. 22, pp. 101-104. The following extract from an abstract of this paper, which appeared in the *Journ. Soc. Chem. Ind.*, vol. 21, 1902, p. 53, appears to show that in this modified process the anodes are inside the barrel as in the Heathfield and the Rawson patents mentioned above, and not as in the Dary process outside the barrel. The abstract runs :—" The special feature of the process is the commutator, which supplies the current to the anode connections inside the bath. Instead of these being permanently connected to the source of the current they are so arranged that, by means of the commutator, the bars not actually touching the articles in the drum are cut out of circuit, so that any leakage of current through them is prevented."

In the absence of drawings the meaning here appears somewhat dark, as it would seem that the time to *disconnect* the circuit from the anodes was precisely when these came in contact with the articles to be plated, whereas precisely the opposite action is stated to occur in this description. The difficulty to be avoided, one would imagine, would be the actual dissolving of the coating on the articles, or the articles themselves possibly if they were permitted to come into contact with the anodes, whilst the question of leakage as ordinarily spoken of would appear not to come into the question, although certainly a short circuit might be caused if the articles under treatment were, as is pre-

sumably the case, connected to the cathode, and then came into contact
with the anodes momentarily.

The Electrolytic Manufacture of Metal Papers.—In the *Electrical Engineer*, vol. 28, 1901, p. 3, the following notice is given of
this process:—" One of the most recent developments in the electrolytic manufacture of metal papers is embodied in a patent recently
issued to Carl Endrumeit, of Berlin, describing a continuous process for
the production of metal papers or paper-backed foils. In this method
an endless belt of metal moves past first a rotary polishing roller, which
removes stains or irregularities, then over a tank containing a dilute
solution of potassium trisulphide, with which the polished face of the
belt is moistened by means of a cylindrical brush. A layer of sulphide
imperceptible to the eye, but sufficient to prevent too close adherence
of the metal afterwards deposited, is thus made to form upon the surface of the metal band.   After wiping and then thoroughly rinsing in
soft water the belt dips into the first plating vat, and receives over the
sulphide a thin coating of nickel, then into a second vat, where this
film is reinforced by a heavier deposit of copper. This is followed by
more brushing and rinsing, and a coating of glue is evenly applied to
the layer of copper, a strip of paper is fed upon it from a roll, and belt,
deposit and paper pass together through rubber compressing rolls and
a drying chamber.   As the deposited film adheres far more strongly
to the paper backing to which it is glued than to the polished and sulphide
faced metal belt, the finished metal-faced paper can be easily stripped
and wound, while the belt passes again to the polishing rolls and the
electro-plating vats. Using a polished cathode belt, it is claimed that
a paper with a brilliant nickeled surface is obtained, and by the addition of mercaptans or other sulphur compounds to the glue the
backing adheres so strongly that the product is thoroughly serviceable."

Electro-deposition of Cobalt.—The following additional remarks
on cobalt plating were made in the Appendix to this work in the 1889
edition (see also pp. 359-362 of this volume) :—In the autumn of 1887
the author, believing that cobalt, as an attractive and serviceable coating
for other metals—but more especially for articles of brass, copper and
steel—deserved more attention than had hitherto been accorded to it by
electro-depositors, pursued a long series of experiments with a view
to ascertain the behaviour of certain salts of cobalt under the action
of the electric current, hoping that the results obtained* would not
only be instructive to the student of electro-chemistry, but also in
some degree useful to those who pursue the art of electro-deposition
as a business. It had often struck the writer as being somewhat
remarkable that cobalt should have received so little attention from

---

* *Vide Electrical Review* Nov. 18th and 25th, and Dec. 2nd, 1887.

those who are engaged in depositing metals upon each other for the various purposes of art, since it is not only whiter than nickel, when deposited under favourable conditions, but it also requires but a very feeble current for its deposition as compared with that required by the former metal. It is true that cobalt is a dearer metal than nickel, but since a cobalt bath requires only about one-third of the quantity of " salts " required to make up a nickel bath, the actual cost of the electrolyte in each case would be about the same. The price of cobalt anodes is also higher than the corresponding plates of nickel, but when we consider the small quantity of metal that is usually required to be deposited upon ordinary work, the relative cost of the respective metals need scarcely be seriously considered. In the series of papers referred to the author endeavoured to show that the adoption of cobalt-plating, or cobalting, would probably be accepted by the public as a novelty, and thus give a fillip to the electro-plating industry, but more especially to that branch of it which is devoted to coating metal work with nickel. On making inquiries as to the probable reason why cobalt-plating had been so little adopted, it soon became apparent that the subject had hitherto been treated chiefly in an experimental way, and that the information furnished by writers on the subject was scarcely sufficient to guide those who were willing to turn their attention to the deposition of this metal as a new industry. Moreover the difficulty of obtaining anodes of cobalt, and reliable cobalt salts, also militated against the adoption of cobalt-plating for practical purposes. Since both these necessaries can readily be obtained from Messrs. Henry Wiggin and Co., of Birmingham, at the present time, there is no practical obstacle in the way of those who may feel disposed to try their hands at cobalting as an addition to their present routine of operations.

Of the many cobalt solutions which the writer tried, the following were found to be the most effective and reliable, namely, the sulphate of cobalt, the double sulphate of cobalt and ammonia, the chloride, the double chlorides of cobalt and ammonium, and the chlorides of cobalt and sodium. These salts were employed in the proportions, and with the results, given below.

Sulphate of Cobalt.—In each gallon of water required to make up a bath, 5 ounces of crystals of sulphate of cobalt are to be dissolved, which is most conveniently done by dissolving the crystals in sufficient hot water, and finally adding cold water to make up the necessary quantity of solution. The bath should be allowed to rest until cold, and be worked with rolled or cast cobalt anodes, either of which may be obtained from the firm above named. The current density should not exceed 2 to 4 ampères per square foot of cathode surface, and the articles to be coated (if required to be bright) should be previously well polished, potashed, and scoured, as for nickel-plating, before being put into the bath. As in the case of nickeling, it is also absolutely necessary that the most scrupulous cleanliness be observed, since carelessness in this respect might cause the work to strip at such parts as may have been badly prepared. Brass, copper, and steel are readily cobalted in this bath.

Double Sulphate of Cobalt and Ammonium.—To prepare a bath from this salt, 5 ounces of the crystals are dissolved in each gallon of water, and the current should be of about the same density as before ; it may, however, be somewhat diminished when coating steel articles. After a time, articles coated in this bath assume a rather dull appearance, but since electrolytic cobalt is somewhat softer than nickel, the dull burr may readily be rendered bright by scratchbrushing, an operation which would have but little effect on a corresponding surface of nickel. If carefully worked, this solution should yield very good results, and may be considered a good practical bath. It must, however, be used with a feeble current, otherwise the metal will deposit too rapidly and consequently be liable to strip off before a sufficiently stout film is obtained. In cobalting with this solution, and indeed with most of them, it is absolutely necessary that the deposit should progress slowly at first ; if the film has "struck" a few seconds after immersion, that is all that is needed to ensure adhesion, and the current may be gradually, but moderately, increased after a few minutes. In other words, if the deposit be allowed to *jump* *on*, it will most assuredly jump off when the film thickens.

Chloride of Cobalt.—A bath may be prepared from this salt by dissolving about 2 ounces, or even less, of the crystals in each gallon of water. With a very feeble current brass and copper readily receive a fine white and bright film of cobalt, which, however, becomes dull after a time. If worked with care, keeping the current low, very good deposits may be obtained from this solution.

Chlorides of Cobalt and Ammonium.—In making up a bath from this double salt, from 4 to 5 ounces of chloride of cobalt crystals are dissolved in each gallon of water, and to the solution is added about 3 ounces of sal-ammoniac, the whole being stirred until the latter is dissolved. This bath must be worked with a still more feeble current than the last, otherwise the deposited metal will be liable to strip from the articles coated. The deposit from this solution is very bright, and of a fine white colour.

Chlorides of Cobalt and Sodium.—To a solution of chloride of cobalt, containing 5 ounces of the salt per gallon of water, 3 ounces of common salt are added and dissolved. With a weak current, articles of brass or copper promptly become coated in this bath, the deposit generally remaining bright for a longer period than in either of the previous solutions. The anodes keep very bright and clean in working this bath.

Double Sulphates of Cobalt and Ammonium, with Chloride of Sodium added.—Well knowing, from his own experience, and that of others, the great advantages which are derived from the addition of common salt to nickel baths prepared from the double sulphates of nickel and ammonium, by which not only is the conductivity of the solution greatly increased, but the character of the deposited metal materially improved, the author determined to ascertain whether similar advantages could be obtained by adding common salt to a bath prepared from the double sulphates of cobalt and ammonium. For this purpose a solution was formed by dissolving crystals of the double salts, in the proportion of about 5 ounces to the gallon of water. About 5 per cent. of common salt was then added and dissolved in the liquor, and a clean brass cathode immersed and allowed to remain until fairly coated. The addition of the salt, however, did not present any apparent advantage ; an additional 5 per cent. was therefore given, after which the deposit—a feeble current being used—became more prompt, and a satisfactory film of cobalt was obtained in about half an hour. A third portion of salt—making 15 per cent. in all—was next added, and a fresh plate immersed, which received a white but somewhat dull film of cobalt, but, however, very uniform in character. Although the deposits obtained in the *salted* solution were much less bright than those from the plain solution of the double sulphates, it may be mentioned that the dull bloom, which had a pearly hue, was readily brightened by the scratch-brush or by scouring with moist silver sand. It would therefore be easily rendered bright by the ordinary process of polishing. When comparing the results obtained from nickel solutions to which common salt has been added, with those in which the saline material has been added to cobalt solutions, it became evident that the advantages derived from the addition of salt were more marked in the case of nickel baths.

In depositing cobalt from its solutions, although it is necessary, in order to obtain a firmly adherent film, to employ weak currents of electricity, the colour of the deposit is never so good—that is, the film never acquires its full degree of whiteness—when the current is below a density that will yield a prompt deposit ; that is to say, the article to be coated should become covered with metal *almost* immediately after immersion.

Professor Silvanus Thompson's Cobalt Process.—This process, for which Professor Thompson obtained a patent in 1887, has for its object " the deposition of cobalt in films of greater tenacity, density, and brilliance of tint, than have heretofore been obtainable with certainty. The solution for depositing cobalt may consist of sulphate or chloride of cobalt, or of the double sulphate or chloride of magnesium, or other suitable soluble salt of magnesium, or a mixed soluble salt of magnesium and ammonium may be added to the solution of cobalt salts. The citrate of magnesium is a useful salt, and it may be formed in the solution by adding citric acid and magnesia, or carbonate of magnesium, to the solution, citrate of ammonium, or simply citric acid may be added to the solution." As an example of one way of carrying out the invention, a bath may be prepared by adding to 10 pounds of pure water* 1 pound of the double sulphate of cobalt and ammonium, $\frac{1}{2}$ pound of sulphate of magnesium, $\frac{1}{2}$ pound sulphate of ammonium, 1 ounce of citric acid, and 2 ounces of carbonate of ammonium. The solution may be used warm or cold, but if heated to at least 35° C. (95° Fahr.) gives a brilliant deposit with greater readiness than when used cold. Another method of carrying out the invention is as follows: —Half-a-pound of sulphate of cobalt is dissolved in 4 pounds of pure water ; $\frac{1}{4}$ pound of sulphate of magnesium is dissolved in 4 pounds of water ; these solutions are mixed, and water added to make up 1 gallon. A quantity of sulphate of ammonium, not exceeding $\frac{1}{2}$ pound, may with advantage be added. In carrying out this process, the author states that good deposits of cobalt are obtained, when either of the above solutions is diluted with an equal bulk of water, but in that case a stronger current in proportion to the surface of deposition must be employed. "The best current density," says the inventor, "I find to be not less than 1 ampère, nor more than 4 ampères per square foot of depositing surface, but a somewhat stronger current may be advantageously employed during the first few minutes of deposition, so as to secure a rapid striking film on the surface, the current density being diminished so soon as the surface is once covered with a film." The work to be coated by this process must be scrupulously clean, and otherwise prepared in the same manner as for nickel-plating.

The Manufacture of Parabolic Reflectors for Search Lights, etc.—Mr. S. O. Cowper Coles has devised a very effective and ingenious method of preparing accurately parabolic reflectors for search lights, etc., by means of electro-deposition.

The older form of search light mirror was made of glass cast and cut as nearly as possible to the shape of a true paraboloid of revolution on both sides and the glass was then silvered by one of the ordinarily employed methods. The drawbacks to this form of reflector were the great expense, the weight, the doubly reflecting surface, and lastly its very brittle nature. A single rifle shot through such a

---

* By pure water, the inventor explains, is meant simply newly-boiled and filtered water, or preferably distilled water.

mirror will shatter it irretrievably, whilst the firing of heavy guns on a man-of-war, on which it is mounted, will it is said sometimes be sufficient to break such reflectors, and the heat of the arc light employed may also crack the glass or damage the silvering. Mr. Cowper Coles' process, which he fully described in a paper read before the Institution of Electrical Engineers (*Journal*, vol. 27, 1898, p. 99), has overcome all these troubles very successfully by means of an electrolytically deposited mirror. This form of mirror is said to be now largely used in the navy.

In the paper above referred to Mr. Cowper Coles says:—"My process is an electrolytic one, and one of the chief features of it is that the surface obtained requires no after polishing or trueing up. When once a true mould has been produced any number of reflectors can be taken from it at a small cost. A glass mould is prepared, the convex side of which is accurately shaped and polished to form a true parabolic or other reflecting surface. As the mould only requires shaping and polishing on the convex side it is comparatively cheap as compared with a glass reflector which has to be ground on both sides. On the prepared surface is deposited a coating of metallic silver, thrown down chemically on the glass, and then polished, so as to ensure the copper backing, afterwards to be applied, being adherent to the silver. The mould thus prepared is placed in a suitable ring and frame (Fig. 125), and immersed in an electrolyte of copper sulphate, the mould being kept horizontal and rotated about a vertical axis, E, at a speed of about fifteen revolutions per minute, and copper is then deposited on the silver by means of an electric current. The copper adheres firmly to the silver and together they form the reflector, which is subsequently separated from the glass by placing the whole in cold or lukewarm water and then gradually raising the temperature of the water to 120° Fahr., when it is found that the metal reflector leaves the glass mould owing to the unequal expansion of the glass and the metal. The concave surface of the reflector obtained is an exact reproduction of the surface of the mould and has the same brilliant lustre and requires no further polish. It was indeed stated in the discussion on the paper, that any attempt at polishing the surface perceptibly damaged the trueness of the metal surface. In order to prevent tarnishing, the surface of the silver is however coated electrolytically with palladium, which gives a good reflecting surface and resists tarnishing and the heat of the arc to a wonderful degree. Mr. Coles states that the cost of palladium is at present about double that of platinum; but that as it has about only one half the density the same area may be plated at about the same cost with either palladium or platinum. It melts at about the melting point of wrought iron and does not readily tarnish, neither is it readily attacked by either sulphuric or hydrochloric acid. In carrying out the manufacture of reflectors by this process it is essential that the glass mould should be perfectly clean and free from grease before the silver coating is applied. It has been found however that if the cleaning is solely

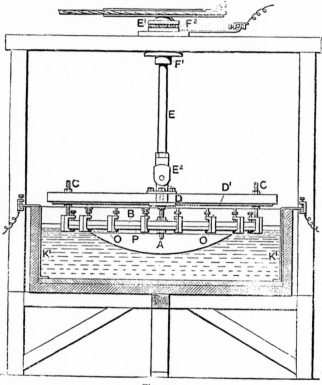

Fig. 125.

affected by chemical means there is a great liability of the silver adhering too firmly to the glass, whereby the mould is in danger of being broken during the removal of the reflector. This difficulty has been overcome by cleaning the glass mould with a suitable paste or powder, such as peroxide of iron, and then removing such paste or powder by washing the glass with a 50 per cent. solution of ammonia. It is necessary that this cleaning operation should be repeated prior to the production of each reflector. After the convex side of the mould has been properly cleansed as described, a thin coating of metallic silver is applied as follows :—Ammonia is added to a solution of nitrate of silver until the precipitate that is first formed is re-dissolved, then re-precipitating by caustic soda, again dissolving in ammonia and then adding glucose to the solution. Excellent results have been obtained with a silvering solution made up of equal parts of solutions of the following strengths :—Silver nitrate, 0·5 per cent ; caustic potash, 0·5 per cent. ; glucose, 0·25 per cent. The surface of the mould to be coated is immediately dipped into the solution face downwards. In from four to five minutes the silver begins to form on the glass mould, the solution changing from pink to dark brown and black, the film thickens quickly and in from thirty to thirty-five minutes a good coating of silver is deposited. Dr. Common has found a good deposit of silver to be equal to a thickness of $\dfrac{1}{28,000}$ of an inch. The silver coating is thoroughly washed and then allowed to dry, and the silver which has been deposited is burnished bright with a piece of cotton wool and peroxide of iron, preferably precipitated by ammonia from a dilute solution of ferrous sulphate. The cost of silvering is found to vary from 2d. to 4d. per inch diameter of the mirror.

During the operations above described the glass mould (which in the case of large mirrors is of considerable weight) is handled by means of a rubber sucker placed on the concave side of the mould. The mould, when silvered and burnished, is attached to the frame figured above, by means of which the silvered surface is electrolytically connected to the terminals of the current generator. The mould is suspended horizontally with its convex silvered face downwards in a tank of the copper sulphate electrolyte. The anode copper plate is at the bottom of this tank. The anode plate is beneath it and is flat, for this shape is found more advantageous than a concave plate because it tends to prevent the formation of copper trees at the edge of the mould, and, moreover, gives a deposit gradually thickening towards the centre of the mirror. In placing the mould in the electrolyte, in the first instance, it is necessary to insert it edge-ways, for if it is put down horizontally the thin silver edges have to carry the whole of the current, which is rather strong at first ; by this gradual edge-ways insertion of the mould in the electrolyte, it causes the surface to be uniformly wetted and a thin film of copper is deposited first at the places of contact of the conductors near the edge of the mould ; the mould is finally allowed gradually to assume the horizontal position in the liquid, and the current is worked at a pressure of about

nine volts for a few minutes, after which the current density
is reduced. The strong current employed at first is in order
to rapidly coat or flash the silver surface with copper, a precaution
which, the inventor states, it is very important to observe, presumably
because otherwise the silver coating would be partially dissolved in
the electrolyte. When the flashing is completed the shaft and
mirror which it supports are rotated about 15 revolutions per
minute, and the operation of depositing the base metal continued
with a current density of about 19 ampères per square foot.
The copper solution employed has the following composition : —
Copper sulphate crystals, 13 parts per cent.; sulphuric acid
(strongest acid), 3 parts per cent.; and water, 84 parts per cent.
by weight. When the copper backing is sufficiently thick the mirror is
detached from the mould by means of warm water as already
described. The concave surface of the mirror is next coated with
a bright coating of electro-deposited palladium. This is performed
by placing the reflector in a solution containing 0·62 per cent. of
palladium ammonium chloride added to 1 per cent. solution of
ammonium chloride. The solution is used at a temperature of 75°
Fahr., and the current density is such, that a 2-foot reflector takes
about 0·5 ampère ; the e. m. f. required at the terminals of the bath
being from 4 to 5 volts. The plating is performed in an earthenware
pan, R, Fig. 126. The anode employed is of carbons, and of only
a few inches cross-section ; it swings in every direction over the
mirror surface. Seventy to 80 grains of the palladium to the superficial foot is found to afford a good protective coating. The silver-
faced reflector, previous to being placed in the palladium solution, is
thoroughly washed with a weak solution of caustic soda. The back
of the reflector is varnished before placing in the palladium solution.
After coating with palladium the reflector is removed from the electrolyte and dipped into boiling water, and then placed in boxwood
sawdust which is kept hot by means of a steam jacket. The reflector
is then complete and ready to be mounted in a suitable ring frame.
Salt water thrown on the surface of these reflectors when much too
hot to touch, results in the water evaporating, leaving a crust of
salt which may be brushed off with a wet cloth and leaves no permanent mark. Although palladium does not give as white a light
as perfectly clean and bright silver, yet owing to the tarnishing of the silver the palladium is finally by far the most satisfactory. Nickel was found to be quite unsuitable. The bluish white
deposit from arc lamp carbons is said to wipe off from the surface of
palladium without damaging it. The weight of the metal mirrors
is not more than one-half or one-sixth of that of glass mirrors
made by different manufacturers.

The reflecting power of several different bright metal surfaces was found to be as follows :

Silver	100
Chromium	100
Platinum	74
Palladium	64

The reflecting power of silver being taken as 100 (S. O. Cowper-Coles, *Electrician*, vol. 44, 1899, p. 267).

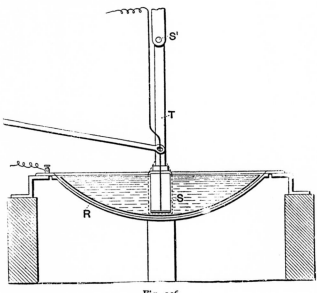

Fig. 126.

" **Arcas** " **Silver-Plating.**—What was a very promising system of electro silver-plating, was brought out in about 1890, under the title of " Arcas " silver-plating. This method of plating consisted in depositing with silver a certain percentage of metallic cadmium. In fact a deposit of an alloy of silver and cadmium was deposited instead of pure silver, as is done in the more usual method of silver-plating. The colour and general appearance of the plated articles was quite as good as that of the best silver electro-plate, but the substitution of cadmium to the extent of 25 per cent. of the silver of course reduces the cost of the metal plating film. The cost of cadmium is about threepence per ounce, whilst silver costs more, nearly thirty pence ; the cost of the cadmium silver alloy containing 25 per cent. of cadmium is therefore about 22 pence per ounce. This economy in the cost of the plating metal therefore offers a reasonable chance of extra profit if the extra care, which is undoubtedly required in plating alloys, is not so great as to unduly raise the cost of labour. The complete patent for this process, which was devised by Mr. S. O. Cowper Coles, is English patent, 1,391, 1892. The Syndicate which worked the process for some twelve or fourteen months advertised the cost of plating by this means in December, 1891, at the following rates :

		£	s.	d.
Spoons and forks, fiddle and old English table . . . . per doz.		0	16	2
Spoons and forks, fiddle and old English dessert . . . . ,,		0	14	5
Spoons, tea . . . . ,,		0	8	5
Dessert knives and forks, with plated handles . . . . per doz. pairs		1	12	3

The prices of a few well known articles are selected from a long list advertised, as indicating the cost yielding a satisfactory profit at which good work could be done. The advertisement concludes that " the prices are for plating in the best manner, and include all ordinary repairs ; a cheaper quality may be had if required. Special quotations offered for quantities."

The solution employed consisted of an alkaline cyanide solution containing about half an ounce of silver per gallon and about 11·5 ounces of cadmium per gallon. This electrolyte is best prepared by dissolving, say an ounce, of metallic silver, and 23 ounces of metallic cadmium in sufficient nitric acid, as described on p. 228. Make up with distilled water or rain water to about two pints, next add sodium carbonate or sodium hydrate solution until a permanent precipitate just begins to form, stirring well all the while with a glass rod or tube. Then add gradually, with constant stirring, sufficient of a solution of pure potassium cyanide until all the metals are precipitated, excess of cyanide must be carefully avoided. Then allow the precipitate to settle, and wash two or three times by decantation as described on p. 230 of this work; finally add gradually sufficient of the cyanide solution to just dissolve all the precipitates, and also add a further excess of the cyanide solution equal to one quarter of that already added which was required to dissolve all the precipitate. The solution is made up to 2 gallons and is worked warm at a temperature of about 50° C. The purest commercial potassium cyanide, such as is employed for gold ore treatment only should be employed, both in this case and whenever cyanide plating solutions are to be prepared.

If, when working this solution, it is found that the anodes become coated, a further amount of the cyanide solution must be added.

The anodes consist of an alloy of silver and cadmium, and contain about six parts of silver to one part of cadmium, that is, about 15 per cent. of cadmium. The alloy of 25 per cent. cadmium, when rolled cold, is somewhat brittle, and breaks up into curious triangular prisms, whose long edges are parallel to the rolls as shown in Fig. 127, which is taken from *Industries* of April 8th, 1892. The 15 per cent. cadmium alloy, especially if rolled at a moderately high temperature, does not show this peculiarity. The patent also claims the plating with alloys of silver and zinc, with or without cadmium, but the employment of zinc was found not to be so satisfactory as cadmium, and was, I believe, employed little or not at all. The difficulties which are experienced in the electro-deposition of alloys and the precautions which it is necessary to take are well known, and are set forth in most works upon electro-plating (see p. 374 of this work). The ordinary brass plater judges of the nature of the deposits which he is obtaining chiefly by its colour, and is in this way guided in his treatment of the bath. Too much copper in the solution and a weak current density are both favourable to the deposition of excess of copper, the more electro-negative element, whilst excess of zinc in the solution and a high current density are favourable to the deposit of excess of zinc. The solution is colourless or nearly so, and in order to regulate the nature of the brass deposited, the plater watches the colour of the deposit on the cathode and regulates his current and composition of electrolyte as this indication dictates. But when electro-plating with an alloy of cadmium and silver the plater is deprived of the valuable assistance which colour affords him in the case of brass.

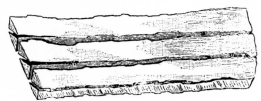

Fig. 127.

An Hotel Silver-Plating Plant.—The Hotel Metropole in New York undertakes its own electro-plating. The plant, which has been in operation for some time, is stated by the Electrical Engineer of New York to have fully repaid the first cost of installation, and has demonstrated the economy and feasibility of this novel scheme. The United Electric Light and Power Company, which supplies alternating current to the hotel, leads its high potential circuit into the sub-cellar, where it is transformed to 220 volts by means of two 1,000 light converters. In close proximity to this place the hotel management has installed the silver-plating plant, which consists of the following apparatus :—A two horse-power 200 volt 2 phase alternate current Tesla motor, running at 1,800 revolutions per minute, belted by means of a countershaft to a buffer running at 3,000 revolutions per minute, and a plating dynamo delivering 75 ampères at a pressure of 5 volts, the necessary fuses, an overload circuit breaker, and the conductors, which in this case are hollow copper tubes leading to the various vats. There are in all four of these vats, holding 25 gallons of liquid, and containing respectively the nickel, copper, silver, and striking solutions. As the operator of this plant is not kept busy constantly, the guests not eating the silver off with sufficient rapidity, he devotes the remainder of his time to the buffing of the silver wire. (The *Electrician*, vol. 41, 1899, p. 487.)

An article giving a useful review of the various conditions which have to be fulfilled in order to obtain an economical working of electro-plating plants is published by C. F. Burgess, in the *Western Electrician* (vol. 22, pp. 195-197, 1898).

Aluminium—It has been frequently stated that aluminium can be electro-deposited from aqueous solutions of its salts in a similar manner to either zinc or copper. Many patents have been taken out for processes of this character, some of which were described in an earlier edition of this book, and practically all of which are described in J. W. Richard's treatise on Aluminium, 1890, published by Carey and Baird, Philadelphia. The present writer has never, however, seen any sample of aluminium so deposited and has never heard of any well authenticated case of it having been so obtained. Even if it were possible to deposit aluminium in this way it is in the highest degree probable that the cost involved would render the process useless commercially. It is, therefore, not necessary to discuss any of these processes here.

Copper and iron may, it is stated, be coated with aluminium by carefully cleaning these metals and plunging them under the surface of molten aluminium in a similar manner to that adopted in the so-called hot galvanising process in which iron is coated with metallic zinc. The writer is not aware that the actual details of this method have been published, that is to say, the best temperature of the melted aluminium bath, the best fluid with which to cover its surface, and the best method of preparing the metallic surfaces which are to be coated with the aluminium has not been stated. Nor has any published statement been made as to the weight of aluminium taken up per square foot of surface and the best treatment of the surface after coating.

Aluminium may be electrically deposited upon any other metal having a higher melting point than aluminium chloride, or cryolite, by passing an electric current through a molten bath of this salt to the metal plate, upon which the aluminium is to be deposited. The deposit obtained hitherto however is powdery and non-metallic in appearance, and probably even if a better looking deposit could be obtained by some modification of this process, the cost would be somewhat high.

Metals have been plated with aluminium by means of a process somewhat similar to the Sheffield silver plating process since a very early date, and it is quite possible that at present, the cost of aluminium having fallen so low, some further application of this process may become commercially successful. Deville, as quoted by Richards, states in 1859, that " M. Sevrard succeeded in 1854 in plating aluminium on copper and brass with considerable perfection. The two metallic surfaces being prepared in the ordinary manner and well scoured with sand, they are placed one on the other and held tightly between two iron plates. The packet is then heated to dark redness at which temperature it is strongly compressed. The veneer becomes very firmly attached and sheets of it may be beaten out. I have a specimen of such work perfectly preserved. The delicate point of the operation is to heat the packet just to that point that the adherence may be produced without fusing the aluminium, for when it is not heated quite near to this fusing point the adherence is incomplete. Experiments of this kind with copper and aluminium foil did not succeed, for as soon as any adherence manifested itself, the two metals combined and the foil disappeared into the copper. In an operation made at too low a temperature the two metals, as they do not behave similarly on rolling, become detached after a few passes through the rolls. Since then the experiments in veneering aluminium on copper with or without the intervention of silver has succeeded very well." Deville stated later in 1862, that Chatel had brought this art to perfection, the veneered plates being used largely for reflectors, etc., in place of silver-plated material.

Dr. Clemens Winckler (" Industrie Blätter," 1873), says: " The coating of other metals with aluminium by the so-called plating method is according to my own experience possible to a certain degree, but the product is entirely useless, every plating requiring an incipient fusing of both metals and their final intimate union by rolling. The ductility of aluminium is however greatly injured by even a slight admixture with other metals, iron makes it brittle and a small percentage of copper makes it as fragile as glass. If, therefore, it were

possible to fuse a coating of aluminium upon another metal an intermediate alloy would be formed between the two metals, from which
alloy all ductility would be gone and which would crumble to powder
under pressure of the rolls, thus separating the aluminium surface
from the metal beneath. But even if it were possible to coat a metal
in this way with a thin plate of aluminium it is doubtful if any
advantage would be obtained. For while compact aluminium resists
oxidising and sulphurising agencies, the divided metal does not. In
powder or leaves aluminium is readily oxidised. In the form of a
coating upon other metals it must necessarily be in a somewhat finely
divided state and hence would probably lose its durability."

With regard to the electro-deposition of gold, silver, and other
metals upon aluminium surfaces, Deville remarks—"that the process
is difficult if good adhesion is desired," he however adds that " M.
Mourey has been very successful in obtaining deposits of gold and
silver upon aluminium surfaces by means of the electric current
passed through baths of aqueous metallic solutions." The nature of
the solutions and details of the process are not however published.
Deville further states that M. Hulot has succeeded in electro-plating
copper upon aluminium from an acid solution of copper sulphate.

Aluminium may also be coated with gold, by the process of simple
immersion, described by Tissier Bros. as follows :—" Eight grammes
of gold are dissolved in aqua regia, the solution diluted with water
and left to digest during twenty-four hours with an excess of lime.
The precipitate with the lime is well washed and then treated with a
solution of twenty grammes of hyposulphite of soda. The liquid
resulting serves for the gilding of aluminium without the aid of heat
or electricity, the metal being simply immersed in it after being
previously well cleaned by the successive use of caustic potash, nitric
acid, and pure water." Very probably analogous methods might be
used to coat aluminium with silver, copper, etc., if such a coating
were desirable.

Aluminium may be coated with gold or silver by a hot plating
process, described by Morin as follows :—" Sheet silver is laid on the
clean aluminium surface, a steel plate is placed over the silver and
the whole is bound into a packet with fine copper wire. Two large
cast iron blocks are heated to a dark red heat, the packet placed
between them, and a pressure of ten tons per square inch is gradually
applied and kept up for fifteen minutes; when removed from the
hydraulic press they can be rolled like silvered copper when brought
to the proper heat. The plating with gold succeeds best if a thin
leaf of silver is slipped between the two sheets of metal, the operation proceeds then exactly as above." Platinum may be plated on
aluminium just as " easily as silver."

Margot's process for coating aluminium with other metals is as follows: — The coating of oxide on the surface of the aluminium, which is the chief cause of any difficulty found in plating this metal, is removed by treating it first by immersion in a solution of an alkaline carbonate and then by rinsing with hot dilute hydrochloric acid (one part of acid to twenty of water), and finally by immersing in a dilute solution of copper sulphate. This last treatment causes the deposition of a film of the copper over and closely adhering to the aluminium surface, and upon this any other metal desired may be satisfactorily deposited by the usual methods.

PART II.—ELECTRO-METALLURGY.

CHAPTER I.

THE ELECTRO-METALLURGY OF COPPER—CHIEFLY HISTORICAL.

Electro-Metallurgy.—The Electric Refining of Copper by Separate Current.—
Dr. Kiliani's Observations on Electrolytic Refining of Copper.—Progress
in Electrolytic Copper Refining up to 1889.—Elkington's Copper Re-
finery.—Wohlwill's North German Refinery at Hamburg.—The Biache
Refinery.—Hilarion Roux's Marseilles Refinery.—The Oker Refinery.—
The Elliott Metal Refining Company's Refinery at Birmingham.—
Electrolytic Refining in America.—Early Attempts at Estimates of Cost
of Refining Copper.

Application of the Term.—The term *Electro-metallurgy*, which
was applied by the late Alfred Smee to the art of electro-deposition of
metals generally, is now more correctly applied to the refining or
purification of metals, and to their separation or extraction from ores
by electrolysis. This important branch of electro-chemistry, the
practical development of which had long been the dream and the hope
of electricians, has during the past twenty years gradually developed
into an art of considerable magnitude, while the great improvements
in magneto and dynamo-electric machines which have been made
within a comparatively recent date have given a stimulus to this field
of enterprise which is likely to render it one of the most important in
its employment of electric machinery. That these great results could
never have been profitably obtained by means of voltaic electricity is
beyond all question.

An early investigation of this subject was made by Maximilian,
Duke of Leuchtenberg, in the year 1847, who proved that impure
copper containing precious metals could be refined so as to yield pure
copper, and leave the precious metals in a condensed form ready for

further treatment. He moreover recognised the great influence which his discovery would eventually have in connection with practical metallurgy. At this time, it must be remembered, electrolytic operations were, with the exception of Woolrich's magneto-electric machine, wholly conducted by means of the current from voltaic batteries, which rendered the following up of this discovery for commercial purposes practically impossible. The introduction of Wilde's magneto-electric machine in 1865 may fairly be taken as the starting-point from which success in this direction became possible as regards the means of obtaining electric power. In the same year Mr. J. B. Elkington introduced a practice process for refining copper by electrolysis, and which, worked by currents from Wilde's successful machines, soon placed the art of refining copper electrolytically upon a sound practical basis. A brief description of this process will be given farther on.

There can be no doubt that the first instance of the application of electrolysis in metallurgy was in the production of what is termed "cement" copper in the wet method of treatment. The drainage water of copper mines is frequently charged with sulphate of copper, due to the oxidation of the sulphide contained in the ore, and it is from these cuprous liquors that the cementation copper is obtained. The wet process is particularly adapted to the treatment of the poorer oxidised ores, especially where fuel is scarce. These ores are treated with acid, either hydrochloric or sulphuric, or with a solution of ammonia, all three of which are good solvents of the oxides of copper. The precipitation is effected in the copper solution by placing iron in it. The action is the result of electrolysis (*Williams*).[*] The cuprous solution being placed in large tanks, fragments of iron are immersed and the copper becomes reduced to the metallic state in the form of a spongy deposit to which the term "cementation copper" is applied. The first patented improvement on the above method, and which may also be considered the first application of a distinctly electrolytic process to the reduction of copper, was due to MM. Dechaud and Gaultier, of France, the patent, dated 1846, being for "Improvements in the extraction of copper from its ores, founded particularly on electro-chemical methods."

Electrolytic Refining of Copper by Separate Current.—This method of obtaining pure copper is now most extensively carried on in various parts of England, on the Continent, and in America, the amount of pure metal annually produced being enormous.

---

* "Mineral Resources of the United States," by Albert Williams. Government Printing Office, Washington, 1883.

I I

One great advantage of this method of refining is that the crude coppers operated upon frequently contain considerable quantities of gold and silver, which valuable metals become entirely removed from the impure metal, and are readily recovered by ordinary refining processes. Another important feature in this system is that when the process is properly conducted the copper obtained is pure—a most important consideration when the metal is required for conducting electricity, as in the case of wire for submarine cables, telegraphs, and the wires employed in the construction of magneto and dynamo-electric machines, and other electrical apparatus. Respecting the presence of gold in copper refined by the ordinary method, or "dry way," we remember the great public excitement that occurred about the year 1844, when it was discovered that the copper coinage of William IV. contained a considerable quantity of gold. So soon as the fact became known, many persons of the Hebrew persuasion became large purchasers of penny pieces, at prices ranging from three-halfpence to twopence each, and many were those who enjoyed the luxury of collecting these coins for the purpose of reaping the advantage of their extra market value. It is well known that by the ordinary refining processes it is practically impossible to extract from copper the gold and silver which not unfrequently exist in this metal in considerable quantities. By the electrolytic method, however, not only are these precious metals recovered, as a natural part of the process, but other impurities—as bismuth, arsenic, iron, manganese, etc.—become separated, and chemically pure copper is obtained, which, from its superior conductivity, realises a higher market value than the best refined copper obtainable in the "dry way."

In the following pages we have given the views and experiences of some of the highest authorities upon the subject of electrolytic refining, from which not only the student, but the practical operator, will glean much that is instructive and useful in this important branch of electro-deposition. A few observations upon the general principles of electro-metallurgy, as applied to the refining of copper more especially, may, however, prove useful to those who have not as yet studied the subject. In the electrolytic process of refining copper, the electrolyte employed is a nearly saturated solution of sulphate of copper, contained in a series of tanks, which are placed in electrical communication with each other by copper connections, as many as forty baths or even more being electrolysed by the current from a single magneto or dynamo-electric machine, which, however, is usually an exceedingly powerful generator of the current, or more properly converter of mechanical into electrical energy.

In most electrolytic copper refining works the anodes consist of cast slabs or plates of *crude* copper, containing not more than from 3 to 4 per cent. of impurities. The cathodes are thin sheets of *pure* copper, presenting the same surface as the anodes. To diminish the resistance of the bath as much as possible, the anodes and cathodes are arranged as close to each other as practicable without danger of coming in contact. When the current passes through the series of tanks, the sulphate of copper solution becomes decomposed, its copper being gradually deposited upon the cathodes, while the liberated sulphuric acid dissolves an equivalent proportion of copper from the anodes, forming sulphate of copper, by which the strength of the solution is kept uniform—that is to say, so far as the impurities of the copper will allow. If pure copper anodes were employed, the solution would keep in a perfectly uniform condition, excepting as regards loss of water by evaporation ; but with impure anodes the bath gradually becomes charged with iron and some other soluble metallic impurities, which in course of time render the bath too foul, if we may use the term, to be further worked, in which case the solution is removed and replaced by a fresh solution of sulphate of copper. The deposit or " mud " which collects at the bottom of the tanks is removed from time to time, and the gold and silver afterwards recovered by the ordinary processes of refining.

In arranging an electrolytic copper refining plant, the resistance of the bath is diminished by increasing the anode and cathode surfaces, by which the cost of the electricity—and consequently of the motive power—is greatly reduced. If, however, this is carried to the fullest extent, it necessarily involves the employment of a costly stock of copper as anodes ; it is therefore preferred by some electro-metallurgists (especially in districts where coal is cheap, or where water power can be obtained) to increase the expenditure of power rather than absorb interest on capital by increasing the quantity of copper in the baths, which would in many cases absorb a large proportion of the profits.

(For a more complete discussion of the questions as to the right area of anode and cathode to be employed in a vat, raised in the above paragraph, as also for an examination of the closely allied and very important question as to the correct value of the current density which should be adopted under any given circumstances in order to obtain the most economical working of a refinery, the reader is referred to the next chapter, in which an attempt is made for, I believe, the first time to place this subject upon a systematic basis.—A. P.)

Dr. Kiliani's Observations on Electrolytic Refining of Copper.—Dr. Martin Kiliani, of Munich, recently published a long paper in the German *Berg- und Hüttenmännische Zeitung*, which in an abridged form is reproduced in *Engineering*,* from which journal we will make the following extracts. Referring to Elkington's process before described, Dr. Kiliani observes, " But however simple this process seems in outline, there are many points which would bring great difficulties to an inexperienced person attempting the use of it, if he desired to get, not only silver and gold, but also a good quality of copper. These points depend on the presence of such impurities as arsenic, antimony, bismuth, &c., and on the necessity of carefully observing certain conditions as to strength of current, composition, and circulation of the solution, &c. Elkington, in his patents, does not deal with those points, and this is perhaps the reason, together with the lack of suitable dynamo machines, why the electro-metallurgy of copper did not make much progress beyond Elkington's works till within the last few years. It is really only during the last decade [more particularly within the past three or four years] that the immense progress of electro-technics has extended also to metallurgy, and enabled great successes to be realised in the working of copper, and opened up the prospect of equal successes in other directions." As to the nature of the process itself, Dr. Kiliani begins by saying that the basis of the whole matter consists in the simple fact that when an alloy of several metals forms the anode in the bath, the electric current does not cause the solution of all the component metals at the same time, but that it makes a selection, and takes one metal after the other in a certain order ; and similarly, when several metals are in solution in a bath, the current selects them in a certain order for deposition on the cathode. A fully satisfactory scientific explanation of these facts cannot be attempted, because the whole matter is even yet too little studied, and the materials for a full explanation have not yet been collected. Even as concerns the order of this solution and deposition there are only full materials published concerning silver, copper, iron, zinc, and lead, and then only so far as concerns some few electrolytes. About those elements which are specially troublesome and important in the metallurgy of copper, arsenic, antimony, bismuth, &c., the published information is very superficial and scanty, and in some instances quite incorrect.

With regard to the selection of the different metals by the current, Dr. Kiliani says that this " takes place, in general, on the principle that as much energy as possible is created (*erzeugt*) and as little energy as possible is consumed ; " that is to say, under conditions that metal will be first dissolved from the anode, the solution of which causes the development of the greatest amount of energy (electro-

---

motive force) ; and that the metal will be first deposited from the solution on to the cathode, the separation of which requires the least consumption of this same energy. A comparative measure of the energy required in these cases is obtained by taking the heat of combination of the metals with oxygen to form oxides or salts. The combination heat of the metals with oxygen to form oxides is taken by the author to form a tabular list in the order in which they are dissolved, as follows:—Manganese, zinc, iron, tin, cadmium, cobalt, nickel, lead, arsenic, bismuth, antimony, copper, silver, gold. Of this list it may be said that all those metals which precede copper, when they are present in the anode metals (not oxides) together with copper, will be attacked by the current before the copper ; whereas silver and gold will only be dissolved after the copper, or if they are present in very small amount, they will fall from the anode as powder, and be found in the " mud " of the bath. In practice this order is fully maintained, and all the above metals dissolve before the copper, and are found in solution in the electrolyte, unless they form insoluble compounds, for example, with lead, when the bath is a sulphate, as in refining copper. When the metals are once in solution, their deposition on the cathode takes place in the reverse order, beginning with gold and ending with manganese. But the correctness of these rules is dependent upon several conditions which must be observed in order that the work may go on in a normal manner. The chief of these conditions concerns the strength of the current, the nature of the electrolyte, the proportions of the metals alloyed together in the anode, and the physical condition of the anode itself.

If the current exceeds a certain strength, all the metals may be dissolved and deposited together. The more neutral the electrolyte is, the more easily will the more electro-negative metals be dissolved, and the more easily will the electro-positive metals be deposited. The same may be said of the electrolyte, the poorer it is in copper solution. If the anode consists of copper containing a large amount of impurities, these will be dissolved more easily than from a copper containing but little impurity. The less dense and compact the anode is, the better the process will go on. This all applies only to copper containing the other metals in the metallic form. If oxides or sulphides are present, the first question is as to their conductivity for the current. Most oxides may be classed as non-conductors under the conditions of the bath, and have nothing to do with the action of the current ; they simply go into the insoluble mud, or are dissolved by the purely chemical action of the electrolyte.

The sulphides are mostly good conductors, but not nearly so good as metallic copper. If, therefore, but a small amount of sulphides is contained in the copper anode, the current will act only on the copper and the sulphides will be found in the mud unacted upon, unless by the acid of the bath : if much sulphide is contained in the copper, the current will be more or less divided between the copper and sulphide, and a portion of the latter will be decomposed, with separation of sulphur. In addition to the above secondary reactions of the bath, there are others, some of which are good and some bad, for conducting the process. The current is always striving to decompose the electrolyte into metal (or oxide) and acid ; whilst the liberated acid is striving to redissolve the deposited metal or oxide. These two forces are always opposed to one another, and under varying conditions either may gain the upper hand. The resolvent action of the acid, in cases where the components of the electrolyte have a strong chemical affinity, may overpower the action of a weak current.

In the case of copper this secondary reaction is not of much importance, copper not being acted upon by dilute sulphuric acid in the absence of air, but still it is quite noticeable in presence of good circulation of the liquor, and more or less access of air in the cathodes. A favourable effect of this secondary action is that any cuprous oxide which may be deposited at the cathode with the copper, owing to weakness of current, is dissolved again.

Respecting the presence of foreign metals and oxides in the anodes, Dr. Kiliani says : " Cuprous oxide, being a bad conductor, is not affected by the current and goes first of all to the mud of the bath. It is then, however, dissolved by the free acid, more or less, according to the time the acid is allowed to remain in the tank. Therefore any cuprous oxide contained in the anodes diminishes the free acid of the electrolyte, and increases the amount of copper in solution." As to sulphide of copper, if it does not exceed in quantity that usually present in " black copper," it deposits in the mud. If there be a high percentage of copper sulphide in the anode, it is decomposed with liberation of sulphur. Gold, silver, and platinum all remain undissolved in the mud when they are not present in considerable quantity, and so long as the electrolyte retains its normal composition as to free acid and dissolved copper. If the liquor becomes neutral the silver dissolves and becomes deposited on the cathode. Bismuth and its oxide go partly to the mud, as insoluble basic salt, and partly into solution, eventually precipitating as basic salt. The presence of metallic bismuth in the anode causes the liquor to become poorer in copper, while the presence of its oxide causes a reduction in the amount of free acid. Bismuth does not become deposited upon the cathode, even when large quantities of the basic salt accumulate in the mud, provided the bath be kept in its normal condition as to copper and acid. Tin dissolves in the bath and after awhile is partly deposited again as basic salt. If the anode contains very much tin, the greater portion remains as basic sulphate, adhering to the anode itself in the form of a deposit of a dirty grey colour while moist, but becoming white when air-dried, increasing rapidly in weight even after long drying at 212° Fahr. : it contains sulphuric acid, and the tin oxide in it is mostly of the variety soluble in hydrochloric acid. The presence of tin, therefore, reduces the amount of copper in the bath without replacing it by any appreciable amount of tin in solution. The tin in solution exercises a surprisingly favourable influence on the copper deposit on the cathode. Copper deposited from a neutral solution of pure copper is rough, irregular, and brittle, but if tin be present the deposits are excellent and tough, even though the deposits give no trace of tin. The resistance of the bath is also much reduced by the presence of tin in the anodes. If arsenic be present in the metallic state, it enters the solution as arsenious acid, and only appears in the mud when the solution is saturated with it. Arsenic in the form of arsenic acid combined with oxide of copper, or other oxides, is at once deposited as mud in neutral solutions, since these oxide combinations are non-conductors. Metallic arsenic thus reduces the amount of copper, and increases that of the free acid in the bath, because it goes into the solution without combining with an equivalent of acid, while at the same time a proportionate amount of copper is deposited with liberation of acid. Arsenic does not enter into the copper deposit in the cathode while the bath remains normal as to copper and free acid ; in a neutral bath, or one in which the copper is insufficient, arsenic is deposited with the copper.

In reference to the above experiments, it must be borne in mind that they were conducted only upon a laboratory scale, at the Technical High School at Munich, and, therefore, should not be accepted as ruling operations conducted on a large commercial scale. We may assume, moreover, without questioning their value as facts, that the results were obtained by means of voltaic batteries, and these we know are not so reliable as dynamo or magneto-electric machines properly constructed to yield large currents of low electro-motive force, suitable for the deposition of copper. In the electrolytic treatment of copper, it is undoubtedly of the greatest importance that not a trace of arsenic, bismuth, or any foreign metal should be present in the deposited copper, since it is well known that even one-fifth per cent. of iron depreciates the conductivity of copper by 25 per cent., while a mere trace of arsenic reduces its conductivity by 66 per cent. The uniform character of the deposit and its absolute purity will depend upon the keen observation of the electro-metallurgist, who, while taking care that the machines he employs yield the exact quality of current necessary for the reduction of copper, will also devote special attention to the condition of his electrolytes, and the removal of the mud as soon as its accumulation in the baths involves risk of partial re-solution.   As far as the nature of the current is concerned, it will be seen, by the foregoing remarks, that the requirements of electrolytic copper refiners have been fully met by those of our electrical engineers who have paid special attention to this subject. There has been much said about the " secrecy " observed at some of the British and Continental works in respect of their electrolytic operations; it must be borne in mind, however, that although the principles of the electrolytic art are common property, *practice* may vary considerably, and special advantages may arise from the suggestiveness or keenness of observation of an expert in one establishment, which, if known to competitors, would reduce their value. These technical advantages often represent extra profit. We need say no more.

Progress of Electrolytic Copper Refining.—It is now about twenty years* since the first practical development of the electrolytic method of refining copper was established by Mr. J. B. Elkington, with the aid of Mr. Henry Wilde's magneto-electric machines, and for many years the process was exclusively practised by this firm.   Since that period, however, the method has been very extensively adopted, not only in this country, but on the Continent, more especially in Germany, Saxony, and France. The North German Refining Works at Hamburg have adopted the electrolytic method for upwards of ten years, the machines of M. Gramme being used. More recently it has been practised at the Oker Foundry, near the mines of Mansfeldt, Germany, with Siemens' large machines ; by MM. Oeschger and Mesdach, at Biache (Gramme machines) ; by M. Hilarion Roux, at Marseilles, (Gramme) ; by MM. Lyon-Allemand, at Paris, and by M. André, at Frankfort (Gramme) ; by the Mansfeldt Mining Company, at Eisleben, and by Messrs. Sterne & Co., at Oker (Wilde's machines). Besides Messrs. Elkington's extensive works at Pembrey, South Wales, where Wilde's large machines are employed, the electrolytic refining process is carried on by Sir Hussey Vivian at Swansea (Gülcher and Crompton machines); by Mr. W. H. Hills at Chester (Gülcher machines) : by Messrs. Williams, Foster, & Co. at Swansea (Crompton machines); Messrs. Charles Lambert & Co. (Gülcher machine); and by the Elliott Metal Refining Company at Selly Oak, near Birmingham, where Wilde's machines are employed. It will thus be seen that the electrolytic method of copper refining is gradually but surely making considerable progress ; indeed, the important improvements which have recently been made in dynamo-electric machines have greatly influenced this result.   From information which has been conveyed to us, we have no doubt that in a very short period the art of electrometallurgy will attain a far greater extent of development, both at

---

* This was written by Mr. A. Watt in 1889.

home and abroad ; and to this end manufacturers of dynamo-electric machines are devoting much attention to the construction of machines specially suited to electrolytic copper refining.

As we have before remarked, there is much difficulty in obtaining information as to the precise methods adopted at the various electrolytic refining works, nevertheless certain points of detail concerning the general system adopted at the respective establishments gradually find their way to the public by some occult means, as is usually the fate of all secret processes sooner or later. The following particulars concerning the principal refineries will give the reader a general idea of the special features of each system of working, and guide his judgment as to which is the most effective and economical. It must not be forgotten, however, that where water-power is not available for driving dynamo machines, the cost of fuel in the various districts is greatly different, often in the proportion of one to four. Besides this, the coppers refined at the various works differ considerably in the nature and proportions of their impurities, by which their conducting power as anodes is greatly influenced, and the cost of electrical energy increased or lessened accordingly.

Elkington's Process of Refining Copper by Electrolysis.— The impure copper, as it comes from the smelting furnace, is cast into slabs or plates, about 18 inches square and $\frac{3}{4}$ inch thick, with lugs projecting from their corners at one end. These plates are placed in troughs, each sufficiently long to take two plates, end to end ; three such rows of plates, or six in all, are placed in each trough, a space of about 6 inches being left between the rows. The lugs of the plates rest on the ends of the trough, and upon a cross-bar fixed midway of its length, to which part strips of copper are attached, and these are all placed in metallic contact with each other. Cathodes of pure thin sheet copper (about $\frac{1}{30}$nd of an inch in thickness) are arranged between the rows of slabs or positive plates, the cathodes being about the same size as the cakes required for the market, or about 12 by 6 inches. There may be four such plates (cathodes) in each row, or sixteen in each trough. These negative plates are each cut with a projecting tongue, by which they are fixed to a frame made of copper rods, the tongue of each plate being lapped round the frame and thus connected and held. The frame has arms at its four corners, which rest on the sides of the trough, on which are copper strips, insulated from the strips at the end of the trough. In this way a series of twenty-five troughs are made, the negative plates of one trough being connected with the positive plates of the next, and so on throughout the whole series, the positive plates being at one end of the series and the negative at the other. Care is taken that all metallic connections are clean and the contacts perfect. The troughs are charged with a nearly saturated solution of sulphate of copper. The negative and positive plates are then connected to the corresponding poles of a large magneto-electric machine, having, say, fifty permanent magnets weighing 28 lbs. each, and fully magnetised. When the positive plates have become so far dissolved and corroded that fragments are likely to fall from them, they are replaced by others, and the old ones recast. The negative plates may be kept in the baths until they are ¾ inch in thickness.

The sulphate of copper solutions are kept at work until they become so charged with sulphate of iron that their further use is inconvenient, when they are changed, and the copper recovered by the usual means. The residue which accumulates at the bottom of the troughs is removed from time to time, and since it frequently contains a considerable percentage of silver, some gold, and also tin and antimony, it has a certain market value, and may be sold to the refiners. Mr. Elkington prefers to work with crude copper known as "blister" or "pimple copper," rather than with that obtained from the earlier stages of the smelting process, which contains higher percentages of impurity.

Copper Refining at Hamburg.—The North German Refinery is under the control of Dr. Wohlwill, who has made the electro-metallurgy of copper his special study. To carry out his system at the above works, six large No. 1 Gramme machines are employed, besides which a still more powerful machine, constructed under his own direction, is used. The principal features of Dr. Wohlwill's method consist in keeping the various baths at a uniform strength, and always at the same temperature, the machines are made to revolve at regular speeds, and are kept in perfect order ; all the coppers to be treated are subjected to analysis both before and after the electrolysis. It is by thus keeping all things equal that he is enabled to produce copper of very fine quality.

The first Gramme machine constructed for Dr. Wohlwill for the chief electrolytic installation at the North German Refinery is provided with two commutators and four brushes : each commutator has twenty sections. The spirals of the bobbins are each composed of seven strips of copper 10 millimetres (0·4 inch) wide and 3 millimetres (or about ⅛ inch) in thickness. Forty groups of copper ribbon correspond to the forty sections of the two commutators, so that each spiral is composed of two identical half spires juxtaposed, and soldered at their extremities to a radiating piece which connects them to one of the sections of the double commutator. The inducted ring is therefore composed of forty partial bobbins, of which twenty are connected to the right-hand side, and a corresponding number to the left-hand side commutator. The total resistance of the armature is ·0004 ohm; when the two parts are joined in parallel this resistance is reduced to ·0001 ohm.

The E.M.F., with a speed of 500 revolutions per minute, is equal to 8 volts for the coupling in series, and to 4 volts for the coupling in parallel. The eight electro-magnets of this machine have iron cores 120 millimetres (or 4¾ inches) in diameter, and 410 millimetres (16·5 inches) in length. On these cores are wound thirty-two turns of sheet copper, corresponding in width to the length of the magnet, and 1·1 millimetre in thickness. The resistance of the eight conductors in one single circuit is ·00142 ohm ; when the magnets are joined in two series their resistance becomes ·00038 ohm. The total resistance of the machine is therefore ·00038 ohm in quantity and ·00182 ohm in tension. The total weight of copper is 1,620 pounds, and that of the entire machine about 49 cwt. The normal output of current of this machine is said to be 3,000 ampères for 4 volts, and 1,500 ampères for 8 volts electro-motive force.

At the North German works there are forty baths arranged in two series of twenty ; the anode surface in each bath is nearly 325 square feet, giving a total of 13,000 square feet of surface for the whole of the baths. The anodes and cathodes, or receiving plates, are arranged at about 2 inches apart. The copper is deposited on the cathodes to the thickness of about ₃₂¹ inch, and at the rate of about 73·3 pounds per hour, in all the 40 baths together, or 1,760 pounds per day of twenty-four hours. The motive power absorbed is about 16 horse power, giving a production of 4·58 pounds of pure copper with the consumption of one horse-power hour. At the same works two other series of baths are employed, the number of which is 120, and these are connected in succession. Each bath is furnished with anodes exposing about 160 square feet of surface, and the entire series of 120 baths has a resistance of 0·1 ohm. The current for these baths is obtained from two No. 1 Gramme machines, connected together in series of 300 ampères, with an electro-motive force of 27 volts. The amount of copper deposited per day of twenty-four hours is 2,000 pounds, at an expenditure of 12 horse-power, or say ½ horse-power per pound of copper per hour. The nature of the electrolyte employed at these works appears to be a secret. It has been affirmed that nitrates are used.

The cost of fuel being an important consideration at Hamburg, Dr. Wohlwill has specially designed his baths to economise motive power as far as possible. After an extensive series of practical trials, he found that considerable advantage was obtained by working with large cathode surfaces, and allowing only thin deposits to take place upon them. In his first installation the rate of deposit is only about 0·005 pound per square foot per hour, or only 0·00012 inch per hour in thickness* ; in the other installations referred to, the deposit is still further reduced, with, as will be seen, a considerable reduction in the

---

* That is, a current density of 1·925 ampères per square foot of cathode.

expenditure of motive power. The value of copper under treatment in one of the installations at Hamburg is said to be equal to about £8,000.

Copper Refining at Biache.—Messrs. Œschger, Mesdach, & Co., of Biache-Saint-Waast, near the English Channel, have an installation in which a large Gramme machine, similar to that constructed for Dr. Wohlwill for the Hamburg works, is employed. Twenty baths are used, from which the daily production of copper is about 800 pounds. The baths are each about 10 feet long, 2 feet 6 inches wide, and 3 feet deep, and are constructed of wood nearly 3 inches thick, and lined with lead. These vats are coupled in series, and are charged with a solution of sulphate of copper maintained at a density of 19° Baumé. Each bath is furnished with 88 anodes and 69 cathodes of equal total surfaces; the anodes, which are 28 inches long, 6 inches wide, and $\frac{1}{3}$ inch thick, are arranged in 22 rows of 4. The cathodes, which are 34 inches long, 7 inches wide, and about $\frac{1}{10}$ inch thick, are placed in 23 rows of 3. The total surface under action represents, therefore, about 10,800 square feet. The copper is deposited upon the cathodes of sufficient thickness to be taken directly to the rolling-mill. The production of copper at these works is about 540 to 800 pounds per day of twenty-four hours, the thickness of the deposit being equal to about ·00012 inch per hour. The silver and gold (if any) is deposited in the "mud" at the bottom of the baths, and this is removed from time to time and washed, and after being dried is fused with litharge or with a reducing agent, the product in the former case being treated as argentiferous lead. When the electrolyte becomes heavily charged with iron and other impurities, it is evaporated and allowed to crystallise.

Copper Refining at Marseilles.—M. Hilarion Roux has an installation at his refinery in Marseilles, in which a No. 1 Gramme machine is employed. There are 40 baths, having a total anode surface of about 10,000 square feet, or about 250 square feet for each bath. There are 115 plates in each vat, each being 2 feet 3 inches long, 6 inches wide, and $\frac{1}{15}$ inch in thickness. Each plate weighs about 26 pounds. The plates are immersed about five-sixths of their length, the anodes and cathodes being placed at a distance of about 2 inches from each other. The total weight of copper under treatment is 54 tons, of which 23 pounds are refined per hour, or about 550 pounds per day, with an expenditure of 530 pounds of coal per day for driving the Gramme machine, which revolves at a speed of 850 revolutions, and absorbs about 5 horse-power. The bath is worked at a density of 16° to 18° Baumé,* and is maintained at a temperature of 25° C. (77°

---

* Or about 18·267 per cent. of sulphate of copper.

Fahr., the deposit being at the rate of 0·0023 pounds per square foot of cathode.*

Electrolytic Refining at Oker.—At these works five large C_1 dynamo-electric machines, furnished by Messrs. Siemens and Halske, of Berlin, are employed, one of which has been at work for upwards of four years. The machines have been constructed to give low internal resistance and great power in ampères, which has been effected by surrounding the armature core and the core of the electro-magnets with copper bars of rectangular section, instead of winding them with wire as in electric-light machines. The armature is provided with a series of bars with the Hefner-Alteneck system of winding; the corresponding bars are joined together by means of large spiral bands, placed on the end of the armature, opposite the commutator. On the anterior face of the core, on the commutator side, the bars are connected to the latter by means of strong copper angle pieces. The armature also consists of a single layer of copper bands coiled by series of seven on each branch, or twenty-eight in all. The bars are insulated by means of asbestos, which, being a non-conductor, allows the machine to become heated without doing harm. Each machine works twelve large vats, and is driven, we understand, by water-power of from four to five horse-power, and deposits 1 kilo-gramme (2·2 lbs.) of copper per hour, or about 300 kilogrammes (6 cwt.) per day.

Electrolytic Refining at Birmingham.—The Elliott Metal Refining Company, of Selly Oak, near Birmingham, employs five large Wilde machines, which refine about ten tons of copper per week. The vats are arranged in five series of forty-eight in each, one Wilde machine being employed for each group of forty-eight baths. The vats are 2 feet 9 inches long, by the same width, and are 4 feet deep; each vat contains 16 anodes, which are each 2 feet long by 6 inches wide, and $\frac{1}{2}$ inch thick, and weigh 26 pounds each. There are only ten cathodes in each vat, each cathode being 1 foot 4 inches long, 22 inches wide, and ·03 inch in thickness, and weighs 2·86 pounds. The total weight of copper per bath is about 450 pounds, and in a series of forty-eight baths about ten tons. The anodes and cathodes are arranged at about $3\frac{1}{2}$ inches apart. The anodes are immersed only to the extent of 20 inches of their length, so that the surface under action, including both surfaces of the anode, is only $1\frac{3}{4}$ square feet, or 30 square feet to each bath. The yield of copper for the forty-eight baths approximates 30 pounds per hour, or about 0·65 pound per bath per hour, which corresponds to a current of 235 ampères. The

---

* This is equivalent to a current density of rather less than 0·9 ampères per square foot of cathode surface.

anodes are replaced every five weeks, and the operation progresses for
156 consecutive hours per week. The temperature of the electrolytic
department is uniformly maintained at 68° F., and the density of the
bath kept up to 16° B. The particulars concerning the improved
Wilde machine adopted at these works are given in Chapter II. of the
1889 edition of this work.

Electrolytic Refining in America.*—It is stated that the Balbach
Refining Works in Newark, N.J., are probably the largest in the
world, the daily production of copper being about six tons. The
current is obtained from four dynamos, furnished by the Excelsior
Electric Company, of Brooklyn, New York. Each of the dynamos is
driven by an independent Westinghouse engine. The three larger
dynamos have an output of 30,000 watts each, while the fourth is
a smaller machine of 15,000 watts capacity, which was put down for
the firm about two years and a half ago, to enable them to ascertain
whether the electrolytic method of refining was remunerative before
embarking in the business upon a large scale. The work goes on day
and night, with a short intermission each day for cleaning and oiling
the engines and dynamos. The foundry for casting the anodes, the
mechanical appliances for handling and transporting them and the
finished plates, are all designed with the object of saving manual
labour as far as possible. There is another large establishment in the
States for the production of electrolytic copper and the separation of
the precious metals, namely the St. Louis (Mo.) Smelting and Refining
Company. Besides the two principal works above referred to, there
are some other less important works, and the great interest taken in
the results obtained shows that electro-metallurgy is making rapid
strides on the other side of the Atlantic, where, of all places in the
world, a successful process is sure to command attention, and when
once taken in hand, pushed forward until it acquires the highest
state of development that enterprise and a due appreciation of the
advantages of labour-saving machinery can bestow upon it.

Cost of Electrolytic Copper Refining.—When it is borne in mind
that the electrolytic method of refining copper is pursued at the various
works under totally different conditions, and that it is evidently not to
the advantage of the respective competitors to make publicly known
the special means by which they secure economy in the cost of their
production, it will be at once seen that any published data in this
connection must be received with caution. It is true that the
capabilities of the leading dynamo-electric machines are well known,
and that a given current will do a certain amount of work; the
coppers operated upon at different refineries, however, vary consider-

* Written by Mr. Watt in 1889.

ably in the nature and extent of their impurities, which of course influences their conductivity; again, the systems of working are different, some employing larger anode and cathode surface than others, while, again, the condition and strength of the electrolytic is varied according to the judgment and experience of each electro-metallurgist, according to the particular metal which he has to treat. That electrolytic copper refining has proved to be a profitable method of obtaining pure copper in large quantities at several large works may be considered as proved, by the fact that the process has been carried on without intermission for a great number of years.

To arrive at an approximation of the cost of electrolytic refining of copper it will be necessary to refer to Mr. Sprague's experiments on the deposition of copper, which were conducted with a bath composed of saturated solution of sulphate of copper, 3 parts ; sulphuric acid, 10 parts, diluted with ten times its volume of water. The source of electricity employed was a Daniell cell, and the current was varied by varying the resistances, so that a given thickness ($\cdot0035$ inch) was obtained in thirty hours as the slowest, and forty-five minutes as the quickest rate. Between a thickness of $\cdot00012$ inch—determined by repeated weighings—and $\cdot00144$ inch deposit per hour,* the deposits were good, up to the limit of the last ; beyond this all quicker rates yielded defective deposits. Mr. Sprague concluded from these results, that the limit of current of 1 ampère to 33 centimetres, or about 5 square inches, could not be exceeded with advantage. This rate, from one to twelve, corresponds to 300 ampères per square metre, or nearly 30 ampères per square foot of anode. These results, however, would not be obtained in practice, from the causes previously indicated (impure anodes, etc.) ; indeed, even under the most favourable conditions and with pure anodes, the deposit of copper in electro-typing seldom exceeds one-third of this rate. The table below shows the thickness of deposit obtained in a working week of 156 hours, including the results obtained by Mr. Sprague and M. Gramme :—

	Inches.
Maximum deposit (chemically pure anodes) .	$\cdot067$
Sprague's results (good deposits) . . .	$\cdot02$ to $\cdot24$
Gramme's results	$\cdot0036$ to $\cdot025$
Hamburg Works	$\begin{cases} \cdot02 \\ \cdot006 \end{cases}$
Biache Refinery	$\cdot02$
Marseilles Works	$\cdot007$
Selly Oak Works, Birmingham . . .	$\cdot06$

The purity of the deposited copper obtained by electrolysis depends upon several important conditions, either of which will greatly influence the character of the deposits ; these are : the strength and

---

* That is, at a current density of from 1·925 to 23·1 ampères per square foot of cathode.

tension of the current; the percentage of impurities in the anodes; the altered condition of the electrolyte after dissolving out impurities from the anodes; the distance between the anode and cathode surfaces; and the temperature and density of the solutions. To these points Dr. Wohlwill appears to have paid special attention, and as a consequence is accredited with producing copper of great excellence. The quality of copper is found to be uniformly pure when Mr. Sprague's limit is not exceeded.

In estimating the cost of electrolytic copper refining,* we have to take into consideration the interest on the capital embarked; the cost of fuel absorbed in driving the dynamos; the cost of labour; the cost of recovering sulphate of copper, etc., from the baths; and the general expenditure of the establishment. Since many refiners are also dealers in the metal, the amount of copper in stock, as anodes, is not of so much consequence, since it is a matter of indifference to them whether it be employed as anodes or otherwise, except in the event of sudden fluctuations in the market prices, when important losses might result. The cost of fuel on the Continent is considerably higher than in Birmingham or Swansea. Upon this point M. Fontaine says: "We can estimate the cost of fuel at 20 francs per ton, although that price would be too high in the case of Birmingham, sufficiently approximate for Hamburg, and quite insufficient for Marseilles. However, as we are only making a comparison, we will maintain an uniform price; it will always be easy afterwards to recalculate the cost, taking as a basis the actual cost of fuel in the locality considered.

"An engine from 4 to 5 horse-power consumes 20 kilogrammes of fuel per hour; the wages of the driver being estimated at 60 centimes per hour, and the necessary expenses of waste, grease, etc., at 40 centimes per hour; the total hourly cost of the motive power is, therefore, approximately 1·60 francs. A 20 horse-power engine consumes 50 kilogrammes of fuel per hour; the driver's wages being about 70 centimes per hour, and the accessory expenses 60 centimes; total 2·30 francs per hour. The cost of maintenance and the wear and tear of the apparatus in use represent a minimum of 10 per cent. of the purchase price; those of the building 5 per cent. The electric conductors which convey the current in the baths do not alter in price. The labour in a factory of forty baths amounts to 75 centimes per hour, or 18 francs per day; it is double this amount in a factory of 120 baths. The general expenditure can be estimated at 100 per cent. of the cost of labour in large installations of 200 or 300 baths, for example, and at 150 per cent. of the cost of labour in installations of only 40 or 50 baths."

---

* For discussion hereon, see the following chapter.

The foregoing figures (which are only approximate) enabled M. Fontaine to compile the following comparative table, which is not to be taken as absolutely correct, but as serving as a basis to a project. The figures "convey an exact idea of the elements which constitute the cost of the electrolytic of refining, and their true interest consists in the comparison which they allow of being established between various factory installations :—

Factories taken as examples.	Expenditure (in francs) per ton of refined copper.					
	Interest on capital.	Motive power.	Main-ten-ance.	Labour.	General expendi-ture.	Total.
Hilarion Roux, Marseilles	78·80	112·00	18	72·00	108·00	388·80
North German Refinery, Hamburg	64·65	39·50	12	40·00	40·00	196·05
Elliott Metal Company, Birmingham . . .	35·95	180·06	30	57·75	57·75	361·45

"The cost of fuel in Birmingham," says M. Fontaine, "is much lower than we have taken as a basis ; but, taking it at 6 francs (5s.) per ton at the works, we find that the motive power still costs 1·20 francs per hour, or 125 francs per ton of copper. If we leave all the other figures unaltered we obtain a total of 306·45 francs, that is to say, a much greater expenditure than at the Hamburg works. The interest on the capital engaged represents a small proportion only of the cost price, whereas at Hamburg it constitutes the main expenditure. As it was easy to foresee, two factories—those of Hamburg and Marseilles—established with the same elements, and on the same lines, give essentially different results in their working, owing to their respective magnitude. At Hamburg, where the operations are conducted on a large scale, the cost price of refining is about 200 francs, whereas at Marseilles, where the works are not of much importance, this cost is nearly doubled. The arrangement of 120 baths in series, and the considerable surface of anodes, is much to be preferred to that of 48 baths of small surface, notwithstanding the enormous capital sunk in the first case. If water-power instead of steam-power were used it would still be necessary, for economically refining the copper, to adopt the disposition in use at Hamburg."

We are quite willing to endorse M. Fontaine's views as to the economical advantage of working with large anode surfaces, no matter from what source the motive power is obtained, and the observations of Keith and others clearly indicate that in this direction is to be found the chief element of economy in electrolytic refining. all other conditions being duly fulfilled. There can be no doubt, however, that a

K K

great deal depends upon the character of the dynamo machines manufactured for this special purpose, and their construction should undoubtedly be based upon the quality of copper which it is intended to refine by their agency. A dynamo that would fulfil all the requirements of electrolysis for one variety of copper would be quite unsuited for refining metal of higher resistance. In taking advantage of this knowledge lies the secret of Dr. Wohlwill's well-known successes.

Respecting the cost of electrolytic refining, the following particulars have been handed to us, from which it will be seen that there is a wide difference when compared with the estimates of M. Fontaine.

Twenty indicated horse-power will deposit 3 tons of copper in 144 hours, the current being generated at a cost of 2 lbs. of coal per horse-power per hour, thus: 20-horse power consumes 40 lbs. of coal per hour; 144 × 40 = 5,760 lb. of coal per 144 hours, or say, less than 3 tons of small coal, which can be purchased anywhere near a coal-pit at 3s. a ton, delivered, or in other districts (as in Birmingham) at, say, 5s. per ton. Thus the total cost of fuel for depositing 3 tons of copper by one large dynamo-electric machine amounts to only about 15s.

An important consideration in the electrolytic method of refining copper is, that the gold and silver—which are often present in considerable quantities in crude coppers—are entirely and easily recoverable, since these metals, during the electrolysis of the impure material, are deposited in the mud at the bottom of the vats, from which they can be readily extracted by the ordinary processes of refining. As evidence of the importance of this process over the dry method of refining copper, by which small traces of gold and silver would not be recoverable—we have been told that in one case, in which a trial sample of 6¾ tons of crude copper were operated upon by the electrolytic method, the mud from the bottom of the baths yielded 5½ ounces of gold and 123 ounces of silver, of the aggregate value of about £54, a sum that would leave a handsome profit after paying the cost of the operation. It is unnecessary to say that in refining this sample of copper by the dry method both the gold and silver would have been practically lost.

Another important consideration in the electrolytic method of refining copper is, that the pure metal obtained, from its high conductivity is invaluable for all electrical purposes, while by its aid dynamo machines may now be constructed of infinitely greater power than would have been practically possible some twenty years ago. In electrotyping, also, the pure metal presents advantages which all practical electrotypists will readily acknowledge. In telegraphy, pure electrolytic copper presents advantages from its high conducting power which cannot well be over-estimated, since even a mere trace of impurity in copper wire considerably reduces its conductivity.

CHAPTER II.

THE COST OF ELECTROLYTIC COPPER REFINING.— CURRENT DENSITY AS A FACTOR IN PROFITS.

Advances in Electrolytic Copper Refining.—Preparing Estimates for Cost of Erection of Refinery.—Cost of Offices, Refinery Buildings, Power Plant, Dynamos, Electrolytic Vats, Electrolyte, Copper Anodes, Stock Copper, Copper Leads, Circulating and Purifying Plant.—Total Capital Invested, Annual Running Costs, Interest on Capital Invested, Depreciation and Repairs, Labour, Melting Refined Copper, Casting Anodes, Fuel, Salaries of Management and Clerical Staff, Rent of Ground.—Importance of Current Density. —Annual Total Profits.—Example of Estimate worked out.—Actual Costs of Electrolytic Copper Refineries.—Further Considerations on Current Density.—Current Density in Copper Conductors in Electrolytic Refineries.

Advances in Electrolytic Copper Refining.—Since the date of the last Edition of this work (1889), an enormous increase has taken place in the amount of copper refined annually by electrolytic processes. Titus Ulke, in a paper appearing in the *Electrical Review* of New York, and reprinted in the *Electrician* (vol. 46, p. 582), states that in eleven of the principal copper refineries in the United States, operated during 1900, the total daily production was estimated to be 516 tons or 188,400 tons annually. The return is made in long tons of 2,240 lbs. In 1897, the world's output of electrolytic copper is stated to have been about 100,000 tons, whilst in 1899 the world's out·put was about 200,000 tons, four-fifths of which was refined in the United States and about one-fifth only in Europe. The *proportion* of the world's output of electrolytic copper which is refined in England is diminishing, notwithstanding the fact that the business of electrolytic copper refining appears to be a very profitable one. This diminution is not due to the fact that less copper is being refined in England, but is caused by the fact that the output of American and European refineries is increasing more rapidly than that of British companies. It is stated that the published annual report of the Boston and Montana Copper and Silver Mining Company, at Great Falls, Montana (which refines 53·6 long tons of 2,240 lbs. each of copper per diem, or 19,296 tons per annum, and produces from the slimes 1,000 ounces of silver and 34 ounces of gold per diem), shows that the sales of copper, gold, and silver produced nearly £1,500,000 and the amount payable to dividend after payment of all expenses was £700,000 (*Electrician*, vol. 44, p. 253), and it may be remarked that if the total value of all the gold and silver extracted (namely under £90,000 per annum) is deducted, the amount available for dividend is still £610,000 per annum. This large profit upon the copper, which works out at £31 12s. 0d. per ton of copper produced and refined, is of course chiefly due to the very high price of copper, at present (1901) over £70 0s. 0d. per ton. The figures given above show that the price per long ton obtained for refined copper was about £73 0s. 0d. It, therefore, appears that the cost of *production* of one ton of *refined* copper by the Boston Montana Company is about £42 per ton, neglecting altogether any profit from the silver and gold present in the copper and obtained as a by-product in the refining process. The extra value of the gold and silver in the copper obtained at these works reduces the cost of producing one ton of copper by about £5. These figures are further confirmed by the Rio Tinto Company's balance sheet, showing that in 1899, an 80 per cent. dividend was paid to its shareholders, whilst large sums were carried forward and to reserve, and J. B. C. Kershaw, remarks (*Electrician*, vol. 46, p. 391), that "this proves the correctness of the opinion that copper can be produced for about £40 per ton."

The Establishment of an Electrolytic Copper Refinery.—The first point to be considered when contemplating the establishment of an electrolytic copper refinery, is the question of the amount of capital available for the undertaking. This important point being fixed the other quantities necessary for making an estimate can be obtained and should be considered in the following order:—

1. Current density required.
2. Area of ground required.
3. Tons of copper to be refined per annum.
4. Profit available per annum.

From the order in which the above quantities are considered it might be imagined that the current density is the ruling factor, and this is undoubtedly the case.

Capital Expenditure.—Each of the four quantities enumerated above will be considered separately further on in this treatise, but in the first place the capital outlay and the annual charges under every head of expenditure in an electrolytic copper refinery must be discussed.

The capital outlay in an electrolytic refinery must be made under the following nine chief headings :—

1. Cost of offices.
2. Cost of refinery buildings.
3. Cost of power plant.
4. Cost of dynamo and switch board.
5. Cost of electrolytic vats.
6. Cost of electrolytic solution.
7. Cost of copper anodes.
8. Cost of copper in stock, with copper leads, etc.
9. Cost of electrolyte circulating and purifying plant.

The annual costs of refining may be considered under the following six chief headings :—

1. Interest on capital invested and depreciation and repairs on plant.
2. Total annual cost of labour required in refinery.
3. Total annual cost of melting refined copper into ingots.
4. Total cost of fuel used for power per annum.
5. Annual salaries of management and clerical staff.
6. Annual rent of ground covered by works.

Cost of Offices.—The cost of offices will to some extent depend upon the output of the works, and if T is the number of tons of copper refined per annum, the cost of the necessary works' offices, may be reasonably represented by the expression :—

$$\text{Cost of offices in pounds} = 150 + 0 \cdot 1 \ T \quad .. \quad .. \quad (1)$$

This would make the cost of offices £250 for a works having an output of 1,000 tons per annum.

Cost of Refinery Buildings.—The area of buildings required for a given output per annum depends upon the current density employed in the works, that is upon the number of ampères per square foot of anode surface, and in fact is inversely proportional to this density. The following data obtained from various sources will enable us to come to a conclusion as to the correct area necessary for a works dealing with an output of T tons per annum,

TABLE I.

PARTICULARS CONCERNING THE AREA COVERED BY VARIOUS COPPER
REFINERIES.

Works.	Authority.	Area square feet.	Output in tons.	Current Density.	Square feet per ton per year.	Square feet per ton per year per unit current density.
Hamburg, North German Refinery	Gore	7,104	330	1·11	21·53	23·68
Oker	,,	861	125	—	6·889	—
Marseilles . . .	,,	3,228	89	0·77	36·27	27·92
Bridgeport, Connecticut . .	,,	12,000	5,720	15 (?)	2·098	31·47 (?)
Stolberg . .	,,	3,498	216	—	16·194	—
Anaconda . . .	*Electrician	141,000	†64,280	10 (?)	2·194	21·94 (?)

In the above table the areas of the works have been taken from
Gore's "Electrolytic Separation of Metals," but the other details,
as to current density, etc., are from Watt's accounts of the same
works. (Watt's *Electro-Deposition*, 1889 edition.)

From the above results it appears that the area of the works required may be reasonably represented by the expression:

$$\text{Area of works in square feet} = 30\frac{T}{D} \quad .. \quad .. \quad (2)$$

Where T = output in tons per annum, and D = the current
density employed in ampères per square foot.

Now if the height of the building is 15 feet, and if it costs four-
pence per cubic foot, the cost of the building over the area $30\frac{T}{D}$

square feet $= \dfrac{30T \times 60}{D \times 240}$ pounds $= \dfrac{30T}{4D}$ pounds. That is, the expres-

sion for the cost of works building is:

$$\text{Cost of works buildings in pounds} = \frac{7·5\,T}{D} \quad .. \quad .. \quad (3)$$

The cost of the works buildings therefore for an output of 1,000
tons of copper per year, and employing a current density of 10
ampères per square foot, is £750.

---

* *Electrician*, vol. 38, p. 144, *et seq.*

† This is not the actual output of the Anaconda Works, but is the
output they are designed to yield.

Cost of Power Plant.—The power required in an electrolytic copper refinery varies directly as the output in tons per annum, and also varies directly as the current density employed.

The following results as to the power required in electrolytic copper refineries are obtained, partly by recalculation, from figures given by Watt, Gore, and others.

TABLE II

PARTICULARS CONCERNING THE POWER PLANT INSTALLED AT VARIOUS COPPER REFINERIES.

Works.	Annual output of copper in long tons.	Output of dynamo in kilo. watts.	Actual H.P. of engines employed at works.*	Calculated B.H.P. of engines required for dynamos.	Current density.	A.	B.
Pembrey	643	38·5	65 I.H.P.	60·362	10	0·01011	0·009387
Hamburg (1)	282·9	11·248	16	17·633	1·925	0·02939	0·03238
,, (2)	321·5	7·263	12	11·387	1·11	0·03459	0·03283
Marseilles	88·41	—	5	—	0·77	0·07345	—
Anaconda Works, 1896†	64,280·	1,620·0	3,000·	2,540·	10 (?)	0·004668	0·003952

In the column marked A in the above table, the figures represent the H. P. (whether indicated, brake or nominal is not stated, except in the details concerning Pembrey), which is actually *installed* at the works, per ton of copper output per annum at unit current density. In the column B the figures given represent the B. H. P. which it would be *necessary* to instal in the works per ton of copper refined per year at unit current density, to obtain the electrical output which it is stated is used. This has been calculated from the kilowatt output of the dynamos, allowing a commercial efficiency of 90 per cent. for the dynamos and 95 per cent. efficiency of transmission for the belt. The dynamos being supposed to be belt driven.

From these results it appears that if gas-engines and gas producer plant is installed, and if the prime cost of supply and erection of this plant is £15 per B. H. P. installed, then the prime cost of the plant may be fairly represented by the following expression:

$$\left.\begin{array}{l}\text{Cost in pounds, including erection}\\ \text{of gas engine and producer plant}\end{array}\right\} = 0\cdot12 \text{ TD} \quad .. \quad (4)$$

---

* With the exception of the Pembrey return it is not stated whether indicated, brake, or nominal H. P is intended.

† See note to Table I. on Anaconda Output.

This amount will amply cover the cost of the necessary power plant. For an annual output of 1,000 tons of copper, and using a current density of 10 ampères per square foot, the cost of power plant would be £1,200.

Cost of Dynamos and Switchboard.—It is quite clear that the output of the dynamos must vary in precisely the same manner as the power it is necessary to instal, and as the electrical power developed may be taken as 85 per cent. of the motive power installed, and as the cost of supply and erection of this plant may be taken at about £11 per kilowatt output, the cost of the dynamos and switch-gear, including erection, may be obtained from the expression :

$$\left. \begin{array}{l} \text{Cost of dynamos and switchboard,} \\ \text{including erection, in pounds} \end{array} \right\} = 0\text{'}056 \text{ T.D.} \quad .. \quad (5)$$

For a works having an output of 1,000 tons per annum, and employing a current density of 10 ampères per square foot, the dynamos would therefore cost £560.

Cost of Electrolytic Vats.—If the anodes and cathodes in an electrolytic vat are packed as closely together as is possible, having due regard to the avoidance of a short circuit, which is the more liable to occur the more closely the anodes and cathodes are permitted to approach each other, it is evident that, however or wherever the vats are set up, provided that they are as closely packed as possible, the yield of copper per cubic foot of vat space must depend only on the current density employed.

The following results obtained from various writers on the electrical refining of copper show how, at the same current density, the output of copper per annum per cubic foot of vat space varies from this fixed value. The variations which are actually observed are due to the greater or less distance between the anodes and the cathodes (the greater the distance the less the copper output per cubic foot of vat space) and the different heights of the lower ends of the electrodes above the bottom of the vats (this greater height is necessitated sometimes in dealing with anodes giving large quantities of mud deposit, and of course reduces the output of the vat per cubic foot. Sederholm states that in American refineries from 1·5 to 2·5 inches space is allowed between the lowest portion of the electrodes and the bottom of the vats. Lastly, the larger the vat the less fraction of the available vat space is lost in clearance between the electrodes and the sides of the vat. The size of the vat must not, however, be too great, or inconvenience in manipulation is found to occur. The smallest annual copper output per cubic foot, at unit current density is, therefore, to be expected from small vats with the electrodes far apart and having the lower ends high above the bottom. Some further remarks upon the most satisfactory size of vat will be subsequently made.

TABLE III.

PARTICULARS CONCERNING VAT SPACE REQUIRED AT VARIOUS
COPPER REFINERIES.

Works.	Authority.	Vol. of each vat in cubic feet.	Number of vats.	Total vat space in cubic feet.	Output of copper in tons per annum.	Current density in ampères per square foot.	Cubic vat, space per ton per year, per unit current density.	Distance between anode and cathodes in inches.
Biache	Watt	75	20	1,500	128·6	1·932	22·54	2·7
Anaconda	*Elect.*, vol. 38	132·4	1,200	158,000	64,280	10 (?)	24·717	2
Elliott	Watt	30·25	240	7,260	578·6	8·888	117·5	3·5
Elkington (Pembrey)	Gore	42	200	8,400	643	10	99·64	2·75
Anaconda	*	140	1,400	196,000	47,250	13	53·93	2·0

From these results it is evident that, with a reasonably satisfactory distance between the anodes, say about 2·7 to 3 inches, a space of 60 cubic feet of vat per ton of copper produced per year, at a current density of one ampère per square foot, should be amply sufficient. The cost of well-made electrolytic vats of pitch pine lined with lead should not be greater than 4 shillings per cubic foot if the vats have a capacity of at least 80 cubic feet, smaller vats cost more per cubic foot, and the amount of capital which should be invested in vats for a given annual output of T tons of copper obtained with a current density D is given by the expression :

$$\text{Cost in pounds of vats} = 12 \, \frac{T}{D} \qquad \cdots \qquad \cdots \qquad \cdots \quad (6)$$

The cost of vats for an output of 1,000 tons of electrolytic copper per annum, if a current density of 10 ampères per square foot is employed, is therefore £1,200.

---

* From paper in the *Electrical World and Engineer*, vol. 37. pp. 186, 187; abstracted in *Science Abstracts*, vol. 4, 1901, p. 423. The output of copper here given is stated to be the *actual* output. The output in the other Anaconda return in this table is that *calculated* from the electrical output of the dynamos installed.

Cost of Electrolyte.—The cost of the electrolyte per cubic foot naturally varies as the cost of copper sulphate. The solution employed contains about 12 lbs. of copper sulphate crystals per cubic foot, and has a density of 17·5° Baumé. Sixty cubic feet of this strength of solution costs, if copper sulphate is £15 per ton, £4 16s. 4d., whilst if copper sulphate costs as at the present date (March, 1901) £25 per ton, the cost of 60 cubic feet of the electrolyte becomes £8 1s. 2d.

The capital it is necessary to put down in electrolyte is given by the expression :

$$\text{Cost in pounds of electrolyte required} = \frac{0\cdot3\ \text{T S}}{\text{D}} \quad \cdot\cdot \quad (7)$$

Where S = price of copper sulphate crystals per ton in pounds. Thus the cost of the electrolyte solution for an output of 1,000 tons of copper per annum, and using a current density of 10 ampères per square foot is £833 6s. 8d. if copper sulphate crystals cost £25 per ton.

Cost of Copper Anodes.—The amount of capital invested in anodes, for a given output of tons of refined copper per annum at unit current density depends upon the thickness of the anodes employed. Of course the thinner the anodes are cast the less amount of capital need be sunk in this form, and it is desirable, therefore, that the anodes shall be as thin as possible down to perhaps 0·25 inch, for the amount of capital invested in this way is usually an important sum. It is interesting to note that a current of 99·54 ampères running day and night for 360 days will deposit an amount of copper weighing one long ton, and the anode *surface* required, if a current density of one ampère per square foot is employed, is therefore 99·54 square feet, or, as both sides of the anode plate are available, a plate having an area of 49·77 or say 50 square feet is required per ton output per year at unit current density. Therefore if the current density is D ampères per square foot of anode surface, and if the output of refined copper is T tons per annum, whilst the cost of anodes is P pounds per ton, and *t* is the thickness of the anode plates in inches the cost of the anodes required is obtained from the expression :

$$\text{Cost of anodes in pounds} = \frac{t\ \text{T P}}{\text{D}} \quad \cdot\cdot \quad \cdot\cdot \quad \cdot\cdot \quad (8)$$

Thus, for example, if the cost of anodes is £70 per ton and the thickness 1·0 inch, whilst the output of the refinery is 1,000 tons per annum and the current density is 10 ampères per square foot, the cost of the anodes required would be £7,000. This is equivalent to 100 tons of copper, that is, the weight of the anodes is 10 per cent. of the annual copper output of the works. At Anaconda the weight of the anodes is 10·85 per cent. of the annual output. (*Elect. World and Engineer*, vol. 37, pp. 186, 187; abstracted in *Science Abstracts*, vol. 4, 1901, p. 423.) It is probable, therefore, that t should be slightly over 1 inch instead of 1 inch as here taken.*

Cost of Copper in Stock, including Cathodes and Copper leads, &c.—Gore states that at least a weight of copper equal to the weight of the anodes employed must be considered as in stock in the form of copper leads, cathode plates, and stock of copper for fresh anodes.

The cost of the copper leads may, however, be estimated with moderate accuracy from the known details of the copper refinery at Anaconda and from the following considerations :—If the output of the refinery and the current density at the anodes is fixed, and the current density per square inch cross-section of the conductors ($= a$) is varied, it is clear that the weight of copper employed in these conductors varies inversely as a. Again, if a is kept constant and the current density D at the anodes is also constant, we may say that the volume of the vats must be increased either (1) by adding more vats of the same size in series, or (2) by setting up a fresh set of the same number of vats in series with a separate dynamo, *i.e.*, duplicating the plant, or (3) by increasing the width of the vats only, their length remaining constant. In all these three cases the weight of copper employed in the conductors must vary directly as the output of copper T. But the volume of the vats may also be increased (4) by increasing the length of each vat only ; it may then be seen that both the cross-section of the copper conductors and their length must be increased in the same ratio, that is the weight of copper employed in the conductors will vary directly as T^2. Lastly, in an intermediate case, (5) if the dimensions of the vats are altered so as to double or treble, etc., their volume and therefore their output, whilst the vats are kept of constant depth and of similar form of plan, then each of the horizontal dimensions of the vats must be multiplied by the square root of the factor by which the output has been multiplied. That is, the length of the conductor must be increased by multiplying it by the square root of this factor, whilst to maintain the current density in the conductors fixed the cross-section of the conductors must be also increased by multiplying them by the factor by which the output is to be increased : this, therefore, means that the weight of the copper

---

* The difficulty with thinner anodes is that they have to be so frequently replaced and the remains of the old ones remelted and recast that the charges for these operations become excessive, and it is therefore more economical to employ thicker anodes, notwithstanding the larger capital thereby necessitated.

employed as conductors must be varied directly as $T^{3/2}$. As for moderately large installations some maximum size of vat is employed for reasons of economy and convenience of handling, and also on account of the cheaper cost per cubic foot of vat and its larger output per cubic foot, it is a reasonable supposition that some form of the first three of the foregoing methods of increasing the vat space will be adopted, and in this case we have the weight of copper employed as conductors varying directly as T.

Lastly the weight of conductors used is independent of the anode current density D, if the output T is kept constant and the current density in the conductors a is also fixed.

The equation for the weight of conductors used along the sides of the vats is therefore of the form $W = K' \dfrac{T}{a}$,

where W is the weight of conductors in tons, and K' is some constant: or if P' is the cost of refined copper per ton in pounds, then :—

The cost of conductors in pounds $= K' \dfrac{P' \, T}{a}$.

I have made an estimate of a probable value of this constant K' from the following details concerning the Anaconda Refinery obtained from a paper in the *Electrician* (vol. 38, p. 144).

At Anaconda a 1,270 kilowatt dynamo of 75 volts and 3,600 ampères runs 200 tanks in series; each tank is 250 cm. long. The total length of tanks is therefore 50,000 cm., and adding one-fifth of this length to allow for inter-connection between the vats and connection to the dynamo, the total length of these leads must be 600 metres each, for the positive and negative, and this length of lead has a cross section of $\dfrac{3,600}{a}$ square inches; its volume is, therefore, $\dfrac{1,200 \times 100 \times 3,600}{2 \cdot 54 a}$ cubic inches, and the weight in tons is $\dfrac{120,000 \times 3,600 \times 62 \cdot 4 \times 8 \cdot 9}{2 \cdot 54 \times 1,728 \times 22 \cdot 40 \times a}$

If it is remembered that 100 ampères per vat yields an output in 360 days of 24 hours each of one ton of copper, we have, using the formula $W = K' \dfrac{T}{a}$ already obtained :—

$$\frac{120,000 \times 3,600 \times 62 \cdot 4 \times 8 \cdot 9}{2 \cdot 54 \times 1,728 \times 22 \cdot 4 \times a} = K' \frac{200 \times 36}{a}$$

Whence K' = 3·39.

Thus if W = the weight of copper in tons used as conductors, T = the annual output in tons of the refinery, and a is the current density per square inch cross section employed in the copper conductors, we have :—$W = 3 \cdot 39 \dfrac{T}{a}$, and if P' is the price per ton in pounds of the refined copper of which the leads are made, then the capital sunk in leads is found from the equation:

$$\text{Prime cost of copper conductors in pounds} = 3\cdot39 \ \frac{P'\,T}{a} \quad \cdot\cdot \quad (9)$$

Thus if $P' = \pounds75$ per ton and the output of refined copper per year is 1,000 tons, whilst the current density in the copper conductors is 250 ampères per square inch, the prime cost of the necessary copper leads would be $\pounds1,017$.

The above estimate only, however, includes the cost of the two positive and two negative copper mains running the whole length of the vats and allowing an extra length of $\frac{1}{5}$th of the total length of the vats for inter-connections of vat to vat and to the dynamo, but it does not include the cost of the cross-bars on the vats from which the anodes and cathodes hang, these cross-bars may be conveniently made of flat iron bars as at the Anaconda Works or old steel rails of small scantling may be used. The cost of this part of the conductors may, therefore, be reduced to a very small amount, and will probably be amply covered by allowing about a fifth of the cost of the copper in the main conductors. The total capital invested in conductors, whether of copper or iron, may be taken as being fairly represented by the expression:

$$\text{Cost of copper and iron conductors} = 4\cdot05 \ \frac{P'\,T}{a} \quad \cdot\cdot \quad \cdot\cdot \quad (10)$$

And thus for 1,000 tons output per annum at a current density in the copper of 250 ampères per square inch, the copper costing $\pounds75$ per ton, would be $\pounds1,215$.

If for the remaining copper in stock for casting fresh anodes, and for the cathodes a third of the amount actually invested in the anodes is allowed, a sufficient provision for copper will have been made. At Anaconda the anodes are renewed in each vat every 34 days, i.e., about one·tenth of the total weight at a time. (*Electrical World and Engineer*, vol. 37, pp. 186, 187.)

Collecting all the amounts of capital for copper anodes, cathodes, stock anodes, and copper and iron conductors together we obtain the expression for the capital invested in these items:

$$K = 1\cdot3 + \ \frac{t\,TP}{D} \ + 4\cdot05 \ \frac{P'\,T}{a} \quad \cdot\cdot \quad \cdot\cdot \quad \cdot\cdot \quad (11)$$

If the thickness of the anodes is 1 inch, and the anode copper costs $\pounds70$ per ton, whilst the output of the refinery is 1,000 tons per annum, and the current density at the anodes is 10 ampères per square foot, whilst the current density in the copper conductors is 250 ampères per square inch, and the copper of which they are made costs $\pounds75$ per ton, the total amount of money invested in the above-mentioned items of anodes, cathodes, stock anodes, and the current carrying conductors from formula (11) £10,548. This amount is half as much again as the amount of capital invested in anodes alone, namely, £7,000, and this is lower than Gore's estimate that the total copper in stock must weigh at least double the amount employed as anodes. Probably if it were desired to reap as far as possible the advantages of a low market in which to buy anodes it might be advisable to hold rather more copper in stock. The question of the rise and fall in price of the copper, however, as it is practically incalculable, has been excluded from consideration in the present investigation.

Cost of Circulating and Purifying Plant for Treating Electrolyte.—The amount of the capital invested under this head is not large per ton output per annum, and in the absence of any very reliable details on this point I have supposed it to be represented by a quantity proportional to the output, but independent of the current density, which must very closely represent the true facts of the case, and the following expression may be employed to calculate out this quantity:

$$\text{Prime cost of circulating plant in pounds} = 0.4\, T \quad \cdots \quad (12)$$

That is, for a refinery having an output of 1,000 tons per annum the cost of the circulating plant would be £400.

Total Capital Invested.—From the above discussion of details, and adding up the items of capital expenditure given in the formulæ from (1) to (12), we obtain the following expression for the total amount of capital, K, invested:

$$\text{Capital invested in pounds} = 150 + \left(4.05\,\frac{P'}{a} + 0.5\right) T + 0.176\, TD$$
$$+ (19.5 + 0.3\, S + 1.3\, t\, P)\,\frac{T}{D} \quad \cdots \quad \cdots \quad \cdots \quad \cdots \quad (13)$$

where
P' = price of refined copper per ton in pounds
P =   „   anode   „   in pounds
T = output of refined copper per annum
D = current density in ampères per square foot at anodes
S = price per ton of copper sulphate crystals in pounds
t = thickness of anode plates in inches
a = ampères per square inch allowed in the copper current conductors,

Or if in a given refinery
P' = £75 per ton
P = £70 „
T = 1,000 tons per annum

D = 10 ampères per square foot
S = £25 per ton
t = 1·0 inch
a = 250 ampères per square inch

Then cost of offices	=	£250	..	..	$150 + 0·1\,T$
„ buildings	=	750	..	..	$7·5\,\dfrac{T}{D}$
„ power plant	=	1,200	..	..	$0·12\,TD$
„ dynamos	=	560	..	..	$0·056\,TD$
„ electrolytic vats	=	1,200	..	..	$12\,\dfrac{T}{D}$
„ electrolyte	=	834	..	..	$0·3\,\dfrac{T}{D}$
„ anodes	=	7,000	..	..	$\dfrac{t\,TP}{D}$
„ conductors	=	1,215	..	..	$4·c5\,\dfrac{P'T}{a}$
„ stock copper	=	2,333	..	..	$0·3\,\dfrac{t\,TP}{D}$
„ circulating plant	=	400	..	..	$0·4\,T$
Total capital invested ..		£11,225		..	K

Annual Cost of Refining.—The next step in our investigation is the examination of the annual costs of running an electrolytic copper refinery. The items of these costs, as has already been pointed out, are :

1. Interest on capital invested and depreciation and repairs on plant.
2. Total annual cost of labour required in refinery.
3. Total annual cost of melting the refined copper into ingots.
4. Total cost of fuel used for power per annum.
5. Annual salaries of management, engineer, clerical staff, etc.
6. Annual rent and taxes of ground covered by works.

Interest on Capital Invested and Depreciation and Repairs of Plant.—The capital invested in copper may be charged for at the rate of p_2 per cent. There is no depreciation in this material except in so far as the market price may vary, but this may either rise or fall, and cannot be very readily taken into account.

If, therefore, the capital invested in copper, and copper sulphate is denoted by K_2, the annual charge for interest on this account will be $p_2\,K_2$.

The remaining capital is invested in material which will depreciate and require repairs. If K_1 is the amount of capital thus invested, and if p_2 is the rate of interest charged on it, whilst p_1 is the percentage rate charged on it for repairs and depreciation, then the annual amount of these charges will be $(p_1 + p_2) K_1$.

The annual charge for interest on capital invested in copper and electrolyte expressed in pounds is therefore :—

$$p_2 K_2 = p_2 \left(1\cdot 3 \; \frac{t \, T \, P}{D} + 4\cdot 05 \frac{P' \, T}{a} + 0\cdot 3 \; \frac{T \, S}{D} \right) \quad .. \quad .. \quad (14)$$

where p_2 is the percentage rate of interest charged on the capital K_2 invested in copper and electrolyte.

The annual interest upon and depreciation and repair charges upon the capital sunk in the remaining plant is in pounds :—

$$(p_1 + p_2) K_1 = (p_1 + p_2) \left\{ 150 + 0\cdot 5 \, T + 0\cdot 176 \, T \, D + 19\cdot 5 \frac{T}{D} \right\} \quad .. \quad (15)$$

Where p_2 is the percentage charge for interest on the capital K_1 sunk in plant and machinery, and p_1 is the percentage charge for depreciation and repairs on the same capital.

The total capital charges to cover interest, depreciation, and repairs are therefore $p_2 \; (K_1 + K_2) + p_1 K_1 = p_2 K + p_1 K_1$. Where K is the total capital sunk in the refinery.

Total Annual Cost of Labour.—The charge for labour may be reasonably fixed from the following considerations :—

At the Anaconda Works (*Electrician*, vol. 38, p. 147), where labour is expensive, and is therefore economised in as much as possible, 25 men deal with an output of 25 tons of copper per diem, and this is in the most recently erected portion of the works, where labour-saving devices are used at every available opportunity. In the older portion of the Anaconda Works, built at an earlier date, 50 men were required to deal with the same output of copper. The average wages of each man at Anaconda is said to be twelve shillings and sixpence per diem, say £4 7s. 6d. per week, which is, of course, extremely high. It is not stated whether this is the complete night and day shifts, *i.e.*, 50 men during the day and 50 during the night. But in the former case the labour would cost £1 5s. per ton of copper, and in the latter double this, or £2 10s. It is therefore pretty clear that 50 men only are employed to run both shifts. In the returns made by the Anaconda Mining Company for the cost of their electrolytic refinery during the year 1897-1898, it is stated that one man at £4 per week is employed to every 140 tons of copper output per annum, that is, the cost of labour is as much as £1 9s. 8d. per ton.

Gore states ("Electrolytic Separation of Metals," p. 223) that probably about 20 men working day and night can completely deal with about 30 tons per week. This, evidently, is meant to represent two shifts of 10 men each, and under these circumstances at £1 0s. per week per man the cost of labour would work out at £0 13s. 4d. per ton of copper produced ; for if this return is meant to represent two shifts of 20 men each, then the cost of labour per ton works out at £1 6s. 8d., which is nearly as high as is the case in the American Anaconda Works, where labour is very expensive. Probably the following expression will give a satisfactory value for the cost of labour required per annum.

$$\text{Cost of labour in pounds per annum} = 0 \cdot 6 \, w \, T \quad \bullet \bullet \quad (16)$$

When w represents the weekly wages in pounds, and T is the number of tons output of refined copper per annum. Thus, for an output of 1,000 tons per annum, and with wages of £1 per week, the annual labour bill will be £666.

Annual Cost of Melting Refined Copper.—The cost of melting the refined copper down into ingots for sale is stated at Anaconda to be as much as 16s. 8d. per ton (*Electrician*, vol. 38, p. 147), but the cost of fuel is high, being £1 2s. 11d. per ton, and labour is, as already stated, very high. It is therefore probable that 10s. per ton will, in most cases, be an ample amount to cover this charge. As, however, this charge varies with the cost of fuel and the price of labour, and as it moreover forms a fairly large percentage upon the total cost of refining a ton of copper, it would be desirable to express the quantity in terms not only of the annual output of refined copper, but also in terms of the cost of fuel per ton and the price of labour.

If m is the cost of melting one ton of the copper, it is probable that the value of m expressed in terms of w, the weekly wages per man in pounds, c the price of fuel in pounds per ton, T the annual output of copper in long tons, would be found from an equation of the form $m = b' \, c \, w - b'' \, T$, when b' and b'' are constants.

However, as I am without sufficiently definite data on the values of the constants involved in such an expression, I limit myself to the following somewhat unsatisfactory estimate.

$$\text{Annual cost of melting refined copper in pounds} = 1 \cdot 25 \, m \, T \quad \bullet \bullet \quad (17)$$

where m is the cost of melting one ton, and the factor $1 \cdot 25$ is to allow for remelting scrap anodes, etc. That is, for an output of 1,000 tons per annum, when m is ten shillings, the charges under this head would be £625.

Cost of Fuel used for Power per Annum.—The cost of fuel employed in the power plant simply depends upon the B. H. P. to be developed, and it is therefore only necessary to calculate out the weight of coal required to develop the B. H. P. employed, in order to obtain a

L L

given output of copper per year at a given current density. The necessary B. H. P. has already been found to be $\dfrac{0\cdot12\ T\ D}{15}$ (see p. 503), and as in a gas-generator plant each B. H. P. hour developed requires always less than 2·5 lbs. of coal, the total weight of coal employed per annum is under 0·07714 T D tons, and if the coal costs c pounds per ton, the annual cost of coal is obtained from the expression :

Cost of fuel used for power per annum $= 0\cdot0771\ TDc$ in pounds .. (18)

That is, the cost of fuel per annum for an output of 1,000 tons of refined copper, at a current density of 10 ampères per square foot of anode, with coal costing £1 per ton, would be £771.

Annual Salaries of Management, Engineer, and Clerical Staff.—These salaries are independent of the current density employed, and may, I consider, be reasonably expressed by the formula :

$$\left.\begin{array}{l}\text{Salaries of management, engineer}\\ \text{and clerks in pounds per annum}\end{array}\right\} = 400 + 0\cdot136\ T \quad .. \quad (19)$$

Thus, with an output 1,000 tons, the annual disbursements under this head would be £536.

Annual Rent and Taxes of Ground Covered by Works.—This annual charge depends upon the area of the works, and the rent per unit area. We have already seen that the requisite area of the works is $30\dfrac{T}{D}$ square feet, and if R is the annual rent and taxes in pounds per square yard, the annual charge under this head is given by the expression—

$$\text{Annual ground rent and taxes in pounds} = 3\cdot33\ \dfrac{TR}{D} \quad .. \quad (20)$$

Thus, if the annual output is 1,000 tons at a current density of 10 ampères per square foot, of anode, and the rent and taxes is as much as 6d. per square yard, that is, £121 per acre, the annual charge for rent and taxes would be £8 6s. 6d. The amount for rent and taxes, even if it is ten times as large as the value here adopted, is a relatively small charge on a ton of copper produced.

The Importance of Current Density.—Having now discussed in some detail the various disbursements which it is necessary to make when conducting an electrolytic copper refinery, we are in a position to observe the great influence which the current density employed has upon the capital and annual outlay per ton of copper refined per year. There is, in fact, some particular current density which will yield the most satisfactory financial results, and this particular best density varies with the price of coal, the cost of melting down each ton of refined copper into ingots, the rate of wages paid, the rate of interest paid upon the capital invested, and lastly, but most important, upon the difference in price *upon the works* between the price paid for anode copper and the price obtainable for the refined copper, together with the value of the silver and gold separated from it (if this value goes to the refiner, which is not always the case).

The following table gives the current density in ampères per square foot of anode surface, as employed at different copper refineries in America, England, Germany and France :—

TABLE IV.— Particulars concerning Current Density employed at Various Copper Refineries.

Works Proprietor.	Location.	Authority.	Current density in ampères per square foot of anode surface.
Wohlwill .	Hamburg, 1st installation	*Watt .	1·925
Wohlwill .	Hamburg, 2nd installation	*Watt .	1·11
Œschger, Mesdach & Co.	Biache	*Watt .	1·932
Hilarion Roux .	Marseilles . . .	*Watt .	0·77
Elliott . .	Selly Oak, Birmingham	*Watt .	8·888
Elkington .	Pembrey (1) . . .	†Gore . .	10·0
Elkington .	Pembrey (2) . .	†Gore . .	8·0
———	Anaconda, U.S.A.	‡Philip .	10 to 15

Two things are evident from these figures, which are, first, that the density employed at different works has varied a good deal ; and, secondly, that although the density employed varies thus over a considerable range, in no case is it over 15 ampères per square foot. In this connection, Mr. S. O. Cowper Coles makes the statement, that " whereas formerly currents of only two to four ampères per square foot were thought permissible, to-day current densities of from 15 to 20 ampères are said to be employed in some establishments, and

* These results are calculated from data given by Watt, " Electro-Deposition of Metals," 1889 edition.

† The density is stated as being employed by Gore, " Electrolytic Separation of Metals," p. 204, 1890.

‡ Calculated from details of these works given in *Electrician*, vol. 38, p. 144, 1896.

the end is not yet in sight " (*Journal of the Institution of Electrical Engineers*, vol. xxix., p. 260, 1900). I am not, however, acquainted with the details of any electrolytic copper refinery, in which a greater current density than about 15 ampères per square foot has been *successfully* employed over any length of time, and I am strongly of the opinion that it cannot pay to employ a greater current density than 16 ampères per square foot at the outside, *if the product of the refinery is merely pure refined copper, and not (as in the Elmore process) some manufactured, or partly manufactured, article of copper, such as sheets, tubes, wire, etc.*

There is a delusive plausibility in the statement that a largely increased current density makes the output of a given plant proportionally greater, and thereby increases the profits obtainable in an equal degree. People advancing such views lose sight of the fact that to double the current density, and therefore the output, the power plant and dynamos must be increased to four times their original size at least.

In order to more clearly bring out the effect which the alteration of the current density employed in electrolysis has upon the cost of refining each ton of copper, let us imagine that a works which has an output of 1,000 tons of copper per annum at a current density of 10 ampères per square foot of surface of anode, alters its density to several other current densities, both smaller and greater than 10. Let us then accept the foregoing figures calculated out on a basis of this output and current density, and then modify them by simply inserting some value of current density other than ten, in the respective formulæ given, and from these calculate the costs of refining one ton of copper at the various current densities selected. It must be remembered then that no factor but the current density is varied in the following calculations. But of course as the refinery is supposed to have the current density varied, the costs of the offices, refinery buildings, electrolytic vats, and rent and taxes remain fixed, whilst the power plant, dynamos, and circulating plant must be increased for an increase in expenditure as the current density is raised, but cannot be reduced when the current density is lowered.

From the results given on this table it is evident that the cost of refining one ton of copper under the particular conditions stated will be least when the current density at the anode is somewhere about 15 ampères per square foot. The direct calculation of the precise value of the most advantageous current density will be dealt with in the following paragraphs.

Annual Total Profits.—The annual total profits obtainable from any copper refinery may evidently be expressed by the following equation (p. 518) :—

TABLE V.—Effect of Varying the Current Density at a given Copper Refinery upon the Available Profits.

The Current Density originally employed at the refinery is supposed to be 10, and the output at this density 1,000 tons of refined copper per annum.

			5	10	15	20
	Current density at anode ..	D	5	10	15	20
K_1	Cost of offices	$150 + 0·1$ T	250	250	250	250
	Cost of refinery buildings . .	$7·5 \dfrac{T}{D}$	750	750	750	750
	Cost of power plant	$0·12$ T D	1200	1200	2700	4800
	Cost of dynamo and switch-board	$0·056$ T D	560	560	1260	2240
	Cost of electrolytic vats . . .	$12 \dfrac{T}{D}$	1200	1200	1200	1200
	Cost of circulating plant . . .	$0·4$ T	400	400	600	800
	Capital invested in plant other than metal and electrolyte	K_1	4360	4360	6560	9640
K_2	Cost of electrolyte	$0·3 \dfrac{T S}{D}$	834	834	834	834
	Cost of copper anodes	$\dfrac{t\,T\,P}{D}$	7000	7000	7000	7000
	Cost of copper and iron conductors	$4·05 \dfrac{P'\,T}{a}$	608	1215	1824	2430
	Cost of copper in stock . . .	$0·3 \dfrac{t\,T\,P}{D}$	2333	2333	2333	2333
	Capital invested in metal and electrolyte	K_2	10775	11382	11991	12597
	Total capital charges . . .	$K_1 + K_2 = K$	15135	15742	18751	22637
	Interest, repairs, and depreciation on capital K_1 at 15% ($p_1 = 9\%\ p_2 = 6\%$) . . .	$K_1(p_1 + p_2)$	654·0	654·0	984·0	1446·0
	Interest only on capital K_2 at 6%	$K_2\,p_2$	646·5	682·9	719·5	755·8
	Annual cost of labour	$0·6\,w$ T	333·3	666·6	999·9	1332·0
	Annual cost of melting copper .	$1·25\,m$ T	312·5	625·0	937·5	1250·0
	Annual cost of fuel for power .	$0·0771$ T D c	193·0	771·0	1737·0	3084·0
	Salaries of management, &c. .	$400 + 0·136$ T	468·0	536·0	604·0	672·0
	Annual rent and taxes . . .	$3·3 \dfrac{T R}{D}$	9·0	9·0	9·0	9·0
	Total annual charges	X	2616·3	3944·5	5990·9	8548·8
	Tons of copper refined annually	T	500	1000	1500	2000
	Cost of refining per ton when 6% is paid on all capital .	$\dfrac{X}{T}$	5·232	3·944	3·993	4·274
	Cost of refining per ton if no interest is paid on capital, but only 9% is charged on the capital K_1 to cover depreciation and repairs .	N	3·415	2·999	3·251	3·607
	Profits available per ton when $P'' - P = £5$.	$5 - N$	1·585	2·000	1·749	1·393
	Total profits under same condition	$(5 - N)$ T	792·5	2000·6	2623·5	2786·0
	Available percentage profits on total capital under same condition	$\dfrac{(5 - N)\,100\,T}{K}$	5·237	12·71	13·99	12·30
	Capital outlay on plant per ton output	$\dfrac{K_1}{T}$	8·732	4·360	4·373	4·820
	Capital outlay on metallic copper and electrolyte per ton output	$\dfrac{K_2}{T}$	21·55	11·382	7·994	6·298
	Total capital outlay per ton output	$\dfrac{K_1 + K_2}{T}$	30·282	15·742	12·367	11·118

Profits per annum in pounds =

$$\text{T} (\text{P}'' - \text{P}) - p_2 (K_1 + K_2) - p_1 K_1 - 0\cdot6\ w\ \text{T} - 1\cdot25\ m\ \text{T}$$
$$- 0\cdot0771\ \text{T}\ \text{D}\ c - 400 - 0\cdot136\ \text{T} - 3\cdot3\ \frac{\text{T R}}{\text{D}} \qquad .. \qquad .. \qquad (21)$$

Where $K_2 = 150 + 0\cdot5\,\text{T} + 0\cdot176\,\text{T D} + 19\cdot5\,\dfrac{\text{T}}{\text{D}}$ see equation (15).

And $K_1 = 4\cdot05\,\dfrac{\text{P' T}}{a} + \dfrac{0\cdot3\,\text{S T}}{\text{D}} + 1\cdot3\,\dfrac{t\,\text{T P}}{\text{D}}$ see equation (14).

If these values of K_1 and K_2 are inserted in the formula (20a), and if it is remembered that $\text{T} = \dfrac{\text{A D}}{30}$, see equation (2), we can obtain an expression for the total annual profits expressed in terms of D, and if this expression is then differentiated with respect to D, and the result equated to zero, a solution for D can be obtained, which gives the value which the current density should have under the given conditions, in order that the available profits on the undertaking shall be a maximum. The particular most profitable value of D depends upon almost all the quantities which we have so far considered. This algebraical solution for D does not, however, give a simple formula, and for this reason the problem of determining the best current density to employ under given conditions has been attacked in a more direct manner. In order to deal with this question it is, however, first necessary to consider

The Preparation of Estimates for, and Design of, an Electrolytic Copper Refinery.—From the various equations developed above, it is now possible to readily prepare estimates for the establishment of electrolytic copper refineries upon a sound basis, and to forecast with considerable accuracy what the profits of working will be.

Let us then imagine that it is proposed to invest, say, £10,000 as capital in an electrolytic copper refinery, and it is desired, as a preliminary, to decide upon the size and cost of the works which should be built, the necessary ground area, the output in tons of refined copper per year and the probable profits which are to be looked for if the venture is carried out.

Firstly, it is necessary to fix upon the most favourable locality, and this must be so chosen that the difference of price between the cost per ton of the refined and the unrefined copper, including the cost of gold and silver present, is a maximum, whilst the cost of fuel and labour are as small as possible. The condition that freightage for markets should be as low as possible is included in the condition that the difference between spot values of the refined and unrefined metal per ton shall be a maximum.

Having then selected our locality we can with very close accuracy know the values of the quantities involved in our formula, namely, cost of coal per ton, cost of labour per man per week, cost of melting one ton of copper, and the difference between the spot values of the refined and unrefined copper per ton. Thus let $K = £10,000$, $P' = £80$, $P = £68$, $P' = £72$, $S = £25$, $t = 1$ inch, $a = 250$ ampères per square inch, $w = £1$, $c = £1$, $m = £0.5$, and let D have successively all values between one and twenty-five ampères per square foot, and let $p = \frac{3}{100}$.

The first step then is to calculate out the value of the area of land required for each current density employed, and this may be done from the formula

$$A = \frac{30\,(K - 150)}{4\cdot05\,\dfrac{P'\,D}{a} + 0\cdot5\,D + 0\cdot176\,D^2 + 19\cdot5 + 0\cdot3\,S + 1\cdot3\,t\,P} \quad \text{.. (22)}$$

This equation (in which A is given in square feet) is obtained from equation (13), the value of T being replaced by $\dfrac{D\,A}{30}$ to which it is equal. See equation (2). The results of these calculations are given in Table VI., column 2 (see overleaf).

The next step is to determine the output of the works in tons of refined copper per annum by means of equation (2), namely, $T = \dfrac{D\,A}{30}$ The results of these calculations for each current density from 1 to 25 are given in Table VI. (see overleaf), lines 27 and 22.

Having thus obtained the output of the works T for each current density employed, we may proceed to calculate out all the various items of capital and annual expenditure by means of the formula already developed in equations (3) to (20). This has been done and the results arranged in Table VI. Finally the annual total profits are calculated out from equation (21), and are given in line 23 of Table VI., when $P'' = £72$, and in line 25 when $P'' = £73$. The percentage profits under each of these conditions are tabulated in lines 24 and 26.

The way in which the percentage of profits varies with the current density is also shown in Fig. 127a in the form of two curves plotted with percentage profits as ordinates and current density as abscissæ. One curve shows the variations of percentage profits with the current density, when the difference between the price of the refined and the anode copper is £4 per ton, and the second when this difference is £5 per ton. In both cases, however, the cost of the anode copper is £68 per ton. The curves, in fact, represent the numerical results in lines 24 and 26 of Table VI. plotted as ordinates, with the current densities in line 1 plotted as abscissæ.

Actual Cost of Electrolytic Copper Refineries.—The foregoing results given in Table VI., and on the curves in Fig. 127a, have been calculated out chiefly in order to indicate the precise line in which the problem of the best design of an electrolytic copper refinery for the establishment of which in a neighbourhood a definite amount of capital is available, should be attacked. Whatever opinion may exist as to the values of the constants adopted, the *form* of the expressions must be closely in accord with the truth, and in my opinion the constants selected cannot be largely in error.

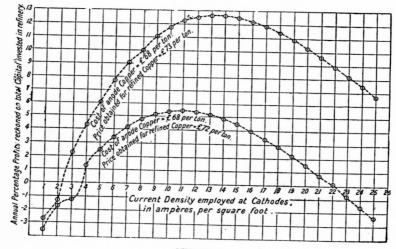

CURVE I.

Showing the variation of the profits obtainable in an electrolytic copper refinery in which a capital of £10,000 is invested when the current density at the cathodes is varied. The details of the various constants involved are given at the head of Table VI. The two above curves are in fact a graphical representation of the two lines 24 and 26 on Table VI.

Fig. 127a.

Besides the general reasoning given in the earlier portion of this chapter, there are two other pieces of independent evidence which point to a considerable accuracy in the method of calculation adopted. Firstly, the results obtained in Table VI. show in line 28 that the smallest cost of electrolytically refining a ton of copper is just over £3, and this is precisely the amount which it is stated to cost from results obtained in the best manufacturing experience (see p. 561).

Secondly, F. R. Badt has published estimates for an electrolytic copper refining plant to give an output of 5,357 long tons of copper per year (*Electrical Engineer* of New York, vol. 13, 1892, p. 598), and he finds that £28,900 would be required for the plant installed, not including the copper anodes and stock copper. Whilst if the current density is assumed to be 12 ampères per square foot of anode (this current density is selected because it is that most generally employed in America, for which country F. R. Badt's estimates are made) it is found that the total capital required as calculated out by the foregoing formula is £29,460. Badt does not state how his results were calculated out, but presumably they represent his experience of some of the better-known American refineries. The results as obtained by Badt, and by calculation by the formulæ here developed, do not differ by 2 per cent., a striking result when one considers that they are quite independently obtained.

I place Badt's estimates and my own side by side :—

Badt's Estimate. (Output of 5357 long tons per annum.)			Calculated from Formulæ $(D = 12 \quad T = 5357)$.		
	£	£		£	£
Buildings	6000		Refinery buildings	3359	
Pavement	400	6800	Office buildings	685·7	4044·7
Rails and travelling tackle	400				
Pipes	800				
Vats	1500		Vats	5357	
Sheet-lead lining	5600	8400	Circulating plant	2142·8	7499·8
Lead burning	300				
Steam injector	200				
Dynamos	6000		Dynamos	3613	
Steam engine and shafting	4600		Power plant	7741	
Electrolyte	900		Electrolyte	3732	
Conductors	2200		Conductors	2830	
Total	£28,900		Total	£29,460·5	

The estimate which Badt made for the necessary anodes and stock copper was only £16,000 and as copper in 1892 had an average value for G.M.B. brands of about £45·5 per ton (according to H. R. Merton and Co.'s lists) the number of tons of copper allowed by Badt for anodes is 351·6 tons, that is a weight of only 6·56 per cent. of the total annual output of the refined copper of the refinery. But the estimate for the copper anodes and stock copper calculated out from the formulæ is £35,828 when copper cost £60 per ton, that is 597·13 tons or 11·147 of the total annual output of the refinery. In my opinion Badt's estimate of the weight of copper for anodes and stock copper was seriously too low, for at Anaconda the weight of anodes alone without stock copper is as already pointed out as high as 10·85 per cent. of the total annual output of the refinery (*Electrical World and Engineer*, vol. 37, pp. 186-187, abstracted in *Science Abstracts*, vol. 4, 1901, p. 423), and it is not at all probable that the total amount of copper anodes and stock copper can be reduced much below this amount unless it becomes economically possible to employ current densities considerably larger than those which Badt's estimates for power plant show that he must have contemplated.

Further Considerations on Current Density at the Cathode in Copper Refineries.—In all the foregoing discussion of the important question of current density, it has been supposed that the only limit to the largeness of the current density to be employed was that imposed by the prime cost, and interest and depreciation of plant, cost of energy, labour, fuel, copper, &c., but besides these important factors there are two other limiting conditions which are as important as any. These conditions are firstly, the alteration of the mechanical character of the electro-deposited copper on the cathode as the current density alters, and secondly, the alteration of the chemical nature of the copper under the same circumstances. In short, it is found that, unless some very special means are used, the copper which is electro-deposited with current densities much above about 14 or 15 ampères per square foot, tends to take a crystalline and loose formation, thus tending finally to long growths, which may short-circuit a vat. Further, if the electrolyte contains much impurity the high current density is favourable to the precipitation of arsenic and antimony with the copper, thus, of course, much reducing its value. My aim has, however, been to call attention to the fact that, without considering any limiting conditions of this kind, the current density must, even under conditions very favourable to the high current density, be kept generally under 15 ampères per square foot, and usually lower than this, if the most profitable working of a refinery is desired. This fact is not popularly recognised, and many efforts have been directed to obtaining a high current density, where such a density is undesirable and cannot be a profitable arrangement. American practice in electrolytic copper refineries is generally to work with a current density not higher than 10 to 15 ampères per square foot, although many attempts have been made to employ higher values.

Current Density in Copper Conductors.—The most economical current density to employ in a bare metal conductor is found by Lord Kelvin's law by the formula

$$a = 1.967 \sqrt{\frac{P\,d\,p}{\rho\,k\,t}}$$

in which a = ampères per square inch cross section of copper conductor, P = the price per ton of the metal in the form of the conductor in pounds, d = the specific gravity of the metal, ρ = the electrical resistance of one cubic inch of the metal at ordinary temperatures in ohms, $p =$ the rate of interest to be charged upon the capital invested in the metal, k is the price in pence of a kilowatt hour or Board of Trade unit of electrical energy, and t is the number of hours the full current is run per annum. If the metal is copper and the value of the Board of Trade unit is one penny, whilst the interest charged on the capital invested in conductors is 6 per cent., and if copper costs £75 per ton, whilst t is equal to 360 days of 24 hours each, $a =$ 164 ampères per square inch, provided that the conductor is used continuously night and day throughout the year, and as this is usually the case in an electric refinery, the current density employed in the copper conductors should not be greater than about 200 ampères per square inch cross section. But if the value of the Board of Trade unit is only one farthing the density should be 328. Usually, however, a higher density than this is employed, the following results from the electrolytic refineries mentioned being given by Gore:—

Milton	833·3 ampères per sq. in.	
Marchese's works at Stolberg .	731·3 ,, ,, ,,	
Boston and Montana Company	454 ,, ,, ,,	
Casarza	305·7 ,, ,, ,,	
Pembrey	260·7 ,, ,, ,,	
Oker	258·0 ,, ,, ,,	

It is of interest to add that the current density in copper conductors usually permitted in electric lighting wires in houses is 1,000 ampères per square inch, which is the normal fire insurance limit, and probably the most economic density in this case is also near this figure, for here we have, not as in the case of an electrolytic refinery, a bare conductor, but an insulated wire, and, moreover, a wire through which, on the average, the current does not flow for more than about 2 hours out of the 24, all through the year. In only one case, however, I am aware of the employment of a current density of as high a value as 1,000 ampères per square inch in a copper conductor in an electrolytic plant. This course is advised in the prospectuses of the Cowper Coles Galvanising Syndicate, for employment when zinc-coating iron. I consider, however, that it is of doubtful economy unless the current is used very intermittently.

The fact that the variation in the price of copper has been rather considerable within the past few years, has, of course, an influence upon the advisability of employing low current density, for the lower the current density the larger the weight of copper in stock in the form of conductors, and consequently the larger the loss in case there is a drop in the price of copper, but conversely the greater the gain in the event of a rise.

CHAPTER III.

SOME IMPORTANT DETAILS IN ELECTROLYTIC COPPER REFINERIES.

Arrangement of Vats.—The vats in an electrolytic refinery may be arranged either in series or in parallel, or in some combination of series and parallel arrangement.

The amount of copper conductor employed *along* the vats is precisely the same *for a given copper output* whatever may be the number or arrangement of vats, or whatever the current density employed in the vats ; but the *connections of copper from the vats to the dynamo* is greater in direct proportion to the number of sets of vats in parallel, and the amount of copper in the form of conductors is also greater per ton output of copper, the greater the number of vats employed for the output because of the extra connections between each vat. Economy in copper employed in conductors is therefore attained, for a given output of copper, by using all the vats in series and having the vats as large as possible. Clearly also the amount of copper employed in the form of current conductors is less the greater the current density employed in these conductors, but this density, as we have already seen, is limited by other considerations of economy (see p. 522).

The electrical energy employed per ton of copper deposited is precisely the same whether the depositing vats are arranged all in series, or all in parallel, or in any combination of these two methods provided the current density at the anodes is fixed, but the loss of energy due to leakage is certainly greater when the series arrangement is adopted than on the parallel system, and in fact these losses may be roughly said to be proportional to the square of the volts employed; that is the losses are roughly proportional to the square of the number of vats in series. The leakage losses can, however, by proper care, be indefinitely reduced, and the advantages obtained by the use of vats in parallel under this head need not therefore be considered as of any importance if proper precautions are employed in insuring sufficient insulation to each vat. At Anaconda, for instance, frames are built up of timber balks. These frames are about 84 feet long, 5 feet wide, and 3 feet deep. The framework is lined with pitch-pine wooden boarding, and is divided up by transverse board partitions into ten tanks, each about $8\frac{1}{2}$ feet long. Every joint of the wood is caulked with insulating material, and all the wood has been soaked in paraffin or similar insulating substance. The vats are supported on wooden beams at a sufficient height above the ground to permit of inspection from below. The sides of the vats are also out of contact with the working floor of the gangways between the troughs. The tanks are all lined with sheet lead, and the cathodes, anodes, and connecting copper conductors are of course carefully kept from all possibility of contact with this leaden sheath, the distance between the edges of the anodes or cathodes being about two inches from the leaden lining of the tanks. The following are details of the arrangement of vats at different refineries :—

Anaconda : six sets in parallel, each set consisting of 200 vats in series.

Boston, Montana : one set of 600 vats in series.

Elkington, Pembrey : one set of 200 vats in series.

Elliott, Birmingham : five sets in parallel, each set consisting of 48 vats in series.

Oker : five sets in parallel, each consisting of 12 vats in series.

Marseilles : one set of 40 vats in series.

Biache : one set of 20 vats in series.

Hamburg : one set of 120 vats in series.

There is no doubt but that modern practice is never to employ less than one hundred vats in series, unless compelled to do so by the fact that the works output is so moderate that a vat of the necessary size to take one hundredth of the annual output would be inconveniently small. American practice is apparently in favour of employing the largest possible number of the largest possible sized vats in series, the limit of vats in series being apparently only limited by the size of the works. This policy as has already been pointed out is the necessary one to pursue, in order that the copper employed in the form of conductors connecting the dynamo to the tanks and inter-connecting tank to tank may be a minimum, and may therefore reduce as far as possible the amount of capital locked up.

Arrangement of Anodes and Cathodes.—The anodes and cathodes in each vat are usually connected in parallel, as shown diagrammatically in Fig. 128.

The Hayden Series System.—In at least two refineries in the United States, namely at Baltimore and at Brooklyn, a special method of arranging the anodes and cathodes in the electrolytic vats is adopted. This arrangement, which has apparently been worked out independently by three investigators, Stalmann, Smith and Hayden, is usually known

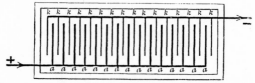

Fig. 128.—Diagram of the Ordinary or Parallel Arrangement of
Anodes and Cathodes in a Copper Refining Vat.
a a, anode plates. *k k*, cathode plates.

under the name of Hayden's Series System. The peculiarity in the arrangement of the copper-plates consists in making the anode and cathode out of one and the same piece of copper : this is done by fixing the anodes in insulating frames, in the form of what may be called diaphragms stretched across the depositing tank, which is thus divided up into a series of compartments, Fig. 129 ; the electric current flows into these compartments in series, entering at the back of one plate where copper is deposited, and leaving this plate at its other side from which therefore copper is removed. One side of a copper-plate is a cathode and the other is an anode. The thickness of the plates remains constant, but the plate gradually changes from crude into refined copper. The great advantage of this method, consists in the fact, that firstly the copper conductors are done away with, and of course the inter-connections from tank to tank, and secondly, as the copper-plates remain the same thickness throughout the refining process, there is no danger of them becoming markedly mechanically weaker, and the plates may therefore be thinner than can be used economically in the more ordinarily employed so-called parallel system of arranging the plates in the vats. The advantages of the system cannot be obtained unless the copper refined by it is unusually pure, for only in this case can the copper-plates be rolled thin. The copper-plates employed at the Bridgeport works, Connecticut, are stated by Swinburne to have been about $\frac{1}{8}$-inch thick, whereas the usual thickness of anodes employed in the so-called parallel system is over $\frac{3}{4}$-inch thick. Copper-plates cannot be successfully cast as thin as $\frac{1}{8}$-inch, and only comparatively pure copper can be rolled. At Bridgeport, one vat required 120 volts across it, and therefore, at a current density of about 10 ampères per square foot, probably contained over 300 plates in series. This refinery has, however, given up this arrangement, and employs the more ordinarily adopted method of arranging the copper-plates in a vat in parallel.

Swinburne states that he saw the Hayden process in America, and that, as the plates were rolled, they were eaten away, and deposited,

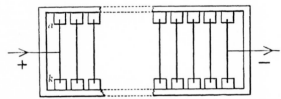

Fig. 129.—Diagram of Series Arrangement of Anodes and Cathodes in Copper Refining Vat. Each line a, k, represents a plate of copper, which acts as a Cathode on its left hand side and an Anode on the right.

very regularly, and the action was continued until very nearly the whole of the copper had been transferred. As the deposited copper on one side of a plate never adhered very strongly to it, the remains of the original anode portion could be picked off the surface of the refined copper. There was, of course, a certain amount of scrap anode formed, which had to be melted and recast.

Swinburne also states that Stalmann, who was for some time manager of, and who designed the Anaconda Electrolytic refinery, devised an ingenious series arrangement in which each plate of crude copper has a sheet of paper stuck on one side. A few copper rivets are put through and the paper is blackleaded. This gives a paper cathode which divides the refined from the crude copper and renders their subsequent separation very easy. In spite of the fact, however, that the Hayden System, or some modification of the method, has been tried at Bridgeport, Anaconda, and at other works, it has subsequently been abandoned, and at the present time is only employed at Baltimore and at Brooklyn.   Titus Ulke states, that the Hayden process is only advantageous when the crude copper to be refined is of a uniformly high grade, and can be rolled out into smooth, long sheets about ¼-inch thick.   As the series system requires a rolling mill, which will consume about one-third of the power of the plant, and as a special refining of the crude blister copper is also necessary in order to adapt it to rolling, the process is not in general use in the United States (Titus Ulke, *Electrician*, vol. 46, p. 582).   It should be added, however, that from whatever reason, such as low current density, etc., or merely because of the merits of the Hayden System already pointed out, the amount of *electrical* energy employed at the two works running this process per ton of copper output is remarkably low.

On the whole it may be said that practical experience seems to be against this series method of arranging the copper in the vats, and in favour of the ordinary parallel arrangement.

Size and Number of Vats.—The size of the vats employed in an electrolytic refinery must not be too small, or their cost per cubic foot and the room which a plant designed for a given output occupies, will become excessive.   On the other hand the size of the vats should not be too great and for two reasons.   Firstly, because the difficulty of arranging and handling the anodes and cathodes becomes greater the greater the size of the vats employed, and secondly, because in the event of a vat either becoming short circuited by accident, or being disconnected from the circuits for cleaning and putting in order, the loss of energy in the first case, or the loss of output in the second case, becomes excessive if the vats are unduly large. It must be remembered in this connection that the output in copper of a vat depends only upon two conditions with a given arrangement of anodes and cathodes, and these two conditions are, firstly, the current density, and secondly, the volume of the vat.   If the current density employed is fixed and the distance apart of the anodes and cathodes is also fixed, the output of a vat depends only upon its volume.   Clearly a large vat is cheaper per cubic foot of contents than a small one, and a good general rule is that the vats must be made as large as possible, the possible largeness being only limited by the inconvenience of the vats of too great dimensions, and the fact that a given factory cannot afford to have more than at most about one-hundredth part of its output wasted at a time, by reason of accidental short circuits occurring in the vats.

Thus, if it is desired to determine the number of vats to employ for a given output of copper, we may firstly decide that at least 100 vats shall be used in series, for then if one vat is being cleaned up or if one vat is by accident short-circuited, the loss of output will only amount to one per cent. of the total output of the factory. If then we divide the annual output of the proposed factory in tons of refined copper by one hundred, we shall obtain the tons output per vat per year, and as has already been shown (p. 505), a safe estimate requires about 60 cubic feet of vat space per ton of copper output per year, at a current density of one ampère per square foot of cathode surface, and therefore if we denote this current density by D, the volume of each vat should be $\dfrac{60\,T}{D\,s}$ cubic feet, where s is the number of vats employed in series, and T is the output of refined copper per annum. For certain outputs this may work out to very small vats, and again under other conditions the size of the vats thus determined will be too large. If it works out too small, say under 25 to 30 cubic feet, a smaller number of vats must be used in series in spite of the inconvenience this entails, as pointed out above. In fact the vats should not under any circumstances be of smaller size than about 25 cubic feet, for if they are the cost of vat per cubic foot becomes excessive, as also does the floor space required and the volume of electrolyte. On the other hand if the vat size calculated out from the above formula is very large, the vats become difficult to manage for shifting anodes, cathodes, etc., and cleaning up.

At Anaconda the vats have an internal capacity of 132 cubic feet each, but in this case the copper is shifted by means of an overhead crane, and it is probable that a vat of a capacity of 80 cubic feet is usually of quite sufficiently large size. If, therefore, the vat size as obtained by the above formula is over 80 cubic feet, it is better to run a larger number in series until vats of a size of about 80 cubic feet are obtained. The sizes of vats at various works are as follows :

	Cubic Feet.	Length. Feet.	Width. Feet.	Depth. Feet.
Biache . .	75	10	2·5	3
*Anaconda . .	132·4	8·2	5	3·25
Elliott . .	30·25	2·75	2·75	4
Elkington . .	42	4	3	3·5
Casarza . .	24·5	6·5	1·25	3

* *Electrician*, vol. 38, p. 144; but in a recent paper (*Elect. World and Engineer*, vol. 37, 1901, pp. 186-187), it is stated that the volume of the Anaconda vats is 140 cubic feet, the dimensions being 8 × 5 × 3·5 feet. The vats are also stated to contain 4,000 litres of electrolyte (= 134·7 cubic feet).

The tanks of the Boston Montana Copper Refining Company have about the same volume each as those employed at the Anaconda Works.

Circulation of the Electrolyte.—Owing to the fact that at the anode copper is passing into solution, whilst at the cathode copper is passing out of solution, the copper electrolyte tends to become weaker, more acid and less dense at the cathode, and stronger and more dense at the anode, and as the dense solution at the anode sinks down to the bottom of the vats, the whole of the solution in the vats tends to become more or less sharply divided into a layer of a dense solution of copper sulphate at the bottom of the vat and a lighter acid solution less rich in copper sulphate floating at the surface of the vat. This separation of the electrolyte into layers of different composition, causes many inconveniences, if it is permitted to take place without any interference. The chief of these inconveniences being : 1st, the unequal corrosion of the anodes, in such a way that the portions of the anodes in the upper layers of more acid electrolyte dissolve more rapidly than the lower ends, which are immersed in the stronger and more neutral copper sulphate solution at the bottom of the vat. This more rapid corrosion at the top eventually causes the lower portion of the anode to break off, probably short-circuiting the whole bath. 2nd, the copper is chiefly deposited at the lower end of the cathode, and as the current density becomes too large there impurities are liable to be deposited with the copper, whilst at the upper ends of the cathodes, hydrogen tends to deposit and polarisation is set up.

Whatever the precise theory of these inconveniences may be, it has undoubtedly been found in practice that a more uniform action of the electric current in depositing the copper is obtained if the electrolyte is maintained in a state of gentle, but steady circulation, whilst deposition is taking place. This circulation is performed by running off the electrolyte at the top of the vats into a reservoir, from which it is pumped back into a second raised reservoir, from which by gravity it flows into the bottom of the vats. The form of pump usually employed is that one known as an injector. The circulating system at Anaconda is arranged as follows : There are six sets in parallel of 200 tanks, all in series, and these 200 tanks are divided up into 20 sets of ten each. These sets of ten tanks are made out of one large wooden girder of beams, which is lined with stout planking, and also divided into ten compartments by means of this stout planking, each compartment forms one tank or vat, and is lined with sheet lead, each tank in this set of these ten tanks then is about one inch higher than its neighbour, and the total fall down the set of ten tanks is, therefore, about nine or ten inches. The liquid runs by gravity from tank to tank, flowing out of one tank at the top by a spout, and falling into a wide lead pipe which carries it downwards, the pipe entering the next tank at the bottom. In this way, at the end of the ten tanks, it finally reaches the collecting pipe which conveys it to the pumps, steam injectors, or air compressor lifts. The pumps in their turn raise the liquid to the upper reservoir, from whence it is delivered to the distributing pipes running to the head of each set of ten tanks. The lead lining and the lead circulating pipes in each vat must not be in electrical connection with the lead of the next vat. The lead linings can, of course, be readily kept separate, and the lead pipes are made discontinuous, the liquid falling from a spout at the top of one tank into a funnel head of a separate pipe running to the bottom of the next tank, the only connection along which electric leakage can take place being, therefore, through the stream of electrolyte running from the top of one vat into the lead circulating pipe of the next, or the circulating pipe of one vat may be joined to that of the next by means of an indiarubber tube. The lead circulating pipes open into the bottom of the depositing vats, through holes pierced in the sides of the pipe which runs along the bottom of the vat. These holes are often covered above by means of screens or hoods of sheet lead, to prevent anode mud settling into them, and thus blocking them up. For circulating purposes, steam injectors or air compressors are preferable to lead pumps, as the latter are liable to get out of order. Steam injectors, of course, both heat the electrolyte and add a certain quantity of water to it; both these actions vary in amount with the temperature, and are not quite readily controlled ; it is, therefore, probably best to use air compressors for raising the liquid, and to employ separate pipes for steam heating or adding water when necessary, as a separate and controllable operation.

Figs. 130 and 131 give views of belt-driven air compressors, made by Messrs. George Scott and Son, of 44, Christian Street, London, E.

Of these, Fig. 131 represents a compressor which, in size A F, runs at 120 revolutions per minute, delivers 12 cubic feet of air per minute at a pressure of 15 lbs. per square inch, a sufficient pressure for air pumps, monte-jus or acid eggs, such as are employed for circulating and aerating electrolytes. The price of this pump is £50. If the pump is run at a higher speed it has a correspondingly increased output. It is desirable on an electro-lytic refinery to drive such pumps by direct connected electro-motors, for the speed is then easily varied, and as the electric plant is necessarily at hand the power transmission is cheaper, more flexible, and in every way more convenient than counter shafts and belting. One B. H. P. is required by the pump shown in Fig. 130 at 120 revolutions per minute, and one B. H. P. is also required by the pump shown in Fig. 131, at the same speed, air-output, and pressure. The latter pump is fitted with a valve which makes it impossible for the air to be delivered at a greater pressure than 15 lbs. per square inch.

Messrs. H. K. Borchers, in their modification of the Siemens depositing vat, Figs. 132 and 133, employ an ingenious system of circulating the electrolyte by means of air. A wide leaden tube L passes downwards into the electrolyte in the vat from the surface and extends along the middle at the bottom of the vat, opening into the electrolyte again at the centre of the bottom of the vat. Into the upper end of this wide tube a narrow glass tube passes downwards, and opens into the wide tube through a narrow drawn out ending. The upper end of this glass tubing passes through an indiarubber cork in a hole in a leaden hood which covers the upper end of the wide lead tube. This hood is intended to catch any spray, and after passing out through the cork, the glass tube is connected, by means of an indiarubber tube I, with a compressed air main, from which air is forced down through the fine open end of the glass tube, and the bubbles rising in the electrolyte in the wide leaden tube cause the column of

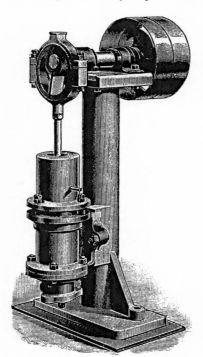

Fig. 130.—Belt-driven Air Compressor. Delivers air at 15 lbs. per square inch pressure.

liquid inside the leaden tube to be lighter than the electrolyte outside it and consequently cause the liquid in this tube to circulate upwards, a constant stream of the denser electrolyte being sucked in at the lower end of the wide lead tube, thus permitting a constant gentle circulation of the liquid in the vat, without disturbing the anode sludge. The lower end of the lead tube opens under a tray or false bottom T at the bottom of the refining vat, and is thus protected from the sludge being drawn in and the electrolyte being rendered turbid thereby. The turbidity of the electrolyte causes the deposit of chemically impure,

Fig. 131.—Belt-driven Air Compressor. Delivers air at 15 lbs. per square inch pressure

and also a mechanically unsatisfactory, cathode copper, and must be avoided. The flow of air through the glass tube can be regulated by means of a screw clip C, which can constrict the indiarubber connection to any desired amount. The depth to which the glass tube can

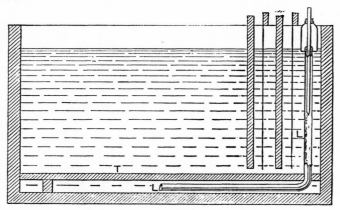

Fig. 132.—Longitudinal Section of Siemens-Borchers Copper Refining Vat, showing two Anodes and two Cathodes in position, and also the Borchers Air Circulating Apparatus.

be pushed into the leaden tube can also be altered, and thus by either or both of these adjustments the rapidity of circulation can be increased or decreased at will. It is stated that the simultaneous oxidation of the liquid by the injected air tends to precipitate any ferrous iron, and also with it any arsenic which may be in the solution as *ferric-arseniate*, the precipitation of the arsenic is, however, disputed by some writers. (See remarks on the use of air for purifying purposes, pp. 543, 544.) When employing this form of air circulator Borchers finds that it is unnecessary to continuously circulate through all the vats by gravity flow as described above, and although from time to time the electrolyte is run off and purified and again returned to the vats, not all at

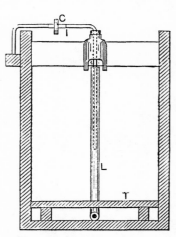

Fig. 133.—Transverse Section of Siemens-Borchers Vat on line x x.

once, of course, but a portion at a time, yet this necessary purification need not be so frequently repeated as is sometimes the case. Borchers' method of air circulation is not patented.*

Heating the Electrolyte.—The copper sulphate solution is usually electrolysed at a temperature of about 25° C., and it is maintained at this temperature by heating the electrolyte, which has been run off from the tanks, before it passes to the pump. The heating is performed by passing the solution over pipes through which steam is blown, or if the evaporation has increased the strength of the solution, the steam may be blown directly through the solution, thus diluting it sufficiently to compensate for the water lost by evaporation. The amount of steam blown through by one method or the other, or both, can be regulated in amount so as to maintain the temperature of the electrolyte in the tanks at a temperature of between 20° C. and 25° C. The amount of steam required for this purpose in summer is, of course less than is necessary in winter. The temperature should be maintained as uniform as possible.

Avoidance of Short Circuits on Vats.—In order that a constant check may be kept upon the condition of every vat in a refinery it is convenient to connect with insulated wires the positive of each vat, and carry all these wires back to one two-pole

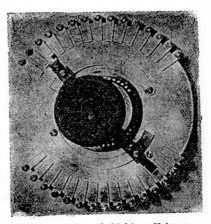

Fig. 134.—Two-pole Multiway Voltmeter Switch, with 15 double contacts.

multiway voltmeter switch in the office. By rotating the handle of such a switch the terminals of each vat may be in turn connected to

---

* There is not room in the present edition of this book for a fuller description of the mechanical details of all the various forms of electrolytic copper refining vats, but full drawings of these, with detailed descriptions, may be found in Borchers' "Electric Smelting and Refining," translated by W. G. MacMillan (London : C. Griffin & Co.), and also in Gore's "Electrolytic Separation of Metals," to which books those persons interested in further details concerning these points are referred.

the terminals of a dead beat low voltage voltmeter reading to a maximum of about 3 volts, and divided in twentieths of a volt, and the volts on each vat thus inspected. Such a voltmeter switch, but only arranged for 15 vats, is shown in Fig. 134. It is made by the British Schuckert Company, and costs 4s. per double contact for large numbers of contacts.

At Anaconda a switch of this character, of course much larger, is rotated by clockwork, readings being recorded automatically for each vat by a recording voltmeter a definite number of times daily. Inspection of the record at once enables the location of a faulty vat to be performed, and any short circuit or other trouble is promptly rectified.

Voltage of Dynamos for Copper Refining.—The voltage of the dynamo employed for copper refining is ruled by the number of vats in series, and the current density employed. If the number of vats in series, as determined from the considerations on pp. 528, 529, is denoted by s, and if the current density in ampères per square foot is denoted by D, then the required voltage V of the dynamo may be obtained with good approximation from the equation

$$V = 0.042\,Ds.$$

Thus, if the current density is 10 ampères per square foot, and there are 200 vats in series, the voltage of the dynamo required will be 75 volts. The dynamo must, of course, be either a shunt wound or a separately excited machine. If a shunt wound machine, it is advisable to have it provided, as is usual in this type of machine, with a rheostat in its field magnet circuit which is capable of varying its voltage up or down by about 20 per cent. This permits of its voltage being kept constant at all loads without altering the speed of the driving engine, and also allows of a certain amount of flexibility in the number of tanks the dynamo can feed in circuit, and the current density may be varied if desired. As the separately excited machine gives a somewhat greater flexibility, it is not unusual to employ such a dynamo, but as it must be run with an exciter dynamo or secondary cells it is not quite so cheap or simple as a shunt machine, unless a refinery is of very large output. At the Boston Montana Refinery, U.S.A., the dynamos are 180 volt separately excited Westinghouse machines having a normal current of 4,500 ampères, but the volts may be raised to 220. The exciters are 125 volt Westinghouse shunt dynamos. The current density in the vats is probably about 14 ampères per square foot of cathode surface.

At the Anaconda Refinery, where 200 vats are worked in series, the dynamos are Westinghouse shunt machines of 75 volts terminal pressure, whilst the current density employed is probably about 12 ampères per square foot of cathode.

The question of the total output and cost of the dynamo required and the cost of the power plant required to drive it has been dealt with on pp. 503–504.

Anode Copper Employed in Refining.—The anode copper employed for refining at the Anaconda Refinery in America is blister copper containing 98 per cent. of copper.* The impurities are arsenic, iron, lead, tellurium, selenium, silver and gold (110 ozs. of silver per ton, and 0·33 oz. of gold). (*Electrician*, vol. 38, p. 146.) At this refinery the anodes are renewed in each vat every thirty-four days.

Titus Ulke states (*Electrician*, vol. 46, p. 582) that the composition of typical anodes of refined blister copper as employed in the United States is as follows :—99·25 per cent. copper ; 0·338 per cent. silver ; 0·001 per cent. gold ; 0·300 per cent. oxygen ; 0·054 per cent. antimony ; 0·033 per cent. arsenic ; 0·009 per cent. lead ; 0·008 per cent. selenium and tellurium ; 0·002 per cent. bismuth ; 0·002 per cent. nickel, and a trace of iron.

Sederholm states that converter copper containing 98 to 99·3 per cent. of copper, and blister copper, containing 99·5 per cent., is largely used for anodes in American refineries (*Journ. Soc. Chem. Ind.*, vol. 14, 1895, p. 756).

Black copper or blister copper, containing as small an amount of copper as 97 per cent., is employed largely for anodes in America. Some works employ an even less pure form of copper for anodes, but general experience and the almost universal custom seems to show that the most economical process is to produce copper of not less than 97 per cent. purity by one or other of the dry methods of copper smelting, before carrying out a further purification by the wet or electrolytic method.

Cleansing and Inspection of Surfaces of Anodes and Cathodes during Refining.—Sederholm (*Journ. Soc. Chem. Ind.*, vol. 14, 1895, p. 756, abstracted from *Dingl. Polyt. Journ.*, vol. 296, 1895, p. 284-288) states that in the North American copper refineries it is the practice to remove each anode and cathode every other day, the anode being well cleaned down from any slime adhering to it, and the cathodes are freed from any excrescences that may have formed. Both plates are then returned unless the anodes are dangerously thin or irregular, when they are replaced by fresh anodes and the old ones are recast. When the cathodes are found to be between one-third and one-half of an inch thick they are replaced by fresh cathode sheets.

Composition of Electro-Refined Copper.—The degree of purity

---

* In a recent paper on the Anaconda Refinery (*Elect. World and Engineer*, vol. 37, 1901, pp. 186, 187), it is stated that the anode copper employed contains 99·6 per cent. of copper.

of electro-refined copper naturally depends upon the purity of the anodes from which it has been prepared, the current density employed, and the precautions taken to keep the electrolyte both chemically pure, and free from turbidity. The following is the analysis of a very favourable sample of electrolytic copper made by the present writer :—

Iron	0·0189
Arsenic	0·0015
Lead	0·0013
Antimony	0·0010
Bismuth	0·0008
Silver $\left.\begin{array}{l}\\\\\\\end{array}\right\}$	
Nickel	Absent
Sulphur	
Copper (by difference)	99·9765
	100·0000

Composition of Anode Sludge.—The anode residues, sludge, mud, or slimes, obtained when the tanks are cleaned up, are washed and dried, and then treated, in order to separate the metal they contain. The chemical composition of this mud must, of course, vary considerably with the nature of the impure copper anodes treated. The extent of this variation may be judged from the following analyses of sludges :—

	Reverberatory Copper.		Converter Copper.	
Silver	53·894	[0·3064]	55·15	[0·3076]
Gold	0·296		0·198	
Copper . . .	11·01		13·82	
Lead	0·91	[0·0093]	2·07	
Bismuth . . .	3·93	[0·0320]	0·34	[0·0035]
Antimony . . .	6·25	[0·0651]	2·44	[0·0510]
Arsenic . . .	2·11	[0·0586]	1·09	[0·0180]
Selenium . .	0·39		0·72	
Tellurium . . .	1·17	[0·0098]	0·89	
Iron	absent		0·80	
Sulphuric Acid (SO₄) .	5·27		10·68	
Water	2·36		2·60	

The column headed "Reverberatory Copper" gives an average analysis in parts per cent. on the air dried material of tank mud or residues obtained during a year's treatment of anodes formed of reverberatory copper; whilst that one headed "Converter Copper" gives an average analysis in parts per cent. on the air dried material of residues obtained during three months treatment of converter copper. The analyses do not add up to one hundred parts, because the oxygen and the water of crystallisation combined with the metals has not been estimated. The numbers in square brackets represent the percentages of the various elements present in the unrefined copper, and therefore indicate the extent of the refining effected by electrolysis. The whole of the silver, gold, selenium and tellurium, contained in the crude copper passed into the residues, whilst of the other elements the percentages of the amounts in the crude copper which passed into the residues is given by the following table:—

	Reverberatory Copper.	Converter Copper.
Copper	0·07	0·08
Bismuth	78·22	60·71
Antimony	61·14	29·9
Arsenic	22·9	37·84

The balance of the impurities, neither passing into the sludge nor into the electrically deposited refined copper, of course becomes concentrated in the electrolyte solution. These figures are taken from a paper by E. Koller (*J. Amer. Chem. Soc.*, 1897, vol. 19, pp. 778-782), an abstract of which is given in the *Journal of the Society of Chemical Industry*, vol. 17, p. 53.

An analysis of anode mud was published in *Erdmann's Journal für Practische Chemie*, vol. 45, 1848, pp. 460-468, by Maximilian, Duke of Leuchtenberg, and is as follows, in parts per cent. on the dried deposit.

Silver	4·45
Gold	0·98
Copper	9·24
Lead	0·15
Bismuth	absent
Antimony	9·22
Arsenic	7·20
Selenium	1·27
Iron	0·30
Tin	33·50
Oxygen	24·82
Sulphur	2·46
Nickel	2·26
Silica	1·90
Cobalt	0·86
Vanadium	0·64
Platinum	0·44
Total	99·69

A. Holland gives the following partial analyses of anode sludge in the *Bulletin de la Société Chémique*, series 3, vol. 19, pp. 470-472. The character of the anode copper treated is not stated.

	I.	II.	III.	IV.
Silver	25·816	36·5210	38·4800	46·5800
Gold	0·0337	0·0768	0·1020	0·1504
Copper	18·475	24·0420	18·5160	18·4750

Gore gives the three following analyses of anode sludge from electrolytic copper refineries.

	No. 1.	No. 2.	No. 3.
Silver	1·815	5·61	0·55
Gold	0·085	0·01	absent
Copper	85·85	19·40	67·90
Lead	0·05	27·70	2·05
Bismuth	0·65	1·25	—
Antimony	0·75	7·35	—
Arsenic	2·48	5·20	—
Selenium	—	—	—
Tellurium	—	—	—
Iron	0·75	0·60	5·55
Sulphuric Acid	1·15	—	—
Water } Oxygen }	4·95	21·05	—
Insoluble Earthy Material	0·95	4·35	3·40
Chlorine	0·25	0·70	—
Nickel	—	0·20	—
Sulphur	—	6·35	18·10
Organic Matter	—	0·20	2·25
Difference	0·02	0·03	0·20
	100·00	100·00	100·00

No statement is made by Gore as to the character of the anodes from which these deposits were obtained.

The anode sludge obtained at the Boston Montana Company's Works is said to be worth £500 per ton.

The silver and gold per ton of electrolytically refined copper produced at the principal American refineries is shown in the following table prepared from the results given by Titus Ulke (*Electrician*, vol. 46, p. 582) :—

Name of Refinery.	Troy ounces of silver per short ton output of copper.	Troy ounces of gold per short ton output of copper.
Raritan Copper Works	58	1·133
Anaconda Mining Co.	80	0·350
Baltimore Smelting and Rolling Co.	80	0·350
Boston and Montana Copper Co.	66	0·150
Nichols' Chemical Co.	16·6	0·566
Guggenheim Smelting Co.	400	3·500
Balbach Smelting and Refining Co.	83	0·433
Bridgeport Copper Co.	60	0·200
Irvington Smelting Co.	80	0·222
Chicago Smelting Co.	20	0·500
Buffalo Copper Works	20	0·400

Weight of Anode Sludge Obtained.—The weight of anode sludge obtained in any given refinery of course depends upon the amount of impurity contained in the anodes employed, and also upon the nature of this metallic impurity. Generally speaking, however, a rough guide is to consider that the weight of dry anode sludge obtained will be twice the weight of the metallic impurities present in the anodes. That is anodes containing 98 per cent. of copper would, in all probability, yield about four tons of dried sludge for every one hundred tons of anodes dissolved. The extra weight of the dried sludge over and above the weight of the metallic impurities in the anodes is, of course, due to the presence of oxygen, water of hydration, and copper, in the deposit.

Treatment of Anode Sludge.—The method of treating the anode sludge must vary with the nature of the deposit, and, as has been shown above, this is very different at different refineries. In a paper by E. Sederholm, abstracted in *Dingler's Polytechnische Journal*, 1895, vol. 296, pp. 284-288, it is stated that concentrated sulphuric acid dissolves most, but not all of the silver, which may then be removed from solution by means of granulated metallic copper. This process is wasteful of acid but the silver obtained is fairly pure (99·5 per cent.). The residual silver undissolved by the sulphuric acid is in combination with selenic and antimonic acids. T. Ulke treats the sludge with *dilute* warm sulphuric acid, through which air is blown during the operation. This dissolves out the copper from the sludge, and leaves a residue containing as much, sometimes, as 90 per cent. of silver.

It is, perhaps, more usual to treat the sludge by a dry process. Thus, it is stated by Gore that the sludge is washed by mixing with water and decantation; the sediment is then dried, sifted to remove fragments of copper, and the resulting powdery residue is melted with litharge, and a reducing flux, and the crude mixture of metals thus obtained is cupelled with a further addition of argentiferous lead, the silver and gold being thus recovered. The dry anode sludge is also sometimes smelted so that the copper passes into the slag, and the metal obtained, consisting largely of silver (60 to 90 per cent.), is cast into ingots which are refined by electrolysis in dilute nitric acid by the Möbius process (see p. 573.) The process for refining the slimes at the Anaconda Mining Company's Refinery is thus described (*Electrician*, vol. 38, p. 144, in an article taken from the *Engineering and Mining Journal* of New York): " The silver mud (this name is given to the slimes) is sent from the refinery in lead lined tank-cars to the silver mill. Arriving there it is hoisted up to the screens, where it is washed with water, and all chips of copper, etc., are taken out. The clean silver mud is then run out into boiling tanks, where it is freed from its copper contents by boiling with acid and steam. From this first set of boiling tanks the silver mud is passed over a filter on which it is thoroughly washed with water. It is then put into the second set of boiling tanks from which the other impurities, notably arsenic and antimony, are taken out. From here the silver mud is again placed on filters, thoroughly washed with water, and dried on large cast iron pans. A subsequent melting in the reverberatory furnace reduces the silver mud to ingots, ready for the parting kettles.

" The silver mud when it goes into the first melting furnace contains only a small amount of impurities. The operation of the furnace consists of a mere melting of the slimes, which is carried on as rapidly as a wood fire will permit, and is not in any way a refining process. The furnace is charged with about two tons of the dried silver mud at a time ; after it is melted it is tapped into moulds which move on a small train in front of the furnace. At Anaconda, the bullion thus obtained is refined by dissolving in strong sulphuric acid and diluting the solution obtained and precipitating the silver from it on copper plates." It may be here remarked that the description given above is not very satisfactory as to the particulars of the compositions of the solutions in which the silver sludge is boiled, possibly because the writer did not know the nature of these solutions. There is some difficulty in obtaining information of this character in America.

The Baltimore Electric Refining Company treat the washed and sifted slimes in lead lined vats with dilute sulphuric acid (1 part acid to 4 parts water), through which air is injected for two or three hours. The solution obtained, which contains arsenic, copper and most of the impurities, is syphoned from the residue, which contains lead sulphate, tellurium, a little bismuth and antimony in conjunction with the silver and gold. This residue is melted on a cupel hearth, at first without any flux, when a brownish slag containing lead and antimony together with some silver beads or prills is skimmed off. The slag after cooling is picked over to separate the silver, and is then added to molten lead in the cupelling furnace, by means of which the last traces of gold and silver are removed from it. After having removed the brown slag from the crude silver as described above, the silver is heated with nitre, which removes the tellurium. The silver on the cupel is now practically pure with the exception of a little copper, and it is cast into bars ready for parting in the usual way, or by Möbius' process. The copper in the acid solution obtained on boiling the sludge with dilute sulphuric acid, is precipitated by means of scrap iron (*Journ. Soc. Chem. Ind.*, vol. 16, p. 49).

Purification of Electrolyte.—The impurities contained in the anodes employed in electrolytic copper refining partly (and, indeed, chiefly) pass into the sludge or slimes, but the remainder dissolve in the solution and gradually, as time goes on, render it more and more impure. The result of this is, that if some means are not taken to repurify the solution, the impurities are electrically deposited on the cathodes, thus rendering the electric refining action less satisfactory, and yielding an electrically-deposited copper of inferior purity.

Indeed, one of the most important points in carrying out successfully the electrolytic refining of copper is to ensure that the electrolyte is kept of a constant composition and strength, and it should be tested chemically every day in order to check its condition. This precaution is now taken in most of the larger refineries.

The means employed to purify and revivify the electrolyte may be considered under three heads :

1st. *Mechanical method.*—This consists in steadily removing some of the impure electrolyte and replacing it with fresh solution of pure copper sulphate of the correct strength and containing the correct percentage of free sulphuric acid. This operation is performed daily. A certain proportion of the liquid running from the depositing vats is not returned by the pumps, but pure fresh electrolyte is substituted for it. By this means the electrolyte in the tanks can always be kept up to a certain standard of purity. The impure solution which is removed is evaporated down, and the copper sulphate crystallised out and employed for forming the fresh solution. The impure mother liquors have the copper removed by scrap iron.

2nd. *Chemical method.*—This method consists in chemically treating the solutions run off from the vats before returning them by the pumps, or partly by blowing air through the electrolyte whilst in the vats (see p. 531). The chemical treatment may vary largely, depending upon the type of copper treated.

3rd. *Electrical method.*—By this method, the impure liquid, after running from the vats and before returning to the pumps, is treated electrically by passing a current of electricity through it with a large current density at the cathode, and employing either pure copper or lead anodes. By this means most, or a large proportion, of the metallic impurities in the solution can be thrown down on the cathode. The purified liquid is then returned to the vats.

Generally speaking, some combination of some or all of the above three methods is employed.

In the Baltimore Smelting Company's Works in the United States of America about one-fifth of the electrolyte is periodically removed and worked up for copper sulphate, for which a ready sale can be obtained, and the mother liquid is treated with scrap iron to recover the remainder of the copper, the composition of the bath being kept as constant as possible by adding freshly-prepared solution. A similar method is followed at the Balbach Works, in Newark.

At Anaconda it is stated that the purifying process consists in passing the impure electrolyte repeatedly through a layer of oxidised copper, so as to partially precipitate the antimony and bismuth present. By this treatment the solution becomes nearly neutral, and saturated with copper, and is then oxidised by passing air through it, so that the iron is partially precipitated as ferric oxide. Titus Ulke states (*Journ. Soc. Chem. Ind.*, vol. 17, p. 160), that one of the best methods for purifying old solution is that in which it is electrolised in special vats, the anodes being of lead and the cathodes of copper. A current density is employed sufficiently great to deposit the arsenic and antimony, but not strong enough to deposit the iron. The solution thus freed from arsenic and antimony is returned to the copper-depositing vats to be used in the ordinary way, and this is repeated until the bath contains so much iron that it is necessary to remove it by crystallising out the ferrous sulphate.

W. Terrill (*Journ. Soc. Chem. Ind.*, vol. 17, p. 466,) states that oxidation by means of chlorine or injected air is often used in the process of purifying the electrolyte, but this leads to the formation of ferric salts and a neutral solution, under which circumstances silver is deposited upon the cathodes with the copper, instead of being deposited in the anode sludge. This method of oxidation should not therefore be adopted except under suitable precautions. Borchers adopted a special method of aëration, for description of which see p. 532. Terrill considers that the best method for keeping the electrolyte in good condition is to gradually renew it, as is done at the Baltimore and Balbach works. The greater part of the copper sulphate crystals obtained can be employed for making up the fresh electrolyte : the surplus necessarily produced must be sold. In a paper on the Anaconda Refinery (*Electrician*, vol. 38, p. 147) it is, however, stated that the impurities in the electrolyte are partially removed on each turn of the electrolyte through the tanks. [No doubt by the chemical treatment mentioned above.] A certain amount of impurity is allowed to stay in the liquid, however; the purifying process being so regulated as to limit the amount of impurity, but not to completely remove it. It is further stated in this paper that the chemical purification employed is extremely simple, requires little attention, and necessitates the use of air and cheap chemicals only. The method is praised as being far preferable to the voluminous and bulky old process of crystallising which is used in almost all Eastern refineries. The actual method of chemical purification adopted at Anaconda is not described in this paper, probably because it is considered to be a secret process. It is, however, as has already been stated, described by Titus Ulke. (See above.)

It will be seen from what has already been said that opinions differ considerably upon the question as to the best process to employ; but in the writer's opinion some process consisting probably of a combination of all three of the general methods mentioned above, could be devised in every particular case which would be most suitable to a given type of anode in a given locality.

Effect of Organic Matter on Copper Deposits.—It is very important that copper solutions shall be kept free from organic matter, as it is found that even very minute quantities present in the electrolyte will seriously affect the character of the deposited metal. F. Förster and O. Liedel have shown that a small quantity of a varnish of caoutchouc in benzene, accidentally present in the copper sulphate solution, caused the deposited copper to become smoother, and to possess a finer structure than the metal deposited from the normal solution. The deposited metal was, however, very brittle, and could be powdered in a mortar. Von Hübl noticed similar effects produced in solutions containing gelatin and other organic substances. The copper deposited contained carbon. It is for this reason necessary to have the electrolytic vats lined with lead, and care must be taken that at no point an opportunity is afforded for the contamination of the electrolyte by turpentine, gelatin, putty, varnish or any oily matter. The steam blown through the electrolyte should either not have been passed through the engines, or should be passed through lead tubes in the electrolyte, and should not be mixed directly with the solution.

Formation of Nodular Growths on Electro-deposited Copper.—Apparently the cause of nodular or tree-like growths upon the surface of electro-deposited metals is caused by the presence of particles of solid matter, anode sludge, etc., becoming stirred up in the electrolyte and then adhering to the surface of the cathode. J. W. Swan, F.R.S., has observed that such nodules always contain a nucleus of some insoluble foreign matter, and he has shown that if the solution is kept free from sediment, the formation of the nodules does not take place. Gore states that the greatest length of the nodule is in the direction of the

N N

greatest density of current, and greatest strength of copper solution ; this observation, however, merely agrees with what might be foreseen. The danger of such growths in copper refining is due to the fact that as they extend more rapidly than the electrodes themselves they may in time short circuit the vat. The formation of these nodules is favoured by a high current density, keeping the anodes and cathodes close together, and turbidity of solution. If they tend to form they may be avoided or reduced by properly altering one or all of these three factors. Cowper Coles has avoided the formation of nodular growths, when using high current densities, by removing any particles which tend to adhere to the cathode surface by means of a rapid rotation of the cathode in the liquid, adhering particles being thrown off by centrifugal force (*Electrician*, vol. 44, p. 288). For a further discussion of this method, see p. 552 of the present volume.

Copper Deposition with High Current Density.—Very many efforts have been directed to the deposition of copper of a satisfactory character with high current density. As is shown on p. 516, if the copper obtained is to be employed as raw copper from which articles will be subsequently manufactured, the employment of a high current density, even if it yielded a mechanically and chemically satisfactory material, would not be economically correct. If, however, the deposited copper is directly formed into a manufactured article as it is deposited, such, for instance, as copper tubes which have a higher value per ton than the pure unmanufactured metal, the current density may economically be increased considerably. Although, generally speaking, the distance apart of the electrodes and the vat space required per ton of copper are quite different when tubes are to be deposited, from what is adopted when electrically refined but unmanufactured copper is to be produced, yet the value of the best current density may be obtained precisely, as on pp. 518–520, different constants being employed however. The chief cause of the increased value of D being clearly due to the fact that P'', the value in pounds per ton of copper tubes, is very much higher than if unmanufactured copper alone were produced, whilst P and P' remain constant.

One of the earlier successful processes for producing copper tubes directly was that of Elmore. The object of this process was, in the first case, the formation of copper tubes directly, without any particular effort to obtain a very high current density, but the highest current density, consistent with a satisfactory deposit, has since been rightly aimed at. Two of the Elmore patents for the electro-deposition of copper, namely, Eng. Pat., 9,214, of July 15, 1886, and 15,831, of December 3, 1886, expired in 1900. The chief patent is 4,499, April 11, 1885. The method consists in rotating a metallic cylindrical mandril B, Fig. 135, horizontally placed in the usual bath of copper sulphate employed for the electro-deposition of copper. The mandril, which in the original process was of iron, was given a preliminary coating with copper in an alkaline copper bath, and then transferred to the acid copper bath where it was employed as the cathode. Whilst it rotated in the solution, an agate burnishing roller passed backwards and forwards along the rotating cylinder, much as the tool on a screw cutting lathe travels. The current density employed was about 15 or 16 ampères per square foot of surface. An early difficulty was found in removing the tubes of copper from the mandril, but this difficulty was overcome by various devices, the first being to coat the iron with melted lead, which completely filled up all pinholes on its surface, and gave a soft surface upon which the copper could be deposited, but to which it did not adhere very strongly. By rolling the outside of the

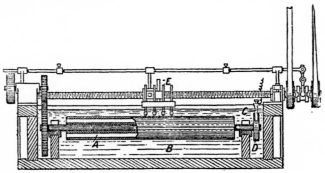

Fig. 135.

tube after the copper deposition was completed, the outer metal tube was sufficiently loosened to be drawn off. Fusible metal cores, and also cores of thin metal expanded by internal pressure, which pressure could afterwards be reduced, thus causing them to shrink, were in turn used. The chief patents taken out over modifications of this process are as follows: Eng. Pat. 12,264, August 25, 1888 (lead and tin coating for mandrils). Eng. Pat. 7,932, May 22, 1890 (fusible metal coated mandrils). Eng. Pat. 5,167, March 23, 1891 (removing tubes from mandrils). Eng. Pat. 10,451, September 3, 1885 (making copper pans, cylinders, etc.) Eng. Pat. 11,800, October 3, 1885 (making tubes, pans, etc.) Eng. Pat. 1,737, February 7, 1885 (cores for tubes). Eng. Pat. 8,707, July 18, 1885 (mandrils for tubes). Eng. Pat. 9,214, July 15, 1886 (making plates, wire, etc.) Eng. Pat. 16,637, December 3, 1887 (copper tubes, etc.) Eng. Pat. 11,778,

August 15, 1888 (making cylinders, tubes, etc.) Eng. Pat. 12,022, August 20, 1888 (making pipes). In view of the importance, if only from the point of view of the amount of capital sunk, of the Elmore processes, it has been considered advisable to give a fairly complete list of the patents involved.

These patents are all taken out by F. E. Elmore. There are some others by A. S. Elmore and J. O. S. Elmore, one of which, Eng. Pat., 21,283, Nov. 9, 1895, claims that high current densities may be employed if a very rapid circulation of the electrolyte between the anode and the cathode is employed. There is, however, apparently no novelty in this discovery. The use of the Elmore process showed that there was a considerable difficulty in avoiding a lamination, exfoliation or scaling of the copper of which the tubes were built up, due no doubt partially to the intermittent action of the burnishers. In the latest work this difficulty is said to be completely overcome, and large quantities of copper tubes are manufactured for both German, French and English naval construction. Nevertheless, from the point of view of the shareholder, the Elmore companies have not yet been a great success, but this seems to be largely due to over-capitalisation, and also, perhaps, to too great an expenditure on the purchase of the patents and promotion. The method itself appears to yield copper tubes at a price for which an article of similar quality cannot be obtained by any of the older processes of tube drawing.

At the annual meeting of the Elmore Patent Company, in June, 1896, it was stated that the Board of Trade had sanctioned the use of Elmore tubes of any diameter in passenger steamers built under their survey, and in 1899 the Elmore companies were amalgamated under the title of the English Electro-Metallurgical Company, with a capital of £700,000. Copper deposited by the Elmore process has shown a breaking tensile stress of 26·5 tons per square inch, with an elongation of 16·5 per cent. on a test piece, whose length, however, is not given; the elastic limit occurring under a tensile stress of 23·3 tons per square inch (*Journal of the Institution of Electrical Engineers*, vol. 29, p. 261). Commercial cast copper breaks at from 8·4 to 11·5 tons per square inch; ordinary wrought copper bolts break at about 14·7 tons per square inch; whilst special samples of non-electrolytic wrought copper have been obtained by working and drawing, with an ultimate breaking-stress of 26·7 tons per square inch. (Anderson's " Strength of Materials," pp. 81-86).

A second process for manufacturing solid copper tubes directly by means of electro-deposition is that patented by E. Dumoulin, Paris, Eng. Pat. 2,709, February, 1897; 2,710, February, 1897; and Eng. Pat. 16,360, 1895. The method adopted under these patents is that of rotating a horizontal mandril in the copper sulphate bath, which mandril acts as the cathode, but instead of employing agate burnishing rolls, animal membranes are hung over the mandril, which rotates beneath them. The pressure employed is only that due to the weight of the membrane. The membranes, which are called impregnators, are made of intestines, bladders, or sheepskin. It is found that the small amount of organic matter introduced into the solution by these impregnators, has a prejudicial effect on the copper deposited (cf. p. 545), and to avoid this the electrolytic bath is worked cold (under 15° C.), to as far as possible prevent solution. The skin is also treated with bi-chromate of potash solution, or with a solution of 10 to 40 per cent. formic aldehyde (Eng. Pat. 13,861, June, 1898), after which it is washed. This preliminary treatment of the impregnators is to render the organic matter in them either insoluble or to previously remove that part which would be soluble in the electrolyte, and it is stated that after this treatment, the skin impregnators may be employed successfully in either a cold or a hot solution. This process is said to be very successful, and under the title of the Electrical Copper Company it carries out the manufacture of copper tubes at Widnes, established in 1897. The capital is £500,000. A description of the Widnes works is given in the *Electrical Review* (vol. 43, 1898, pp. 561-562), by J. B. C. Kershaw. He states that there are 30 depositing vats, which are shallow lead lined troughs, each provided with a revolving mandril, 12 feet long and 1 foot 4 inches in diameter. The mandrils are half immersed in the electrolyte which contains 7 per cent. (*i.e.*, about 12 ounces per gallon), of free sulphuric acid and 40 ounces of copper sulphate per gallon. The solution is cooled and filtered after each passage through the depositing vats. The anodes consist of thick plates of raw copper bent into a semi-cylindrical form. Under normal conditions, 44 lbs. of copper are deposited on one mandril producing a tube of deposited copper weighing 14 ozs. to the square foot. The current density is from 35 to 40 ampères per square foot and the e. m. f. required is 1·6 volts per vat, at first, rising however, as the anodes become eaten away. Large tubes of copper made at Widnes, are cut longitudinally, and converted into sheets, having a superficial area of 48 square feet. The capacity of the plant in 1898 was equal to 60 such sheets per day (*i.e.*, two sheets per vat), but it was to be increased by the erection of depositing vats for the production of boiler tubes.

Fig. 136 shows a longitudinal section of Dumoulin's apparatus, Fig. 137 is an end elevation, and Fig. 138 is a plan. The apparatus is thus described in patent 16,360, 1895, by the patentee :—" *c* is a vessel or tank placed between two cast standards *b*, *b*, each of which carries a shaft, *e*, *e*, of brass, which is adapted to actuate the mandril and to conduct the current to the same.

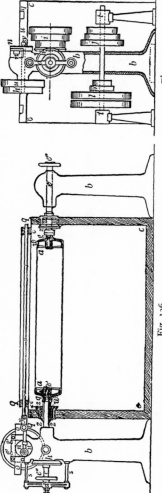

One end of each of these shafts is provided with a hand-wheel, e'', e'', and the other is provided with a screw-head, e^1, e^1; the said shafts pass into the vessel or tank through stuffing-boxes, f, f, which are allowed necessary play (about one centimeter in the vessel e, as shown at v, v). Of these two shafts, which slide freely in the standards b, b, one is free and the other strikes against an adjustable stop d, carried on two supports s, s, fixed to one of the standards b."

"This shaft carries a bevel wheel h, which is driven by another bevel wheel fixed to the step pulleys i, adapted to drive the mandril at three different speeds: the said pulleys are driven by the step pulleys j keyed to the driving shaft t."

"The mandril consists of a steel tube a, provided at its ends with brass heads w, w, provided with axial holes which are screwed to receive the ends e^1, e^1 of the two screwed shafts. In cases where the said heads are situated at some distance apart, as, for instance, in the manufacture of tubes, I provide in the axis of the mandril or tube a a metal rod connecting the two heads, the said rod being a good conductor, which distributes the current evenly along the length of the tube by means of copper wires arranged in bunches (*en tête de loup*)."

" The apparatus is employed as follows :—The ends of the tube a are placed opposite the two shafts e, e, and are screwed to the same by means of the hand wheels e'', e'' ; the stop d is then put in position, and the apparatus is ready to work."

" For the purpose of moving the impregnators to and fro, I arrange on the standard which carries the mechanism for actuating the mandril a small shaft u, u, which carries a step pulley k, which is actuated from another step pulley m keyed to the driving shaft. To the other end of this shaft u I attach a grooved plate n, n, in the groove of which the head of a connecting rod p slides. This connecting rod is fixed at the other end to a slide o, o, which moves in a guide g^1, g^1, so as to ensure the rod y sliding in a straight line : this movement is transmitted to the bar r, which carries the impregnators, and is supported by the frame q."

" By this means the movement is effected quite regularly : the stan-

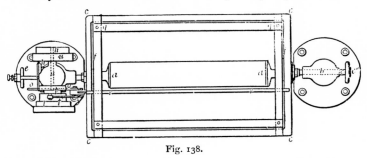

Fig. 138.

dards b, b, which are not connected with the tank and the stuffing boxes, have no metal deposited on them, which would very soon render them useless. The mandril is fixed to the shafts by the screw thread e^1, but the said shafts are protected against deposit by a ring of india-rubber g, which covers the end of the mandril, and is firmly clamped by rings of insulating material z, z."

A recent patent taken out by Dumoulin, Eng. Pat. 7,918, April 15, 1899, is for the formation of spiral bands on a revolving mandril by his process, presumably for subsequent manufacture into wire.

The theory favoured by the inventor of this process as explaining the satisfactory mechanical character of the copper deposited at such high current density, is that any rough particle of copper sticking out from the revolving surface scrapes a portion of non-conducting organic matter from the skin impregnator, and is thus protected from further growth by copper deposition until the remaining surface of the cathode is at the same height. There is, I consider, some doubt as to the satisfactory nature of this explanation, and it seems probable that all the methods which have been so far devised for the electro-deposition of copper in the form of tubes, etc., at high current density, and yielding copper of exceptionally good mechanical properties may be explained in one and the same way, an explanation which will be considered later. (See p. 559.)

For further details of later modifications of plant employed by Dumoulin, the reader is referred to the original patents cited above, the space limits of this work making it impossible to include any further details here.

Mr. Cowper Coles has devised a third method for depositing copper electrolytically at high current density in the form of tubes. The copper obtained has very satisfactory mechanical properties. A paper on this method was read by Mr. Cowper Coles, before the Institute of Electrical Engineers in January, 1900. The method consists in rotating mandrils with vertical axes in a bath of copper sulphate of the usual composition. No burnishers or rubbing contacts on the copper surface are employed, but the mandrils are rotated at a very high angular velocity. The arrangement of the cathode and anodes was described in the original paper as follows :—

" The apparatus employed for the centrifugal process consists of a wooden vat ' A,' Figs. 139 and 140, in which are placed anodes composed of crude copper. The cathode is a hollow mandril ' B,' made of brass, which is supported on a revolving shaft ' C.' The shaft is brought through the bottom of the vat to the top of the cell, and is protected from the acid copper sulphate solution by a lead-cased wrought-iron column. The shaft is caused to revolve by gearing placed beneath the depositing cell, the speed varying with the size of the mandril employed. The mandril at one end is fitted with an eye bolt ' D ' for lifting purposes, and a circular brass casting ' E,' against which the contact brushes ' F ' rub for collecting the negative current. The brushes are fitted to arms ' G,' which are made to turn back to allow of the easy withdrawal and insertion of the mandril. ' H ' is a baffle plate made of an insulating material, placed at the bottom of the mandril to prevent the formation of copper ' trees ' or ' nodules.' The electrolyte is briskly circulated through the cells by means of an acid proof pump or air pump, the solution being forced or pumped to a reservoir, where it passes through a filter to rid it of all impurities in suspension. When it is desired to obtain a sheet of copper, the tube, after being removed from the mandril, is cut, flattened out, and annealed ; when a wire, a piece of insulating material is wound round the mandril in the form of a spiral, and copper is deposited between the threads until the thickness of copper is equal to

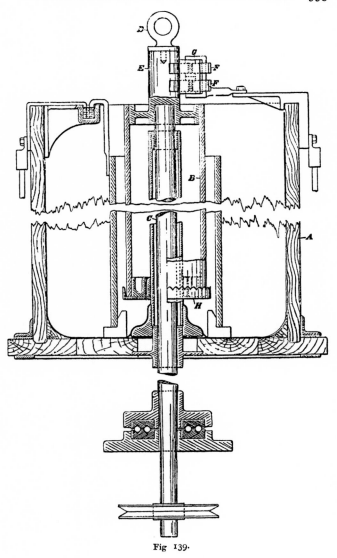

Fig 139.

the distance between the threads, the spiral is then removed, and after annealing is drawn out in the usual manner.   The mandril is given a small taper, and is slightly greased to facilitate the removal of the tubes.   Figs. 141 and 142 show a modified arrrangement for driving the mandril from above instead of from below.''

The inventor apparently advises working the solution at a temperature of about 65° C., with a peripheral speed of surface of the cathode

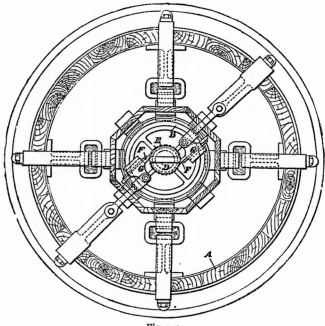

Fig. 140.

of about 500 feet per minute, and a current density of from 100 to 200 ampères per square foot.   The solution found satisfactory by the inventor had the following composition :—

Copper sulphate crystals, 32 ozs. per gallon.
Sulphuric acid (H_2SO_4) 12·6 ozs. per gallon.
Made up to one gallon with water.

The copper obtained by this process (unannealed), had a breaking tensile stress of 22·1 tons per square inch, and had, in the form of drawn wire, a breaking strain of 29 tons per square inch, whilst annealed wire broke at 20 tons per square inch. The electrical conductivity was 99 per cent. (presumably of the Mathiessen standard). Tubes 12 inches in diameter have been produced.

In the experiments described, the inventor stated that there was a critical speed of surface of the anode below which the good character of the copper deposited at high current densities was not maintained. The critical speed was stated to be not less than 1,000 feet per minute.*

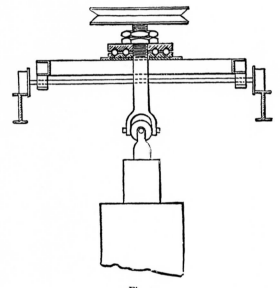

Fig. 141.

This invention is covered by Eng. Pat. 16,210, Aug. 9, 1899, and Eng. Pat. 21,197, 1898. An abstract of the former given in the *Journal Soc. Chem. Ind.*, vol. 19, p. 671 is as follows:—

---

* This appears to be an error in Mr. Cowper Coles' paper, for he also speaks of normally using a surface speed of 500 feet per minute and obtaining excellent results.

" Manufacture of thin tubes or sheets of copper or other metal by electro-deposition :—Thin tubes or sheets of copper or other metal are formed by successively depositing coatings or layers of metal on a cathode or mandril whilst it is rotated at high speed, and a thin coating of a greasy matter, or of an oxide, or sulphide, or the like is applied before each coating or layer of metal is deposited. The rotation may be effected by means such as those described in Eng. Pat. 21,197, 1898."

The large value per ton of copper tubes of high class mechanical property, of course, justifies the employment of far larger amounts of energy than could be used with economy if only raw electrolytic copper requiring further manufacturing operations were employed, but it appears to me somewhat doubtful, if even these advantageous conditions for a high current density can economically allow of a current density of 100 to 200 ampères per square foot, and at the same time

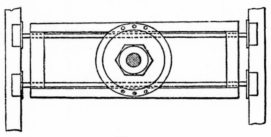

Fig. 142.

permit the further large consumption of energy necessary to maintain a surface speed of cathode of 500, 600, and even as the author states 1,000 feet per minute (*Journ. Inst. Elect. Eng.*, vol. 29, p. 284), whilst the electrolyte is pumped through the cell at such a rate that it is renewed every five minutes (*loc. cit.*, p. 266). No doubt time and the operation of this process on the manufacturing scale will finally settle this question. In my opinion, if the method is finally successful, the current density employed will probably be not at all higher than 100 and the peripheral speed will be kept as low as possible, whilst the rate of renewing the electrolyte will be reduced. Prophecy is, however, usually a rash and always a thankless office. Mr. Cowper Coles attributes the favourable results obtained with his process, firstly to the skin friction between the copper and the liquid burnishing the metal, and secondly to the centrifugal force driving off all solid particles and gas bubbles.

Many other inventors have devised processes for either rotating the cathodes or the electrolyte, or for rapidly circulating the electrolyte, or continuously rubbing the deposited copper surface, but none of these processes have to my knowledge been described in any detail, although several of them are said to be in successful use for the direct manufacture of copper tubes and sheet, thus saving the cost of remelting, rolling, etc.

Mr. Henry Wilde, F.R.S. patented (Eng. Pat. 4,515, 1875), a process in which both the liquid and the cathode are rotated at a high speed, by means of which he claimed, a sound copper deposit could be obtained at an exceptionally high current density. Mr. Wilde states that his process has been successfully operated in Manchester and elsewhere for more than 20 years (*Journ. Soc. Elect. Eng.*, vol. 29, p. 272).

Mr. Z. T. de Ferranti states that somewhere about 1893 or 1894, he saw a demonstration in Paris of a process for obtaining sheet copper, in which horizontal rollers half immersed in copper sulphate electrolyte, were rotated at a very high speed, and copper deposited on them at a high current density. The cylinders were from 2 feet 6 inches to 3 feet in diameter. The copper sheet deposited was beautifully smooth and regular (*Journ. Soc. Elect. Eng.*, vol. 29, p. 280).

Mr. Sanders, who was at one time manager of the Elmore works, also produced at Eastbourne very fine deposits of copper without burnishing, by rotating a horizontal cathode cylinder three quarters immersed in the copper sulphate electrolyte at a speed of about 100 revs. per minute, and employing a very high current density. Mr. Sanders appeared to wish to apply this method to the production of wire, for he wound a wire on the cylindrical cathode before depositing the copper, and a most beautiful spiral deposit of copper was obtained on this wire, which grew until it was large enough to be uncoiled from the mandril. Mr. Ferranti states that the wire thus obtained after being drawn through a few dies was of most excellent character. This method was not, however, employed commercially, because of the prevailing idea that it is not possible to produce copper wire more cheaply than by the old-fashioned method, the prevailing feeling being that the absolutely lowest possible cost price has been already reached. (*Journ. Soc. Elect. Eng.*, vol. 29, p. 280.)

Mr. Swan also has shown in a lecture at the Royal Institution, that very smooth and excellent copper deposits may be obtained at very high current density, by moving the cathode plates at a high surface speed in the electrolyte. (*Journ. Soc. Elect. Eng.*, vol. 29, p. 281.)

A process was patented by Paul David in February or March, 1894, and was taken over and used by La Société des Cuivres de France, 3, Rue Cambon, Paris. The centrifugal process in this patent is claimed

" for the fabrication of tubes and other objects in copper, by means of simple rotation in the electrolyte." (*Journ. Soc. Elec. Eng.*, vol. 29, p. 283).

It was stated in an article on the Anaçonda Copper Refining Co. (*Electrician*, vol. 38, Nov., 1896, p. 147), that after six years' attempts, M. Thofehrn, the manager, had developed a process for rapid copper deposition, by which the metal was obtained in the finished state, either as sheets for rolling, or as tubes. The process was as follows :—" A hollow cylinder about eight feet long and three feet in diameter, is immersed in the electrolyte and forms the cathode, upon which the copper is electrically deposited with a current density of 50 to 100 ampères per square foot. The anodes are common converter pigs, ingots, shot or scrap copper or whatever is at the disposal of the refiner. It is, indeed, stated that white metal containing only 75 to 80 per cent. of copper has been used satisfactorily. It is an advantage that this copper in its crude market state can be used without transforming it previously by melting into plates for anodes, and by avoiding this preliminary casting as much as 12/6 per ton is saved. The cylinder which is used as a cathode revolves in a tank at a slow speed, and the copper that precipitates upon it, is in the shape of extremely fine crystals assuming the form of hexagonal needles or hairs which can only be seen through a powerful microscope. In order to produce a good and dense deposit of copper, these microscopic needles must be interwoven, felted and compressed. This is attained by the action of numerous small jets of electrolyte directed under pressure against the revolving cylinder. While apparently to the naked eye no action whatever seems to follow from the jets, the final result is remarkable. The copper deposited is to the full extent of its thickness thoroughly dense ; even after continuous annealing and hammering no foliation whatever can be obtained. The foliation is one of the main drawbacks of the Elmore process, rendering its use for the manufacture of tubes, etc., unsafe. The method of directing a stream of fresh electrolyte on the surface of the cathode has another advantage. The whole cylinder is surrounded by the fresh electrolyte coming from the purifying tanks and no matter how impure the anodes are, and how large the amount of impurities in the liquid, the cathodes are fully surrounded by clean electrolyte only. The course of the liquid is as follows :—From the epuration tanks it is directed against the cathode, surrounding it with a layer of about ¼ inch thick. . It goes from there to the anodes, and then to the collecting tanks previous to its return to the epuration tank. After one inch thickness of copper is deposited upon the cathode, the sheet of metal is taken off by opening the cylinder by means of a small hydraulic jack, especially devised for this purpose. The plates of the deposit on the cylinder are not continuous. A small seam being left in the cylinder for the attachment of the hydraulic jack. After opening the plates sufficiently to slide out the cylinder, it is placed on a cast iron bed plate, held fixed at one edge, and a heavy roller is inserted, and pulled forward by a hydraulic ram, thus transforming the plate into a flat sheet. The plates so produced are ready to go to the rolling mill without preliminary melting or annealing. The quality of the copper produced by this method is higher than that produced in the usual way by melting, etc. The conductivity is generally 100 per cent, Mathiessen's standard. The tensile strength is about 33·48 tons per square inch,* and the elongation is about 2 per cent., whilst the number of twists in six inches of No. 12 wire is about 100. These tests are all made on the hard drawn wire." And yet—and yet, it is announced in 1900, only four years after this extremely laudatory description, that this method had been abandoned (*Journ. Elec. Eng.*, vol. 29, p. 283), or to be more precise the projection of the electrolyte on the cathode had been abandoned, and although I am under the impression that the whole of the extra density current deposition process has been discarded at Anaconda, I do not think I could produce written or printed evidence to that effect. What is the reason of this abandonment? Does it not also seem to point to the fact that the cost of the energy at such large current density, and with so much pumping of solution, is too high, in spite of the fact that it is stated at the end of the paper above quoted that the cost of refining one ton of copper by the high current density method described is *estimated* to be not more than £3 6s. 8d. per ton. Was the estimate correct?

Reason for Possibility of Using High Current Density with Rotating Cathodes.—In the discussion on Cowper Coles' paper on his centrifugal process, Mr. Alan Williams pointed out that probably the true action of Dumoulin's so-called impregnators, and also the high surface velocity in the Cowper Coles process, were both the same, and consist in the wiping off from the surface of the cathode the impoverished electrolyte from which the copper is so rapidly removed by the high current density employed. Mr. Williams said, "My theory of this process, and of all other processes of this character, which have produced satisfactory results is this : In an ordinary depositing cell immediately the current is switched on, copper is deposited on the cathode and a thin layer of electrolyte touching the cathode becomes very considerably weakened in consequence. Before more deposition can take place this layer of exhausted electrolyte must be removed. *If left to itself*

---

* This appears enormous, but even if the load is given in short tons of 2,000 lbs. it is still equivalent to an ultimate tensile strength of 29·88 tons avoirdupois, which is surprisingly large.

*it will naturally break up in a sort of granular formation, being perforated
at a number of points, and allowing copper-bearing electrolyte to have access to
the cathode through the perforations, and permitting a number of nodules
of copper to be formed* ; an irregular deposit then grows on these nodules.
To prevent this action from taking place it is necessary to remove the
film of exhausted electrolyte immediately it is formed, either by
friction, violent circulation, or rotation of the cathode.   Consequently
the process which most effectually removes this film, and which at the
same time gives the most uniform distribution of current and the most
uniform strength of electrolyte over the whole surface of the cathode,
should be one to produce the best deposits of copper and to allow of the

COST AND OUTPUT OF COPPER.

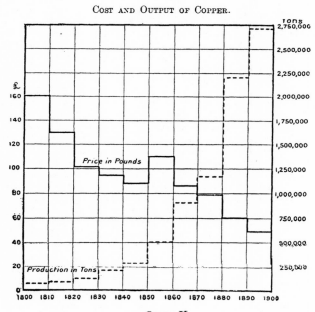

CURVE II.

Average price per ton of English Tough Copper during each decade of
a century, shown ─────────
Total Copper production during each decade,   ·······················
Fig. 142a.

highest current density." That portion of the above remarks here given in italics, was not so indicated in the original, but I have so marked it to point out that this is the only portion of Mr. Williams' excellent explanation which does not quite commend itself to my judgment. I fancy that the bad deposit under the condition of a layer of electrolyte weak in copper, close against the cathode, is due to the deposition of hydrogen, which must occur in a weak solution simultaneously with the copper, thus causing bubbles of gas to be interspersed with particles of copper, rendering the deposit loose and flocculent.

The Price of Copper and its Fluctuations.—A very complete treatise on the price of copper and its annual output for a hundred

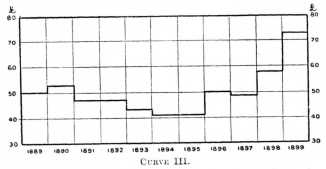

CURVE III.

Curve showing the price of G. M. B.'s per ton on December 31 of each year from 1889 to 1899 (October 1).

Fig. 142b.

years back has been recently produced by Messrs. N. Brown and C. C. Turnbull (London : Effingham Wilson, 1899, 2s. 6d.), and in a review of this pamphlet appearing in the *Electrician*, vol. 44, 1899, p. 225, Curves II. and III. shown here are given, which practically sum up the more important features in these variations.

The G. M. B. copper is of 96 per cent. purity, and the " tough " copper is about 99·5 per cent. purity, the " tough " costing from £2 to £3 per ton more than the good merchantable brands (G. M. B.).

The price of electrolytic copper is from £2 to £3 per ton more expensive than the " tough " copper.

The prices per ton of various brands of copper given in the *Western News* Metal Market on August 8th, 1901, were as follows :—
G. M. B.'s, £65 5s. : English tough, £71 ; best selected, £72 ; and strong sheets, £83. Probably electrolytic would be £73 to £73 10s.

The Cost of Refining Copper Electrolytically.—The cost per ton of refining copper electrolytically, when the anodes and cathodes are arranged in the usual parallel method in the vats, is stated by Peters* to be £3 5s., whilst when arranged in the series method, as in the process adopted by Hayden, Stalmann, and others, Peters states the cost of refining to be about £3 8s. per ton. As is stated on page 559 of this volume Trofehrn estimated the cost of his special high current density process of refining to be £3 6s 8d. per ton, whilst when using the ordinary process, at Anaconda, he considered that the cost was £2 18s. 6d. per ton. These figures do not allow anything for the value of the silver and gold recovered, which is an extra source of profit.

T. Ulke states (*Elect. Rev.*, N. York, vol. 38, 1901, p. 85 and 101—103) that the cost of refining one ton of anode copper is less than £1 13s. 4d. at the Raritan Copper Refinery at Perth-Amboy, which is now the largest in America, with an output of 150 tons of copper per day. To reduce the costs of refining to this very low value automatic machinery is employed for casting the anodes, and every possible labour-saving device is adopted. I find it difficult to believe that this last estimate can possibly cover all the costs of production. For I am of the opinion that if the cost of labour and the salaries for management, etc., were reduced to zero the cost of refining one ton of copper electrolytically could not be less than about £1 10s.; that is, if the value of the gold and silver separated from the anode copper is not considered as a set off to part of the cost of refining.

---

* "Modern Copper Smelting." Peters.

CHAPTER IV.

ELECTROLYTIC GOLD AND SILVER BULLION REFINING.

Electrolytic Refining of Gold Bullion (Wohlwill's Process). —Pforzheim Process of Recovering Gold and Silver from Complex Jewellery Alloys.— Electrolytic Silver Refining (Möbius' Process).

The Electrolytic Refining of Gold Bullion.—The only process employed for the electrolytic refining of gold bullion is that devised by Dr. Wohlwill, which has been employed, it is said, since as early as 1880, at the Norddeutsche Affinerie, at Hamburg (E. Wohlwill, *Zeitschrift für Elektrochemie*, vol. 4, pp. 378—385, 402—409, and 421—423). Three good resumés of Dr. Wohlwill's papers may be found in English in the *Journal of the Society of Chemical Industry*, vol. 17, 1898, p. 585 ; in the *Electrician* at the end of a paper by Titus Ulke, vol. 46, p. 582 ; and in a paper by J. B. C. Kershaw, on "Electrolytic Methods of Bullion Refining," in the *Electrician*, vol. 41, p. 187.

The electrolyte employed by Wohlwill consists of an aqueous solution of auric chloride, containing from 5 to 30 grams of gold per litre, mixed with from 20 to 50 cubic centimetres of fuming hydrochloric acid (s. g. $= 1\cdot19$) per litre. Distilled water should be employed for the solution. Instead of the hydrochloric acid from 21 grams up to as much as nearly 100 grams of sodium chloride per litre may be used. The amount of sodium chloride must not, however, exceed 100 grams per litre, because silver chloride is, as is well known, soluble in strong solutions of common salt, and under the condition of the presence of an excess of sodium chloride in the electrolyte the silver chloride formed at the anode, instead of remaining insoluble and becoming detached and falling into the anode sludge, dissolves, and is finally deposited with the gold at the cathode. The temperature at which the solution is worked is from 60° to 70° C.

The anodes consist of bars of the impure gold bullion $0\cdot16$ inch thick, and containing lead, silver, platinum, palladium, iridium, and osmium as impurities. When first inserted the anodes weigh $8\cdot8$ lbs. each, and have a total surface (on both sides) of $1\cdot1128$ square feet.

The cathodes are made of thin rolled sheets of electrolytically deposited gold of the same length as the anodes, but are narrower, for during the course of the electrolytic action they rapidly grow in thickness in every direction. There is no formation of arborescent or loose crystalline growths at the cathode even when the current density is as high as 100 ampères per square foot. The deposited gold is in fact dense and coherent.

The distance between the anodes and cathodes need not be greater than about 3 centimetres, and there is then no danger of a short circuit occurring due to metallic growths on the cathodes. It is a curious fact that at the commencement of the process, whilst the electrolyte is a comparatively pure solution of auric chloride with hydrochloric acid or sodium chloride, the cathode deposit is looser and less dense than it is after the soluble impurities from the anode have impregnated it with chlorides of lead, platinum, etc. Palladium, however, has a prejudicial effect in this respect, and the solution must not be permitted to contain more than 5 grams per litre of this metal. At the starting of a new bath of solution it is on this account usual to keep the anodes and cathodes farther apart than the 3 cm. prescribed above, and the refiner changes the anodes more frequently in order to avoid the danger of a short circuit.

The electrolytic tanks employed consist of porcelain or stone-ware vats, upon which wooden frames are supported carrying the positive and negative copper conducting leads. On these leads there are resting copper cross-bars, about nine per vat, from which four rows of anodes and five rows of cathodes hang. The tanks are arranged in series. The depositing vats in the Frankfort Gold and Silver Refinery in Hamburg, where a current density of 95 ampères per square foot is employed, have an output of 1,650 lbs. of gold per 24 hours (that is 265·3 tons per year of 360 days of 24 hours each), and yet only occupy an area of 64 square feet. (The area occupied by a copper refinery running at a current density of about 12 ampères per square foot of cathode surface for a similar daily output is about 581·9 square feet. This figure refers to the area of the whole works at Anaconda. The *vat space* at Anaconda for the same output is 182·1 square feet. If, however, the current density used at Anaconda were the same as that employed in the Hamburg Refinery, namely, 95 ampères per square foot, the works area for the same output would be 73·52 square feet, and the area of the vats would be 23 square feet, see pp. 502 and 505.)

The impurities present in the anodes partly pass into the solution and partly are deposited as anode sludge.

The anode sludge contains all the silver as chloride together with about ten per cent. of its weight of gold. The gold thus deposited in the anode sludge is in the metallic state, but its presence is not due to the mechanical disintegration of the anode. Dr. Wohlwill finds that it has quite a different mechanical condition to the gold of which the anode is formed, and he considers that its presence is due to the formation of aurous chloride at the anode, which aurous chloride is immediately converted by a secondary reaction into auric chloride and gold, thus :—

$$3 \, AuCl = AuCl_3 + 2 \, Au,$$

the gold being precipitated in a very finely divided form.

A high current density at the anode, and a high temperature of the electrolyte, are unfavourable to the formation of the aurous chloride, and these conditions are, therefore, conducive to a small amount of gold present in the anode sludge, and must, therefore, be as far as possible attained. With a current density above 136 ampères per square foot, and a temperature of 65° to 70° C., the gold present in the anode sludge is reduced to a minimum, whilst, on the other hand, with a current density of 0·09 ampères per square foot apparently only aurous chloride is formed at the anode by the direct action of the current, a correspondingly large amount of gold appearing in the sludge.

Bismuth present in the anode is converted into bismuth oxychloride, and if sufficient hydrochloric acid is not present to hold it in solution it is precipitated with the anode sludge.

The greater part of the iridium and other metals of the platinum group, except platinum and palladium, remains undissolved and is precipitated with the sludge.

The lead present in the anodes is at first dissolved in the solution as lead chloride until the electrolyte is saturated, when it crystallises out with the sludge.

If a large amount of lead is present in the bullion anodes, both electrodes and the internal surface of the tanks and the surface of the electrolyte become coated with crystals of lead chloride, and it is found advisable under these circumstances to avoid the inconvenience of the formation of these coatings of lead chloride crystals by adding to the electrolyte some free concentrated sulphuric acid about equal in volume to the hydrochloric acid present. The lead then forms an insoluble sulphate, which passes into the anode sludge. If the anode bullion is rich in silver or lead, or both, it may become so coated with a covering of silver chloride and lead chloride or sulphate as to seriously interfere with the electro-chemical action, and under these circumstances the anode surface must be kept clean by an automatic scraper, which comes into action periodically, at intervals depending upon the rapidity with which the coating is formed.

Part of the lead and bismuth, and all the platinum and palladium, pass into solution as chlorides, but they are not precipitated with the gold, and indeed their presence, as has already been stated (with the exception of the palladium, which must never reach a higher value than 5 grams per litre), is favourable to the formation of a dense and pure deposit of gold.

The platinum is recovered from the electrolyte from time to time, when sufficient has dissolved, by first precipitating the gold as metallic gold by means of ferrous sulphate, and then the platinum as ammonium platinum chloride by adding ammonium chloride to the solution. The palladium may finally be precipitated as iodide by the addition of iodide of potassium.

The cathode gold obtained has a fineness of over 999·8 parts of gold per 1000. The impurity present is chiefly silver, a small amount of the anode silver being dissolved in the hydrochloric acid or alkaline chlorides present in the electrolyte.

Potassium chloride is said to be inconvenient for use instead of sodium chloride with the auric chloride electrolyte, on account of the precipitation of the platinum chloride which would thereby occur, but it appears to the present writer that this might probably prove advantageous; for the platinum would thereby be automatically removed into the sludge, and could be easily re-obtained, when desired, by a simple treatment, apparently more readily than by the method already referred to, by which it is removed periodically from the electrolyte.

Clearly the formation of chlorides by the impurities in the anode bullion, and the fact that only gold is precipitated at the cathode, must gradually render the electrolyte weaker and weaker in auric chloride. This reduction of the amount of auric chloride present is made good from time to time by the addition of fresh amounts of this salt in solution.

The anode sludge is collected, drained, and washed with distilled water from time to time, and the washings from the sludge are returned to the electrolytic vats to make up the loss of water which is constantly occurring, due to the high temperature at which the electrolysis is conducted.

As it is necessary to maintain the electrolyte at a uniform level, notwithstanding the loss due to evaporation, a closed reservoir containing fresh electrolyte is arranged above the electrolytic tanks, and is provided with two long tubes, the mouth of one of which (A, Fig. 143) passes just beneath the surface of the solution in the tanks, and the mouth of the other, B, passes from the bottom of the reservoir to nearly the bottom of the electrolytic vat. Immediately the evaporation of the electrolyte uncovers the mouth of the first tube, air passes up it into the reservoir, and liquid flows from the reservoir down the second tube into the electrolytic tank until its level rises so much as to once more close the mouth of the first tube, and thus stop the entry of the air, and therefore the further flow from the reservoir. Fig. 143 shows this arrangement somewhat diagrammatically.

With an anode 0·16 inch thick, and a current whose initial density is 40 ampères per square foot, the anode is so far consumed in twenty-four hours that the remainder, about one-tenth of the original weight,

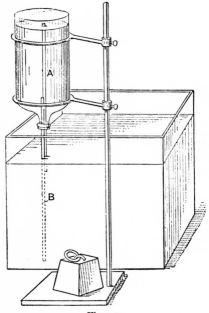

Fig. 143.

can be removed, washed, remelted, and used in making fresh anodes. The final current density with solutions containing a satisfactory amount of foreign salts is from about 92 to 95 ampères per square foot, and the voltage required at this current density is 1 volt.

If the sodium chloride or the hydrochloric acid are not present in the electrolyte chlorine is evolved at the anodes, and very little gold is dissolved. This fact, ascertained by Dr. Wohlwill, is somewhat surprising, and this gentleman has shown that undoubtedly the compound which is really decomposed electrolytically in this process is a double salt of auric chloride and sodium chloride, or auric chloride and hydrogen chloride, as the case may be, and, in short, may be any compound of the form

$$\text{AuCl}_3\text{MCl,}$$

where M is any monovalent radical such as Na, H, K, etc.

Finally, Dr. Wohlwill has shown that if low current density is employed a larger weight of gold is deposited at the cathode per ampère hour than if greater current densities are employed. This he finds to be due to the fact that with low current densities we have to deal with aurous compounds, whilst with high current densities auric compounds with only one-third of the electro-chemical equivalent are being treated. It would therefore appear at first sight that it might be preferable to work at low current densities, but there are two reasons against this ; one is that when low current densities are employed and aurous chloride is formed, a far larger amount of gold is precipitated with the anode sludge, as has already been explained; and the second reason is that the low current density would necessitate the locking up of an unnecessarily large weight of gold in the anodes and cathodes, for a given output of gold, thus increasing the cost of refining.

The weight of anode and stock gold employed at Hamburg, at the current density of 92 ampères per square foot, is from about $\frac{1}{100}$ to $\frac{1}{300}$ of the weight of the annual output of the refinery.*

Dr. Wohlwill's gold bullion refining process is employed com. mercially at the Frankfürter Gold und Silber Scheide Anstalt in Hamburg, where the output is 265·3 tons per annum, and also at the U. S. Mint, Philadelphia, Pa., where the estimated daily output is 1,000 oz. troy (= 11·04 tons per annum). The process here installed is the Wohlwill process slightly modified by Dr. D. K. Tuttle and Mr. H. J. Schlacker.

The English patent for Dr. Wohlwill's process is No. 7,783, 1896.

The fact that this process involves practically no cost in acid, and also that it has been so recently adopted by the U.S.A. Mint after careful inquiry, points to the fact that it must entirely supersede the older chemical acid parting methods of bullion refining.

Electrolytic Recovery of Gold, Silver, etc., from Complex Alloys.—A very large quantity of low-grade gold and silver alloys is employed in jewellery and other metal plate work, and the precious metals, and indeed the copper, may be profitably recovered by electro-

---

* The weight of copper in stock at Anaconda in the form of anodes is about 10·85 per cent., and as cathodes and stock anodes is about 3·5 per cent. of the total annual output of the refinery, that is a total of about 14·5 per cent. (See p. 507 and p. 509.

lysis from miscellaneous scraps and cuttings, and from old or damaged articles made of these complex alloys. The ordinary Möbius silver refining process is not suited to this type of work, for the electrolyte becomes rapidly loaded with copper, zinc and nickel salts, the copper especially tending to become precipitated with the silver at the cathode, and in such cases it is advisable to employ the electrolytic process which has been successfully employed at Pforzheim, in the Gold und Silber Scheide Anstalt there, ever since 1893. This process is described by A. Dietzel (*Zeitschrift für Elektrochemie*, vol. 6, pp. 81-85, July 27th, 1899. An abstract of this paper appears in *Science Abstracts*, vol. 2, 1899, p. 776). The alloys which are treated at Pforzheim consist of cuttings from various sources, and the average composition is as follows :—

Gold	5 to 7 per cent.
Silver	22 to 50 ,, ,,
Copper	40 to 65 ,, ,
Lead, Zinc and Tin . . : .	About 5 ,, ,,

Whilst traces of cadmium, iron, nickel and platinum are also present.

The electrolyte which is employed is a solution of copper nitrate ; the copper and nickel pass into solution and the copper is deposited electrolytically at the cathode, but the silver remains in solution and is deposited subsequently by chemical means outside the electrolytic vat. The electrolyte must be maintained acid by the addition of nitric acid. (Presumably this is to prevent the co-deposition of the silver with the copper, and it is rather astonishing that in any case this does not take place to an inconvenient extent.) The chemical method of removing the silver from the electrolyte from time to time consists in shaking up the electrolyte in special cylinders filled with scrap copper, which displaces the silver by chemical substitution. The silver in solution can be reduced to 0·03 per cent. by this means. The iron in the solution is removed, or rather is partially removed, by aeration and filtration. The insoluble anode sludge contains gold, silver, platinum, copper, tin and lead, and is treated by chemical methods for the recovery of the gold. It is stated that under favourable working conditions the value of the copper deposited in the electrolytic vat covers the whole cost of the process. Zinc, aluminium, cadmium and nickel when present in the alloys pass into solution with the copper and silver; and as these impurities cannot be removed economically, the electrolyte becomes in time surcharged with impurities and has to be discarded. The apparatus employed at Pforzheim is fully described and illustrated in the original paper. About 120 to 130 pounds avoirdupois of alloy are treated daily in the refinery.

Electrolytic Silver Refining.—" It has been truly said that industrial electro-chemistry and electro-metallurgy are rich in resurrections of forgotten processes " (J. B. C. Kershaw, *Electrician*, vol. 38, p. 605), and in illustration of this remark the writer pointed out that English Patent, 13,755, 1851, by Watt, describes a method of electrolytic silver refining which is essentially the process which is the most successfully employed on a commercial scale at the present day.

This successful electrolytic silver refining process is that patented by Möbius (English Patent, 16,554, Dec. 16th, 1884). This process consists essentially of employing anodes of impure silver, containing about 95 per cent. of metallic silver, in a bath of dilute nitric acid, the silver is deposited upon sheet silver cathodes in the form of loosely adherent crystals, the current density employed being as high as possible without producing inconvenient heating of the electrolyte. The current density employed at the works of the St. Louis Smelting and Refining Company is stated to be rather over 32 ampères per square foot of cathode surface (C. Schnabel, " Abstracts Proc. Inst. C. E.," vol. 116 [ii] 1894, p. 86-87). At the St. Louis works the silver, which contains 28·8 grains of gold per lb. troy (=5 parts per thousand), is cast into plates measuring 10 inches by 8 inches and 2½ inches thick. Two such plates placed together in a linen bag form one anode, and for a cathode a rolled plate of fine silver is employed. The dissolving vessels are pitch pine vats, each divided into 7 cells and rendered impermeable to the electrolyte by a coating of bitumen. Each cell contains 10 pairs of electrodes, the plates in one cell are in parallel, whilst the cells are arranged in series and 10 vats containing 70 cells, arranged in series, require 100 volts at a current density of 32 ampères per square foot, that is about 1·5 volts per cell.* The anodes are completely dissolved in from 30 to 40 hours. The electrolyte used in the first instance is water containing one-tenth per cent. of nitric acid, and as the process progresses the nitric acid strength of the solution, which is reduced by the formation of cupric nitrate, is kept up by the addition of a very weak solution of silver nitrate. The reduced silver separates in a crystalline form shooting across from the cathode to the anode, so that short circuiting would frequently occur if it were not that the loose crystalline deposit of silver on the cathode is scraped off at intervals. The deposited silver so removed falls to the bottom of the bath into boxes with double bottoms perforated and covered with linen, for collecting the precipitate which is removed once a day. It is then washed, pressed, dried, and melted. The finely divided gold is kept back in the linen anode bags, where it is allowed to accumulate for a week before cleaning out. The gold is then boiled with nitric acid, washed, dried, and melted with a little

---

* The current is supplied by a 100 volt, 200 ampère dynamo.

sand or borax and yields a bullion 999 fine. The silver crystals, after washing and drying, are found to be free from gold, and assays 999·5 fine. The silver nitrate obtained from the final treatment of the gold is used to keep up the strength of the bath as described above. At the St. Louis works 30,000 ounces of silver are parted by this method daily, and at Pittsburg a still larger plant of 40,000 ounces capacity is in use. At the latter place the finely divided gold is melted with a small quantity of silver before being subjected to the final boiling with nitric acid, this is no doubt to facilitate the nitric acid parting.

Möbius states that it is necessary that nitrate of copper should be in the electrolyte to ensure that all the lead present in the anodes may be converted into peroxide. The linen bags which surround the anodes are saturated with coal oil, linseed oil, and paraffin, in order to protect the fibre from the action of the dilute nitric acid. Under these conditions they are very little affected.

The anode precipitate contained in the linen bags round the anodes contains all the gold, platinum, lead (as lead peroxide), and antimony, and frequently some silver as peroxide. The copper, zinc, and iron accumulate in the electrolyte, and when the copper becomes too concentrated it is removed from the solution by passing, firstly a feeble current with carbon anodes, by means of which the silver is removed from the solution, and then the silver cathode is replaced by a copper one, and the whole of the copper is deposited from the solution as a powdery crystalline precipitate upon the cathode by means of a current of high density. The liquid thus nearly or completely freed from copper is then used once more, either for making up new electrolyte or for replacing liquid lost by evaporation in the course of the process.

In silver and in gold refining by electrolysis it is desirable, on account of the high price of the metal under treatment, that the weight purified per diem should be as large as possible for a plant of a given size; it is, moreover, quite unimportant in what state of cohesion the metal is deposited, for both gold and silver can be readily melted down without loss or any other inconvenience, even when they are in a very fine state of division. This peculiarity is due to the fact that oxygen does not act upon them when in the melted state, a characteristic which is not shared by the other metals which are more usually met with. In devising his process Möbius recognised this fact, and therefore employed the high current density stated, and arranged that a mechanical device to scrape off the precipitated silver on the cathodes should act at short intervals in order to prevent the danger of a short circuit caused by the silver crystals from the cathode bridging across to the anode. In the latest form of this silver purification process, however, Möbius employed a moving con tinuous band of silver as a cathode, which, passing through the electrolyte, received a deposit of silver crystals, and then, as it moved out of the solution, it passed under scrapers outside the vats, which removed the silver deposit from it. When this modification of Möbius' patent was first employed at Perth Amboy it was a failure, and was finally abandoned, chiefly on account of the difficulty of satisfactorily scraping off the adhering silver deposit from the moving cathode belts, but by oiling the belts, a device of Mr. G. Nebel, it was found that the close adherence of the silver deposit to the cathode was much reduced, and that the crystals might then be readily scraped off. The silver cathodes, whether in the form of plates or belts, must be removed from the electrolyte when the current is not passing, or they will be found to gradually dissolve. (Titus Ulke, *Electrician*, vol. 46, 1901, p. 583.)

A further description of the Möbius process, as conducted in 1888, is given by Courtenay De Kalb (*Eng. and Min. Journ.*, vol. 45, p. 452), an abstract of which I quote here (*Journ. Soc. Chem. Ind.*, vol. 7, 1888, p. 571) :—

"The Möbius process for separating gold and silver has now been put into practical operation and with excellent results. A small plant has been erected in Kansas City, and another in the Pennsylvania Lead Company's Works near Pittsburg. Twenty thousand ounces of silver bullion are being refined daily, and the product is said to be the best ever offered to the United States Mint in Philadelphia, the bars running from 999 to 999·5 fine. H. G. Toney has secured the patent rights for New York City. The success of the method has there received fresh demonstration, one lot of 1,000 ounces having turned out 1,000 fine. Refining by this method, it is said, can be done in New York for three-fourths of a cent per ounce (that is, £56 per ton), whilst the minimum charge that will leave a profit to the refiner using other methods is one cent per ounce."

The principle upon which the process depends is that when silver to be refined is made the anode in a weak nitric acid bath, the silver will pass into solution as nitrate through the action of an electric current from a dynamo, and will be re-deposited in the metallic state upon a silver plate which constitutes the cathode. In practice the bullion is cast in plates ½ inch thick and about 14 inches square. These plates are inserted into muslin bags intended to retain the anode deposit. The plates in the bags are suspended in the tank of acid from copper rods. Alternating with these are the silver plates for the deposition of the refined silver. The current employed is 150 ampères. One volt is required per tank. The tanks can be arranged in series or in parallel at will. In each tank is a tray for catching the silver as it is scraped from the cathodes by automatic brushes, which move constantly backwards and forwards across the plate, thus preventing the silver from accumulating. For convenience in cleaning the trays, they are arranged so that, after the electrodes have been removed, they may be lifted out of the tanks, and the bottom of each tray is divided in the middle and hinged to the sides of the frame of the tray, so that upon removing a silver pin these sections swing down and outwards, dropping the silver into receptacles placed to receive it. The electrolyte contains no more than one per cent. of nitric acid, and will not for a very considerable period produce any noticeable effect upon the muslin bags. Under the influence of the electric current the silver rapidly dissolves, and any copper in the bullion dissolves also, but remains in solution.

In a paper on " Electrolytic Copper Refining in North America," by Sederholm (*Teknisk Tidskrift Afdeler för Kemisch Metallurgi*, 1895, I., and *Ding. Polyt. Journ.*, 1895, vol. 296, pp. 284-288 ; abstracted in the *Journ. Soc. Chem. Ind.*, vol. 14, 1895, p. 756), it is stated that the crude silver obtained from the anode slimes by smelting (see p. 542) is a metal containing from 60 to 90 per cent. of silver. It is cast into anodes and refined by Möbius' method. At the St. Louis Smelting and Refining Company and at another large Pittsburg works using the Möbius process aluminium wire is used to suspend the anodes in the dilute nitric acid electrolyte, and it is said that the aluminium is not attacked. At the St. Louis works 70 silver refining vats are stated to be connected in series and a current of 180 ampères sent through them, the voltage required being 100 volts. Fourteen kilos (30·8 pounds) of silver are yielded from each vat in twenty-four hours, giving, therefore, a total output of 980 kilos. per diem, that is, 2,156 pounds avoirdupois or nearly one long ton. The current density employed at this voltage of 100 for 70 vats is stated to be 300 ampères per square metre, that is, rather under 28 ampères per square foot.

A more recent description of Möbius' process, as practised at the Pennsylvania Lead Company's Works, was published by G. Faunce (*Oesterr. Zeits. für Berg. und Hüttenw.*, 1896, vol. 44, p. 30), and the following abstract of this paper is taken from the *Journal Soc. Chem. Ind.*, vol. 15, 1896, p. 361 :—

The silver to be refined is first treated by ordinary well-known metallurgical processes to reduce the quantity of other metals present, such as lead, copper, bismuth, etc., to at most about two per cent. It is then cast in sheets, measuring about 18 × 10 × ½ inches, and weighing 28·5 to 33 pounds each. These plates serve as anodes. The cathodes are formed of thin rolled sheets of pure silver 13 × 22 inches in size. The electrolyte is a solution of the nitrates of copper and silver, to which 0·5 to 1 per cent. of nitric acid is added to prevent the deposition of the copper. Four cathodes and three anodes are placed in each cell, with a distance of nearly two inches between the cathodes and anodes. The anodes are enclosed in muslin bags for the purpose of intercepting the undissolved matters which fall from them as the action proceeds. These consist of gold, bismuth, the principal portion of the lead as dioxide, and a little silver and copper.

A sheet of woollen cloth stretched on a frame near the bottom of each cell catches the silver as it is removed from the cathodes by a mechanically moved wooden scraper.

The current density employed is 18 ampères per square foot of cathode surface.

The silver is collected from each cell at intervals of two days, the gold once a week. The silver is washed with water and then melted in graphite crucibles capable of holding nearly 1,300 lbs. each, and is thus obtained of a fineness of 999 to 999·5.

The residue of gold, etc., after being melted, granulated, and treated with acid, gives gold of a fineness of 996 to 998.

In conducting this process the writer states that care must be taken that the amount of copper in the electrolyte does not exceed 4 to 5 per cent , for otherwise the silver is not obtained in a pure state.

Titus Ulke states that the cost of parting Doré silver by the Möbius electrolytic process, with careful management, should not exceed 20 to 30 cents per 100 ounces (that is, £15 to £22 per ton), (*Electrician*, vol. 46, p. 583, 1901)* ; whilst, as has already been stated, C. De Kalb gives three-quarters of a cent per ounce (£56 per ton) as being the price at which the electrolytic refining by this process was carried out in New York in 1888 ; refining by non-electrical methods being stated to cost at least one cent per ounce, or £70 per ton.

The following table, given by Titus Ulke, gives details of all the electrolytic silver refineries in the United States operated in 1900 :—

ELECTROLYTIC SILVER REFINERIES IN THE UNITED STATES OPERATED IN 1900.

No.	Name of Company and location of Works.	Material treated.	Estimated daily output.
1	Guggenheim Smelting Co., Perth Amboy, N.J.	Doré bullion	100,000 ounces troy = * tons
2	Pennsylvania Lead Co., Pittsburg, Pa.	Doré bullion	35,000 ounces troy = * tons
3	Globe Smelting and Refining Co., Denver, Colorado	Doré bullion	25,000 ounces troy := * tons

* See also *Elect. Rev.*, New York, vol. 38, 1901, p. 85 and pp. 101-103.

Although Titus Ulke states that Gustave Nebel first suggested the device of oiling the cathode travelling silver band to prevent the adhesion of the silver crystals to it, and thus permit their easy removal by a scraper (*loc. cit.*), yet there appears to be a little uncertainty as to the precise way in which this improvement was introduced ; because Möbius' patent for the moving band cathode is dated 1895, while the device of oiling the stationary cathode is described in the 1890 edition

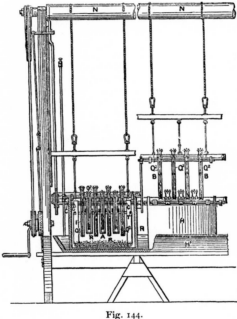

Fig. 144.

of Gore's "Electrolytic Separation of Metals," p. 240, where it is stated that "the cathodes are made of sheet silver slightly oiled to prevent adhesion of the deposited metal." I have not been able to see a patent on this point either by Möbius or by G. Nebel, and it is a matter of some interest.

The official abstract of Möbius' first patent (English Patent 16,554, 1884) is as follows :

" Silver, etc.—refining ; gold, etc.—obtaining; electrodes ; electrolytes.—Relates to an electrolytic process for refining silver, etc. The
metal to be refined is cast into blocks A, Fig. 144, fixed to cross-bars
resting on longitudinal bars which may be raised or lowered. These
form the anodes. The cathodes, in the case of silver, consist of silvered
copper plates B. These anodes and cathodes are suspended from
rollers N so that they can be raised from the troughs H, which contain
the solution. Each anode is enclosed in a bag, the whole series of
electrodes (*anodes and cathodes.*—A.P.) being enclosed in a larger bag.
These contain the metals, etc., which drop from the plates.

"The liquid in the vessels is agitated by bent strips of metal to equalise
the densities (of the solution.—A.P.) at different levels and prevent
local action. The electrodes are provided with brushes Q_3 on each side
to remove all deposits. The solution ordinarily consists of silver and
copper nitrates and nitric acid, the copper being required to cause
the lead present to be deposited on the anode as peroxide along with
gold, platinum, and other metals which may be present. The ions
(*crystals of silver and anode slime* ?—A.P.), are removed by the
brushes, and fall into the separate bags. Any copper which may
have been deposited with the silver at the cathode is re-dissolved by
the acid solution. The contents of the gold bag are treated for the
separation of the metals. The excess of copper is removed from
the exciting liquid in the battery by first suspending copper plates,
etc., in the solution (whereby the silver remaining is deposited) and
then precipitating the copper by an electric current."

The abstract of Möbius' second patent (English Patent 469, Jan.
8th. 1895) is given in the *Journal of the Society of Chemical Industry*
(vol. 14, p. 371) as follows :

" Improvements in method of, and apparatus for, separating metals.
B. Möbius, New York, U.S.A.

" Between two rollers placed horizonally beneath the surface of the
electrolyte is stretched an endless band of silver sheet which is kept
in constant movement by turning the rollers. Horizontal trays,
having porous bottoms, are placed just above the horizontal top surface of the silver band, and contain the bullion to be treated. The
bullion is made anode and the silver band cathode. A second endless band of rubber, cloth, canvas, or other material is stretched
between a roller beneath the liquid and ,another placed above it and
beyond the end of the bath : when both bands are in movement the
latter presses on the silver, where it passes round one roller, and
rubbing off the loose crystalline deposit, carries it beyond the tank
and deposits it in a special receptacle. A strong bath of sodium
or potassium nitrate is preferred, with sufficient acid to keep some
silver and all the copper in solution. A current of small electro-

motive force is used (*whatever that may mean, presumably that the volts required per tank, at the current density employed, are not high.*—A.P.). The silver collected is ready for the melting pot, the gold, platinum, and lead remaining in the porous trays. The movement of the cathode and its constant cleaning by friction with the rubber band prevent crystalline growths and consequent short circuiting."

CHAPTER V.

THE ELECTROLYTIC TREATMENT OF TIN.

The Electrolytic Refining of Tin.—The Recovery of Tin from waste Tin Plate by Acid Processes.—The Recovery of Tin from waste Tin Plate by Alkaline Processes.—Properties of Iron contaminated with Tin.—Cost of old and new scrap Tin and Quantity Available.

The Electrolytic Refining of Tin.—It is stated that Messrs. Bolton, at Froghall, Staffordshire, have refined tin electrolytically for several years by a process similar in principle to that in general use for the electrolytic refining of copper. The tin is, however, deposited in the form of coarse crystals, and is remelted. (J. B. C. Kershaw, *The Electrician*, 1897, vol. 38, p. 693.) The details of this refinery have not been published. The cost of refining cannot well be less than about £3 per ton, and it does not seem probable that such a method of refining can be economically successful on any large scale, unless it becomes apparent that there is a corresponding opening for specially pure metal. At present no such opening exists in so far as I am aware. The refined tin referred to above is stated by Kershaw to be used in Messrs. Bolton's wire factory, whether for coating iron or copper wire, or whether for making tin wire for fuses is not stated.

It was due to the fact that a great increase in the conductivity of ordinary copper was found to be caused by the removal of very small quantities of impurities from it that enabled a high price to be paid for specially purified copper for electrical, and chiefly, in the first place, for telegraphic purposes, and thus the electrolytic refining industry became possible, an industry which has to-day reached such enormous proportions. It does not seem at present in any way probable that a sufficiently high price can ever be offered for specially refined tin to render its electrolytic refinement a paying industry.

The Recovery of Tin from Tin Plate.—The problem of the successful recovery of the metallic tin from scrap tin has long possessed a great fascination for the inventor, and since 1857 no less than 50 to 60 English patents have been taken out on this subject. The difficulties, however are considerable, and are unexpected in their nature. Tin plate, the manufacture of which consumes the greater part of the world's tin output at the present time, contains from 3 to 9 per cent. of metallic tin, covering, with or without an admixture of lead, the surface of wrought iron plates. Dr. J. H. Smith states that, out of a large number of samples of tin plate which he examined, the above figures represented the amounts of tin present, and he found that on the average the tin plates contained 5 per cent. by weight of metallic tin. The recovery of tin from tin plates can, perhaps, only be considered as a process of refining tin by a stretch of the imagination, yet it is the only process for preparing tin by electrolysis which has had any commercial success, and it will be considered here. The raw material for this industry, namely, scrap tinned plates and old broken tins, exists in enormous quantity. It is reckoned that in Paris alone 3,000 tons of tinned iron scrap cuttings are produced per year (J. H. Smith, *Journ. Soc. Chem. Ind.*, 1885, vol. 4, p. 312). The cost of collection and supply of this waste tinned iron is in many cases, however, very high.

In the discussion on Dr. Smith's paper referred to above, Mr. Gatheral stated that the mere cost of collection of the waste tins in England was often as high as from 25 to 30 shillings per ton, whilst to this amount the cost of the carriage to the electrolytic works must presumably be added, and this latter, on account of the bulky nature of the material, must be somewhat high ; this means that if the tinned iron collected contains 5 per cent. of metallic tin—that is one hundredweight of tin per ton of scrap—and with tin at £79 per ton (this was its price in 1885, the date of the paper), the cost of collection and delivery cannot at most leave more than £2 3s. per ton to pay the cost of the electrolytic recovery. The tinned scrap obtained by Dr. Smith, however, only cost 2 francs per ton, and this allowed of over £3 16s. per ton to pay the cost of treatment. (For further remarks upon the price of tin scrap and cost of collection, see p. 593 of this volume.) It must be added that the very high price of tin at present (1901), over £116 per ton, would appear to offer a particularly favourable condition for the undertaking of this process ; for the value of the tin in a ton of the scrap is probably on the average over £6. The fact, however, that the price of tin fluctuates a good deal, probably due to market manipulations, would necessitate some careful consideration before embarking upon an undertaking for recovering tin from scrap. Under special conditions there can be little doubt that the process can be made to pay ; if, for instance, the tin cuttings in a factory, or a group of factories, are as large in amount as from four tons upwards per week, and the supply can be relied upon as being constant.

Swinburne, in his Cantor Lecture III., in 1896, on Applied Electro-

Chemistry (*Electrician*, vol. 37, p. 759), in the course of his remarks on the removal of tin from tinned iron stated that there are supposed to be thousands of tons of scrap tin wasted every year, and that it is even employed for road mending, but that when he tried to purchase it in quantity none was forthcoming, and that it seemed as if the whole idea of there being an important industry in the recovery of tin from scrap was erroneous; the difficulty was to get the scrap. He adds: "makers of large tin goods sell their clippings to people who do smaller work, and finally the scrap is distributed in such a way that it cannot be collected profitably. The recovery of tin and lead from old cans might be carried on on a very small scale by corporations, but it is questionable if it would pay owing to the (small) quantities available."

The methods which have been proposed for the recovery of tin from tin scrap by electrolysis are very many. In every case the tin scrap is made the anode in some electrolyte. The chief electrolytes which have been proposed are as follows :—Dilute sulphuric acid, a mixture of dilute nitric and hydrochloric acids, ferric chloride solution, stannous chloride solution, caustic alkaline solutions, solutions of caustic alkali mixed with common salt. There can be little doubt but that the alkaline electrolytes are the most satisfactory, for the consumption of chemicals is smaller, the tin is obtained in a pure state, and the metallic iron is, to a large extent, recoverable. But, as Dr. Smith has written a very detailed paper describing an actual and successful installation which he planned, erected, and conducted for recovering the tin electrolytically, employing an acid bath, I propose to consider the methods in which acid or non-alkaline electrolytes are used first, and to reproduce his paper at some length.

Methods for Recovering Tin when using Acid Solutions.— Dr. Smith says: I am only aware of three methods for recovering tin from tinned iron plate which have been employed on any large scale, viz: dissolving in a mixture of hydrochloric and nitric acids, and subsequent precipitation of tin by metallic zinc ; treatment with caustic soda and litharge, forming stannate of soda ; and, lastly, the formation of tetrachloride of tin by the action of chlorine gas.

The last-named process is worthy of special mention. It has been described by Prof. Lunge in his report of the Chemical Section of the Swiss National Exhibition, held in Zürich, in 1883.

Some time ago I was requested to recommend some practical method for the utilisation of these tin cuttings in a district where they could be collected in large quantities. (This was in Milan.) After reading all the literature at my disposal, relating to the subject, I was induced to make some experiments, and amongst others, actuated by a knowledge of certain electrolytic processes employed for the precipitation of metals from their ores, I tried the effect of a current of electricity in passing through an acid solution in which a quantity of cuttings were suspended, forming the anode, a copper plate serving as cathode.

These experiments proved that not only is tin dissolved at the anode, but that it is also deposited in a pure form at the cathode very soon after the commencement of the action. . . . I recommended that a trial should be made of this process ; this was undertaken, the ceremony of obtaining a patent was duly observed, and I was intrusted with the supervision of the erection of the plant, an account of which, and of the results obtained therefrom, I now propose to give you in a brief form.

Although ignorant at the time, I am now aware that at least four English patents have been granted for the same object effected by very similar means; still, as there are important differences in the process about to be described, and especially as, to the best of my knowledge, it was under my direction that the first plant was erected for the employment of an electrolytic method on any large scale, I venture to hope that the following details may interest the members of our Society :—

General.—The cuttings with which we had to deal varied considerably in value. Some of the thicker ones contained little more than 3 per cent. of tin, while some of the thinner kinds contained 8 or 9 per cent. I considered 5 per cent. as about an average. The quantity (of tin scrap) obtainable was calculated at about 6 tons per week, and the plant was designed to accommodate that quantity, charging twice a day. The iron was designed to be converted into sulphate, a large quantity of which could be disposed of at a high price ; the remainder was to be converted into "iron mordant" ; the tin was to be converted into stannous chloride, and other salts of tin employed as mordants, and largely used by the dyers of that neighbourhood.

Dynamo.—A Siemens-Halske dynamo, No. C 18, similar to that used for copper deposition at Oker, was employed. It was a shunt-wound machine, giving 240 ampères at 15 volts, and requiring 7 B.H.P. to drive it.

Electrolytic Baths.—These were made of wood lined with india-rubber, and had the following dimensions : 5 feet long, 2 feet 4 inches wide, and 3 feet 3 inches deep ; the thickness of the wood was 2 inches, and the rubber was about $\frac{1}{16}$-inch thick. As a matter of fact, one large tank, measuring 10 feet long and 4 feet 8 inches wide, was divided up by one transverse and one longitudinal partition into four tanks of the dimensions above stated. The sides of the tanks outside had long iron bolts and nuts firmly clamping them on to the wooden ends, the form of construction so well known and often used for wooden depositing vats (*see p.* 289). The vats were placed end to end, and thus arranged exactly occupied the width of the building, and were fixed at its end, with their bases at a height of 3 feet 3 inches above the floor. In front was a platform, at one side of which was a door for the admission of the cuttings after being washed and packed. The dynamo was situated in the engine room just behind the baths, and electrical communication was made by two copper cables passing through a hole in the wall. At either side wall, and on a level with the baths, was a dissolving tank, capable of accommodating half the cuttings after the removal of the tin. A little farther on, and nearly on a level with the ground, were the evaporating down tanks. The crystallising tanks were situated beneath the ground level, so that the solution could be run off from one stage to another without any pumping arrangement.

Anodes.—These were of course composed of the tin scrap. Baskets were obtained to pack the cuttings in. These at first were made of wicker-work, but as they were too flexible and soon rotted by the action of the acid, their place was supplied by strong wooden baskets, whose sides were formed of stout upright wooden bars $\frac{3}{4}$ inch thick, with spaces between them sufficient to allow the solution to circulate freely, while preventing the exit of the scrap. The internal dimensions of the baskets were: Length, 3 feet 11 inches; breadth, 1 foot; and depth, 2 feet 9 inches. Great care was required in packing the scrap, because if it were packed too closely the metallic surfaces thereby united, and thus preventing the circulation of the electrolyte between them, necessarily retained their coating of tin. These baskets held from 132 to 154 pounds of the scrap, the eight being capable of accommodating about half the total quantity required (*i.e.*, for a treatment of six tons of scrap per week). Long and narrow strips of tinned iron were employed to complete the connection of the anode scrap with the copper conductors. As the resistance of the iron to the current is comparatively great, a large number of these were required in order to prevent excessive heating. (An allowance of 7 to 8 square inches cross section of iron per 1,000 ampères transmitted should at least be made.) At one end these strips were soldered together and connected with the copper conductor by means of binding screws; the other extremities were distributed throughout the scrap.

Cathodes.—Copper plates were employed. These had a thickness of about $\frac{1}{12}$ inch, were 3 feet 11 inches long, and were 3 feet 1 inch wide. There were sixteen cathode plates in all, two being in each bath, and they were arranged one on either side of the baskets. To keep these thin plates of copper as flat as possible, each was surrounded by a framework of copper rod of square section. The copper of the cathodes was coated with tin to prevent corrosion, as well as to avoid solution by an accidental reversal of the current. (Not a very efficient

protection.) These cathode plates rested in grooves at the sides
of the tanks, placed at a distance of 4 inches from the sides of the
baskets. They were provided, as also were the baskets, with india-
rubber rollers extending to the sides of the baths, enabling them to be
raised out of the same with ease, and without injury to the india-
rubber coating.

Electrolyte.—Dilute sulphuric acid formed the electrolyte. This
was employed not only on account of its comparatively small resistance to the current, but also because it was convenient to turn the
solution into the sulphate of iron tanks as soon as it became saturated
with that salt and all the tin had been precipitated from it. (I do not
quite follow the latter reason here given.) Commercial acid of 60°
Baumé was diluted with nine volumes of water. (Of course the acid is
poured into the water.)

Above the tanks was a pulley arrangement for raising the baskets
and plates out of the baths as required ; there was also an arrangement
of levers and eccentrics constructed, whereby the baskets were kept in
gentle motion in the baths, thus exciting circulation in the liquid and
tending to prevent polarisation. The horizontal axis upon which the
eccentrics were disposed made about two revolutions per minute,
thereby raising the baskets a distance of about two inches. Levers
were fulcrumed into the wall. These passed over the eccentrics, and
at their extremities ropes were fixed communicating with the baskets.

The current conductor consisted of a thick copper wire of several
plies, and although already coated, they were enclosed in indiarubber
tubing as an additional protection.

The eight baths were arranged in series, and took a current of rather
under 240 ampères at 15 volts (*i.e.*, about 1·7 volts per bath).

Quality of Tin deposited.—This was at the first of a spongy nature,
owing to the great acidity of the bath. Soon, however, it began to be
precipitated in a more dense, extremely fine, granular and partially
crystalline state, which, indeed, was preferable, as it fell to the bottom
of the bath and was not in danger of forming a communication with
the anode. The tin obtained was not chemically pure, but it was purer
than commercial tin, and when thoroughly washed contained no trace
of iron. It fused readily and almost completely, and that without any
addition, provided it had been thoroughly washed and dried. The
rapidity with which the powdery electro-deposited tin dissolved in
hydrochloric acid was not to be compared to the slow action of that
acid upon granulated tin, and this rendered the product peculiarly
suited for the manufacture of stannous chloride.

Quantity of Tin deposited.—Rather over one-half of the weight
of tin was deposited which should have been obtained theoretically
as calculated from the current employed. Dr. Smith adds that the

discrepancy was due to part of the current being absorbed in dissolving the iron as well as the tin, as soon as the former began to get bare. This, together with the natural solution of the iron in the acid led to the rapid accumulation of sulphate of iron in the baths. (No doubt hydrogen was deposited at the cathode in place of the missing tin.) The acid employed in one charge of the eight vats took about seven weeks to become saturated. (It is not stated if the baths were worked day and night.) On analysis the baths were found to vary in a very remarkable manner, first one and then another containing the largest quantity of iron. The tin, on the contrary, remained very constant in amount, both in the individual baths and in the total average ; it amounted in the average to 1·5 grams per litre. Pure tin was deposited until the acid was saturated with iron salts and all the tin had been removed from solution, then hydrate of iron began to form. This could be avoided for a time by the addition of more acid but it was better to run the acid into "green vitriol" tanks and add fresh solution. It was found not to be at all necessary to continue the action of the current until all the tin had been removed ; in fact after a certain time the action on the iron was even stronger than that on the tin. It was found in practice that after the passage of the current for the space of 5 or 6 hours the quantity of scrap referred to was sufficiently free from tin to be dissolved in the sulphate of iron tanks with the greatest ease ; the tin remaining unacted upon in the presence of the large excess of iron always provided for, and it was not difficult to recover that tin and utilise it with the rest. (The above remarks show that if one charge could be removed from the vats after a treatment for 5 to 6 hours with the current, and as each charge weighs half a ton, the vats must be capable of dealing with about 2 tons per diem of 24 hours, or about 1 ton per diem of 12 hours ; and as the plant dealt with 6 tons of scrap per week it is fairly clear that the plant was run six 12-hour days per week.)

Cost of Plant and Expenses of Working.—"These, of course, must depend upon the neighbourhood fixed upon and other circumstances, but in any case they would be little compared with the value of the tin capable of being recovered." (This statement is disappointingly vague and begs the whole question.) "I think I have given sufficient details for any one to be able to calculate approximately their amount for any particular locality."

"Generally the scrap is obtained for a mere nominal sum, frequently just for the expense of transport.

"*Labour required.*—One stoker and two or three labourers would be quite sufficient to work three tons per week, using one of the small dynamos described above. The three hundredweight of metallic tin obtained therefrom (*i e.*, from three tons of scrap), at £3 18s. per cwt., will compare favourably with the cost of fuel necessary to maintain the annual output of 7 or 8 B.H.P., wages of workmen, interest on plant, and for occasional repairs." (I consider this doubtful, for the actual cost of obtaining a ton of metallic copper by electrolytic deposition must certainly, under the best conditions of large output, etc., be as much as £2, and there are many reasons why the cost of electrolysing one ton of tin scrap should cost more than this ; as, for instance, the larger size of vats per ton of anode necessary for the tin process, the large amount of energy consumed, namely, one kilowatt hour to every 3·42 lbs. of tin produced, and also the amount of acid consumed, which is large, whilst it is practically zero in the case of copper refining.   However, Dr. Smith undoubtedly worked under very favourable circumstances, firstly, because he only had to pay 2 francs per ton for the scrap tin, and secondly, because he was able, as he states, to sell the iron sulphate produced " at a high price."   It is difficult, however, to understand how so large a quantity of iron sulphate as 28 tons, and 6 cwt. of iron sulphate crystals ($FeSO_4 + 7\ H_2O$), the amount produced from 6 tons of scrap tin containing 5 per cent. of tin, could be disposed of per week over any long period of time.) Dr. Smith further adds, " If the process were worked on a larger scale, and the tin and iron further worked up into salts, as in the case described above, the profits would be increased in a much greater ratio.   When we consider that in Paris 3,000 tons of this scrap are produced annually, it is no unimportant matter to determine the best method of utilising it."   In conclusion, Dr. Smith suggested that a solution of stannous sulphate would answer better as an electrolyte than the dilute sulphuric acid which he employed.

The following accounts of three or four other processes of treating scrap tin by acid solutions were given by Watt in earlier editions of this work, and they are retained, not because the processes appear to be of great promise, but because the descriptions are fairly complete, and on account of their historical interest.   In so far as the present writer is aware, none of these processes, which involve the use of electrical energy, have been employed on a commercial scale, and it is in the highest degree improbable that they could be successful if their employment were attempted.

Mr. Watt writes :—The recovery of the tin from tin scrap, preserved fruit and meat tins, and other waste of a similar character, has long occupied the attention of experimentalists, both at home and abroad, and several ingenious processes have been devised and patented, and, we believe, carried out with some success.   When experimenting upon this subject a few years since, the author tried, amongst other solvents, perchloride of iron—a material which, for another purpose, he had prepared in very large quantities, and was therefore well acquainted with the most economical method of preparing it. In stripping tin from tinned iron scrap with the perchloride of iron, the solution of the salt was employed hot, and the scrap then introduced and kept constantly in motion until the whole of the tin was dissolved from the surface of the fragments, which was generally effected in a few minutes. After a time the characteristic orange-yellow colour of the perchloride disappeared, leaving a green solution of protochloride of iron, in combination with the corresponding salt of tin. It was now found necessary to again peroxidise the iron solution, which was effected by trying various ordinary agents in succession, nitric acid being found to answer the purpose readily. Since the perchloride of iron as a de-tinning agent has somewhat recently formed, in part, the subject of a patent, it will be well to give the inventor's description of his method of applying it. The process referred to, which is given below, does not, however, depend exclusively upon the employment of perchloride of iron, as will be seen from the abridged details given.

Garcia's Process.—This process, for which a patent was granted in 1886, consists more especially of "an improved method of treating cuttings, or waste, of tin plate in order to remove the metallic tin, but is also applicable to the separation of tin combined with metals, such as copper, zinc, or lead, or in combination with oxygen. The process consists first in making solutions of tin, preferably protochloride of tin, slightly acid, and afterwards extracting the tin in a metallic state by means of the simple apparatus used in the ordinary process of electro-typing." This consists in applying the "single cell" arrangement, as it is called, in which a plate of zinc is immersed in a porous cell filled with a saline solution, or acidulated water, and which is placed in a larger vessel containing a solution of tin in lieu of the sulphate of copper, as in electrotyping. The zinc plate is connected by a suitable conducting wire to sheets of tin plate, placed round the porous cell, upon which plates the tin becomes deposited. "Supposing," says the inventor, "cuttings of fresh tin plate are to be used, as often happens, I place them in vessels of material unaffected by the chemical agents employed, and keep them in movement by any convenient mechanical means, so as to prevent them from adhering together, since if left at rest they would remain adherent, and to a great extent escape the action of the acid liquids. These acid liquids may vary according to the character of the tin cuttings, or fragments, but I prefer to use a mixture of hydrochloric and nitric acids, in about the following proportions, namely, 250 to 300 kilogrammes of hydrochloric acid, and 8 to 12 kilos., or even less, nitric acid." The vessels employed for this purpose may be constructed of stoneware, glass, slate, etc., and the mixed acids diluted with about 1,000 litres of water, the acid liquor being heated by steam passing through coils, or otherwise, to the temperature of about 80 to 100 degrees. [Centigrade?] "While the process of solution is going on a certain proportion of iron is also dissolved in removing the last particles of tin, the fragments being removed when the further solution would cease to be economical." It is stated that the foregoing proportions of acids and water are sufficient to effect the separation of the tin from one ton of tin scrap, as, for example, preserved food cans. "Instead of nitric acid," says the patentee, "perchloride of iron may be used, by which the tin is effectually separated. The perchloride is preferably produced by dissolving 80 to 90 kilogrammes of peroxide of iron in 300 to 350 kilos. of hydrochloric acid, this quantity being sufficient to separate the tin from a ton of tin plate, the quantity of water being the same as before, so as to completely cover the metal from which the tin is to be removed." The patentee next states that "the use of perchloride is new for the purpose and in the manner described, and forms an important part of the invention." As to the *novelty* of perchloride of iron being applied as a "stripping" solution for tin, we can scarcely agree to this, since this salt has long been used for the purpose by jewellers and others in common practice, and indeed has been recommended by various writers as a useful solvent not only of tin, but also of lead and copper.

Montagne's Processes.—The first of these two processes—both of which are protected by patent—consists of a method and apparatus for separating tin from tinned sheet metal, such as empty tin boxes, tin scrap, etc. "Broadly speaking," says the inventor, "the method consists in conveying hydrochloric acid gas into a closed vessel containing the material the separation of the tin from which it is desired to obtain. After the closed vessel containing the tinned sheet metal has become completely saturated [charged?] with hydrochloric acid gas, whereby the latter combines with the tin, a shower of water is allowed to fall over the sheet metal, and instantaneously the said gas is converted into liquid protochloride of tin ; the tin, entirely removed from the sheet metal, is dissolved in the protochloride of tin." The solution of tin is afterwards drawn off from the closed vessel and the metal precipitated either by means of zinc, "lime wash," or otherwise. It is obvious, however, that the metal would be more economically recovered by electrolysis. In the accompanying drawing, A is a retort into which is placed a charge of sulphuric acid and chloride of sodium for producing, with the application of heat, hydrochloric acid gas in the usual manner. The gas is conveyed by a pipe, B', into a closed vessel, B, containing a charge of tinned sheet metal scrap or cuttings, the latter material being heated to a temperature of 150° to 160° Fahr. by a jet of steam supplied by a generator, D', and pipe, D. After the tinned sheet metal cuttings, or scrap, have become saturated with hydrochloric acid gas, a shower of water, supplied by a pump, F, and a pipe, C, is conveyed into the closed vessel, B, by a perforated pipe, E, so as to completely wash over the tinned metal scrap. The solution thus produced is protochloride of tin, which is drawn off into a pan, H, by a tap, G. The tin, precipitated by zinc or otherwise, is placed under a press, M, to remove the liquid attached to it, and is afterwards melted in a crucible, P, and cast into ignots.

By a second patent, dated May 2nd, 1887, M. Montagne introduces an improvement on the foregoing process, which may be thus briefly

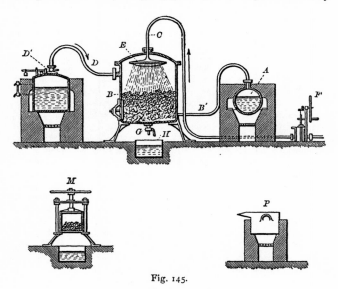

Fig. 145.

described :—The tinned scrap is placed in a series of brickwork chambers lined with any suitable material capable of resisting the action of acids, and sufficiently thick to retain heat for a long time, Fig. 145. These chambers are put into communication by a system of piping, provided with suitable valves capable of allowing the acid gas to pass successively into several chambers before passing to the exit or chimney, completely exhausted by its passage through material less and less attacked by acid. "Let it be supposed for the sake of clearness," says the inventor, "that three chambers are employed, and that the gas passes successively through chambers 1, 2, and 3. The gas first reaches chamber 1 (through a pipe and valve), in which the extraction [dissolving of the tin] will be soonest completed. When the extraction has been effected, chamber 1 is completely shut off by closing the valve of the gas supply pipe, and the valve which puts chamber 1 in communication with chamber 2 is opened. The acid gas will then first enter chamber 2 by its valve and be led into chamber 3, and thence to the chimney. While chamber 1 is shut off a spray of water is allowed to fall into the same from a perforated coil or otherwise. At the same time, in order to reach the parts which are least accessible, a jet of steam is supplied from a generator by pipes, the same being cooled by the spray of water, and forming vapour which settles on all the surfaces covered with protochloride of tin, and carries it off in solution. The condensation of all the hydrochloric acid gas is thus certain to be effected, and chamber 1 can be opened for removing the metal deprived of its tin, and recharging the same with a fresh batch of tinned scrap without being inconvenienced by the acid vapours. When this is done the chamber is closed, and placed again in circuit by opening the valve, and thus making this chamber the last in the series; that is to say, the gas entering through chamber 2 passes into chamber 3, and then into chamber 1, whence it escapes through the chimney. The order of circulation thus becomes reversed, as will be readily seen, and it is obvious that the number of chambers may be varied at will." To recover the tin from the solutions thus obtained, the inventor employs zinc or iron, or otherwise lime; but the electrolytic method would doubtless be more economical.

Methods for recovering Tin when using Alkaline Solutions. —As has already been stated the methods for recovering tin from scrap by means of either chemical or electrical processes which involve the employment of alkaline solutions appear much more promising than those in which an acid solution is adopted. The following is a complete lists of patents, with their dates, granted for the removal of metallic tin from tin scrap by means of hot alkaline solutions, with or without the assistance of an electric current, between the years 1877 and 1888 inclusive.

W. D. Walbridge, April 6, 1877, 1,346.—This inventor employs a boiling hot solution of either (1) caustic soda and sodium nitrate, (2) caustic potash and potassium nitrate, or (3) caustic soda and sodium chloride as electrolyte, with tin scrap anodes.

R. S. Laird, August 15, 1883, 3,952.—The process consists of a "single cell" deposition method, using tin scrap anodes and an electrolyte of caustic soda or potash.

E. L. Cleaver, October 23, 1883, 5,033.—Tin plate scrap is cleaned and boiled either in hydrochloric acid or caustic potash or soda bath. The solution obtained is electrolysed. Or the removal of the tin may be hastened, and the tin at the same time deposited, by using the tin scrap as anode and a metal cathode.

A. P. Price, December 22, 1883, 5,847, and *January* 25, 1884, 2,119.— Practically identical with Cleaver's patent, but hydrochloric acid is not to be employed.

W. Beatson, September 18, 1885, 11,067.—The waste metal is cut into small pieces, and enclosed in a cylinder of wire gauze ; the cylinder is slowly rotated in a hot bath of soda or potash, and a current of electricity is passed through the solution ; the cylinder forms the anode. A plate or arrangement of rollers forms the cathode. Instead of the cylinder shallow gauze boxes may be used, and shaken by hand or otherwise. Instead of a special cathode, the iron vat containing the alkali may be used for receiving the tin. An alkaline cyanide may be added to the caustic alkaline electrolyte. The use of the cyanide was afterwards disclaimed as useless and expensive.

Mr. James Swinburne, in his Cantor Lecture on Applied Electro-Chemistry in 1896 (*Electrician*, vol. 37, p. 759), stated that he had worked out a process for removing tin from scrap which was not electric, and consisted in dipping the scrap in baskets into hot caustic. If old tins were used this treatment also melted out the solder and corroded labels and varnishes, so as to get at and remove the tin under them. The resulting stannate of soda was to be sold in that state. The scrap coming out of the caustic had a good deal of the caustic sticking to it. It would on a large scale, therefore, be washed in successive baths, getting weaker and weaker, and ending in water; the liquid being gradually moved up into the fused vat to replace that removed as stannate. (A process which would be more safely conducted on paper than in any other way.) By this means the expense of evaporating down all the caustic that adhered to the tin could be kept down. This process works fairly well, but it is difficult to get the caustic to take up enough tin. The stannate thickens it very quickly, and the resulting mixture never contains enough stannate to be a commercial article. The tin could be easily removed from the caustic by electrolysis, and came down well. Mr. Swinburne adds : This process seemed fairly promising, but at this stage it was found that it had been tried before, and a company had been formed to work it. Whether or not the company was successful the lecturer was not aware. A fresh attempt at the solution of the problem was therefore made. "Curiously enough " (I quote Mr. Swinburne's own words) "if sheet tin is used as anode in caustic soda solution " (presumably cold) " the tin does not go into solution, but oxygen comes off. I had tried a number of experiments in 1882, with sheet tin as a backing, for peroxide plates in alkaline solutions for secondary batteries. If sheet tin is boiled in caustic lye the tin is not attacked. If clean iron is dropped in bubbles of hydrogen come off the iron, and the tin is dissolved." (This, of course, means that where iron and tin are in *contact* in boiling caustic alkalies the tin dissolves.) " Strangely enough, however, this process does not act, because very soon the clean iron gets coated with tin, and the action stops. It is very odd that tin should be eaten off one piece of iron and then deposited from the solution on to another piece. This gave the key to the electrolytic process. All that is needed is to heat the caustic bath. The anode no longer gives off oxygen, and the tin is stripped off perfectly and deposited nicely on the cathode. This process seemed perfectly successful ; the next thing was to try it commercially, but when we tried to get the scrap tin none was forthcoming." I have quoted Mr. Swinburne at some length, not because the results he obtained were new even at the time he published them, but because a frank statement of results obtained in endeavouring to hammer out a new process is always instructive, and the more so that the chemical and electrical method decided upon as most satisfactory was identical with that of Cleaver's patent of 1883, and those of Beatson of 1885 and 1890.

Dr. W. Borchers (Borchers' *Electric Smelting and Refining*, 1897, translated by McMillan (Griffin), p. 323) describes some experimental results which he has recently obtained. He found that hot solutions of caustic alkalies or of sodium stannate did not give satisfactory results when they were electrolysed with tin scrap anodes, because of the precipitation of stannic oxide in the bath, which stannic oxide is of course difficult to collect. Dr. Borchers, however, found that a 12 to 15 per cent. solution of common salt, to which 5 per cent. of sodium stannate had been added, gave much better results, provided that the solution was kept distinctly alkaline. Hot alkaline hydroxides, however, in the presence of iron and tin in contact dissolve the tin without the aid of the electric current, and the bath therefore tends to become richer in stannite. Dr. Borchers, therefore, commences the process with an electrolyte made of a solution of sodium chloride mixed with a few parts per cent. of caustic soda, which latter becomes converted into stannate as the action proceeds. The e. m. f. required per vat (the current density and arrangement of the bath is not stated, and therefore the e. m. f. required is somewhat meaningless) at the commencement of the electrolysis is very low, and tin dissolves practically unaided (this is a somewhat vague statement) if the solution is warmed to from 40° to 50° C. But as the tin coating is gradually removed from the surface of the tin, the current density per unit area of the remaining tin increases (this statement is not clear to me), and the power absorbed becomes higher in consequence. (The only explanation I can suggest for this is that the surface of the exposed iron becomes oxidised in the alkaline solution, and thus setting up a back e. m. f. on the iron would concentrate the current on the tin.) The e. m. f., rising at the end of the process to 3 volts, even when a low current density is employed, calculated upon the cathode area, which remains constant throughout. The difference of potential between the electrodes at the commencement of the work is much lower than 3 volts even when the current density amounts to several hundred ampères per square metre. (This is equivalent to several tens of ampères per square foot.) In the experiments that were made, the necessary e. m. f. averaged 1·5 to 2 volts per bath when the current density did not exceed 15 ampères per square foot of cathode, and providing that the current density was reduced to one-half by joining in parallel two baths, which had been charged at the same time, directly the bulk of the tin had been dissolved from the anodes. Borchers further states that the iron remaining in the anode basket after treatment is so pure that it may be applied to any purpose for which good sheet iron scrap is used. The powdery tin obtained on the cathodes is washed and pressed or dried in a centrifugal machine, and may be melted and cast into ingots quite satisfactorily (see Dr. Smith's experience on this point, p. 583 of this work).

The advantages which Dr. Borchers claims for his process are :—

1. The complete stripping of the iron, which may therefore be employed as sheet-iron scrap.

2. The recovery of the pure tin free from iron.

3. The possibility of using iron cathodes and vats.

4. The possibility of constructing the anode baskets of iron, which forms a very durable material for the purpose.

It must be remarked that the employment of a hot solution of sodium chloride with a mixture of caustic soda was, as has already been pointed out, patented by W. D. Walbridge on April 6, 1877, English patent 1,346, for the removal of tin from tin scrap by electrolysis, the scrap forming the anode. Therefore Dr. Borchers' process is only a revival of an old method, no doubt independently arrived at. It is, however, evident that no valid patent can exist for it.

Iron Contaminated with Tin.—If the cleaned scrap iron from which the tin has been recovered is to have any commercial value, it must be quite free from tin, for even very small amounts of tin—0·19 part per cent. in wrought iron—cause it to be very cold, short, and prevents its welding. Percy, in fact, states that the iron was absolutely worthless (Percy's *Metallurgy of Iron and Steel*, 1864, p. 161).

The present writer saw some iron obtained by melting down tin scrap in a blast furnace in Birmingham in 1891, which was cast and employed in making window sash-line weights. An enormous amount of white fume was produced in the process, and it was subsequently abandoned on account of difficulties with this fume, and for other reasons.

Borchers, as has been stated above, considered the iron cleaned from tin to have been quite free from that metal, but he does not state whether the iron scrap was used to produce good wrought iron, the only test which is of any importance, and, on the other hand, in an article (*Industrie Electro-Chimique*, vol. 2, pp. 2, 3, 1898, abstracted in *Science Abstracts*, vol. 1, 1898, p. 434), it is stated that caustic soda solution, when employed as an electrolyte to remove the tin from tinned iron, removes the outer metallic tin coating, but fails to attack the alloy of tin and iron, a layer of which exists under the tin uniting it to the iron; and the iron left is stated to have but little commercial value owing to its contamination with tin.

An abstract of a paper by A. Zugger (*Chem. Zeit. Rep.*, 1901, vol. 25, series 16, p. 143), on the Influence of Tin on the Quality of Iron and Steel, appeared in the *Journal of the Society of Chemical Industry*, vol. 20, 1901, p. 583, which states that observations on the product of a basic Siemens-Martin furnace, in which tin was accidentally present, showed that a metal containing

Tin	0·55 parts per cent.
Antimony . .	0·015 ,, ,
Arsenic	0·030 ,, ::
Copper . . .	0·182 ,, ,,

rolled quite satisfactorily to plates; on further rolling to sheet, however, cracks began to develop at the edges of the sheet, but they could not, as it was proved, be attributed to red shortness. Bars rolled from the ingot iron folded completely over, under the cold-bending test, and showed a tensile strength of 40 kilos., with an extension of 31 to 34 per cent. (The length of the test piece upon which this elongation was measured is not stated in this abstract.) Hence it appears to be proved that the presence of 0·55 per cent. of tin does not affect malleability, tenacity, or extensibility; the welding qualities at most are impaired. Properly stripped tin can in fact apparently be employed for steel making, but not for the manufacture of wrought iron.

The Cost of Old and New Scrap Tin.—I have been at some pains to ascertain the price at which both new and old scrap tin can be obtained, and also the quantities available. Dr. J. H. Smith, in his paper before the Society of Chemical Industry, given *in extenso* above, states that he paid 2 francs per ton for the cuttings of scrap tin he treated, and he kindly wrote me recently, in August, 1901, stating that this price was paid for cartage only as the scrap was actually given away in large quantity by several very large button manufactories in the neighbourhood of Milan, where the tin recovery works were situated. Dr. Smith states that owing to an insufficiency of capital to keep the factory at Milan working and bring it to a successful issue it became bankrupt, and it was said to have been sold and converted into another business. Dr. Smith adds: "I was convinced that with capital behind it, so necessary to every form of manufacturing business, the result would have been entirely satisfactory, and I still believe that under the conditions then existing the process would have turned out to be a very profitable one." With regard to this remark I must add that it has been my fortune to see something of the struggles of two syndicates to make a profit out of the fascinating recovery of tin from tinned scrap by different processes, neither, however, electrolytic, and both of these, after a gallant fight, went under, one with a loss of over £3,000 in less than eighteen months. Nevertheless, I feel convinced that this recovery of tin can be performed with financial success, but in my opinion it must be by an electrolytic process such as I have sketched above.

My friend, Mr. F. W. Harbord, A.R.S.M., of Cooper's Hill, who has some considerable experience in attempts at the recovery of tin from scrap on the commercial scale, writes to me that in Birmingham in about 1891 he could purchase new scrap at the rate of ten tons per week at a price at the works of the makers of from 12s. 6d. to 15s. per ton, and another 5s. had to be paid for collecting and cartage. Old scrap could be purchased at the same date in Birmingham for 2s. 6d. per ton, with an extra 5s. per ton for collection. The amount of old scrap available was about from five to seven tons per week. Mr. Harbord further states that new scrap varies very much in price. Mr. T. Twyman, F.I.C., of the Talbot Continuous Steel Company, who has also had a large experience in this matter, has kindly given me a good deal of information, and he tells me that in London he was able to obtain old tin scrap at from 6s. to 7s. per ton delivered, i.e. carted a short distance, carriage being the principal factor determining price. He found that he could get about 60 tons per week of this old tin scrap from firms in the East-end of London alone ; the trade being largely in the hands of one man. Tin cuttings, i.e. new scrap tin, fetched a far higher price, namely, about 25s. per ton in London in 1899. These cuttings are nearly all shipped abroad to Hamburg.

Another correspondent, in the metal trade in Birmingham, wrote in September, 1901, stating that he believed that new tin scrap at that date in Birmingham varied from £1 to £1 5s. f.o.r. Birmingham. This correspondent also states that he considers the percentage of tin in average new tin scrap to be nearer 3 per cent. than the 5 per cent. found by Dr. Smith in the scrap obtained in Milan. (See p. 581.) A few further details from other sources concerning the price and available amount of tin scrap will be found on p. 579 of this book.

CHAPTER VI.

THE ELECTROLYTIC REFINING OF LEAD.

Keith's Electrolytic Lead Refining Process for Base Bullion—Tommasi's Electrolytic Lead Refining Process.—Formation of Spongy Lead.— Richly Argentiferous Lead Treated by Tommasi's Process.—Refining Argentiferous Lead in Lead Nitrate Solution.

Electrolytic Refining of Lead.—Keith's Process.—Professor Keith, of New York, in 1878, devised a process for the electrolytic refining of impure lead, with the object of extracting the silver and at the same time separating the lead in a pure metallic state. The process, so far as the arrangements of the baths and the disposition of the anodes and cathodes, is the same as in Elkington's copper-refining process. In base bullion the chief constituent is lead, which forms at least 90 per cent. of the mass. To separate this from the silver, antimony, arsenic, &c., by electricity, many electrolytes were tried, including nitrate of lead ; but since the nitric acid set free would also dissolve the silver from the bullion anodes, this was not found to be a suitable electrolyte ; sulphuric acid, which dissolves lead in small quantity, but the sulphate of lead formed became precipitated in the solution, while no metallic lead was deposited upon the cathode. After having tried all the known salts of lead in varying combinations, Professor Keith finally obtained several solutions which would serve as more or less perfect electrolytes for lead, and it was to his success in this direction that the apparent practicability of his processes was due.[*] The electrolyte which he found most successful, and which was accepted as the best for the electrolytic treatment of lead, consisted of acetate of soda, about $1\frac{1}{2}$ pounds to the gallon, in which $2\frac{1}{2}$ to 3 ounces of sulphate of lead were dissolved. The bullion-plates or anodes were thinner than those used in electrolytic copper refining, being only from $\frac{1}{8}$th to $\frac{3}{16}$ths of an inch in thickness. A plate of bullion 15 by 24 inches of this thickness weighs about 20 pounds. Before being put into the vats a muslin bag is drawn over each bullion-plate, the object being to prevent the residues (silver, &c.) from falling to the bottom

---

[*] This process was carried on by the Electro Metal Refining Company, New York, but has since been discontinued.

of the vat. The lead is intended to accumulate on the bottom of the vats, and by the above arrangements the impurities are prevented from mixing with the deposited metal.

The muslin employed to enclose the anodes is fine enough to prevent the fine particles of residue from being washed through by agitation of the solution, while the liquid is freely diffused between the inside and outside of the bags, and the resistance not perceptibly increased. The diffusion of the liquid by agitation is absolutely necessary, since the lead becomes dissolved as sulphate by electrolysis, and this sulphate must diffuse itself in the liquid to replace the sulphate decomposed at the cathode to deposit metallic lead. If the diffusion were to take place too slowly, in a short time the amount of lead present in the solution outside the bags would be too little to satisfy the depositing power of the current, and hydrogen would be evolved, followed by polarisation. If the solution be constantly agitated, therefore, a much stronger current may be employed without danger of polarisation. Heating the solution also favours the diffusion, while at the same time it materially reduces the resistance; it is, therefore, usually heated to about 100° Fahr. During working the solution is maintained in a neutral state; if allowed to become alkaline, the alkali (in this case soda) becomes decomposed, furnishing oxygen to the anode and per-oxidising the lead thereon, while hydrogen is evolved at the cathode and produces polarisation, irrespective of the diffusion of the solution.

In working the above solution there is no polarisation, provided the liquid is kept in a normal condition; the lead is dissolved from the anodes, and an exact equivalent of metal is deposited on the cathodes, and gathers as a crystalline coherent layer. In some cases, after the layer becomes sufficiently thick, it rolls off the surface of the cathodes and falls to the bottom of the vat; sometimes, however, the cathodes require to be gently scraped to remove the deposited metal. When the lead is all electrolysed from the bullion-plates (anodes), the impurities alone remain in the bags, and these usually constitute about 5 per cent. of the whole mass. If the bag containing the anode be carefully removed, it will be found that the anode has changed but little in appearance; it has a bright metallic aspect with a display of iridescent colour, and appears as if it had undergone no action; if the plate be touched, however, it yields to the finger like blue clay, to which it bears some resemblance. This is the residuum, in which may be seen here and there small fragments of lead which had become detached from the anode as it grew thinner. The residuum is immersed in water, and the scraps are washed from the "mud," which then deposits in a clayey mass, leaving the water perfectly clear; the scraps are afterwards re-melted and recast into plates. The deposited lead is drawn out from the vats at convenient intervals, and after washing to free it from the solution which attaches to it, it is dried quickly, and either pressed into "slugs," or otherwise melted at a low heat and cast into pigs. The residue, which contains the silver and gold besides the impurities, as arsenic, antimony, &c., is treated by a special process, which is thus described by Professor Keith :—

"In laying out our plan of procedure, we must first consider the conditions and liabilities. These may be formulated thus :—

"1. It is a wet powder, and must be dried.

"2. The oxidisable constituents must be oxidised.

"3. It must be mixed with fluxes and fused.

"4. Antimony and arsenic are volatile, and carry off in vaporising, mechanically or otherwise, silver, and perhaps gold. It is absolutely necessary to get all the gold and silver, and as pure as possible, though they may be alloyed together. It is obvious that drying the powder and roasting it in a reverberatory furnace will cause a great loss in silver from volatilisation with arsenic and antimony, besides loss of powder carried off by the draught. Its roasting needs most careful treatment, as from the easy fusibility of antimony, masses of alloy may be formed which cannot be practically oxidised. Recognising these conditions and difficulties, the plan of proceeding is this : After having removed the powder from the filters while it is still wet, it is mixed with the proper quantity of nitrate of soda, when it may be dried without loss of dust, as the nitrate cements the whole together. When sufficiently dry it is placed in crucibles for fusion. These are cautiously heated : the nitrate decomposing gives oxygen to the antimony, arsenic, copper, iron, &c., thus forming teroxide of antimony, arsenious acid, and oxides of copper, iron, &c. The soda combines with the teroxide of antimony and the arsenious acids, forming antimoniate of soda and arsenite of soda, which are fusible ; a little borax added makes the slag more liquid when the oxides of iron and copper are present. A button of pure gold and silver collects at the bottom of the crucible. Now, though antimony, arsenic, and arsenious acid are volatile, antimoniate of soda and arsenite of soda are not, so there can be no loss from their volatilisation. Nitrate of potash may be substituted for the soda salt with the same effect. This slag of antimoniates and arsenites can be utilised in the following manner : When treated with hot water the arsenite of soda or potash is dissolved, and the antimoniate remains undissolved, together with the oxides of copper and iron. The arsenite of soda or potash is obtained by crystallisation, and finds its use in dyeing, colour-making, &c. ; or metallic arsenic may be obtained from it by sublimation. Antimony may be obtained from the residue by mixing it with charcoal and melting in a crucible. No copper or iron need be reduced with the antimony with proper care, but if they are, they may be removed by subsequent fusion with some teroxide of antimony. Perhaps it will not be found profitable to carry the utilisation farther than to save the antimony and arsenic."

In this process the anode plates are cast thin, because the speed at which the electrolysis can be pushed is limited by the rate of diffusion of the sulphate of lead through the bags, as before explained. In practice it was found that with plates 15 by 24 inches, the rate of the electrolytic transfer of lead was from $1\frac{1}{2}$ to 2 ounces of lead per hour, and a plate of this size, weighing 20 pounds, would therefore require from six to eight days. With plates twice as thick, or $\frac{1}{4}$ of an inch, it would last twice as long, and to make a given return per day, the amount under treatment would require to be greater. Larger and thinner plates would be electrolysed more rapidly. The electrolyte does not become changed by continued use. Iron and zinc, if present in the bullion, become dissolved and remain in solution, but this does not impair its efficiency. A small quantity of sulphate of lead added to the bath will correct any defect from this cause ; moreover, the sulphate of iron becomes gradually oxidised, and sesquioxide of iron rises to the surface, which may be skimmed off.

The chief object in the treatment of lead base bullion is the separation of the silver from the lead, which by the old or dry method is not only costly but imperfect, while the presence of antimony and other impurities greatly influences the facility of such treatment. Indeed, some varieties of bullion, containing antimony in large proportions, are so refractory that in many cases they cannot be separated with profit. By this process, however, the silver and lead are directly separated, whether antimony be present in the bullion in large or small quantities, while at the same time this metal is also saved, and can be sold at the market value. All the lead is recovered with the exception of a trifling percentage lost by vaporisation, oxidisation, &c., because the bullion is only heated sufficiently to be melted and cast into plates, instead of being repeatedly heated, as in the old process. Again, all the gold and silver are saved, while the lead is obtained in an almost perfectly pure state—an important advantage in the electrolytic method of treatment. A sample of Keith's electrolytic lead was found on analysis to contain only ·000068 per cent. of silver, or ·02 ounce per ton, while only traces of antimony and arsenic could be detected, though a large quantity was used for analysis ; there was no copper, though there had been some in the bullion. The presence of the small quantity of silver was believed to be due to carelessly handling the bags covering the anodes, by which small particles of the residue washed out into the bath, and finally deposited with the lead at the bottom of the vat. The bullion from which this lead was electrolysed was also submitted to analysis, with the following result :—

Lead	96·36
Silver (161·7 ounces per ton) . .	·5544
Copper	·315
Antimony	1·07
Arsenic	1·22
Traces of zinc and iron, undetermined	
matter, and loss	·4806

100·0000

Although this ingenious process is not now being worked, there is much in its details that is instructive, and may serve as a guide to the experimentalist in his researches in the electrolytic treatment of lead—a successful process for which we may yet hope will be discovered. Mr. Williams* thus describes the arrangements formerly adopted for carrying out the process by the Electro Metal Refining Company of Rome, New York, who were the owners of Professor Keith's patents. The process was first developed on a comparatively moderate scale as follows :—" In this plant there were 4 vats 10 feet long, 2 feet wide, and 3 feet 6 inches deep, made of wood, and covered with pitch without and within to make them watertight. Copper rods 1 inch square, resting on the edge of each tank, served as conductors for the current. The anodes and cathodes rested on these conductors by means of hooks projecting from their upper margin. A piece of paper was placed between the hook of each anode and the conductor on one side, so as to prevent contact, while the cathodes were also insulated in like manner from the other conductor. There were about 40 anodes and as many cathodes in each tank. The tanks were connected in series to a Weston dynamo-electric machine for electro-deposition, having an electromotive force of about 1 volt, and a very low resistance. A circulating and heating apparatus was also provided as follows : The solution was allowed to run off from a gutter at one end of each tank, and was thence conveyed to a tub, from which it was pumped up into a cask placed higher than the tanks. From the bottom of this cask was a delivery-pipe which subdivided into four smaller pipes, one extending along the bottom of each tank. These small pipes were perforated with numerous holes, through which the solution entered the tank. In the cask a copper still-worm was placed, which was heated by steam. Thus the solution was agitated and heated at the same time. The weight of each plate (15 by 24 by $\frac{1}{8}$ inches) being about 20 pounds, the amount under treatment was consequently 160 (plates) × 20 = 3,200 pounds. At the rate of $1\frac{1}{3}$ ounces per plate per hour, which was the average of working, the

* " Geological Survey of the Resources of the United States."

deposit was 360 pounds per day of twenty-four hours. It will be readily understood that the machine can be worked by night as well as by day. At this rate a plate would be exhausted in somewhat less than nine days. In practice the plates are not all exhausted, however; there always remain small pieces which become detached from the rest of the plate. The weight of these scraps would average about 1 pound, though in this case many of the plates were cracked to begin with, and did not hold out well for this reason.

The company subsequently fitted up a more extensive plant at Rome, New York, which is thus described: "The works are in a one-storey building, 150 feet by 50 feet, of which the working capacity (three tons per day) requires only one-third the space. The casting of the bullion-plates is done by means of a casting machine, or system of mechanical moulds rotating around a centre, and passing successively under the spout of the melting furnace. There are twelve moulds, each holding at its upper part two thin strips of copper perforated with holes. When the lead is poured into the mould it fills these holes, and the strips form suspension lugs and connections at the same time. At the side of the revolving system opposite from the furnace, the plates are taken away by a boy, who replaces other copper strips and closes the mould again for another round. A man and a boy will make 180 plates per hour. Each plate is 24 inches by 6 inches by $\frac{1}{8}$ inch, and weighs 8 pounds. The plates are hung from a frame and carried by an overhead railway to the vats. There are thirty circular vats, made of a kind of concrete mixture. Each vat is 6 feet in diameter, 40 inches high, and has a central core or pillar 2 feet in diameter, and equal in height to the vat.

"The cathodes consist of thirteen circular hoops or bands of sheet brass, two feet high, and arranged concentrically two inches apart. The plates of bullion are lowered between these circular cathodes. The anode frame or bullion carrier has twelve consecutive rings of brass, 2 inches wide and $\frac{1}{8}$ inch thick, also arranged two inches apart. Rivet heads of copper project from these rings, and the bullion-plates are suspended to these by the eye-holes in the suspension lugs. Each frame will receive 270 bullion-plates, making a total weight of bullion about 2,100 pounds per vat, or slightly over one ton. The carrying power of the overhead railway is 3,000 pounds. The solution is allowed to overflow from the vats by a small gutter to the floor, which is of concrete, and grooved with gutters that lead to cisterns at the end of the building, which have a capacity of 3,000 gallons, whence it is pumped by a centrifugal pump to an overhead tank, where it is heated by a system of steam pipes to 100° Fahr., automatic electrical regulation of the temperature being secured by a special device. From this tank the solution is distributed to the vats by a system of pipes. An Edison dynamo-electric machine, constructed specially for this purpose, is used to furnish the current. This machine has an electromotive force of ten volts, and an internal resistance of ·005 ohm, and produces the enormous volume of current of 2,000 ampères. This current will nevertheless be entirely safe to the employés on account of its very low electromotive force.

" The vats are connected in series, and the power used by the machine does not exceed 10 horse-power for 30 vats. The vats are charged in rotation, three per day, and on the tenth day the first three are renewed, after which the renewal is kept up in the same order. In this way three tons are put under treatment every day, and three tons refined and returned. The anode carriers can be rotated around the core of the vat as a centre, and they carry mechanical fingers which scrape the surface of the cathodes by the motion. By removing a plug, each vat may be rapidly emptied, and the crystalline lead shovelled out. This lead is washed in water and placed in a centrifugal dryer, after which it is melted under oil or other reducing material, to expel the remaining traces of moisture without oxidation, and it is then ready for the market. In this establishment there are ovens and muffles, &c., for assay purposes and for reducing residues. These residues are washed in water, and the water run through a sieve to take out the scraps of bullion ; they are then allowed to settle, after which the water is decanted and the residuum dried."

Tommasi's Process of Lead Refining.—This process was first described by Dr. Tommasi in *Comptes Rendus,* 1896, vol. 122, p. 1476, and the following description is taken with a few modifications from an undated pamphlet describing it, which was circulated somewhere about 1898 by a French Syndicate, who were running the process commercially. The address of this Syndicate was 6, Rue des Immeubles Industriels, Place de la Nation, Paris.

"The Tommasi Electrolyser (Figs. 146, 147, 148, and 149), consists of a rectangular tank d, in which is placed two anodes p p, and between them the cathode c, the latter being a metallic disc mounted on a gun-metal spindle, and capable of being rotated. Only a part of the disc is submerged in the bath, so that as it rotates it passes through the electrolyte and then out into the air. The part of the disc which emerges from the liquid passes between two scrapers, o, whose duty it is, not only to scrape off the spongy deposit, but also to depolarise the surface of the disc. (This depolarising action is assumed to consist in the removal of deposited metallic lead, and with it any hydrogen which may have been deposited with it on account of a high current density being employed. The actuality of this claimed advantage appears to me to be more than doubtful.—A. P.)

The anodes can be either in the form of plates or of coarse powder. When used in a granular state the compounds are simply packed in a perforated receptacle, in the centre of which a metallic plate, to act as a conductor, has previously been placed. To make sure that the contact is good between the granular pieces and the conductors, a similar metallic plate is placed at the bottom of the receptacle in which they lie. In the larger form of electrolysers the anodes are made up of two or more parts to permit of their being withdrawn and replaced when necessary, without having to disturb the disc cathode. The disc is made all in one piece, when the metal to be recovered is deposited in a spongy or loose crystalline state; it is, however, made in several interchangeable sections when the metal is deposited in a compact state.

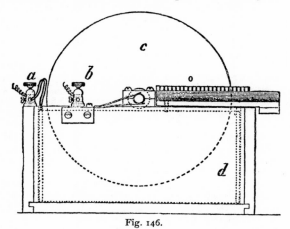

Fig. 146.

The sections of which the cathode is built up are fixed to the spindle by means of bolts and screws which can be unfastened when it is desired to remove or replace any of the parts. When the sections are coated with a sufficiently thick layer of metal, they are withdrawn by a travelling block and fall, suspended above the electrolysers, and are carried over a bath of molten metal, similar to that deposited on the sections, and into this they are dipped; the electro-deposited coating is melted off into the bath, and the sections can then be once more returned to the electrolysing vats in order to receive a fresh deposit. (The Tommasi electrolyser, as will be understood from the above description, is intended for the treatment of all kinds of metal, but it is clearly more adapted to the treatment of lead deposits than to other metals, and the pamphlet from which this description is taken chiefly refers to its use for this low-melting-point metal.—A.P.) The principle upon which the Tommasi process, when applied to the refining of metallic lead, is based, consists in electrolysing a solution of a lead salt which not only has a very low electrical resistance, but also is of such a character that the formation of lead peroxide on the anode is prevented. It is of course well known that the electrolysis of lead solutions nearly always causes the deposition of peroxide of lead on the anode, thus polarising the electrolytic cell, and giving rise to a waste of energy whilst the current is passed, which otherwise would not occur. That is, for a given current density in the electrolytic vat a larger e. m. f. must be supplied by the dynamo per vat, if the lead peroxide is formed, than would be required if this did not happen. The electrolytic solution used consists of a mixture of acetate of lead with acetate of sodium or potassium dissolved in water, and to this is added " *certain substances, the nature of which is kept secret.*" (The function of these secret materials is undoubtedly that of preventing the formation of the lead peroxide on the anodes, and the material or materials probably consist of some reducing organic compound which reduces the lead peroxide (or rather prevent its ever being formed), and is itself oxidised. The importance of this action was recognised by Keith many years

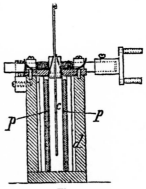

Fig. 147.

previously, and he found that if organic acid salts of lead were used with moderate current density and constant circulation of the electrolyte no polarisation of the anode due to the formation of lead peroxide took place, but nevertheless the organic acid present, namely, acetic acid, was gradually oxidised into water and carbonic acid, and thus removed from action, a loss of acid which naturally had to be made good later. Whatever may be the nature of the substance employed by Tommasi for preventing the formation of lead peroxide, it appears that it must be unavoidably consumed in the course of the electrolysis, and the cost of its renewal must therefore be considered in the calculation of expenses involved in this process of refining lead.—A. P.)

Tommasi, in describing his method of refining, further states that the lead anodes become dissolved by the current, and the metal is deposited in the form of spongy crystals on the cathode disc, whereas all the silver in the alloy, being insoluble in the bath, falls to the bottom of the electrolyser, and is caught in a perforated receptacle placed there for the purpose. In the event of there being antimony and arsenic present with the silver, they may be separated from it by fusion in a crucible with a mixture of borax and sodium nitrate. The antimony and arsenic are converted into antimonate and arsenate of sodium respectively, and all the silver remains in the metallic state.

The details of working are described by Tommasi as follows :—The

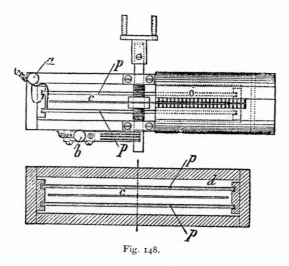

Fig. 148.

argentiferous lead is melted and cast into suitable moulds. The half anodes (each anode is made in two parts) are attached to four metal rods placed at the top of the electrolyser, each provided with a screw and bolt. These are in connection with the positive pole of the dynamo. The anodes are so held that it is possible to keep them at a fixed distance from the cathode disc, but at the same time it is possible to gradually bring them closer and closer to the cathode, as the anodes are gradually dissolved away. In some cases the distance between the electrodes can be reduced to as little as 2 or 3 cms. Contact with the cathode disc is made by means of a metallic brush rubbing on the spindle. The cathode disc, which makes from one to

two revolutions per minute, is for large plants as much as 3 metres
(= 9·84 ft.) in diameter, and is made of aluminium, bronze, or of
nickelled iron. It may also be made of copper or even of sheet iron.
As soon as the current is turned on, the lead commences to be
deposited on the disc in the shape of spongy crystals. When the
deposit of lead is considered to be of sufficient thickness, and it is
thought advisable to remove it, the current is switched off, and the
scrapers are applied. (From this statement it appears that the process
is not strictly continuous, although from other statements which have
been published concerning the methods employed, the lead is apparently continuously scraped away from that portion of the anode disc

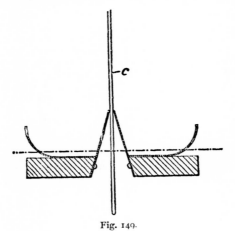

Fig. 149.

above the vat, whilst it is continuously deposited upon the portion of
the disc in the electrolyte, and the disc is kept slowly rotating
throughout the whole process. Which of these methods is actually
employed, or if both, is of some importance in arriving at a correct
estimate of the cost of working.—A. P.)

The scrapers which are employed to remove the lead consist of two
brass or aluminium-bronze plates, which by a simple mechanism can
be brought close up against the disc or withdrawn as is desired. They
are shown in Figs. 148 and 149. By the action of the scrapers the lead
is detached and falls into suitable gutters or channels, by which it is led
to a sieve of metallic gauze. (The mechanism by which this leading off
of the metallic scrapings along the gutters is effected is not described

in any way.—A. P.) The lead is then washed with distilled water and submitted to great pressure, and the liquid which exudes is collected with the washings, which are then all evaporated down to a solution of a density of 30 degrees Baumé. It is then allowed to cool, and is pumped back into the electrolysers. The compressed lead is heated in a crucible with 2 to 3 per cent. of charcoal powder, and is cast into ingot moulds. When the anodes are dissolved they can be replaced by fresh ones, and the silver which has fallen to the bottom of the tank can be collected. To do this the disc is raised by suitable means (It must be remembered that this disc of the size recommended weighs well over a ton and a half.—A. P.), and the perforated receptacle or tray, which has been placed at the bottom of the tank at the beginning of the operation, is raised, and the deposited silver sludge is cleaned out.

Tommasi continues :—Theoretically, the extraction of silver from argentiferous lead, or, as it is most generally called, the refining of lead by means of the electric current, does not demand any consumption of energy. (This is only approximately true ; firstly, if there is no back e. m. f. from the electrodes, i.e., if both the anode and cathode have surfaces of identically the same chemical constitution ; and secondly, if the current density is excessively small, say a small fraction of an ampère per square foot of cathode surface. This, of course, would involve, for a given yearly output of lead, an enormously large factory space and an immense capital outlay in vats, solution, lead anodes, and refinery buildings.—A. P.) The amount of heat evolved by the solution of a molecule of lead as acetate is exactly the same as that absorbed by the setting free of a molecule of lead from the same salts. The energy which is employed, if there is no back e. m. f., is used solely in the resistance of the electrolyte and leads. In short, if it were possible to do away with the electrical resistance of the electrolyte, and suppress the polarisation of the electrodes, the electrolysis of any salt using a soluble anode made of the same metal as that contained by the salt in solution would take place with practically no expenditure of energy.

And the writer of this pamphlet adds : " To the important advantages of this process already mentioned there is one other, none the less important, to be added, viz., the composition of the electrolyte. As a matter of fact it is not easy to choose a good bath, particularly for refining lead, for that bath must satisfy certain conditions on which depend to a great extent the success of the operation. These conditions are :—(1) The resistance must be very low ; (2) During the progress of the electrolysis there must be no crystalline or amorphous deposit on the electrodes ; (3) The lead deposited must be in such a state as to adhere sufficiently so as not to fall off, yet not so strongly as not to be taken off by means of the scrapers ; and (4) It should dissolve

out the lead only without attacking the silver contained in the alloys
These conditions are fulfilled by the use of an electrolyte, the composition of which has been given above." (It would be more correct to say
"the composition of which has *not* been given above."—A. P.)

The following figures are stated by Tommasi to represent the best
results of a large number of experiments. Presumably the first set
refer to a small installation, although the details of cost, etc., are not
worked out.

DETAILS OF SMALL INSTALLATION (?)

Number of disc cathodes . .	1
Diameter of disc	1metre (= 3·28 feet).
Thickness of disc	2 mm. (= 0·08 inch).
Height of disc above electrolyte .	40 cm. (= 1·312 feet).
Surface of the disc submerged .	2,910 sq. cm. (= 451 sq. in.).
Material of which disc is made .	Copper or sheet-iron.
Revolutions of disc . . .	One per minute.
Number of anodes	2
Thickness of anodes . . .	1 cm. (= 0·4 inch).
Composition of anodes . . .	Argentiferous lead.
Distance between anodes . .	4 cm. (= 1·6 inch).
Distance between anode and cathode	19 mm. (= 0·8 inch nearly).
Composition of electrolyte . .	Double acetate of lead and potassium.
Current	75 ampères.
Resistance of bath	0·00387 ohm.
E. M. F. required	0·29 volts.

It may be added that from these figures the current density employed
at the cathode is 23 ampères per square foot. The total weight of the
lead anodes must be at least 73 lbs., and judging from the form of
cells employed, they are more likely to be over 150 lbs. in weight.

The following are the details given by Tommasi of an installation
which would be capable of desilverising electrolytically from 25,000
to 30,000 long tons of argentiferous lead per annum by his process:—

Number of baths in series . . .	500
Cathode discs per bath	1
Diameter of disc cathode . . .	3 metres (= 9·84 feet).
Thickness of disc cathode . . .	2 cm. (= 0·8 inch).

(Each of these discs must at least weigh over a ton and a-half if made
of brass or gun metal of a density of 8, even if the shaft is not reckoned
in. It is possible that (owing to a misprint in the original pamphlet)
the thickness of the cathode is given as 2 cm. instead of 2 mm. If
this is so, the weight would be only a tenth of that stated.—A. P.)

Nature of disc . . . . . Aluminium bronze.
Anodes in each bath 2
Thickness of anodes . . . . 5 cm.
Nature of anodes . . . . . Argentiferous lead.

(The weight of this argentiferous lead must be five tons per bath or 2,500 tons for the whole installation.—A. P.)

Distance between anodes and disc . 2 cm. (= o·8 inch).
Revolutions of disc . . . . One per minute.
Composition of electrolyte . . . Double acetate of lead
              and potassium.
E. M. F. per bath . . . . . o·75 volts.
Total E. M. F. for the 500 baths (a) . 375 volts.
Current . . . . . . 1,800 ampères.
Conductors consist of 400 copper wires of 6 sq. mm. cross section each.
Total length of conductor, go and return . . . . . . 500 metres (= 547 yds.)

(The current density in the conductor is, therefore, 484 ampères per sq. inch, and there is just about one yard of conductor connecting vat to vat.—A.P.)

Resistance of conductor . . . o·0035 ohm.
Pressure drop in conductor (b) . . 6·3 volts.
Total E. M. F. required (a and b) . . 382 volts.
Power required . . . . 687,600 watts.
Output of dynamo . . . 721,980 watts.
Motive power required . . 980 B. H. P.

(If the output of the dynamo is to be 721,980 watts then, with 95 per cent efficiency of conversion, the output of the engine required is 1,002 B. H. P.—A. P.)

Cost of 1,000 B. H. P. engine . . 200,000 francs (= £8,000).
Cost of one dynamo of 721,980 watts,
 or five each of 144,326 watts . 80,000 francs (= £3,200).
500 electrolysers, including the
 electrolyte, at 1,000 francs . . 500,000 francs (= £20,000).

The cost of a brass disc weighing one and a-half tons at £40 per ton is £60, and the cost of the five tons of lead in each vat must be at least £40. And on to this, the further cost of the vats, mountings, scrapers, solution, etc., must be added. The estimate under this head is very much too low, and the 500 electrolysers, if made as is specified, cannot, in my opinion, be less than £50,000.—A. P.)

Total capital cost, as per Tommasi
   estimate .   .   .   .   . 780,000 francs (= £31,200).
Copper conductors .   .   .   . 20,000 francs (= £800).

Interest at 10 % on 780,000 francs . 78,000 francs (= £3,120).
Interest at 5 % on 20,000 francs   . 1,000 francs (= £40).
Coal, assuming 800 grams per horse-
   power hour, at 15 francs per ton . 86,400 francs (= £3,456).

(This estimate allows 1·76 lb. of coal per B. H. P. of engines.—A.P.)

Oil and sundries .   .   .   . 10,000 francs (= £400).
Two stokers .   .   .   .   . 5,000 francs (= £200).
Two electricians .   .   . 5,000 francs (= £200).
Thirty labourers .   .   . 30,000 francs (= £1,200).

Total annual outlay .   .   . 215,400 francs (= £8,616).

Tommasi adds the following further details :—

In each bath per hour   .   .   . 7,020 grams of lead are
                                     obtained.

(This is the full theoretical yield from 1,800 ampères.—A. P.)

And this gives in a year of 300 days of 24 hours each 25,272 long tons. (This estimate presumes that not only the theoretical yield of lead is obtained whilst the vats are running, which may indeed very closely represent the facts of the case, but it also necessitates the vats being run continuously, which according to Tommasi's description is not the case. (*See p.* 605.)

From the above results, by dividing the supposed annual cost of 215,400 francs by the still more supposed annual output of 25,272 tons of lead, Tommasi states that the cost of refining one ton of lead is 8·6 francs when a steam-driven plant is employed, and by deducting nearly the whole annual cost of coal per ton of lead output, namely 2·8 francs, he obtains 5·8 francs as the cost of refining one ton of lead if a water-power plant is used. This is, however, an absurdly incorrect method of calculation, for it allows practically nothing further for the capital cost of installation of the water power. That is to say, that the total cost of installing the water-power, including lease of power, turbines, and all other requisites complete is reckoned at £8 per B. H. P. output of turbines. This is certainly at least 20 per cent. below the average price at which water-power can be installed, although it is true this cost varies enormously with the district.

From the sketch plans given and the stated dimensions, it is evident

R R

that each vat in this installation must be about 12 feet long, 2·5 feet wide, and about 5 feet deep. That is, the number of cubic feet of vat space per ton output of lead per annum, at a current density of one ampère per square foot, cannot be less than 60 cubic feet, which is just about the same amount of space as is necessary for copper refineries. (*See p.* 505.)

The current density employed by Tommasi in the installation described above (if it has ever been actually run on anything like so large a scale) is about, and is certainly not less than, 20 ampères per square foot, and if a higher current density is actually employed for the size of vats here described and the output stated, the vat space per ton per year per unit current density will work out still larger than 60 cubic feet.

According to Tommasi's figures each horse-power hour yields 7·865 pounds of metallic lead, whilst at Anaconda the practical yield of copper is stated to be 6·67 lbs. per hour per B.H.P. installed in the works, and it must be remembered that the B.H.P. installed in Tommasi's estimate is the theoretical power required for the dynamos only, when their efficiency is taken as 95 per cent., and it allows no margin for power required for rotating the cathodes, circulating the electrolyte, and shifting metal, nor is any allowance made for the necessary discontinuity of the operations. Further it must be remembered that Tommasi's estimates do not include such very large and important items as the interest on the large weight of metallic lead in stock and in the form of anodes, nor the weight of gun-metal as cathodes. Nor are the costs of buildings included in the capital charges, nor rent and taxes in the annual. In short the whole of Tommasi's estimate carries evidence of want of actuality, and I cannot believe that the electrolytic refining of a ton of lead by this or any other known vat method can be carried out successfully for a less amount than about 20 shillings per ton ; probably this is a good deal too low an estimate.

Spongy Lead.—Tommasi's process is, however, probably, as he suggests, a satisfactory method of obtaining cheaply a pure form of spongy lead for making secondary battery plates. Tommasi calculates that the formation of a ton of spongy lead by the chemical method of precipitating metallic lead from lead acetate solution by means of metallic zinc to be as much as £54 per ton, whilst by his own process he considers that far purer spongy lead might be formed for about 258 francs (= £10 6s. 8d.) per ton. I do not, however, think that the particular chemical method, which is the alternative suggested by Tommasi is the cheapest which could be devised, whilst I do not consider that the Tommasi process could produce lead, if the anodes cost £10 per ton, at much under £11 10s. to £12 per ton. However there is not much doubt that if spongy or crystalline lead can obtain any large market, it can probably be prepared most cheaply by Tommasi's process or some variant of it.

Tommasi has also suggested the employment of the pure crystalline lead obtained by his process for the starting point in the preparation of pure litharge, red lead and white lead at a cheap rate and of high quality. What is evidently a reprint of the pamphlet from which the above descriptions of Tommasi's process has been obtained, may be found in the *Electrician*, vol. 41, 1898, p. 591.

Richly Argentiferous Lead Treated by Tommasi's Process.—The richly argentiferous lead obtained in the process of the de-silverisation of lead by Pattinson's process is stated to be, and may possibly be, treated economically by Tommasi's process, for in this case the refinery product may be considered to be the silver whilst the desilverised lead obtained on the cathode forms a by-product. The silver, as has already been pointed out, is obtained in the anode sludge. This anode sludge is washed, dried and fused with sodium nitrate and a little borax, the arsenic and antimony are thus oxidised and fluxed off, leaving a pure silver; for further remarks on this point, see p. 542. Rich Pattinson-ised lead contains from 400 to 600 ounces of silver per ton, *i.e.*, about 1·11 to 1·67 per cent. of silver, which is two to three times the silver content of the base bullion treated by Keith's process, see p. 599. After examining carefully Tommasi's description of this process and also the published descriptions of the Keith's process, I think there can be little doubt that if such richly argentiferous lead can be treated successfully by Tommasi's process, it can be treated more cheaply still by Keith's method.

Cost of Refining Base Bullion by Parkes' Process.—In a recent paper M. W. Iles (*Engineering and Mining Journal*, 1900, vol. 70, series 7, p. 185), makes a very detailed examination of the costs of refining base bullion by the Parkes' non-electrolytic process, and he gives the following figures obtained from actual working in the Globe Works, Denver, Colorado, from Jan. 1895 to June 1896 :

1. Cost of labour per ton	.	.	1·968 dollars	
2. Cost of spelter	. . .	.	0·861 ,,	
3. Cost of coal .	. .	.	0·496 ,,	4·135
4. Cost of coke	. .	.	0.521 ,,	
5. Supplies, repairs, and general expenses	. .	.	0·289 ,,	
6. Interest	. . .	.	1·317 ,,	
7. Parting and brokerage	.	.	2·121 ,,	
8. Reworking by-products	.	.	1·492 ,,	
9. Expressage (carriage) .	.	.	1·085 ,,	
			10·150 ,,	

Out of this total amount of £2 2s. 3½d. per ton, the cost of actually *refining* the base bullion which is included in the first five of the above items is from 3 to 5 dollars per ton, a cost which varies from time to time owing to the irregularity of the bullion supply and the consequent effect of this on the work of the plant (abstract *Journ. Soc. Chem. Ind.*, vol. 19, 1900, p. 900).

The above prices, were obtained with the Parkes' process of refining lead by means of zinc (a well known process which is described in all text books of metallurgy). The lead which is started with is refined to lead free from silver, and it is this process which costs from 3 to 5 dollars per ton of pure lead produced; besides this, however, a further quantity of a zinc-silver-lead alloy is produced, and the treatment of this to obtain all the silver costs another 5 dollars per ton of lead treated. It therefore seems very probable that the highly argentiferous zinc-lead-silver alloy could be very profitably parted electrolytically, the process resolving itself into—1st, a Parkes' process refining of the lead, and production of a certain amount of the lead-zinc-silver alloy : this part of the process costing, say, 4 dollars per ton of lead refined. 2nd, an electrolytic refining of the zinc-lead-silver alloy, costing say 15 dollars per ton of this alloy refined, and a ton of this alloy is about equivalent to 20 tons of lead, and hence the cost per ton of base bullion refined would be 4·75 dollars per ton to yield all the lead and silver in the refined state.

Refining of Argentiferous Lead in Lead Nitrate Solutions.— L. Glaser (*Zeitschrift für Elektrochemie*, 1900, vol. 7 [24] 365-369 and [26] 381-386) states that argentiferous lead, containing as much as 5 per cent. of metallic silver, may be refined by employing it as the anode in solutions of lead nitrate saturated with lead chloride. The lead obtained at the cathode is in a compact state, and is free from silver. A small amount of free hydrochloric acid in the electrolyte may be present without any disadvantage. The e. m. f. required per vat is stated to be only 0·05 to 0·2 volts, but in the abstract from which this account is taken (*Journ. Soc. Chem. Industry*, vol. 20, 1901, p. 259), it is not stated what current density is obtained when these voltages are employed. This process, in as far as it is possible to judge from the very sketchy details given, would seem to offer a possibility of a commercially successful refining method. Its success, however, must depend upon the good conductivity of the electrolyte, the deposition of the lead in the massive condition, the absence of a polarising back e. m. f., the absence of silver and other foreign metals in the cathode lead, and last, what includes two of the foregoing conditions, the lowness of the required electrolysing e. m. f. at a high current density. The details concerning all these points are wanting in the abstract referred to above.

CHAPTER VII.

ELECTROLYTIC PRODUCTION OF ALUMINIUM AND THE ELECTROLYTIC REFINING OF NICKEL.

Properties of Aluminium.—The most remarkable properties of the metal aluminium are, firstly, its extreme lightness when compared with any other metal employed in commerce. The weight of one cubic centimetre of aluminium is from 2·6 to 2·8 grammes, whilst the weight of an equal volume of rolled copper is about 8·9 grammes. Secondly, the expansion of aluminium with heat is for equal volumes and equal rises of temperature, although somewhat lower than that of metallic zinc, yet about twice as great as that possessed by iron, and more than one-third greater than that of copper. Thirdly, aluminium, although more readily superficially oxidised by the air at ordinary temperatures than iron, does not corrode anything like so rapidly on exposure to moist air as the latter metal, the reason probably being that there is not any intermediate oxide lower than that formed on the exposed surface; the continuous rusting of iron being largely due to the fact that the red ferric oxide on the surface hands on oxygen to a layer beneath, when a lower oxide of iron is formed, and at the same time itself takes up a further amount from the air. In the case of aluminium, however, this action cannot take place, for only one oxide exists, and this cannot therefore give oxygen to the metal beneath it, but protects it from further action of the air. Fourthly, the colour of the oxides of iron and copper are markedly dark red, brown, orange, and black, whilst the sulphides of iron and copper and silver are all black. On the other hand, the oxide of aluminium is colourless and transparent, and so also is its sulphide. For this reason no discolouration of an aluminium surface due to atmospheric action takes place. Lastly, the conductivity of pure commercial aluminium for electricity is as high as about 0·6 of that of copper, although, if the aluminium contains very small quantities of copper, iron, and other impurities, its conductivity is more nearly 0·5 that of pure copper.

Effect of Mercury on Aluminium.—Mercury amalgamates readily with the surface of aluminium, cleaned with either dilute acids or with caustic alkali solutions. The bright metallic patch of mercury amalgamated aluminium, however, when exposed to the air, rapidly heats and throws out branches and crusts of white aluminium oxide, due to the rapid oxidation of the aluminium metal when dissolved in mercury and capable of circulation, fresh surfaces of metal being constantly exposed to the air. The appearance of this action is very remarkable, and it is clear that all aluminium surfaces must be carefully guarded from the simultaneous action of mercury and air. Yet it is quite possible that a surface of aluminium quicked with mercury salt, and then, without exposure to the air, plated with another metal, may offer a very satisfactory resistance to oxidation by the air.

Electrolytic Aluminium Smelting.—Many processes have been devised during the past fifteen years or so for the production of metallic aluminium from its compounds, by electrolysing these substances whilst in the condition of igneous fusion. Of these various processes, although many have been more or less largely worked, only three remain at the present date in active work, and one of these may still be considered as perhaps not having emerged from the experimental stage.

By far the most important processes are those known as Heroult's and Hall's process. These two processes are identical in principle, and, as far as can be judged, are carried out in essentially the same manner. They were curiously enough patented at almost identically the same dates, the one by Heroult in France first. In America he applied for a patent on May 22nd, 1886, but Hall, who had been developing the same process in America, was granted the patent on July 9th, 1886. Heroult's English patent was dated May 21st, 1887, No. 7,426. Apparently at present the processes are amalgamated, for they are stated to be employed indifferently at most of the works manufacturing aluminium. The English patents will normally expire this year (1901), but an appeal for an extension is to be made by the British Aluminium Company.

The Heroult or Hall process, as worked by the British Aluminium Company, is described as follows by R. W. Wallace (*Journal of the*

Society of Chemical Industry, vol. 17, p. 308) in April, 1898. The Heroult cell, as used by the British Aluminium Company, consists of a carbon-lined iron box, itself forming the cathode ; the anode is a bundle of carbon rods suspended within it and reaching nearly to the bottom. The bath consists of molten cryolite holding alumina in solution, and it is constantly fed with fresh pure aluminium oxide, as that already present in the bath is removed by the action of the electric current. The exhaustion of the alumina in the box is immediately indicated to the furnace attendant by a sharp rise in the volts required to keep the current steady; or, what is the same thing, the current rapidly drops if the volts are kept constant. The action goes on smoothly and continuously. The temperature of the fused bath has been measured, and has been found to be between 750° C. and 850° C. An electric pressure of from 3 to 5 volts is ample per bath to maintain the working temperature, as well as to effect the electrolysis. The current density employed at this voltage is about 700 ampères per square foot of cathode surface, or, as there are about $11\frac{1}{2}$ square feet of cathode surface per cell, the total current taken per cell is about 8,000 ampères. The practical yield is 1 pound of aluminium per 12 electrical horse-power hours. That is, a dynamo of 12 electrical horse-power output, or about 9 kilowatts, will yield 12 pounds of aluminium per hour. At this rate in one of the cells employed by the British Aluminium Company described above as taking 8,000 ampères at say 5 volts, the yield should be about 45 pounds of aluminium per hour.

The raw materials employed in this process consist of alumina and carbon. The fused cryolite which merely serves as a solvent for the alumina, is itself but little consumed if the current density is kept within the limits stated above. But if the current density is permitted to rise too high, or, what is the same thing, if the volts are raised, the decomposition of the cryolite and escape of fluorine will take place. The alumina used by the British Aluminium Company is prepared from bauxite, an impure oxide of iron and alumina, which is dug at Glenravel, in Ireland, thirty-five miles from the company's alumina purification works at Larne Harbour. Here it is chemically treated by a method described in the paper (*loc. cit.*) from which this description is abridged, and the pure alumina is transported by boat through the Caledonian Canal to the reduction works at the Falls of Foyer. The cost of this pure alumina is about £12 per ton to the company.*

---

* Mr. Claude Vautin states that he has seen a process working by means of which it is claimed that alumina can be obtained from felspar at a cost of from £3 to £4 per ton, and as two tons of pure alumina are consumed per ton of aluminium manufactured a reduction in the cost

The current density in the furnace at the carbon rods or anodes, is very much greater than over the bottom of the carbon-lined cell, which forms the cathode. The cathode density, as has already been stated, is about 700 ampères per square foot. But the anode current density is 35 ampères per square inch, that is to say, as much as over 5,000 ampères per square foot. Theoretically, two parts by weight of carbon should be consumed for every three parts by weight of aluminium separated, but practically the weight of carbon consumed is equal to the weight of aluminium obtained, and as all the impurities the carbon anodes contain must fall into the bath, it is necessary that the material should be exceptionally pure. It is equally important that the anode rods shall not disintegrate rapidly under the dense current they carry, or short circuiting of the bath by masses of carbon may occur, and thus large waste of energy will be caused. The British Aluminium Company manufacture the carbons they employ at a factory they have started in Greenock. The cryolite employed by this company is imported from Greenland, but arrangements have recently been made for the manufacture at their Larne works of an artificial fluoride which will be employed instead of cryolite as the solvent for the alumina.

Professor Chandler, in a discussion on a paper on aluminium production, read before the New York Section of the Society of Chemical Industry (*Journal*, vol. 16, p. 223, March 1897), describes the Hall process as he saw it carried out by the Pittsburg Reduction Co. as follows : " This seemed to be as simple and beautiful an electro-chemical process as could possibly be carried out. There were iron vessels there some 6 feet by 8 feet cross-section and 2 feet deep, lined with carbon. These vessels constituted the cathode. There were some 30 anodes of very compact carbon hanging from copper rods in 3 rows of about 10 carbons each. The carbons were about 3 inches in diameter. The cryolite bath had previously been fused in another pot and the impurities electrolysed out so that the cryolite that was actually employed for the bath, from which the aluminium was to be produced by electrolysis, was free from silicon and any metal which might contaminate the aluminium as it was reduced. The pure alumina was fed into this pure cryolite bath and the electrolysis proceeded regularly. Whenever there was a short supply of alumina the resistance of the bath increased, and an electric meter showed at once that the alumina had been consumed, and workmen simply shovelled

---

of the alumina of £8 or £9 per ton would reduce this cost of the aluminium by from £24 to £27 or more. As the present price of aluminium is under £136 per ton this would mean a very substantial reduction in its cost of manufacture, and in time in its selling price.

in some more. The process consisted chiefly in shovelling in alumina and ladleing out aluminium, day in and day out." The above accounts show that the Hall and Heroult processes as now conducted are identical.

A. Minet (*Comptes Rendus*, vol. 128, pp. 1163-1167, 1899), points out that during the past few years great advances have been made in the commercial production of pure aluminium, and in support of this statement he gives the following analyses :

	Silicon.	Iron.	Aluminium.
Aluminium in 1890 contained	0·90	0·40	98·70
„ 1893 „	0·25	0·40	99·35
„ 1897 „	0·02	0·12	99·86

A purity of product which compares not unfavourably with that of commercial copper.

The following table showing the enormous increase in output of aluminium and the correspondingly large decrease in its price is given by Mr. Wallace, and is of great interest :—

WORLD'S ANNUAL MAKE OF ALUMINIUM AND PRICE PER TON.

Year.	Tons per annum.	Price per ton. £
1890	40	1,083
1891	200	504
1892	300	308
1893	530	298
1894	1,200	186
1895	1,800	160
1896	2,000	155
1897	2,500	148
1899	—	136

The world's output of aluminium for the year 1900 has been stated to be about 6,000 tons, and the price per ton is about £170 for manufactured wire, rod, tubes, etc., in bulk.

At the present high price of copper and tin the selling price of aluminium is cheaper, bulk for bulk, than either of these metals, and it has lately been replacing copper for the manufacture of electrical conductors, several rather large bare metal distributing mains having been installed which are made of this metal. Whether it will stand the test of time must largely depend on the question of its corrosion by atmospheric influence, and also on the constancy of its mechanical properties. Both of these important points appear to be closely connected with the purity of the metal. If pure aluminium, or if, at any rate, aluminium of a constant and desired known composition, can be turned out regularly and in large quantities at the present prices, it appears highly probable that it may altogether displace copper in many important applications. This seems to be the more likely, as the patents for the only commercially successful processes will expire this year (1901), unless a special extension is granted, and even in that case the lapse of the patents cannot be long delayed. The cost to the British Aluminium Company of alumina, per ton of aluminium produced, is stated by Wallace to be about £25, and the cost of electrical energy by the Heroult process is about £12. If, therefore, the aluminium is sold in bulk at £136 per ton the cost of carbons, labour, depreciations, interest on capital and profits must be nearly £100 per ton ; a price which seems to admit of large profits.

The Minet aluminium reduction process appears, as far as can be seen, to be essentially identical with the Hall process, and has, indeed, been worked in conjunction with it at St. Michel in France. Perhaps the simplest way of understanding the actual differences between the Heroult, the Hall, and the Minet processes is to consider those differences to consist in legal rights as defined by dates of patents, etc.

There is one other process for the preparation of aluminium by electrolysis which may be said to be still only in an experimental stage. This was devised by Bucherer, and published in 1892 (*Zeitschrift für Ang. Chem.*, 1892, pp. 483, 484). The process is really identical with those of Heroult, Hall, and Minet, except that instead of the alumina employed in those processes aluminium sulphide is used. It is dissolved, like the alumina, in a bath of aluminium fluoride or cryolite, and then electrolysed in the same way. It is claimed that the voltage and, therefore, the horse-power required is less. The cost of aluminium sulphide is, however, at present said to be not less than £28 per ton, and as it contains only 36 parts per cent. by weight of aluminium, whereas pure aluminium is stated by Wallace to cost the British Aluminium Company only £12 per ton,[*] and contains over 52·9 parts per cent. by weight of aluminium, it is clear that unless some considerable reduction is made in the future in the cost of aluminium sulphide, this process is not likely to compete very seriously with those at present in commercial use. Apparently sulphur, in a more or less available form, is obtained as a by-product in this process, and possibly this fact, and the fact that the consumption of the carbon anodes should be much less than is the case in the alumina electrolysis processes (where it is stated by Wallace to be as high as 2 tons per ton of aluminium obtained) may put the sulphide process on a sound financial footing eventually.

The chemical reaction involved in the Heroult process is as follows :

$$2 \, Al_2 \, O_3 + 3 \, C = 2 \, Al_2 + 3CO_2$$

---

[*] It should be added that the cost of pure alumina in America is stated to be as high as £37 6s. 8d. per ton.

Whilst the reaction in Bucherer's process is given by the equation

$$Al_2 S_3 = Al_2 + S_3$$

if the temperature of reaction is low, or if it is higher, the sulphur combines with the carbon anodes, carbon bisulphide being formed, an action represented by the equation

$$2\ Al_2 S_3 + 3C = 2\ Al_2 + 3\ CS_2$$

The Cowles Process.—One of the earliest processes for the electrolytic production of alloys of aluminium was the Cowles' process, which process had a very considerable commercial success. A description of this process which was contained in the 1889 edition of this work is, therefore, retained in full here, and is as follows :—

An account of the methods adopted by the Cowles Electric Smelting and Aluminium Company, of Cleveland, Ohio, was given in the New York *Engineering and Mining Journal*,* and will be read with considerable interest. The Company referred to is, it appears, carrying on electric smelting commercially, but at present are chiefly devoting themselves to the production of aluminium and silicon bronze. The system, however, may hereafter be extended to other metals, and the operations be conducted upon a more extended scale. The Messrs. Cowles have succeeded in greatly reducing the market value of aluminium and its alloys, and thereby vastly extending its uses ; they are said to be the largest producers in the world of these important products. As described in their patents, the Cowles process consists essentially in the use, for metallurgical purposes, of a body of granular material of high resistance or low conductivity, interposed within the circuit in such a manner as to form a continuous and unbroken part of the same, which granular body, by reason of its resistance, is made incandescent, and generates all the heat required. The ore or light material to be reduced—as, for example, the hydrated oxide of aluminium, alum, chloride of sodium, oxide of calcium, or sulphate of strontium—is usually mixed with the body of granular resistance material, and is thus brought directly in contact with the heat at the points of generation. At the same time the heat is distributed through the mass of granular material, being generated by the resistance of all the granules, and is not localised at one point or along a single line. The material best adapted for this purpose is electric light carbon, as it possesses the necessary amount of electrical resistance, and is capable of enduring any known degree of heat, when protected from oxygen, without disintegrating or fusing ; but crystalline silicon or other equivalent of

---

* *Engineering and Mining Journal.* New York. August 8th, 1885.

carbon can be employed for the same purpose. This is pulverised or granulated, the degree of granulation depending upon the size of the furnace. Coarse granulated carbon works better than finely pulverised carbon, and gives more even results. The electrical energy is more evenly distributed, and the current cannot so readily form a path of highest temperature, and consequently of least resistance, through the mass along which the entire current or the bulk of the current can pass. The operation must necessarily be conducted within an air-tight chamber or in a non-oxidising atmosphere, or otherwise the carbon will be consumed and act as fuel. The carbon acts as a deoxidising agent for the ore or metalliferous material treated, and to this extent it is consumed, but otherwise than from this cause it remains unimpaired.

Fig. 150.

Fig. 150 of the accompanying drawings is a vertical longitudinal section through a retort designed for the reduction of zinc ore by this process, and Fig. 151 is a front elevation of the same. Another form of the electric furnace, which is illustrated by Fig. 152, is a perspective view of a furnace adapted for the reduction of ores and salts of non-volatile metals and similar chemical compounds. Figs. 153 and 154 are longitudinal and transverse sections respectively, through the same, illustrating the manner of packing and charging the furnace. The

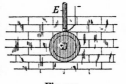

Fig. 151.

walls and floor, L L, of the furnace are made of fire-bricks, and do not necessarily have to be very thick or strong, the heat to which they are subjected not being excessive. The carbon plates are smaller than the cross-section of the box, as shown, and the spaces between them and the end walls are packed with fine charcoal. The furnace

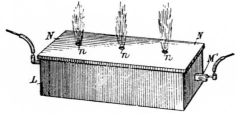

Fig. 152.

is covered with a removable slab of fire-clay, N, which is provided with one or more vents, n, for the escaping gases. The space between

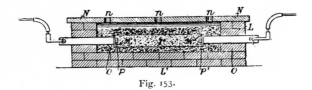

Fig. 153.

the carbon plates constitutes the working part of the furnace; this is lined on the bottom and sides with a packing of fine charcoal, o,

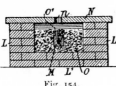

Fig. 154.

or such other material as is both a poor conductor of heat and electricity—as, for example, in some cases silica or pulverised corundum or well-burnt lime—and the charge, P, of ore and broken, granular, or pulverised carbon occupies the centre of the box, extending between the carbon plates. A layer of granular charcoal, o, also covers the charge on top.

The protection afforded by the charcoal jacket, as regards the heat, is so complete that, with the covering slab removed, the hand can be held within a few inches of the exposed charcoal jacket; but with the top covering of charcoal also removed, and the core exposed, the hand cannot be held within several feet. The charcoal packing behind the carbon plates is required to confine the heat, and to protect them from combustion. With this furnace aluminium can be reduced directly from its ores, and chemical compounds from corundum, cryolite, clay, etc., and silicon, boron, calcium, manganese, magnesium, and other metals are in like manner obtained from their ores and compounds.

In a paper read before the American Institute of Mining Engineers at Halifax,* Dr. T. Sterry Hunt, who had devoted two entire days at the experimental works at Cleveland, furnished some additional particulars concerning this interesting and valuable process, from which we make a few extracts, feeling confident that this important application of electricity will receive the fullest attention on this side of the Atlantic. Dr. Hunt says, "If alumina, in the form of granulated corundum,† is mingled with the carbons in the electric path, aluminium is rapidly liberated, being in part carried off with the escaping gas, and in part condensed in the upper layer of charcoal. In this way are obtained considerable masses of nearly pure fused aluminium and others of a crystalline compound of the metal with carbon. When, however, a portion of granulated copper is placed with the corundum, an alloy of the two metals is obtained, which is probably formed in the overlying stratum, but at the close of the operation is found in fused masses below. In this way there is got, after the current is passed for an hour and a half through the furnace, from 4 to 5 lbs. of an alloy containing from 15 to 20 per cent. of aluminium and free from iron. On substituting this alloy for copper in a second operation, a compound with over 30 per cent. is obtained. . . . . The reduction of silicon is even more easy than that of aluminium. When silicious sand, mixed with carbon, is placed in the path of the electric current, a part of it is fused into a clear glass, and a part reduced, with the production of considerable masses of crystallised silicon, a portion of this being volatilised and reconverted into silica. By the addition of granulated copper, there is readily formed a hard brittle alloy holding 6 or 7 per cent. of silicon, from which silicon bronzes can be cheaply made.

"The direct reduction of clay gives an alloy of silicon and aluminium, and with copper, a silico-aluminium bronze that appears to possess properties not less valuable than the compound already mentioned. Even boric oxide is rapidly reduced, with evolutions of copious brown fumes, and the formation, in presence of copper, of a boron bronze that promises to be of value, while, under certain

---

* *American Engineering and Mining Journal*, September 19th, 1885.

† *Corundum* is a very hard genus of aluminous minerals, to which the gems sapphire, ruby, salamstein, and adamantine spar belong. *Emery* is an impure, compact, amorphous, and opaque variety of corundum, and consists, according to Tennant, of alumina, 80; silica, 3; iron, 4 parts.

conditions, crystals of what appear to be the so-called *adamantoid boron* are formed. In some cases also crystalline graphite has been produced, apparently through the solvent action of aluminium upon carbon."

By another improvement in the Cowles furnace, the copper or other metal used for the alloy is in the form of rods running across the furnace, it having been found that where grains of copper were used, they sometimes fused together in such a manner as to short circuit the current. The new electric smelting furnace is shown in Fig. 155.

Since the publication of the earlier editions of this work, the Cowles electric smelting process has been (as we anticipated would be the case) introduced into this country, with the result that the Cowles Aluminium Syndicate have established extensive works at Stoke-upon-Trent, where operations for the manufacture of aluminium bronze were commenced, twelve or thirteen years ago, with a

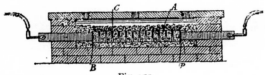

Fig. 155.

dynamo electric machine built by Messrs. R. E. Crompton & Co., of Chelmsford, of a capacity of 300,000 watts. With this powerful machine the Cowles Syndicate reduce from its oxides, in a run of about one hour, upwards of 20 pounds of metallic aluminium in its most valuable alloyed state—that of aluminium bronze—at a cost far below that at which it could be produced by any other known means. The raw material generally used by the Cowles Company, at their works in Lockport, U.S.A., is corundum, but for carrying out the manufacture in this country it is found that they can use not only corundum, but also bauxite, wochleinite, and other aluminous ores which are plentifully distributed in various parts of the world.

In Fig. 156. a view is shown of a series of electric furnaces erected at the new works at Stoke-on-Trent, the current being supplied by the large Crompton dynamo referred to.

The production per machine of the capacity above named at the Stoke works is stated to be considerably above one ton of ten per cent. aluminium alloy per day. With this machine, which is rather more than one-fourth larger than the Brush " Colossus " machine running at the Lockport works in America—the production is proportionately very much greater, and evidences also that the Cowles process is capable of operating dynamos of almost any magnitude. We understand that a second dynamo, having a guaranteed capacity of 402,000 watts, is in course of construction and will shortly be put down at the Stoke works. The production of aluminium alloys by this process has recently been much simplified, with a proportionate reduction in cost, and it is believed to be probable that these alloys will soon be marketable at prices competing with the common metals and alloys.

Fig. 156.

The Cowles Syndicate are also making a ferro-aluminium alloy, with a new continuous furnace, from which the product runs direct in its fluid state.

When we consider the enormous importance of aluminium alloys—aluminium bronze, perhaps, more especially—and the numerous purposes for which they are particularly applicable, a few details concerning the alloys manufactured by the Cowles furnace may not be uninteresting. One of the most important purposes to which aluminium bronze — if obtainable in large quantity and at a moderate cost—could be applied, would be in the manufacture of heavy ordnance. Indeed, Mr. J. W. Richards, Instructor in Metallurgy in the Lehigh University, Bethlehem, Pa., U.S.A., has stated that " the most recent action of the Italian Government in the artillery line is the replacement of 4,000 steel field-pieces by bronze or gun-metal pieces, the advantage of the latter being that while as safe from bursting, and as effective as steel guns of equal weight, they can be cast at much less expense and to greater perfection than steel guns. Such being the advantages in using ordinary gun-bronze, let me ask if it is not in the power of our Government to take a step in advance of the best and most recent artillery practice by undertaking to cast its heavy guns of aluminium bronze?" Following up Mr. Richards' sensible suggestion to the American Government, may we express a hope that our own Government will seriously consider the advisability of turning its attention to the same alloy for our own naval and military ordnance? Respecting the tensile strength in lbs. per square inch of Cowles' aluminium bronze, as compared with that of some other metals and alloys, the following data are given by the manufacturers :—Cast copper, 24,000 ; gun-bronze of copper and tin (cast), 39,000 ; cast gun-metal (U.S. Ordnance), 30,000 ; steel plates (rolled), 81,000 ; cast steel, average Bessemer ingots, 63,000 ; Cowles' aluminium bronze, *in castings*, 100,000 ; the "special" grade of the same has, *in castings*, a tensile strength up to 130,000. It should be mentioned that in the latter grades of this alloy a small percentage of silicon—about one-tenth the percentage of the aluminium—is combined with the alloy, that is to say about 11·3 per cent. of aluminium and silicon are added to pure Lake Superior copper to form this special alloy. Messrs. Cowles state that every ingot of alloys made at their works has a test bar made from it, which is tested in a machine built by Tinius Olsen & Co., of Philadelphia, a trustworthy machine, and one used in many of the colleges and ironworks in the States, and the record of all tests is carefully preserved. The grade, tensile strength, and elongation are stamped upon each ingot.

The beautiful golden-yellow colour of the ten per cent. aluminium bronze renders it highly suitable for the manufacture of art metal work, since it has not only a richer colour than brass, but it is also much less liable to corrosion. We have seen some very pleasing specimens in the form of salvers, candlesticks, sugar-bowls, salt-cellars, &c., manufactured from this bronze, which by contrast would really form very pretty table ornaments when mingled with articles of silver or electroplate.

The Electrolytic Refining of Nickel.—Frankly there is, at present, no known and commercially workable process for nickel refining by electrolysis, but it is stated that such a process is conducted secretly at the Balbach Smelting and Refining Company's Works at New Jersey, U.S.A., whilst the Orford Copper Company showed some thick plates of electro-deposited nickel at the 1901 Glasgow Exhibition ; and in England Messrs. Thomas Bolton and Sons., of Cheadle, are credited with being able to successfully refine nickel electrolytically ; but again the method employed is kept scrupulously secret. With regard to these secret processes it is interesting to note that Titus Ulke, in a paper published in the *Electrical Review*, of New York, early last year (vol. 38, p. 85 and pp. 101-103, 1901), states that the Balbach Works, which were estimated in 1900 to have a daily output of 6,000 lbs. of metallic nickel, *have probably ceased to operate electrically*. Dr. Mond states that the total world consumption of metallic nickel was about 9,000 tons per annum in 1900, but the world's production in 1896 was 4,603 metric tons (*Engineering and Mining Journal*, October 16, 1897, p. 452).

There can be no doubt whatever that a really successful method for electrolytically refining nickel on a commercial scale would yield a very large profit.

The chief difficulty in nickel refining is caused by the fact that dense and reguline deposits of the metal cannot be obtained at the cathode; after the deposit attains a certain and very moderate thickness it flakes or peels off. It would be fairly simple to obtain electrolytically deposited nickel in the form of scales and flakes, but this material would require remelting, and the high melting point of nickel, together with its oxidisability, would render such a remelting process very costly.

An interesting paper on nickel, by F. Foerster, appeared in the *Zeitschrift für Elektrochemie*, vol. 4, pp. 160-165, and was abstracted in the *Journal of the Society of Chemical Industry*, vol. 16, 1897, p. 808. The author found that with soluble nickel anodes, and an electrolyte consisting of an aqueous solution of commercial nickel sulphate (10 parts by weight of water, and 1·5 parts by weight of nickel sulphate), if the electro-deposition were carried out at the ordinary temperature thin flakes of non-adherent metal could only be obtained, but if the temperature of the bath was raised to from 50° to 90° C. tough cakes of metallic nickel could be obtained of any desired thickness from either the sulphate solution, or from the chloride slightly acidified with 0·25 per cent. hydrochloric acid. The best deposits were obtained with a current density of 200 to 250 ampères per square metre (= 18·5 to 23 ampères per sq. ft.), a silver white deposit being obtained. If the current density were reduced as low as 50 ampères per square metre (4·6 ampères per square foot), the deposited metal, although still compact, had a dull grey coloured surface. The higher the current density employed the smoother and brighter coloured was the deposit.

The analysis of the deposited nickel obtained by Foerster showed it to contain all the iron and cobalt which had been present in the anode, so that no very satisfactory refining action had occurred. It is, however, probable that if some organic acid were present the separation of the iron, at least, might be attained, but in this case the iron would collect in the electrolyte, from which, however, it might probably be separated from time to time as a basic organic salt by oxidising, neutralising, and then boiling the solution. Foerster found that the presence of organic matter from vegetable parchment or muslin bags surrounding the anodes which were acted upon by the solution had a very prejudicial action upon the metal deposited, rendering it brittle and non-adherent. This action of the solution and the current on the muslin and vegetable parchment was much more marked with the solution of nickel chloride than when the sulphate was used. The somewhat high current densities employed caused a good deal of evolution of hydrogen at the cathode, no doubt accompanied by the usual drawbacks of polarisation and reduced yield of metal. In order to avoid the roughening of the cathode by the bubbles of hydrogen evolved, Foerster had recourse to vigorous stirring. It appears to the present writer that if with the nickel sulphate solution free from nitrates an admixture of sodium chloride and boric acid, as recommended for nickel plating (see pages 303, 333 and 334), were used as electrolyte, and if acetic acid or sodium acetate, or perhaps a tartrate, were present in the solution, the iron would probably be retained in the solution, and could afterwards be removed by neutralising with ammonia, boiling and filtering, and more satisfactory results than those obtained by Foerster might be forthcoming. He also made experiments upon the electrolysis of solutions of nickel chloride containing about 100 grams of nickel per litre, with $\frac{1}{4}$ per cent. of free hydrochloric acid, at a temperature of about 80° C., employing plates of carbon for anodes. With a current density of 18·5 ampères per square foot, excellent deposits of nickel were obtained at first, but in time, due to the action of the hot solution and the current upon the carbon anodes, the electrolyte became contaminated with soluble organic materials, which caused the deposited nickel to deteriorate very much in quality, in fact to be useless for any commercial employment.

Titus Ulke states (*Engineering and Mining Journal*, 63 [5], pp. 113, 114) that the nickel anodes employed at Balbach in 1896, and which are purchased from the Orford Copper Company's Works, at Constables Hook, N.J., had the following composition:

Nickel	95 per cent
Copper	0·55 „
Iron	0·75 „
Silica	0·25 „
Carbon	0·45 „
Sulphur	3·00 „
	100·00

Five ounces of platinum per ton is also present.

The electrolytically refined cathode plates obtained measure about 20 inches by 30 inches, and are said to be 0·375 inch thick; they are very tough and elastic, and will not break or crack under the hammer. As stated above this firm exhibited some fine plates of electro-deposited nickel at the 1901 Glasgow Exhibition. Thicker cathode plates of nickel are difficult to obtain, owing to the scaling and flaking which occurs, and which is said to be due to surface oxidation. The chemical composition of the refined nickel obtained is as follows:

	I.	II.
Nickel.	99·5	99·7
Copper	0·1	0·2
Arsenic	0·03	0·03
Sulphur	0·02	0·02
Iron	0·1	0·1
Platinum	trace	trace
	99·75	100·05

About 1,000 lbs. of this pure nickel was produced per diem, in 1896, at the Balbach Works (*Electrician*, vol. 39, p. 337, 1897).

Titus Ulke states that the electrolytic refinement of the nickel at Balbach is effected in an alkaline cyanide solution, but as the method employed has been carefully kept secret, and as nickel is practically never deposited from cyanide solutions in electro-plating work,[*] there appears to be a good deal of doubt as to the accuracy of Ulke's guess. The Balbach Works are now said to have given up the electrolytic process, see p. 626.

The Canadian Copper Company, of Cleveland, Ohio, in 1897 are stated to have erected an experimental plant for operating a method of nickel refining devised by T. Ulke. This process is designed to treat 1,000 lbs. per diem of a bessemerised nickel matte containing on the average:—

Copper	43·0 parts per cent.
Nickel	40·0 ,, ,,
Iron	0·3 ,, ,,
Sulphur	13·7 ,, ,,

With about 7 ounces of silver and 0·1 ounce to 0·2 ounce of gold per ton. It is stated (*Zeits. für Elektrochemie*, June 5, 1897; and abstract *Electrician*, vol. 39, 1897, p. 337), that if a separation of the nickel and copper is desired this bessemerised matte cast in the form of slabs is employed as anode material in a bath of dilute sulphuric acid. The copper is first deposited in an acid solution, and then after making the bath neutral or slightly alkaline, the electrolysis is continued and a deposition of the nickel is obtained. The whole of this

---

* See p. 305.

statement is, however, very vague, and very wanting in detail. It is difficult to see how the method proposed could be run commercially.

It has been sometimes suggested that the unsatisfactory condition in which electrolytic nickel deposits are usually obtained is due to the partial oxidation of the surface of the deposit, and consequent want of adherence of the subsequently deposited metal. This idea is, however, supposed to be negatived by the experiment made by Bischoff and Thiemann in 1895. These investigators were endeavouring to prepare pure nickel by electrolysis from its pure salts in order to furnish Winkler with material on which to make a determination of the atomic weight of nickel; and they found that the deposited metal peeled off the cathode in flakes, and to test it as to whether or no it contained oxygen, a considerable weight of the deposit was heated to redness in a current of purified hydrogen, but no alteration of weight took place, and hence the freedom from all oxygen was assumed. For my part, however, I do not think that a gravimetric experiment is really capable of deciding whether or no the mechanical faults of an electrolytic deposit are due to the presence or absence of a material such as oxygen. The question is, I think, still an open one. It is at least as likely that the trouble is due to hydrogen, for, as has already been noticed, if a considerable quantity of sodium chloride is present in the solution (which would tend to prevent the direct deposition of hydrogen at the cathode), a more reguline and adhesive deposit is obtained (p. 333) than when this addition is neglected.

B. Neumann (*Zeitschrift für Elektrochemie*, vol. 4, pp. 316-322, and 333-338, 1898) concludes that it is not possible to separate nickel and copper on the commercial scale because, when the solution becomes weak in copper, a very unsatisfactory type of copper deposit is found to be deposited, and he suggests that the best method is to partially remove the copper from solution by means of the electric current, and then when the solution gets so weak in copper that there is only about 0·01 grams of copper per cubic centimetre of solution the electrolysis should be stopped, and the copper precipitated by sulphuretted hydrogen, and the nickel then electrolysed out of the filtered solution. Neumann also states that it is useless to attempt to remove iron from nickel solutions by blowing air through them whilst they are kept heated, for even if the treatment is continued for nineteen hours 25 per cent. of the iron originally present still remains in the solution. It must be confessed that Neumann's suggested method for separating nickel and copper on the commercial scale by partial electrolytic and partial chemical treatment smacks somewhat of the laboratory, with beakers and stirring rods for manufacturing plant. An abstract of this paper appears in *Science Abstracts*, vol. 2, p. 217.

Mond's Nickel Refining Process.—Dr. Ludwig Mond's very extraordinary discovery in 1890 of a volatile compound of metallic nickel and carbonic oxide, from which compound the pure nickel can again be removed by heat, and the carbonic oxide gas employed once more for removing fresh amounts of nickel from the ore is, of course, in no way an electrolytic process, but it offers such an extremely simple method of obtaining pure metallic nickel (if the costs of working on the commercial scale are not prohibitive) that it must always be considered when coming to any conclusion as to the value of any electrolytic process for nickel refining. The experimental Mond works at Smethwick have manufactured over 50 tons of nickel by this method, and very satisfactory reports as to the quality of the metal were made by the consumers to whom it was sold. The process has now been taken up by a company asking for a capital of £600,000 (May 14, 1901), but whether it will prove as successful as has been hoped must be settled by the future. No less than 70 patents have been taken out, of which six are for Great Britain, and the dates and numbers of these English patents are as follows :—

12,626, 12th August, 1890; 21,025, 24th December, 1890; 8,083, 11th May, 1891; 23,665, 10th December, 1895; 23,665A, 10th December, 1895; 1,106, 14th January, 1898.

The feature which, on the surface, appears most unfavourable to the Company is the fact that the earliest patent is already eleven years old. The Company has, however, acquired the mining rights over no less than 4,913 acres of nickel bearing, and reputedly nickel bearing, land at Denison, Garson, and Blezard, in the Sudbury district, Ontario, Canada. The nickel ore is to be smelted for a bessemerised matte at Sudbury. The matte will contain about 40 per cent. of nickel and 40 per cent. of copper, and will be transported to Clydach, near Swansea, in Wales, where it will be treated by the Mond process. At the works erected there an annual output of 1,000 to 1,500 tons of nickel can be obtained. The Company state that the present (1901) price of refined nickel per ton is £165, whilst its average price for the past five years has been about £125.

(In concluding this section on the commercial electrolytic methods of obtaining nickel, or rather their absence, I would counsel any one who contemplates experimenting in order to place this branch of electrolytic refining upon a more satisfactory basis to read carefully the sections of this volume written by the late Mr. Alexander Watt on the subject of the electro-plating of nickel (pp. 288–338 and 460–464), which contain a large amount of useful information on this matter not usually treated in such full detail, and owing not a little to that writer's long personal experience of the practical details of these operations.—A. P.)

CHAPTER VIII.

ELECTRO-GALVANISING.

Galvanised Iron.—The galvanising, or plating of iron with zinc, by electro-deposition has already been described in the 1889 edition of this work, pp. 345-347, and this description of galvanising from Watt's Alkaline Solutions and from Hermann's Sulphate Solutions is retained in the present edition, pp. 353-355. From May, 1891, the present writer undertook a prolonged investigation of the best means of depositing a sound and adhesive coating of zinc upon iron by means of electrolysis, and for this purpose the behaviour of a very large number of solutions of zinc salts was examined at different temperatures, and employing different current densities. The general result of these investigations was as follows :—

1st. *Nature of Solutions.*—Aqueous solutions of zinc sulphate, and of this salt mixed with about molecular proportions of sodium sulphate, potassium sulphate, ammonium sulphate, aluminium sulphate, and magnesium sulphate, all gave electrolytes from which good and adherent deposits of metallic zinc could be obtained by electrolysis, but on the whole a solution of zinc sulphate and magnesium sulphate in molecular proportions, and containing about 30 ounces avoirdupois of zinc sulphate per gallon, was the solution which yielded the most satisfactory results. Zinc deposited from this solution did not contain more than a very small trace of magnesium, and it is quite possible that the amount detected (0·028 parts per cent.) may have been due to the small traces of the magnesium salt dissolved in the electrolyte adhering to the deposited metal.

2nd. *Temperature of Solutions.*—In almost all the solutions mentioned above the high temperature of the solution was an advantage for two reasons: Firstly, the resistance of the baths was decreased and therefore the power required was diminished, and, secondly, and this is of far greater importance, the nature of the deposit obtained was more regular and coherent. Any temperature up to 95° C. may be used with advantage, but probably a temperature of about 50° C. will be most useful, as the higher temperature causes greater inconvenience due to loss of electrolyte from evaporation and cost of fuel for heating.

3rd. *Current Density.*—It was found with these solutions that in confirmation of Kiliani's results (p. 355 of this book) very low current densities, and also small strength of zinc salt, gave bad deposits, but that as the current density increased better results were obtained. With a current density of 30 ampères per square foot of the surface being plated with zinc very good deposits were obtained from hot solutions of the magnesium and zinc sulphate solution described above, and the current could even be increased to as high as 40 ampères per square foot of surface plated, and satisfactory coatings be deposited, but the best results were obtained with a current of from 20 to 30 ampères per square foot. Good deposits are obtained at any lower current density down to 6 ampères per square foot. Above 45 ampères per square foot the coatings were bad.

4th. *Voltage required.*—The volts required to run a zinc depositing vat in which the electrodes were two equal-sized rectangular plates, at a distance of six inches apart, and with a current density of 30 ampères per square foot of surface, at a temperature of about 50° C., was found to be 4·5 volts; with a larger distance between the electrodes the required e. m. f. under the same conditions would of course be correspondingly increased, and if, instead of a zinc plate anode, a lead or carbon anode is used, a further increase of voltage is required to overcome the back e. m. f., due to the employment of insoluble anodes.

5th. *Weight of Zinc actually deposited per hour.*—If a current density of 30 ampères per square foot of surface plated is employed, the weight of zinc deposited per hour should theoretically (if nothing but zinc is liberated at the cathode and if there is no leakage) weigh 1·282 ounces per square foot of surface. As a matter of fact, however, it was found that, due to secondary reactions, leakage, etc., the actual weight of zinc deposited was not much greater than about 75 per cent. of the theoretical amount which should have been obtained, that is, a deposit made at a current density of 30 ampères per square foot of surface plated would in an hour weigh about 0·9615 ounce avoirdupois, instead of 1·282 ounces. If a current density of 20 ampères per square foot of surface is employed, a coating of zinc weighing $1\frac{1}{4}$ ounces of zinc per square foot of surface of the iron would be deposited in about two hours.

6th. *Protective Effect of Electro-deposited Zinc.*—Sir W. H. Preece has prescribed a test for examining the character and thickness of metallic zinc deposits, which is known as the Preece test. This test, as modified by the present writer, consists in immersing the zinc-coated iron in a saturated solution of copper sulphate at a temperature of 15° C. for one minute, and then immediately removing the object from the copper solution, it is placed under a rapidly running stream of water from a tap, in which it is well shaken ; this treatment will remove any of the loose flocculent deposit of copper which has been formed on the surface of the zinc by the zinc displacing the copper in the copper sulphate solution, but if the zinc has been so far removed as to expose the surface of the underlying iron to the action of the copper solution, a much more coherent deposit of bright looking copper is deposited on the iron, *which is not removed by the shaking in the stream of rapidly flowing water from the tap.* The number of successive times, therefore, that a zinc-coated piece of iron will withstand this treatment by Preece's test is a measure of the thickness and regularity of the protective zinc coating. It must be noted here that although copper deposited, as above described, upon a wrought-iron or mild steel surface will adhere so firmly that it cannot be detached by briskly rubbing with the surface of the finger under water, yet if the iron is very steely in character, *i.e.*, contains a large amount of carbon, as for instance is the case in what is known as plough steel wire, the copper deposited on the steel surface, although bright, is readily removed by rubbing with the surface of the fingers, but it is not removed by shaking the article under a rapid stream of water from an ordinary water service tap ; whilst, finally, copper deposited upon a *zinc* surface is quickly washed off by this treatment. It is, therefore, necessary to carry out tests by this means with caution, or misleading results will be obtained. The saturated copper solution should be kept stored in a large bottle, and only small portions should be taken out for each dip in a small beaker, and should not be again used or returned to the bottle, but should be thrown away. The number of times a zinc-coated piece of iron, which has been coated by zinc by some given method, will withstand Preece's test, is proportional to the amount of the zinc per unit of surface of the iron covered by it ; but the protective effect also depends upon how the zinc coating has been applied, and I have found that undoubtedly the same weight of zinc per unit of surface of iron has a greater protective action against the

Preece test when it has been deposited electrolitically than is the case when it has been deposited by the ordinary hot galvanising process of dipping the iron surface into the melted zinc. The following are some results illustrative of this fact :—

Zinc coated wire.	S. W. G.	Weight in ounces per square foot of surface.	Result of Preece Test.	Ounces of zinc per square foot of surface per dip.
Electro-plated, sample A . No.	11	0·997	failed at 6 dips	0·166
,, ,, B . ,,	19	0·997	,, 6 ,,	0·166
Bullivant's hot galvanised . ,,	11	1·200	,, 3 ,,	0·400
,, ,, ,, . ,,	19	0·7442	,, 3 ,,	0·248

Apparently with zinc coatings obtained by the old-fashioned hot galvanising method the amount of zinc required to protect an iron surface so that it will withstand one one-minute immersion in the saturated copper sulphate solution at $15°$ C. is about 0·248 ounce avoirdupois per square foot, but owing (especially in the case of wires) to the irregularity of the thickness of the zinc coating, the amount may become as great as 0·4 ounce per square foot, whilst in the case of electro-deposited zinc as little as 0·166 ounce per square foot of surface will afford the same protective effect. The reason of this difference is possibly due to the fact that the greater purity of the coating zinc, when electro-deposited, renders local action, and hence corrosion, smaller than is the case with the less pure zinc employed in the hot galvanising methods.

Since 1891, the date at which the results described above were obtained by the present writer, and also to a smaller extent before that date, a large amount of ironwork covered with electro-deposited zinc has been turned out on a commercial scale, but precise details of the works where this was done were not published, in so far as I am aware, until comparatively recently. I have, however, examined several samples of iron and steel wire and other articles zinc-coated electrically on a commercial scale before 1891. I believe that the Warrington Wire Rope Company galvanised wire by this electro-plating process, as also Messrs. Ramsden, Camm and Company, and Messrs. Siemens Brothers. Elmore had brought out a patent for electro-zincing iron wires, and at the same time burnishing their surface, Eng. patent 9,214, 1886. In all these cases samples of wire supplied by these firms had a thin coating of zinc which would not in any case stand as many as two immersions by Preece's test, although they would all stand one immersion. The adhesion of the zinc to the underlying iron was excellent. They also all showed the peculiarity of having the deposit of zinc uniformly thicker along one side of the wire, as though indeed a straight line had been ruled along the side of the wire. This appearance cannot be observed in wire galvanised by the old-fashioned hot galvanising method, and is no doubt due to the fact that in these cases, as the wire runs through the electrolytic galvanising troughs, the anodes are arranged beneath it and at its side, but not above it, as should also have been done.

In 1893 Richter electro-deposited zinc upon iron tubes and still-worms by the following method, described in 1895 (*Zeitschrift für Elektrotechnik und Elektrochemie*, 1895, pp. 79-82, and pp. 98-103, and *Journ. Soc. Chem. Ind.*, vol. 14, 1895, p. 874) :—Wrought iron tubes 6 metres long were coated both internally and externally with zinc from a solution of zinc sulphate in water having a specific gravity of 1·2 (= 50 ounces of zinc sulphate per gallon). Spirals or stillworms of tube as much as 2 metres in diameter, and containing 300 metres of tubing, were also coated. The deposit of zinc was 0·05 millimetres thick, and was deposited in ten minutes. The current density was from 20 to 30 ampères per square foot. At this current density, and up to as high a value as nearly 70 ampères per square foot, the deposit obtained was found to be adherent, ductile and silvery white in colour. Powdery and loose deposits were only noticed when the current density was permitted to become too small. The author states that in order to obtain successful deposits of zinc two things are necessary ; (1) a very careful cleansing of the surface upon which the zinc is to be deposited ; and (2) a uniform current density. Richter employs the following routine for cleansing the ironwork. First, grease is removed by dipping the article in a 10 per cent. solution of caustic soda heated nearly to its boiling point. After draining, the articles are next placed in a pickling bath containing dilute sulphuric acid (7·5° Baumé). Great care must be taken to avoid getting grease upon the surface of the ironwork by handling it after it has been treated with the caustic alkali solution, and it should be moved by means of pincers and crane only. After pickling, the surface is scoured by means of sand and chopped straw, and any cavities or depressions not readily get-at-able are treated by means of the sand blast. After pickling, the acid must be removed as quickly and thoroughly as possible by first washing it with a powerful water spray, and then soaking the iron in clean water. Oxidation is further prevented by adding 0·2 per cent. of ammonia to the soaking water. The articles are next placed in the plating vat, and after plating they are washed in warm water, and then dried in warm sawdust. Richter states that baths for coating 7-inch tubes will take a current of about 2,000 ampères, and that at this current density a pressure of 5 volts will suffice for two baths in series.

A very similar process for electro-deposition of zinc upon iron has been used by S. O. Cowper Coles, and was described in the *Journal of the Society of Chemical Industry* in 1896 (vol. 15, pp. 414-417). The solution employed was a neutral solution of zinc sulphate containing 40 ounces avoirdupois per gallon of solution, specific gravity = 1·156, The anodes employed were of lead, and the solution was kept neutral, or nearly so, by constantly pumping it off from the electrolytic vat, where the zinc was removed from it by electrolysis, and forcing it to run through scrubbers of coke containing zinc dust or zinc oxide, but zinc dust was preferred to zinc oxide on account of the comparatively high price of zinc oxide if pure. The scrubbers are in duplicate and are used alternately, fresh zinc being added to the one which is for the moment unemployed. The scrubbers are termed regenerators by Mr. Cowper Coles, and with regard to their efficiency, he states that a zinc sulphate solution containing 12·59 per cent. of free acid after passing through a filter-bed containing 10 per cent. of zinc dust (and the rest presumably coke) contained only 0·68 per cent. of free acid. In a subsequent paper, read before the Society of Engineers, October 3rd, 1898, this experimenter states that he employs a solution of 40 ounces of zinc sulphate per gallon, together with 5 ounces of ferrous sulphate per gallon, and he adds that the ferrous sulphate gradually becoming oxidised to ferric sulphate by the air takes up acid from the bath and tends to keep it neutral. The ferric sulphate is of course once more reduced to ferrous sulphate by the zinc dust as the solution is pumped through the scrubbers. Mr. Coles considers that the presence of the ferrous sulphate tends to prevent the formation of the powdery deposits which are ascribed by different writers to the presence of a hydride of zinc, or, as I think, with far greater probability, to the presence of an oxide or hydrate. The great advantage which Mr. Coles claims for this process is the employment of zinc dust in coke or sand scrubbers for revivifying the solution, but I must remark that this method was proposed by me to Mr. Cowper Coles in March of 1893, the description of the method sent to Mr. Coles being as follows :—

Draft of Specification for a Provisional Application for a Patent for Improvements in the Electro-metallurgy of Zinc.

The London Metallurgical Company, Limited, and Sherard Osborne Cowper Coles, Engineer, of 80, Turnmill Street, London, E.C., and also Arnold Philip, Metallurgist, of 43, Onslow Road, Richmond, Surrey, do hereby declare the nature of this invention to be as follows:—

In order to prepare metallic zinc by means of electricity, or to electroplate articles with metallic zinc, we take excess of artificially prepared zinc oxide, zinc hydride, or zinc oxide carbonate, or any mixture of all or any of these compounds of zinc, with or without the addition of zinc dust, and treat them with an aqueous solution of any acid which will form a zinc salt soluble in water. (We employ by preference an aqueous solution of sulphuric acid.) When all the free acid has thus become saturated, we remove the solution from the excess of the remaining undissolved zinc compound by means of filtration, settling or other well-known method. The aqueous solution of zinc thus obtained is next submitted to electrolysis ; as the zinc is thus electrically deposited the solution becomes acid and is pumped away over a further quantity of the zinc compound, and thus resaturated. with zinc. The method is thus made continuous, for a continuous stream of an aqueous solution of a zinc salt containing a small quantity of free acid is withdrawn from one end of the electrolytic bath and resaturated with zinc by mixing with a further quantity of the zinc compounds, whilst a constant stream of fresh electrolyte, which contains practically no free acid, enters at the other end of the bath, thus keeping it level. To facilitate the solution of these compounds we may mix them with any substances which are practically insoluble in the dilute acids used, as for instance sand, burnt clay, pebbles or coke.

It is therefore clear that the only portion of Mr. Coles' patent, 2,999, 11th Feb., 1895, which is novel is the admixture of the ferrous sulphate with the zinc sulphate solution. In connection with the above statement it is interesting to note that the Cowper-Coles Galvanising Syndicate warn galvanisers and others against the employment of zinc dust, either in their zincing baths or in regenerators, except under licence from the Syndicate. The great advantages that are claimed for the employment of zinc dust in the way described above is that no scale or sludge is formed in the electrolytic bath, the formation of which is very difficult, in fact one may say impossible, to avoid when using metallic zinc anodes or even granulated zinc scrubbers, and moreover the cost of zinc dust per ton is considerably lower than the cost of an equal weight of zinc in the form of any other pure material such as oxide, hydrate, carbonate, etc. A second advantage of this method is that by its means the electrolyte can be readily kept neutral or nearly so, which cannot be done if solid zinc anodes alone are employed.

Another method of keeping the zinc solution saturated, however, is, instead of running the solution over zinc dust or metallic zinc, to pump it hot through scrubbers containing zinc and copper or zinc and carbon in intimate contact, the electric couple thus formed setting up local action and thus assisting in the neutralisation of the acid present, or the scrubbers may contain zinc electrodes, with an electric current flowing from them through the acid electrolyte. These methods have not been patented and may therefore be freely employed by any experimenter.

Zinc dust is a very fine powder containing from 75 to 90 per cent. of metallic zinc. It cannot be melted down owing to the finely divided zinc oxide mixed with it. It usually contains some cadmium, and is obtained as a by-product in the manufacture of zinc by the Belgian process.

In the papers quoted above the following estimates for the cost of an electrolytic galvanising plant are given :—

ESTIMATE FOR ZINCING 20 TONS OF PLATE IN A WORKING WEEK OF 52 HOURS, THE AVERAGE THICKNESS OF THE IRON BEING $\frac{4}{10}$ OF AN INCH, AND THE ZINC COATING BEING 1 OUNCE PER SQUARE FOOT OF SURFACE.

	£	s.	d.
Labour, piecework, at 1s. per cwt.	20	0	0
Incidental expenses, at 15 per cent. on labour	3	0	0
Yard labour, 10 per cent. on labour	2	0	0
Zinc, 448 lb. at £17 per ton (1898)	3	8	0
Royalty, at 2s. 6d. per ton	2	10	0
Electrical energy, at 1d. per E.H.P.	2	18	0
Pickling, at 5s. per ton	5	0	0
Rent of building, at £2 per week	2	0	0
Interest on capital	2	17	8
Depreciation of plant	3	17	0
	£47	10	8

Cost per ton = £2 7s. 6½d.

It may be remarked concerning this estimate that only 3·4 shillings per ton of iron treated represents the total cost of the zinc, and therefore if granulated or rolled zinc plates were employed in contact with copper or lead, as suggested above, instead of zinc dust, the cost would only be increased by the extra cost at most, as claimed by Mr. Cowper Coles, of £7 per ton on the cost of the metallic zinc, or, that is, at the ratio of about 17 to 24 ; that is, the cost of galvanising per ton of iron plate, for the zinc alone, would be increased from 3·4 shillings, as above, to 4·8 shillings per ton; that is, the total cost would be increased by 1·4 shillings per ton, for the cost of electro-zincing per ton using zinc dust is £2 7s. 6½d., but using cast zinc it costs less than £2 8s. 11½d. per ton ; but allowing the cost of royalty at 2s. 6d. per ton would bring down the cost of electro-zincing this class of iron from £2 7s. 6d., by zinc dust process, per ton, to £2 6s. 5d. per ton when not using zinc dust.

ESTIMATE FOR A GALVANISING PLANT CAPABLE OF ZINCING 7,200 SQUARE
FEET OF SURFACE PER WEEK OF 54 WORKING HOURS WITH A THICK-
NESS OF ZINC EQUAL TO 1 OUNCE PER SQUARE FOOT.

	Price.			Approx. weight.	
	£	s.	d.	tons	cwt.
1 dynamo to give 3,000 ampères at 6 volts, speed 800 revs. per minute	200	0	0	2	0
1 switchboard with measuring and regulating instruments	59	0	0	0	½
1 galvanising tank, 12 ft. × 5 ft. × 3 ft. . .	40	0	0	0	19½
1 pickling tank ,, ,, ,, . .	74	0	0	1	10
1 washing tank ,, ,, ,, . .	40	0	0	0	19½
2 circular regenerating tanks, with fittings .	62	10	0	1	0
1 air compressor for circulating electrolyte .	21	0	0	0	2
1 complete set of anode and cathode suspension bars for zincing tank	130	0	C	0	17
1 special arrangement for zincing tubes inside and outside up to 6 feet in length . .	150	0	0	0	8
	£776	10	0		

An engine to give about 30 *Indicated* H. P. is stated to be required to
run the above plant, but to keep the dynamo running at full load, if it
has a 95 per cent. efficiency and is belt driven with a 5 per cent. transmission, an engine of 26·74, say 27 *Brake* H. P., would be required,
and allowing 3 B. H. P. for the air compressor, etc., an engine of
30 *Brake* H. P. would appear to be necessary. The solution is stated to
cost 1¾d. per gallon, containing 35 ounces of crystallised zinc sulphate
per gallon. That is, a cost of 10½d. per cubic foot containing 6½
gallons.

ESTIMATE FOR A GALVANISING PLANT CAPABLE OF ZINCING 4,800 SQUARE
FEET OF SURFACE PER WEEK OF 54 WORKING HOURS WITH A THICK-
NESS OF ZINC EQUAL TO 1 OUNCE PER SQUARE FOOT OF SURFACE.

	£	s.	d.
1 dynamo to give 2,000 ampères at 6 volts . .	150	0	0
1 switchboard, with measuring and regulating instruments	50	0	0
1 galvanising tank, 9 ft. × 5 ft. × 3 ft. . .	30	0	0
1 pickling tank ,, ,, ,, . .	60	0	0
1 washing tank, 9 ft. × 5 ft. × 4 ft. . .	30	0	0
2 circular regenerating tanks, with fittings . .	50	0	0
1 air compressor for circulating electrolyte . .	21	0	0
1 complete set of anode and cathode bars for zincing tank	130	0	0
	£521	0	0

ESTIMATE FOR A GALVANISING PLANT CAPABLE OF ZINCING 2,400 SQUARE FEET OF SURFACE PER WEEK OF 54 WORKING HOURS WITH A THICKNESS OF ZINC EQUAL TO 1 OUNCE PER SQUARE FOOT OF SURFACE.

	£	s.	d.
1 dynamo to give 1,000 ampères at 6 volts . .	110	0	0
1 switchboard, with measuring and regulating instruments	45	0	0
1 galvanising tank, 6 ft. × 5 ft. × 3 ft. . . .	25	0	0
1 pickling tank, 6 ft. × 5 ft. × 3 ft.	45	0	0
1 washing tank, 6 ft. × 5 ft. × 4 ft.	25	0	0
2 circular regenerating tanks, with fittings . .	40	0	0
1 air compressor for circulating electrolyte . .	21	0	0
1 complete set of anode and cathode bars for zincing tank	110	0	0
	£421	0	0

Figs. 157 and 158 show the general arrangement of an electrolytic zincing plant as given by Mr. Cowper Coles.

In another paper Mr. Coles states that the cost of a galvanising plant having a capacity of 6,700 gallons (30 ft. × 6 ft. × 7 ft.) is about £600, which he says is but little more than that of an old-fashioned molten zinc galvanising plant having a bath capacity of only 10 ft. × 4 ft. × 4 ft. 6 in. outside dimensions. Such a hot galvanising tank would hold 28 tons of zinc, which at £15 per ton amounts to £420. To keep this large quantity of zinc melted entails a heavy expenditure in fuel, the thickness of the iron of the bath averaging more than one inch, whilst the iron baths are a constant source of annoyance and expense, for the iron is dissolved by the zinc, ultimately destroying the tank, and besides this the zinc is gradually rendered useless by absorbing iron, not only from the tank itself, but also from the iron plates immersed in it.

The great advantages of using the electric process are that a better adhering coating is obtained, in which a given weight of zinc has a greater protective action. (See p.635.)

The thickness of the zinc coating can be made of any desired value.

The iron is not weakened if thin plates are zinced, either by the zinc eating into the iron, or by the high temperature drawing the temper of steel, both of which troubles occur in the old process.

The electrolytic plant permits of plates of very large dimensions being very cheaply treated, and also permits small articles, or articles of intricate form, being rapidly and cheaply zinced.

There is no waste of materials or expense incurred in renewal of the bath. All the zinc is consumed, no dross being formed.

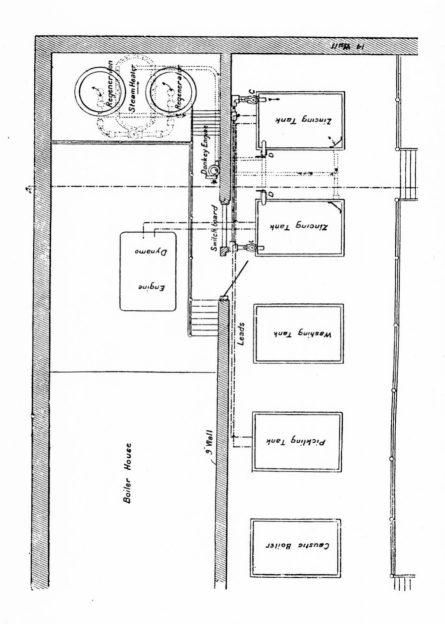

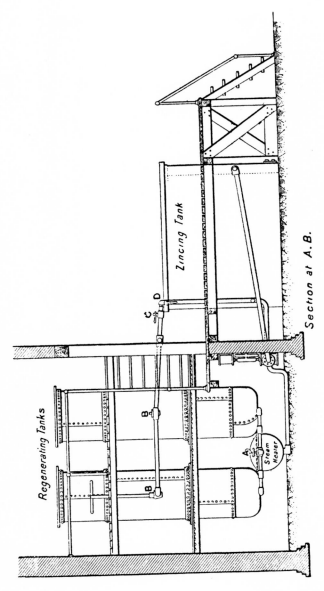

Fig. 157.— Ground Plan of Electro-Zincing Plant.

Regenerating Tanks

Zincing Tank

Steam Heater

Section at A.B.

Fig. 158.—Sectional Elevation of Electro-Zincing Plant.

[*To face p.* 640.

The zinc coating obtained by the electrolytic process takes paint better than ordinary galvanised iron.

The surface of the electrolytically zinced iron can be cut and worked far more readily than ordinary galvanised iron, which is very hard and brittle.

Large structures, such as girders, propeller shafts, pillars, etc., can be easily zinced, either completely or partially.

The cost of the electrolytic is less than that of the old hot galvanising process.

The work does not require to be dried after cleaning and pickling, but can be passed at once into the electrolytic vat.

The electro-galvanising of iron shows up very clearly any faults or defects in the iron, and I have been informed by Mr. W. Weston, Chief Chemist to the Admiralty, that the zincing of iron work in boiler tubes is carried out by this method with a very thin coating, not with the idea of protecting the iron from corrosion, for which such a coating in these situations has little or no value, but rather that any latent defects in the tubes may be brought out and rendered readily apparent.

Electro-galvanising has been largely employed in this country and abroad during the past ten or fifteen years for coating boiler tubes, sugar machinery, wire, stillworms, the frames and plates of torpedo boats, wire gauze, and other articles of iron too numerous to mention.

Pickling and Cleansing Iron.—In order that a satisfactory and adhesive coating of zinc should be obtained by electro-deposition, it is of the highest importance that the iron shall be most carefully cleaned, not only from grease but also from all scale and oxide of iron. A general description of the process adopted by Watt is given on p. 354 of this work, and the method used by Richter on p. 635. Cowper Coles employs similar methods. The Admiralty specify that all steel steam pipes, boiler and collector tubes, and all plates for boilers are to be pickled in a liquid consisting of 19 parts of water and 1 of hydrochloric acid until the black oxide or scale formed during the manufacture is completely removed. Plates have to be placed in the pickling vat on edge, and not laid flat. When taken out of this dilute acid all the surfaces are to be well brushed and washed to clean off the loose scale. They are then to be placed in a bath, filled and kept well supplied with fresh water, or must be thoroughly washed with a hose as may be found necessary, and then placed on end to dry ; but for obtaining a good adhesion of electrolytically deposited zinc, sulphuric acid gives better results than hydrochloric acid. The objection to using only sulphuric acid, however, is that the iron does not stand so good a mechanical test after pickling in this acid as is the case if hydrochloric acid is employed. On account, therefore, of these two conflicting requirements Mr. Weston has advised that after pickling and cleaning by the Admiralty hydrochloric acid method, the plates should then be dipped in weak sulphuric acid for about a quarter of an hour, or for a few minutes in a stronger acid immediately before they are transferred to the electro-zincing vats. Probably the electrolytic cleaning in baths of neutral salts, described on p. 646, may overcome both the drawbacks found with hydrochloric acid on the one and sulphuric acid on the other hand. In all probability the unsatisfactory mechanical condition of the iron after pickling in sulphuric acid is due to occluded hydrogen, which is well known to make iron very brittle, but precisely why it is absent when hydrochloric acid is used it would be difficult at present to say. Cowper Coles states, in one of the papers already quoted, that sand blasting is often used instead of pickling, and for many purposes it is found to give better results, especially for cast iron work, from which it is very difficult to remove the last traces of acid. The cost of sand blasting per square foot, under the most favourable conditions, including the cost of labour, sand, and power, is one-tenth of a penny, but the actual cost varies considerably with the nature of the work. When using quartz sand of the best quality, the loss is about 10 per cent. (presumably of the sand employed) each time it is passed through the machine ; the loss when using chilled iron sand is very small, the waste of the material replacing more than nine-tenths of the apparent loss.

The process which Mr. Cowper Coles employs in pickling iron is described in *Engineering*, December 30, 1898, as follows :—The usual practice is to place the iron in a solution containing one part of hydrochloric or sulphuric acid to ten parts of water for a period varying from half an hour to 24 hours. The amount of acid consumed in the pickling of close annealed sheets, such as are used for roofing purposes, varies from 3 cwt. to 7 cwt. of muriatic acid per ton of sheets, but if sulphuric acid is employed a considerably smaller weight of acid is required. Muriatic acid is preferred by many manufacturers because it is cheaper bulk for bulk, and also because sulphuric acid is said to act too quickly on the "skin" of the metal and so tends to eat it away. In Worcestershire alone there are about 1,053 carboys (hundred-weights) of acid, corresponding to some 3,000 cwt. or 27,000 gallons of waste pickle, on an average every week, which have to be got rid of. One of the first experiments made by Mr. Cowper Coles was to make the iron to be pickled the anode in an acid bath, but this was found to pit the plates, for that portion of the iron which was not protected by the mill scale was more readily dissolved than that which was so covered. To overcome this difficulty the current was periodically reversed, the iron being alternately made anode and cathode. To quicken the process and reduce the electrical resistance of the solution it is advantageous to heat the pickling solution or electrolyte, which is done by passing exhaust steam through lead pipes in the pickling vat. A tray or false bottom is placed in the vat, which is capable of being raised out of the vat bodily by flotation; it is normally anchored at the bottom, and it is allowed to float to the top at the end of each day and the heavy mill scale which has become detached from the plates and fallen to the bottom is thus removed from the acid and a corresponding economy in acid is obtained. Another device employed to remove the mill scale deposits from the vat is to circulate the solution, by means of a small pump, through a lead-lined box or chamber, behind which are placed electro magnets, the result being that as the solution flows past the magnet poles the iron scale in suspension is attracted and retained and can be moved from time to time. The author further adds that in America an electrical pickling process has been tried, sulphate of soda (Bisulphate waste from nitric acid retorts?—A. P.) or sulphuric acid solutions being employed, and the iron plate is made the cathode or negative electrode. The hydrogen set free is said to reduce the oxide on the surface to metallic iron and decompose any grease present. Mr. Cowper Coles claims that the method which he employs to remove the loose scale from the pickling tanks effects a very considerable saving of acid, as the scale after removal from the work under treatment is not allowed to remain in the bath to be further acted upon by the pickle. He states that it has been found by experiment that as much as 30 per cent. of the scale usually allowed to fall to the bottom of the pickling tank is dissolved in a week of 168 hours in a 1 per cent. solution of sulphuric acid. It is also claimed that less time is required to pickle the work, as the solution is kept undiluted (presumably this means unneutralised), and does not deteriorate so rapidly. The apparatus can be attached to any ordinary pickling vat, and does not require skilled labour.

In a paper read by Mr. Cowper Coles before the British Association (Section G, 1899), it is stated that the scale is in many cases $\frac{1}{64}$ inch in thickness and very adherent, and that although by the ordinary pickling process only 30 minutes is required for close annealed sheets, yet as long as 24 hours is required for plates and forgings such as are employed for ship and bridge building. Another form of electro-magnetic scale collector is described in this paper, which consists of an electro-magnet encased in copper, Fig. 159. This is placed in any convenient position in the tank, and the current required to work it is about 10 ampères with an e. m. f. of 6 volts. In the case of large tanks two or more scale collectors can be employed or only one may be used, this being moved about to different parts of the bath at suitable intervals. From time to time it is taken out and the adhering scale removed. Fig. 160 shows a scale collector in position in a lead-lined wooden pickling vat.

In 1892, finding some difficulty in obtaining a good adherence of fairly thick coatings of zinc upon high carbon steel wire (plough steel), the present writer adopted, among other devices, the expedient of cleaning

Fig. 159.—Cowper Coles' Magnetic Scale Collector, showing scale adhering to the copper jacket of the electro-magnet.

it as it ran through the depositing vat, by causing the current to run from the wire as an anode in a preliminary vat of zinc sulphate solution immediately before it ran into the zinc sulphate solution, in which it acted as a cathode receiving the zinc deposit. Previous to this electrolytic cleaning the wire had passed through a hot caustic vat to remove grease, and then through an acid pickle, and finally it had been washed in water. The adhesion obtained by this device was good. Recently (*Electrician*, vol. 44, 1900, p. 434), Mr. Cowper Coles has employed a modification of this method for cleaning iron in the electro-

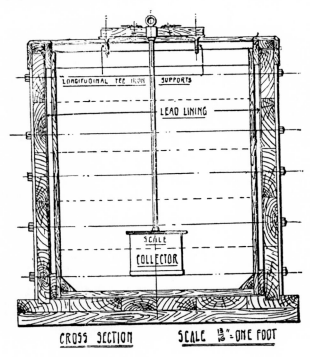

Fig. 160.—Magnetic Scale Collector in tank.

lytic vat immediately before the zinc is deposited upon it. The operation is described as follows:—" Before galvanising (but after cleaning in the usual manner) the plates were placed in the zincing bath, and the current caused to flow in the opposite direction to that required to deposit zinc upon them for a period of $2\frac{1}{2}$ minutes. The current was then reversed and the zinc deposited on the plates until a calculated weight of 1¼ oz. was obtained. They were then removed, washed, and dried, and were subjected to the same bending tests as a similar set of plates coated without the preliminary reversal of current. The adhesion of the zinc to the plates, in which the preliminary reversal of current had taken place, was found considerably better than was the case with the plates treated in the more ordinary way. It was also found that plates which had been "flashed" with zinc from a zinc sulphate solution containing $\frac{1}{10}$ ounce of free sulphuric acid per gallon, and then coated in a neutral zinc sulphate solution, gave much better adhesion than was the case with plates completely coated in a neutral solution. Mr. Cowper Coles considers the good adhesion in both these cases to be due to the acid on the surface of the plates preventing the formation of thin and invisible films of oxide, which may occur after the iron is cleaned with acid and whilst it is being washed with water. Deposits of zinc obtained from perfectly neutral zinc sulphate solutions are smoother and more even than those obtained from solutions containing $\frac{1}{10}$ ounce of free sulphuric acid per gallon, but they were non-adhesive, whilst those deposited from the slightly acid solution were perfectly adhesive, but were less uniform and smooth on the surface.

The following remarks upon the electrolytic cleansing of metallic surfaces are of interest. They form the substance of a paper by F. Loppé published in *L'Electricien*, vol. 20, 2nd series, 1900, p. 106, and translated into the *Electrical Engineer*, vol. 28, 1901, p. 329, by the present writer.

" In ordinary practice metallic surfaces are cleansed by immersion in either alkaline or acid solutions. Acid solutions are employed for cleansing the surface of iron, copper, and alloys of copper, whilst alkaline solutions are used to clean aluminium or zinc. These methods of cleaning metallic surfaces by simple immersion are lengthy and costly; the solutions used gradually become saturated with metal and lose their cleansing power; and if, as in the case of copper, the metal dissolved in the cleaning process is valuable, it is necessary to recover it from solution, a recovery which is not always easily carried out. Moreover, the liquids employed are corrosive, and therefore, if their action is not carefully checked, the metal surfaces may be damaged, or the workpeople may be affected by their corrosive or poisonous nature, whilst, finally, there is often considerable difficulty in getting rid of the waste pickling solutions owing to local regulations concerning drain and water contamination. Attempts have been made to overcome some one or more of the above mentioned difficulties by employing electrolytic methods of cleaning, such as that used by Cowper Coles. These methods have been applied to the cleaning of iron surfaces, such as boiler tubes, etc., but the employment of an acid solution has always been advocated. Recently, however, the Vereinigte Elektrizitäts Aktiengesellschaft, of Vienna and Budapest, have patented a process (French patent 292,333 of 1899) for the electrolytic cleansing of metallic surfaces, which has yielded remarkably satisfactory results, both from the point of view of rapidity and of economy. This new process may be applied to any kind of metallic surface, and the cleansing solutions employed do not become deteriorated in their properties by use. Any metal removed from the surfaces which are under treatment may be readily recovered if the value of the metal renders such recovery desirable, and, further, the solution being neutral and practically harmless, it may, when necessary, be run off into the drains, etc., without creating a nuisance. The electrolyte used consists of an aqueous solution of an alkaline salt, and one of the electrodes is formed of the metal whose surface it is desired to clean, whilst the other is formed of carbon, or of some metal which is not acted upon by the solution (or, as will be pointed out later, both electrodes may consist of the metal whose surface is under treatment).

In the case of iron or copper, or alloys of copper, the metal surface is made the anode, and the metallic oxide which is dissolved from the anodes is reprecipitated from the solution by the alkali which is generated at the cathode. The electrolyte is thus continuously regenerated. When it is wished to clean zinc and aluminium surfaces, these metals are employed as the cathodes, and alkaline aluminates and zincates are formed, from which, however, the zinc and aluminium oxides are subsequently precipitated by the acid liberated at the anode.

This form of electrolytic cleansing process may, moreover, be used for the cleaning of metallic surfaces from greasy and fatty matters, and for this purpose the metal must always be used as the cathode, the alkaline hydrate liberated at this electrode saponifying and dissolving the grease on the metallic surface. If it is wished to use the process for the preparation of iron plates, etc., which are to be subsequently tinned or electro-galvanised, it is carried out as follows :—The electrolyte employed consists of an aqueous solution containing 20 per cent. of sodium sulphate, such as is produced as a waste product in some factories, and both of the electrodes are formed of the iron whose surface is to be cleaned. The current is first of all passed in one direction during a certain period, and the plates acting as anodes are cleaned of metallic oxide by the acid set free at their surface, whilst the plates acting as cathodes are cleansed from grease by the alkali. The anode plates are then removed and replaced by fresh uncleaned plates, and the current is then reversed so that the plates which were previously cleansed from grease as cathodes in the first operation are now cleansed from oxide as anodes, and the freshly introduced plates now acting as cathodes are in their turn freed from grease. When this operation is completed the current is again stopped, and the anode plates are replaced by fresh plates, and the current is once more started in the reverse direction. The process is then continued in this way, the plates being successively cathodes and cleansed from grease and then anodes and cleansed from oxide, and are finally removed from the cleaning vats. The duration of the process naturally depends upon the current density employed. With a current of 60 to 120 ampères per square metre—*i.e.*, rather under 6 to 12 ampères per square foot— of the metallic surface, each operation lasts about half an hour, and as each surface undergoes a double treatment, first as cathode and then as anode, the complete treatment requires from 60 to 120 ampère-hours per square metre of surface. The mean voltage required is about four volts, and the energy required per square metre of surface cleaned is therefore from 240 to 480 watt-hours, which is a very small expenditure. When the articles cleaned are iron plates they are slipped into frames made of lead-covered iron wire, which are not attacked by the solution.

The space inside the frames is about 2 c.m. wide, and the sheets are slipped in either through an opening at the top or at the side, and are so adjusted that they come into contact with the frames at several points, thus obtaining a good electric contact. The frames are arranged parallel to each other, but are insulated from one another by wood separators. At one side of the vat the alternate frames are connected together to one conductor, whilst the intermediate ones forming the second electrode are connected together to the other conductor. The electrolytic vats are usually made of concrete, and the whole of the frames can be lifted from the vats together in order to remove and replace the plates. When the current is passed there is a considerable evolution of gas, and the solution becomes turbid. The solution also becomes heated, and a flocculent precipitate of the reddish-brown ferric hydrate is formed, which partly falls to the bottom of the vats and partly floats at the surface. The electrolyte is circulated by a pump, which forces it through a filter, from whence it returns to the vat. Each vat requires a voltage of about four volts, but if a higher voltage is available several vats may be arranged in series with advantage. The process has been employed at the large plate factory at Teplitz, in Bohemia, and a still larger installation is now being erected."

Effect of Pressure upon Electro - deposition of Zinc.— Mr. Cowper Coles has studied the effect of depositing zinc electrolytically upon iron in enclosures in which a hydrostatic pressure is maintained (*Electrician*, vol. 44, 1899, p. 183). The pressure employed is not stated. The solution contained 35 ounces of zinc sulphate and $\frac{1}{10}$th ounce of sulphuric acid per gallon, and a small quantity of zinc dust was added. The anode was of zinc and the solution was not circulated. Current densities up to 100 ampères per square foot were employed, and at 50 ampères per square foot excellently adhesive and beautifully smooth, close grained deposits were obtained. The author states that in a great number of cases the plates had round spots upon them which appeared to be uncoated, and were probably due to gas bubbles clinging to the surface. There does not appear to be any advantage obtainable by electro-deposition under pressure at the present date.

Zinc Sponge Deposits.—With many solutions of zinc salts, both acid and alkaline, and especially if the solution is dilute and the current density small, powdery, spongy deposits of zinc, which appear often almost black in colour whilst wet, are obtained on the cathode. It is a peculiarity of this zinc sponge that solutions which will perhaps for half-an-hour yield quite satisfactory regular electrolytic deposits of zinc will, when a certain thickness of deposit has been reached, commence to deposit the spongy and powdery metal, and this occurs not only in neutral or acid solutions but also in alkaline solutions. Two theories have been advanced as to the cause of this unsatisfactory deposit, namely, one that it is due to the formation of a hydride of zinc and is due to the liberation of free hydrogen with the zinc at the cathode, and the other, and to my mind the more probable, states it to be due to the formation of zinc hydrate. The formation of spongy zinc may, however, apparently be avoided in practice by employing strong solutions of zinc sulphate with only small amounts of acid present and using high current densities (about 15 and better from 20 to 30 ampères per square foot of surface to be coated), and lastly by employing a rapid circulation of the electrolyte. The formation of the zinc sponge from solutions which to commence with yield good reguline deposits is probably due to the impoverishment in zinc salt of the layer of electrolyte in contact with the cathode. Zinc sponge is only partially soluble in pure mercury and the insoluble portion consists of zinc hydrate.

Sources of Zinc Dust Supply.—The United States Consul at Liège, Belgium, stated (1898) that the export of zinc dust collected at the several zinc works in Liège had largely increased. During the year 1896 the declared value of zinc dust exported for the American market was 35,826 dollars. In 1897 it amounted to 105,000 dollars. For the first quarter of 1897 the exports of this material had a value of 14,621 dollars, whilst for the same quarter in 1898 the invoiced value was 28,456 dollars. Zinc dust is said to be employed in America for the manufacture of paints.

Price of Zinc and Zinc Dust.—Cowper Coles states that zinc dust containing 75·5 per cent. of metallic zinc, and usually a trace of cadmium, costs less than virgin spelter. It is not stated whether this applies to the unit of metallic zinc, or simply to the ton of the two materials.

Messrs. Paul Speier, of Breslau, state that sheet zinc in February, 1900, rose from 47 marks to 53 marks per 100 kilos. (£23 8·7s. to £26 8·5s. per ton), whilst crude zinc (spelter) rose in the same month to 23 marks per 100 kilos. (£11 9·4s. per ton), but in April and afterwards the price was as low as 18·75 marks per 100 kilos. (£9 6·9s. per ton). The price of zinc dust from May to July of the same year was 56 marks per 100 kilos. (£27 18·5s. per ton). These figures are taken from an article entitled "The Zinc Market in 1900," published in *The Electro-Chemist and Metallurgist* (January, 1901, vol. 1, p. 29), but do not bear out Cowper-Coles' statement that zinc dust cost from £7 to £8 per ton less than ordinary rolled zinc anodes (*Journ. Soc. Chem. Ind.*, vol. 15, 1896, p. 416). Ordinary spelter was quoted at £16 12s. 6d. on the London market, August 8, 1901, whilst the highest price of spelter on the English market in 1900 was £22 10s., and at the commencement of 1901 its price was £18 15s. per ton c. i. f. Hull net.

USEFUL TABLES.

TABLE I.—ELEMENTS, THEIR SYMBOLS AND ATOMIC WEIGHTS.

Name.	Symbol.	Atomic Weight.	Name.	Symbol.	Atomic Weight.
Aluminium	Al.	27·5	Mercury	Hg.	200·
Antimony	Sb.	122·	Molybdenum	Mo.	96·
Arsenic	As.	75·	Nickel	Ni.	59·
Barium	Ba.	137·	Niobium	Nb.	97·5
Bismuth	Bi.	210·	Nitrogen	N.	14·
Boron	B.	10·9	Osmium	(s.	199·
Bromine	Br.	80·	Oxygen	O.	16·
Cadmium	Cd.	112·	Palladium	Pd.	106·5
Cæsium	Cs.	133·	Phosphorus	P.	31·
Calcium	Ca.	40·	Platinum	Pt.	197·
Carbon	C.	12·	Potassium	K.	39·1
Cerium	Ce.	92.	Rhodium	Ro.	104·3
Chlorine	Cl.	35·5	Rubidium	Rb.	85·
Chromium	Cr.	52·5	Ruthenium	Ru.	104·2
Cobalt	Co.	59·	Selenium	Se.	79·5
Copper	Cu.	63·5	Silicon	Si.	28·
Didymium	D.	96·	Silver	Ag.	108·
Erbium	E.	(?)	Sodium	Na.	23·
Fluorine	F.	19·	Strontium	Sr.	87·5
Gallium			Sulphur	S.	32·
Glucinum	G.	9·3	Tantalum	Ta.	138·
Gold	Au.	196·6	Thorium	Te.	129·
Hydrogen	H.	1·	Tin	Tl.	204·
Indium	In.	113·4	Thorinum	Th.	119·
Iodine	I.	127·	Tin	Sn.	118·
Iridium	Ir.	197·	Titanium	Ti.	50·
Iron	Fe.	56·	Tungsten	W.	184·
Lanthanum	La.	92·	Uranium	U.	120·
Lead	Pb.	207·	Vanadium	V.	137·
Lithium	L.	7·	Yttrium	Y.	(?)
Magnesium	Mg.	24·3	Zinc	Zn.	65·
Manganese	Mn.	55·	Zirconium	Zr.	89·5

TABLE II.—RELATIVE CONDUCTIVITY OF METALS.

By L. Weiller.

Names of Metals.	Conduc-tivity.	Observations.
1. Silver, pure	100·	These experiments have
2. Copper, pure	100·	been conducted with a
3. Copper, pure super refined and		series of bars especially
crystallised	99·9	prepared for the pur-
4. Silicium bronze (telegraphic) . .	98·	pose. These said bars have
5. Copper and silver alloy at 50 per cent.	86·65	been molten at a uniform
6. Gold, pure	78·	diameter of about 13 mil-
7. Silicic copper (with 4 per cent. of		limètres. They have been
silicon)	75·	cut so as to show the
8. Silicic copper (with 12 per cent. of		grain of the metal, and
silicon)	54·7	the detached portions
9. Aluminium, pure	54·2	have then been drawn
10. Tin, containing 12 per cent. of		into wires.
sodium	46·9	It is on the wires so
11. Silicium bronze (telephonic) . .	35·	obtained that the said
12. Plumbiferous copper, with 10 per		experiments have been
cent. of lead	30·	carried out, and of which
13. Zinc, pure	29·9	the results are given in
14. Phosphor bronze (telephonic) . .	29·	the table.
15. Silicious brass, with 25 per cent. of		As regards those alloys
zinc	26·49	which can neither easily
16. Brass, with 35 per cent. of zinc .	21·15	be drawn nor rolled, such
17. Phosphide of tin	17·7	as certain phosphides or
18. Gold and silver alloy, 50 per cent. .	16·12	silicides, the measure-
19. Swedish iron	16·	ments have been taken
20. Pure tin of Banca	15·45	direct from the bars ac-
21. Antimonous copper	12·7	cording to the method of
22. Aluminium bronze, 10 per cent. .	12·6	Sir W. Thomson.
23. Siemens' steel	12·	The measurements have
24. Platinum, pure	10·6	been taken by means of
25. Amalgam of cadmium, with 15 per		a Wheatstone bridge with
cent. of cadmium . . .	12·2	a sliding index, a diffe-
26. Mercurial bronze, Drosnier . .	10·14	rential galvanometer and
27. Arsenical copper, with 10 per cent. of		a battery of four cells.
arsenic	9·1	
28. Lead, pure , ...	8·88	
29. Bronze, with 20 per cent. of tin .	8·4	
30. Nickel, pure	7·89	
31. Phosphor bronze, with 10 per cent.		
of tin	6·5	
32. Phosphide of copper, with 9 per cent.		
of phosphorus	4·9	
33. Antimony	3·88	

TABLE III.—SPECIFIC RESISTANCE OF SOLUTIONS OF SULPHATE
OF COPPER.

By FLEEMING JENKIN.

Sulphate of copper.	Temperature. Fahrenheit.	57°	61°	64°	68°	75°	82°	86°
	Water.							
8 parts.	100 parts.	45·7	43·7	41·9	40·2	37·1	34·2	32·9
12 „	100 „	36·3	34·9	33·5	32·2	29·9	27·9	27·0
16 „	100 „	31·2	30·0	28·9	27·9	26·1	24·6	24·0
20 „	100 „	28·5	27·5	26·5	25·6	24·1	22·7	22·2
24 „	100 „	26·9	25·9	24·8	23·9	22·2	20·7	20·0
28 „	100 „	24·7	23·4	22·1	21·0	18·8	16·9	16·0

TABLE IV.—SPECIFIC RESISTANCE OF SOLUTIONS OF SULPHATE
OF COPPER AT 50° FAHR.

By EWING and MACGREGOR.

Density.	Specific resistance.	Density.	Specific resistance.
1·0167	1·644	1·1386	35·0
1·0216	1·348	1·1432	34·1
1·0318	9·87	1·1679	31·7
1·0622	5·90	1·1823	30·6
1·0858	4·73	1·2051 (satu-	29·3
1·1174	3·81	rated).	

TABLE V.—TABLE OF HIGH TEMPERATURES.

Degrees. Fahr.	Description.	Degrees. Fahr.	Description.
977	Incipient red heat.	1700	An orange red heat.
980	A red heat.	1873	A bright red heat.
1000	A dull red heat visible in daylight.	1996	A dull white heat.
		3000	A white heat.
1140	Heat of a common fire.	3300	Heat of a good blast furnace.
1200	A full red heat.		
1310	Dull red heat.		

TABLE VI.—COMPARATIVE FRENCH AND ENGLISH THERMOMETER SCALES.

French, or Centigrade. Cent. or C.	English, or Fahrenheit. Fahr. or F.	Cent.	Fahr.	Cent.	Fahr.
Degrees.	Degrees.	Degrees.	Degrees.	Degrees.	Degrees.
0	32	33	91·4	67	152·6
1	33·8	34	93·2	68	154·4
2	35.6	35	95	69	156·2
3	37·4	36	96·8	70	158
4	39·2	37	98·6	71	159·8
5	41	38	100·4	72	161·6
6	42·8	39	102·2	73	163·4
7	44·6	40	104	74	165·2
8	46·4	41	105·8	75	167
9	48·2	42	107·6	76	168·8
10	50	43	109·4	77	170·6
11	51·8	44	111·2	78	172·4
12	53·6	45	113	79	174·2
13	55·4	46	114·8	80	176
14	57·2	47	116·6	81	177·8
15	59	48	118·4	82	179.6
16	60·8	49	120·2	83	181·4
17	62·6	50	122	84	183·2
18	64·4	51	123·8	85	185
19	66·2	52	125·6	86	186·8
20	68	53	127·4	87	188·6
21	69·8	54	129·2	88	190·4
22	71·6	55	131	89	192·2
23	73·4	56	132·8	90	194
24	75·2	57	134·6	91	195·8
25	77	58	136·4	92	197·6
26	78·8	59	138·2	93	199·4
27	80·6	60	140	94	201·2
28	82·4	61	141·8	95	203
29	84·2	62	143·6	96	204·8
30	86	63	145·4	97	206·6
31	87·8	64	147·2	98	208·4
32	89·6	65	149	99	210·2
		66	150·8	100	212

TABLE VII.—BIRMINGHAM WIRE GAUGE FOR SHEET COPPER AND LEAD.

Thickness by B. W. G.	Thickness in Inches.	Weight per Square Foot. Sheet Copper.	Sheet Lead.	Thickness by B. W. G.	Thickness in Inches.	Weight per Square Foot. Sheet Copper.	Sheet Lead.
No.	inch.	lbs.	lbs.	No.	inch.	lbs.	lbs.
0000	·454	20·566	26·75	19	·042	1·93	2·48
000	·425	19·252	25·06	20	·035	1·61	2·04
00	·380	17·214	22·42	21	·032	1·47	1·89
0	·340	15·6	20·06	22	·028	1·29	1·65
1	·300	13·8	17·72	23	·025	1·14	1·47
2	·284	13·	16·75	24	·022	1·01	1·30
3	·259	11·9	15·26	25	·020	·918	1·18
4	·238	11·	14·02	26	·018	·826	1·06
5	·220	10·1	12·98	27	·016	·735	·945
6	·203	9·32	11·98	28	·014	·642	·826
7	·180	8·25	10·63	29	·013	·597	·767
8	·165	7·59	9·73	30	·012	·551	·708
9	·148	6·8	8·72	31	·010	·480	·600
10	·134	6·16	7·90	32	·009	·420	·532
11	·120	5·51	7·08	33	·008	·370	·472
12	·109	5·02	6·42	34	·007	·323	·413
13	·095	4·37	5·60	35	·005	·262	·309
14	·083	3·81	4·90	36	·004	·194	·236
15	·072	3·31	4·25				
16	·065	3·00	3·83				
17	·058	2·67	3·42				
18	·049	2·25	2·90				

TABLE VIII.—NEW LEGAL STANDARD WIRE GAUGE
ISSUED BY THE STANDARDS DEPARTMENT OF THE BOARD OF TRADE. CAME INTO FORCE MARCH 1ST, 1884.

Descriptive No. B. W. G.	Equivalents in parts of an inch.	Descriptive No. B. W. G.	Equivalents in parts of an inch.	Descriptive No. B. W. G.	Equivalents in parts of an inch.
7/0	·500	13	·092	32	·0108
6/0	·464	14	·080	33	·0100
5/0	·432	15	·072	34	·0092
4/0	·400	16	·064	35	·0084
3/0	·372	17	·056	36	·0076
2/0	·348	18	·048	37	·0068
0	·324	19	·040	38	·0060
1	·300	20	·036	39	·0052
2	·276	21	·032	40	·0048
3	·252	22	·028	41	·0044
4	·232	23	·024	42	·0040
5	·212	24	·022	43	·0036
6	·192	25	·020	44	·0032
7	·176	26	·018	45	·0028
8	·160	27	·0164	46	·0024
9	·144	28	·0148	47	·0020
10	·128	29	·0136	48	·0016
11	·116	30	·0124	49	·0012
12	·104	31	·0116	50	·0010

TABLE IX.—CHEMICAL AND ELECTRO-CHEMICAL EQUIVALENTS.

Name of Element.	Symbol.	Chemical Equivalent Weight.	Electro Chemical Equivalents.	Weight deposited by one ampère flowing for one hour.
			mgr.	grammes.
Hydrogen . . .	H	1	0·010352	0·03726
Aluminium . . .	Al	9·16	0·09479	0·3413
Antimony . . .	Sb	40·6	0·4305	1·5120
Arsenic	As	25	0·2587	0·9313
Barium	Ba	68·5	0 7089	2·5520
Bismuth . . .	Bi	70	0·7244	2·6080
Boron	B	3·65	0·03778	0·1360
Bromine . . .	Br	80	0·8279	2·9810
Cadmium . . .	Cd	56	0·5795	2·0860
Calcium . . .	Ca	20	0·20704	0·7452
Chlorine . . .	Cl	35·5	0·3674	1·322
Chromium(ic) .	Cr	17·5	0·1811	0·6519
Cobalt . . .	Co	29·5	0·3053	1·099
Copper (Cuprous) . .	Cu	63·6	0·6583	2·370
,, (Cupric) . .	Cu	31·8	0·3291	1·185
Fluorine . . .	F	19	0·1966	0·7244
Gold (Aurous) . . .	Au	196·6	2·0334	7·310
,, (Auric) . . .	Au	65·5	0·6778	2·440
Iodine . . .	I	127	1·314	4·732
Iridium . . .	Ir	48·3	0·4998	1·799
Iron (Ferrous) .	Fe	28	0·2898	1·0430
,, (Ferric) .	Fe	18·6	0·1449	0·6929
Lead	Pb	103·5	1·0710	3·8560
Magnesium . .	Mg	12·2	0·1263	0·4545
Manganese . .	Mn	27·5	0·2845	1·0240
Mercury (Mercurous) . .	Hg	200	2·0704	7·4500
,, (Mercuric) .	Hg	100	1·0352	3·7250
Nickel . . .	Ni	29·5	0·3053	1·0990
Nitrogen . .	N	4·6	0·04761	0·1714
Oxygen . . .	O	8·0	0·082816	0·2981
Palladium . .	Pd	26·6	0·2753	0·9910
Phosphorus . .	P	10·3	0·1038	0·3837
Platinum . . .	Pt	44·3	0·4584	1·6500
Potassium . .	K	39·1	0·4047	1·4560
Selenium . . .	Se	39·5	0·4088	1·4720
Silver . . .	Ag	108	1·118	4·0240
Silicon . . .	Si	7·0	0·07246	0·2608
Sodium . . .	Na	23	0·2380	0·8569
Strontium . .	Sr	43·7	0 4523	1·6290
Sulphur . . .	S	16	0·16563	0·5961
Tin (Stannous) .	Sn	59	0·6106	2·1990
,, (Stannic) .	Sn	29·5	0·3053	1·0990
Zinc	Zn	32·5	0·3363	1·2110

TABLE X.—SPECIFIC GRAVITIES OF METALS.

Metal.			Sp. Gr.	Metal.			Sp. Gr.
Iridium	.	.	22·4	Cast Iron .	.	.	7·62
Platinum	.	.	21·5	Manganese	.	.	7·52
Gold	.	.	19·5	Tin	.	.	7·29
Mercury	.	.	13·6	Zinc	.	.	7·15
Thallium	.	.	11·8	Antimony.	.	.	6·71
Palladium .	.	.	11·4	Tellurium	.	.	6·40
Lead	,	.	11·3	Arsenic (Crystals)	.	5·72
Silver	.	.	10·5	Aluminium	.	.	2·65
Bismuth	.	.	9·85	Strontium	.	.	2·54
Copper	.	.	8·93	Magnesium	.	.	1·74
Nickel	.	.	8·90	Calcium	.	.	1·57
Cadmium .	.	.	8·66	Sodium	.	.	0·97
Cobalt	.	.	8 62	Potassium	.	.	0·87
Wrought Iron	.	.	7·82	Lithium	.	.	0·59

XI.—TABLES OF WEIGHTS AND MEASURES.

APOTHECARIES' WEIGHT.

1 pound	. .	*equals* .	. .	16 ounces.
1 ounce	. .	,, .	. .	8 drs. (480 grains).*
1 drachm	. .	,, .	. .	3 scruples.
1 scruple	. .	,, .	. .	2 grains.

TROY WEIGHT.

1 pound	. .	*equals* .	. .	12 ounces.
1 ounce	. .	,, .	. .	{ 20 pennyweights (dwts.) or 480 grains.*
1 pennyweight	.	,, .	. .	24 grains.

IMPERIAL MEASURE.

1 gallon	. .	*equals* .	. .	8 pints.
1 pint	. .	,, .	. .	20 ounces.
1 ounce	. .	,, .	. .	8 drachms.
1 drachm	. .	,, .	. .	60 minims.

One cubic foot of water weighs 62·5 lbs., and contains 6·25 gallons. One gallon weighs 10 lbs.

FRENCH OR METRICAL SYSTEM.

French Weight.

Kilogramme, 1,000 grammes	.	*equals* .	2 lbs. 3¾ ozs. *nearly.*
Gramme (the unit)	. .	,, .	15·432 grains.

French Measure of Volume.

1 litre (the unit)	. . .	*equals* .	34 fluid ounces *nearly.*

Long Measure.

Mètre (the unit)	. . .	*equals* .	39·371 inches.
Decimètre (10th of a mètre)	.	,, .	3·9371 ,,
Centimètre (100th of a mètre)	.	,, .	0·3937 ,,
Millimètre (1000th of a mètre)	.	,, .	0·0393 ,,

* An ounce Avoirdupois is only 437·5 grains.

TABLE XII.—SPECIFIC GRAVITIES CORRESPONDING TO DEGREES OF BAUMÉ'S HYDROMETER FOR LIQUIDS HEAVIER THAN WATER. (WATER = 1·000.)

Degrees Baumé.	Specific gravity.	Degrees Baumé.	Specific gravity.	Degrees Baumé.	Specific gravity.	Degrees Baumé.	Specific gravity.
0	1·000	20	1·152	40	1·357	60	1·652
1	1·007	21	1·160	41	1·369	61	1·670
2	1·013	22	1·169	42	1·382	62	1·689
3	1·020	23	1·178	43	1·395	63	1·708
4	1·027	24	1·188	44	1·407	64	1·727
5	1·034	25	1·197	45	1·420	65	1·747
6	1·041	26	1·206	46	1·434	66	1·767
7	1·048	27	1·216	47	1·448	67	1·788
8	1·056	28	1·225	48	1·462	68	1·809
9	1·063	29	1·235	49	1·476	69	1·831
10	1·070	30	1·245	50	1·490	70	1·854
11	1·078	31	1·256	51	1·495	71	1·877
12	1·085	32	1·267	52	1·520	72	1·900
13	1·094	33	1·277	53	1·535	73	1·924
14	1·101	34	1·288	54	1·551	74	1·949
15	1·109	35	1·299	55	1·567	75	1·974
16	1·118	36	1·310	56	1·583	76	2·000
17	1·126	37	1·321	57	1·600		
18	1·134	38	1·333	58	1·617		
19	1·143	39	1·345	59	1·634		

TABLE XIII.—SPECIFIC GRAVITIES ON BAUMÉ'S SCALE FOR LIQUIDS LIGHTER THAN WATER.

Degrees Baumé.	Specific gravity.	Degrees Baumé.	Specific gravity.	Degrees Baumé.	Specific gravity.	Degrees Baumé.	Specific gravity.
10	1·000	23	0·918	36	0·849	49	0·789
11	0·993	24	0·913	37	0·844	50	0·785
12	0·986	25	0·907	38	0·839	51	0·781
13	0·980	26	0·901	39	0·834	52	0·777
14	0·973	27	0·896	40	0·830	53	0·773
15	0·967	28	0·890	41	0·825	54	0·768
16	0·960	29	0·885	42	0·820	55	0·764
17	0·954	30	0·880	43	0·816	· 56	0·760
18	0·948	31	0·874	44	0·811	57	0·757
19	0·942	32	0·869	45	0·807	58	0·753
20	0·936	33	0·864	46	0·802	59	0·749
21	0·930	34	0·859	47	0·798	60	0·745
22	0·924	35	0·854	48	0·794		

Table XIV.— Degrees on Twaddell's Hydrometer and the Corresponding Specific Gravities.

[Note the degrees of Twaddell's hydrometer are converted into their corresponding specific gravities by multiplying by 0·005 and adding 1·000.]

Degrees Twaddell.	Specific gravity.	Degrees Twaddell.	Specific gravity.	Degrees Twaddell.	Specific gravity.
1	1·005	8	1·040	15	1·075
2	1·010	9	1·045	16	1·080
3	1·015	10	1·050	17	1·085
4	1·020	11	1·055	18	1·090
5	1·025	12	1·060	19	1·095
6	1·030	13	1·065	20	1·100
7	1·035	14	1·070		

Electrical Units.—One ampère for a second is called an ampère-second, or one coulomb. There are 3,600 coulombs in an ampère-hour.

The watt is the unit of power or activity, and is the same kind of unit as the horse-power : 746 watts = one horse-power. The activity or power of a dynamo is obtained in watts by multiplying its terminal pressure in volts by its current in ampères. This activity is usually expressed in kilowatts. The kilowatt = 1,000 watts = about 1·34 horse-power.

Electrical energy is measured in kilowatt hours or (as they are usually called) Board of Trade Units. One Board of Trade Unit = 1·34 horse-power hours.

SUBJECT INDEX.

For INDEX OF NAMES *see pp.* 67○-58○.

ACCOUTREMENT, army, gilding, 193
Accumulators or secondary batteries, 22-33
Acierage, 348, 446
Adams' process for metallising moulds, 139
   ,, Mr. I., process for nickel-plating, 300, 460
Air-bubbles, to prevent the formation of, 101
Albert chains, gilding, 185, 186, 191
Alumina, cost of, 615, 618
Aluminium, cost of, 617
   ,, cost of sulphide, 618
   ,, plating other metals with, 368, 476, 477
   ,, plating with other metals, 478, 479
   ,, properties, 613
   ,, profits on producing, 618
   ,, alloys, production by Cowles' process, 619
   ,, Electrolytic smelting of, 614
   ,, world's output, 617
   ,, action of mercury upon, 614
Alloys, aluminium, 619
   ,, electro-deposition of, 374, 387, 398, 399, 400, 453, 473
   ,, new white metal, 399
   ,, refining complex jewellery, 568
Amalgam gilding, 210
Amalgamating zincs, 19, 100
American copper refineries, 494, 499, 500, 521
Ammeters, 38
Ampère or practical unit of current, 6

Animal substances, copying, 124
Anodes, 78
   ,, brass, 386
   ,, carbon, 370
   ,, carbon rod, 615
   ,, charcoal iron, 350
   ,, copper, gilding with, 221
   ,, German silver, 397
   ,, gold, 179, 186, 191, 563
   ,, lead, 366, 599, 602
   ,, mud from, see Sludge
   ,, nickel, 337, 399, 436, 627
   ,, palladium, 362
   ,, platinum, 195, 357
   ,, ,, wire, 186
   ,, silver, 229, 233, 570-573
   ,, ,, and cadmium, 475
   ,, ,, and platinum, 399
   ,, steel, 349
   ,, tin, 398
   ,, tin and chromium, 398
   ,, brass cleaning, 400
   ,, bullion, 563-577
   ,, cast cobalt, 465
   ,, cast nickel, 292, 436
   ,, cleansing and inspection of during refining, 537
   ,, cobalt, 465
   ,, copper employed in refining, 537
   ,, dirty, 400
   ,, gold, worn, 219
   ,, iron, 349
   ,, iron wire, 351
   ,, rolled cobalt, 360
   ,, ,, nickel, 337, 436
   ,, ,, silver, 282
   ,, sludge from, see Sludge
   ,, worn, 282
   ,, zinc, 353, 354

INDEX OF NAMES.

680

N

680   NAME INDEX.

SALZEDE, see De Salzede
  Sanders, 557
Schlaeker, H. J., 568
Schlumberger, 150
Schnabel, C., 570
Schottlaender, 129
Schuckert Co., British, 532
Scott, George & Son, 532
Sederholm, 537, 541, 573
Sevrard, 477
Siemens, 493, 535, 634
Sire, 354, 386
Slater, 398
Smee, 5, 14, 97, 141, 442
Smith, Dr. J. H., 346, 579, 580, 593, 594
Smith, 398, 526
Société des Cuivres de France, 557
Speier, P., 650
Spence, 346
Spencer, 82
Sprague, 455, 495
Stalmann, 526, 527, 561
Steele, 172, 345
Swan, J. W., 545, 557
Swinburne, 462, 527, 579, 590, 591

TANGYE, 72
  Terrill, W., 544
Thiemann, 629
Thofehrn, 558
Thom, 441
Thomas, 370
Thompson, Prof. S., 467
Thompson, L., 456
Tilley, 370
Tissier Bros., 478
Tommasi, 601-611

Toney, H. G., 572
Tuck, 233
Tallis, J., 59
Tuttle, Dr. O. K., 568
Twynam, T., 594

ULKE, Titus, 528, 537, 540, 541, 544, 562, 563, 574, 627, 628
Umbreit and Matthes, 3, 11
Unwin, 302, 332

VAUTIN, Claude, 615
  Von Hübl, 545

WAHL, 131, 141, 279
  Walbridge, W. D., 587, 592
Walenn, 157, 352, 382
Wallace, 617, 618
Watt, 184, 353, 396, 414, 442
Webber, 74, 75
Weil, 158, 342
Weston, E., 303
Weston, W., 641
Wiggin, Messrs. Henry, 399, 437, 465
Wilde, Dr. Henry, 151, 488, 557
Williams, Alan, 481, 559
Williams, Albert, 599
Winckler, Dr. Clemens, 384, 477
Wohlwill, 490, 496, 498, 563, 565, 567, 568
Wollaston, 189, 250
Wood, 177, 380
Woolrich, 371, 380
Wray, Messrs. L. and C., 51
Wright, 96

ZINGSEM, 462, 463
  Zugger, A., 593